T0350419

More information about this series at http://www.springer.com/series/4318

René Erlín Castillo • Humberto Rafeiro

An Introductory Course in Lebesgue Spaces

René Erlín Castillo
Universidad Nacional de Colombia
Bogotá, Colombia

Humberto Rafeiro
Pontificia Universidad Javeriana
Bogotá, Colombia

ISSN 1613-5237 ISSN 2197-4152 (electronic)
CMS Books in Mathematics
ISBN 978-3-319-30032-0 ISBN 978-3-319-30034-4 (eBook)
DOI 10.1007/978-3-319-30034-4

Library of Congress Control Number: 2016935410

Mathematics Subject Classification (2010): 46E30, 26A51, 42B20, 42B25

Printed on acid-free paper

This Springer imprint is published by Springer Nature
The registered company is Springer International Publishing AG Switzerland

A la memoria de Julieta mi mamá-abuela
A mi esposa Hilcia del Carmen
A mis hijos: René José
Manuel Alejandro
Irene Gabriela y
Renzo Rafael
R.E.C.

à Daniela pelo seu amor e paciência
H.R.

Preface

This book is not a treatise on Lebesgue spaces, since this would not be a feasible work due to the extension of their usage, e.g., in physics, probability, statistics, economy, engineering, among others. The objective is more realistic, being the introduction of the reader to the study of different variants of Lebesgue spaces and the common techniques used in this area. Since the Lebesgue spaces measure integrability of a function, they can be seen as the father of all integrable function spaces where more fine properties of functions are sought.

We can find many books where the subject of Lebesgue spaces is touched upon, for example, books dealing with measure theory and integration. In the literature, we can also find some books dealing with Sobolev spaces and they dedicate, in general, not more than one chapter to Lebesgue spaces. A book dedicated solely to Lebesgue spaces is unknown to the authors. With this in mind, we decided to write a book devoted exclusively to Lebesgue spaces and their *direct* derived spaces, viz., Marcinkiewicz spaces, Lorentz spaces, and the more recent variable exponent Lebesgue spaces and grand Lebesgue spaces and also to basic harmonic analysis in those spaces. We think this will be a welcome to any serious student of analysis, since it will give access to information that otherwise is spread among different books and articles, as well as more than two hundred problems.

For example, one of the attractiveness of Lorentz and weak Lebesgue spaces is that the subject is sufficiently concrete and yet the spaces have fine structure and importance in applications. Moreover, the area is quite accessible for young people, leading them to gain sophistication in mathematical analysis in a relatively short time during their graduate studies. These features, among others, make the subject particularly interesting.

We think it is appropriate to comment on the choice of the writing style and some peculiarities. In the first part dealing with function spaces, we tried to be as thorough as possible in the proofs, although this could sound prolix for some readers. Another aspect is the inclusion of proofs of classical results that deviate from the standard ones, e.g., in the proof of the Minkowski inequality, we used the classical approach via Hölder's inequality for the Lebesgue sequence space and the less-known direct approach for the Lebesgue spaces. We also decided to include a chapter briefly touching upon the so-called nonstandard Lebesgue spaces, namely, on variable exponent Lebesgue spaces and on grand Lebesgue spaces since these are areas where intense research is being made nowadays. The topic of variable exponent spaces became very fashionable in recent years, not only due to mathematical curiosity, but also to the wide variety of their applications, e.g., in the modeling of electrorheological fluids as well as thermorheological fluids, in the study of image processing, and in differential equations with nonstandard growth. Grand Lebesgue spaces attracted the attention of many researchers and turned out to be the right spaces in which some nonlinear equations in the theory of PDEs have to be considered, among other applications.

This text is addressed to anyone that knows measure theory and integration, functional analysis, and rudiments of complex analysis.

Part of the content of this book has been tested with the students from *Universidad Nacional de Colombia* and also from the *Pontificia Universidad Javeriana* in the classes of advanced topics in analysis and also in measure theory and integration.

Since many of the results were collected in personal notebooks throughout the years, a considerable number of exact references were lost. We want to emphasize that the content is NOT original of the authors, except maybe the rearrangement of the topics and some (hopefully a small number) mistakes. If the reader finds misprints and errors, please let us know.

H.R. was partially supported by the research project "Study of non-standard Banach spaces", ID-PPTA: 6326 in the Faculty of Sciences of the Pontificia Universidad Javeriana, Bogotá, Colombia.

Bogotá, Colombia René Erlín Castillo
Bogotá, Colombia Humberto Rafeiro
October 2015

Contents

Chapter 1
Convex Functions and Inequalities

All analysts spend half their time hunting through the literature for inequalities which they want to use but cannot prove.
HARALD BOHR

Abstract Inequalities play an important role in Analysis, and since many inequalities are just convexity in disguise, we get that convexity is one of the most important tools in Analysis in general and not only in Convex Analysis. In this chapter we will introduce the notion of convexity in its various formulations, and we give some characterizations of convex functions and a few applications of convexity, namely, some classical inequalities as well as not so known inequalities.

1.1 Convex Functions

The concept of convex function is a relatively new one, introduced in the beginnings of the 20th century, although it was implicitly used before. The importance of convex functions, among other reasons, steams from the fact that amid nonlinear functions, they are the ones closest in some sense to linear functions.

We now introduce the concept of convexity in a more analytical way, namely

Definition 1.1. Let f be a real-valued function defined on an interval (a,b). The function is said to be *convex* in (a,b) if, for every $x,y \in (a,b)$, it satisfies

$$f\left(tx+(1-t)y\right) \leqslant tf(x)+(1-t)f(y) \tag{1.1}$$

for all $0<t<1$. Similar definition can be given for a closed interval. *Strict convexity* is when we have strict inequality in (1.1). Inequality (1.1) is sometimes called *Jensen's inequality*, although this is also reserved for the more general case (1.22). A function is called *concave* if $-f$ is convex. If f is both convex and concave, f is said to be *affine*. $\oslash$

The notion of convexity introduced in the Definition 1.1 has a very nice geometric interpretation, namely, the graph of the function f is below the secant line that passes through the points $(a,f(a))$ and $(b,f(b))$, as shown in Figure 1.1.

R.E. Castillo, H. Rafeiro, *An Introductory Course in Lebesgue Spaces*, CMS Books in Mathematics, DOI 10.1007/978-3-319-30034-4_1

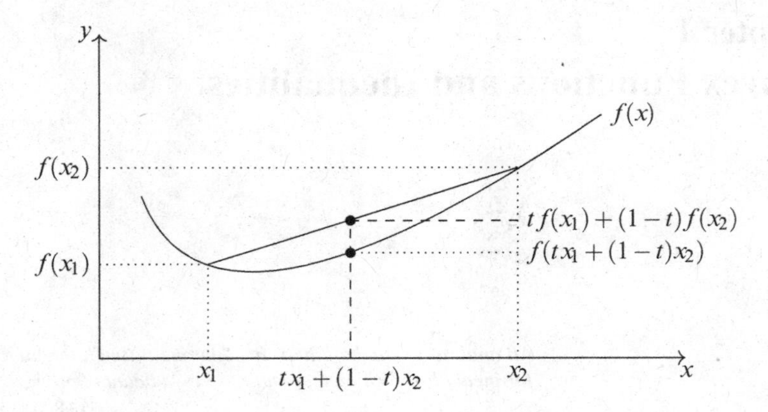

Fig. 1.1 Convexity defined by having the graph below the secant line.

It is possible to define convexity in a more geometric fashion, namely, a function f is *convex* if its epigraph

$$\operatorname{Epi}(f) := \Big\{(x,y) \in \mathbb{R}^2 : x \in (a,b), \quad y \geq f(x)\Big\}$$

is a convex set (epigraph is also designated as *supergraph*). Both definitions of convexity are equivalent, see Problem 1.15.

We can characterize convexity in a number of forms.

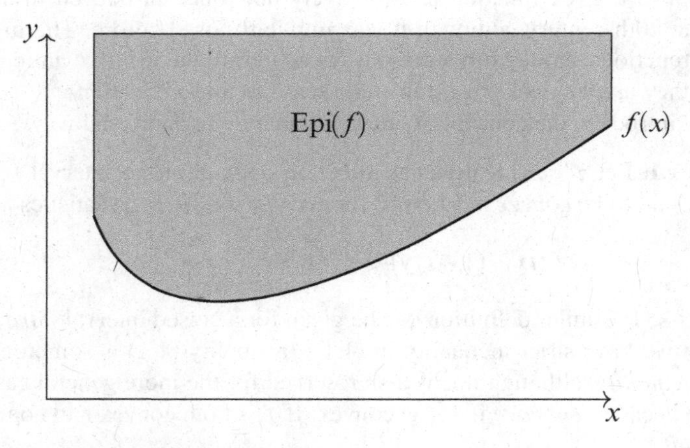

Fig. 1.2 Convexity defined via convex epigraph.

Theorem 1.2. *Let f be a real-valued function defined on $[\alpha,\beta]$. Then f is convex in $[\alpha,\beta]$ if and only if*

$$\frac{f(x)-f(a)}{x-a} \leqslant \frac{f(b)-f(a)}{b-a} \tag{1.2}$$

with $a<x<b$ in $[\alpha,b]$.

Proof. Taking the convexity of f and $t=\frac{x-a}{b-a}$ in (1.1), we get (1.2).

Conversely, let $a<x<y<b$ and $0<t<1$, then $0<1-t$, then clearly that $tx<ty$ and $(1-t)x<(1-t)y$ and so

$$x = tx+(1-t)x \leqslant tx+(1-t)y < ty+(1-t)y = y$$

i.e.

$$x < tx+(1-t)y < y,$$

which, from hypothesis, we have

$$\frac{f(tx+(1-t)y)-f(x)}{tx+(1-t)y-x} \leqslant \frac{f(y)-f(x)}{y-x}$$

and, as before, we get

$$f(tx+(1-t)y) \leqslant tf(x)+(1-t)f(y).$$

□

As a consequence of the previous theorem we get the inequalities in (1.3) which are quite important, and are sometimes referred to as the *fundamental inequalities*.

Corollary 1.3 *Let f be a convex real-valued function defined on $[\alpha,\beta]$ and $a<x<y<z<b$ in $[\alpha,\beta]$. Then*

$$\frac{f(y)-f(x)}{y-x} \leqslant \frac{f(z)-f(x)}{z-x} \leqslant \frac{f(z)-f(y)}{z-y}. \tag{1.3}$$

Proof. Taking x,y,z such that $x<y<z$, then by Theorem 1.2 we have

$$\frac{f(y)-f(x)}{y-x} \leqslant \frac{f(z)-f(x)}{z-x}. \tag{1.4}$$

On the other hand $x<y<z$ implies that $-z<-y<-x$ or equivalently $0<z-y<z-x$, then we obtain $0<\frac{z-y}{z-x}<1$. Taking $t=\frac{z-y}{z-x}$, we get

$$y = \frac{y-x}{z-x}z+\frac{z-y}{z-x}x$$

then

$$f(y) = f\left(\frac{y-x}{z-x}z + \frac{z-y}{z-x}x\right)$$

and now from the convexity of f it follows that

$$f(y) \leqslant \frac{y-x}{z-x}f(z) + \frac{z-y}{z-x}f(x)$$

and we obtain the inequality

$$\frac{f(z)-f(x)}{z-x} \leqslant \frac{f(z)-f(y)}{z-y},$$

which, together with (1.4), proves the fundamental inequalities (1.3). □

Definition 1.4. Let f be a real-valued function defined on an interval (a,b). The function is said to have *midpoint convexity* in (a,b) if, for all $x,y \in (a,b)$, it satisfies

$$f\left(\frac{x+y}{2}\right) \leqslant \frac{1}{2}f(x) + \frac{1}{2}f(y). \tag{1.5}$$

Sometimes this convexity is also called *convexity in the Jensen sense*, *J-convex* and *midconvex*. ⊘

The following theorem tells us that we can obtain convexity just from continuity and *midpoint convexity*. Therefore, for nice functions, the two concepts are equivalent.

Theorem 1.5. *Let f be a real-valued continuous function defined in (a,b) such that satisfies* (1.5). *Then f is convex in (a,b).*

Proof. Proceeding by induction, let $n \in \mathbb{N}$. For $n = 1$, we have

$$f\left(\frac{1}{2}x + \left(1-\frac{1}{2}\right)y\right) = f\left(\frac{x+y}{2}\right) \leqslant \frac{1}{2}f(x) + \left(1-\frac{1}{2}\right)f(y).$$

Suppose that, for $0 < k < 2^n$, we have

$$f\left(\frac{k}{2^n}x + \left(1-\frac{k}{2^n}\right)y\right) \leqslant \frac{k}{2^n}f(x) + \left(1-\frac{k}{2^n}\right)f(y).$$

Now consider

$$x_0 = \frac{k}{2^n}x + \left(1-\frac{k}{2^n}\right)y, \quad y_0 = \frac{k+1}{2^n}x + \left(1-\frac{k+1}{2^n}\right)y$$

$$x_0 + y_0 = \frac{2k+1}{2^n}x + \left(2-\frac{2k+1}{2^n}\right)y, \; \frac{x_0+y_0}{2} = \frac{2k+1}{2^{n+1}}x + \left(1-\frac{2k+1}{2^{n+1}}\right)y$$

and for all $h \leq 2^{n+1}$ choose k such that $h = 2k+1$, then

$$
\begin{aligned}
f\left(\frac{h}{2^{n+1}}x+\left(1-\frac{h}{2^{n+1}}\right)y\right) &= f\left(\frac{x_0+y_0}{2}\right) \\
&\leqslant \frac{1}{2}f(x_0)+\frac{1}{2}f(y_0) \\
&\leqslant \frac{h}{2^{n+1}}f(x)+\left(1-\frac{h}{2^{n+1}}\right)f(y).
\end{aligned}
$$

Let $\lambda \in (0,1]$, $n \in \mathbb{N}$ then there exists $k_n \in \mathbb{Z}^+$ such that

$$\frac{k_n}{2^n} \leqslant \lambda \leqslant \frac{k_n+1}{2^n}$$

from which we obtain that

$$\lim_{n\to\infty}\frac{k_n}{2^n} = \lambda.$$

From the continuity of f we have

$$
\begin{aligned}
f(\lambda x+(1-\lambda)y) &= f\left(\lim_{n\to\infty}\frac{k_n}{2^n}x+\left(1-\frac{k_n}{2^n}\right)y\right) \\
&= \lim_{n\to\infty} f\left(\frac{k_n}{2^n}x+\left(1-\frac{k_n}{2^n}\right)y\right) \\
&\leqslant \lim_{n\to\infty}\frac{k_n}{2^n}f(x)+\lim_{n\to\infty}\left(1-\frac{k_n}{2^n}\right)f(y)
\end{aligned}
$$

and we obtain $f(\lambda x+(1-\lambda)y) \leqslant \lambda f(x)+(1-\lambda)f(y)$. □

One consequence of convexity is continuity, as stated in Theorem 1.6. Another approach to prove continuity of convex functions is showing the existence of finite left-hand and right-hand derivatives, see Problem 1.45.

Theorem 1.6. *Let f be a real-valued convex function defined on (a,b). Then f is continuous on (a,b).*

Proof. Let x_0 be any point in (a,b) and δ a positive real number such that the closed ball of center x_0 and radius δ is contained in (a,b), i.e.

$$B_\delta(x_0) = \left\{x \in \mathbb{R} : |x-x_0| \leq \delta\right\} \subset (a,b).$$

Let $M = \max\left\{f(x_0-\delta), f(x_0+\delta)\right\}$. We note that any $x \in B_\delta(x_0) = (y_1,y_2)$ where $y_1 = x_0-\delta$ and $y_2 = x_0+\delta$ can be written in the following manner

$$x = \frac{y_2-x}{y_2-y_1}y_1+\frac{x-y_1}{y_2-y_1}y_2$$

and we obtain

$$f(x) \leqslant \frac{y_2 - x}{y_2 - y_1} f(y_1) + \frac{x - y_1}{y_2 - y_1} f(y_2)$$

which implies that $f(x) \leqslant M$, and this means that f is bounded in $B_\delta(x_0)$.

For $x \neq x_0$, define $u = \delta\left(\operatorname{sgn}(x - x_0)\right)^{-1}$, then x has two possibilities:

(a) x is in $(x_0, x_0 + u)$, or
(b) x is in $(x_0 - u, x_0)$.

Consider the possibility (a), i.e., $x_0 < x < x_0 + u$. From this we obtain

$$0 < \frac{x - x_0}{u} < 1$$

writing

$$t = \frac{x - x_0}{u} = \frac{|x - x_0|}{\delta}$$

we get

$$x - x_0 = tu \tag{1.6}$$

operating properly in (1.6) we obtain

$$x = t(x_0 + u) + (1 - t)x_0, \qquad x_0 = \frac{x}{1+t} + \frac{t}{1+t}(x_0 - u).$$

Since f is convex

$$f(x) \leq t f(x_0 + u) + (1 - t) f(x_0), \; f(x_0) \leq \frac{1}{1+t} f(x) + \frac{t}{1+t} f(x_0 - u) \tag{1.7}$$

considering that f is bounded on $B_\delta(x_0)$ and operating properly in (1.7) we obtain

$$-t\left(M - f(x_0)\right) \leqslant f(x) - f(x_0) \leqslant t\left(M - f(x_0)\right)$$
$$\left|f(x) - f(x_0)\right| \leqslant t\left(M - f(x_0)\right).$$

Since $t = \frac{|x - x_0|}{\delta}$ we have

$$\left|f(x) - f(x_0)\right| \leqslant \frac{\left(M - f(x_0)\right)}{\delta} |x - x_0|. \tag{1.8}$$

When considering (b) also gives (1.8), concluding that f is continuous at (a, b). □

We cannot drop the hypothesis in Theorem 1.6 that the interval is open.

Theorem 1.7. *The following claims, for $f : [a, b] \to \mathbb{R}$ a differentiable function, are equivalent:*

(a) f is convex;
(b) $f' : [a, b] \to \mathbb{R}$ is a nondecreasing monotone function;

(c) For all $x, y \in [a,b]$ we have $f(x) \geq f(y) + f'(y)(x-y)$, in other words, the graph of f lies above its tangent lines at each point.

Proof. $(a) \Rightarrow (b)$: Let $\alpha < x < \beta$, then by the fundamental inequalities (1.3) and making $x \to \alpha+$ and then $x \to \beta_-$, we obtain that

$$f'_+(\alpha) \leq \frac{f(\beta) - f(\alpha)}{\beta - \alpha} \leq f'_-(b).$$

$(b) \Rightarrow (c)$: Let $y < x$. By the mean value theorem, there exists $z \in (y, x)$ such that $f(x) = f(y) + f'(z)(x-y)$. Since f' is a nondecreasing monotone function, we get that $f(x) \geq f(y) + f'(y)(x-y)$. For $y > x$ the proof is analogous.

$(c) \Rightarrow (a)$: Let $x_1 < x < x_2$, then by (c) we have

$$f(x_1) \geq f'(x)(x_1 - x) + f(x) \quad \text{and} \quad f(x_2) \geq f'(x)(x_2 - x) + f(x).$$

The convexity of f now follows multiplying the first inequality by $\lambda = (x_2 - x)/(x_2 - x_1)$, the second inequality by $1 - \lambda$ and taking the sum. □

The next corollary is quite useful since it characterizes in an easy way the set of convex function which are "regular" enough.

Corollary 1.8. *Let f be a real-valued function defined in (a,b) which is twice differentiable. Then f is convex if and only if $f''(x) \geqslant 0$, for all $x \in (a,b)$.*

1.2 Young Inequality

In this section we study Young's inequality, or to be more precise, different versions of Young's inequality. We give an alternative analytic proof of Young's inequality different from the usual one which is based upon geometric considerations, see Problem 1.40 for the classical proof of (1.9).

Theorem 1.9 (Young's inequality). *Let $y = f(x)$ be a strictly increasing function on $[0, \infty)$ with $f(0) = 0$. Suppose $f \in C^1[a,b]$ with a and b nonnegative real numbers. Then*

$$ab \leqslant \int_0^a f(x)\,\mathrm{d}x + \int_0^b f^{-1}(y)\,\mathrm{d}y, \tag{1.9}$$

where $f^{-1}(y)$ is the inverse function of f. The equality holds if $b = f(a)$.

Proof. If $f \in C^1[a,b]$, then by integration by parts we get

$$\int_a^b f(x)\,\mathrm{d}x = bf(b) - af(a) - \int_a^b xf'(x)\,\mathrm{d}x. \tag{1.10}$$

Let $y = f(x)$, then $\mathrm{d}y = f'(x)\mathrm{d}x$, and $x = f^{-1}(y)$. Replacing in (1.10) we obtain

$$\int_a^b f(x)\,\mathrm{d}x = bf(b) - af(a) - \int_{f(a)}^{f(b)} f^{-1}(y)\,\mathrm{d}y. \tag{1.11}$$

Now, if $r \leqslant x \leqslant a$, then $f(r) \leqslant f(x)$, so

$$(a-r)f(r) \leqslant \int_r^a f(x)\,\mathrm{d}x$$

$$af(r) - rf(r) \leqslant \int_r^a f(x)\,\mathrm{d}x. \tag{1.12}$$

Note that

$$\int_r^a f(x)\,\mathrm{d}x = \int_r^0 f(x)\,\mathrm{d}x + \int_0^a f(x)\,\mathrm{d}x$$

$$= -\int_0^r f(x)\,\mathrm{d}x + \int_0^a f(x)\,\mathrm{d}x.$$

By (1.12), we have

$$af(r) - rf(r) \leqslant -\int_0^r f(x)\,\mathrm{d}x + \int_0^a f(x)\,\mathrm{d}x. \tag{1.13}$$

By (1.11)

$$\int_0^r f(x)\,\mathrm{d}x = rf(r) - \int_0^{f(r)} f^{-1}(y)\,\mathrm{d}y. \tag{1.14}$$

Replacing (1.14) in (1.13) we get

$$af(r) - rf(r) \leqslant -rf(r) + \int_0^{f(r)} f^{-1}(y)\,\mathrm{d}y + \int_0^a f(x)\,\mathrm{d}x$$

which entails

$$af(r) \leqslant \int_0^a f(x)\,\mathrm{d}x + \int_0^{f(r)} f^{-1}(y)\,\mathrm{d}y.$$

If $0 < b < f(a)$, we can choose $r = f^{-1}(b)$. Thus,

$$ab \leqslant \int_0^a f(x)\,dx + \int_0^b f^{-1}(y)\,dy.$$

Finally, by (1.14) for $r = a$ and $b = f(a)$ we have

$$\int_0^a f(x)\,dx = af(a) - \int_0^{f(a)} f^{-1}(y)\,dy$$

then

$$\int_0^a f(x)\,dx = ab - \int_0^b f^{-1}(y)\,dy$$

giving

$$ab = \int_0^a f(x)\,dx + \int_0^b f^{-1}(y)\,dy.$$

□

For a particular choice of f in Theorem 1.9, we can obtain such inequalities as (1.15) and (1.16) which are also called Young's inequality.

Corollary 1.10 (Young's inequality) *Let $1 < p < q < \infty$ be such that $\frac{1}{p} + \frac{1}{q} = 1$. Then*

$$ab \leqslant \frac{a^p}{p} + \frac{b^q}{q} \tag{1.15}$$

for a and b positive real numbers. The equality holds if $a^p = b^q$.

Proof. First, if $a^p = b^q$, then $a = b^{q-1}$. Thus, $ab = b^q\left(\frac{1}{p} + \frac{1}{q}\right)$, i.e., $ab = \frac{a^p}{p} + \frac{b^q}{q}$. Now, consider $f(x) = x^\alpha$ with $\alpha > 0$ and $f^{-1}(y) = y^{1/\alpha}$, note that f satisfies the hypothesis of Theorem 1.9, then

$$ab \leqslant \int_0^a x^\alpha\,dx + \int_0^b y^{1/\alpha}\,dy = \frac{a^{\alpha+1}}{\alpha+1} + \frac{b^{1/\alpha+1}}{1/\alpha+1}.$$

If $p = \alpha + 1$ and $q = \frac{\alpha+1}{\alpha}$, thus $ab \leqslant \frac{a^p}{p} + \frac{b^q}{q}$, since $1/p + 1/q = 1$. □

It is noteworthy to mention that inequality (1.15) can be demonstrated in several conceptually different ways, see Problem 1.38 for another approach.

Corollary 1.11 (Young's inequality) *For $a > 0$, $b > 0$, we have that*

$$ab \leqslant a \log^+ a + e^{b-1}, \tag{1.16}$$

where

$$\log^+ x = \begin{cases} \log x, & \text{if } x > 1; \\ 0, & \text{if } 0 \leqslant x \leqslant 1. \end{cases}$$

Proof. Let φ and ψ be functions defined by

$$\varphi(x) = \begin{cases} \log x + 1, & \text{if } x > 1; \\ 1, & \text{if } 0 \leqslant x \leqslant 1. \end{cases}$$

and

$$\psi(y) = \begin{cases} e^{y-1}, & \text{if } y > 1; \\ 0, & \text{if } 0 \leqslant y \leqslant 1. \end{cases}$$

Now, let us define

$$\varphi^*(x) = \varphi(x+1) - 1 \quad \text{and} \quad \psi^*(y) = \psi(y+1) - 1.$$

Note that φ^* is of class C^1 on $[0, \infty)$ and $\varphi^*(0) = 0$. Note also that ψ^* is the inverse function of φ^*, then by Theorem 1.9 for $a > 1$ and $b > 1$ we have

$$\begin{aligned}
(a-1)(b-1) &\leqslant \int_0^{a-1} \varphi^*(x)\,dx + \int_0^{b-1} \psi^*(y)\,dy \\
&= \int_0^{a-1} \varphi(x+1)\,dx - (a-1) + \int_0^{b-1} \psi(y+1)\,dy - (b-1) \\
&= \int_1^a \varphi(u)\,du + \int_1^b \psi(t)\,dt - (a+b) + 2.
\end{aligned}$$

Then

$$\begin{aligned}
ab - (a+b) + 1 &\leqslant \int_1^a \varphi(u)\,du + \int_1^b \psi(t)\,dt - (a+b) + 2 \\
ab &\leqslant \int_1^a \varphi(u)\,du + \int_1^b \psi(t)\,dt + 1 \\
ab &\leqslant \int_1^a (\log u + 1)\,du + \int_1^b e^{t-1}\,dt + 1
\end{aligned}$$

$$= a\log^+ a + e^{-1}[e^b - e] + 1$$
$$= a\log^+ a + e^{b-1} - 1 + 1$$

finally

$$ab \leqslant a\log^+ a + e^{b-1},$$

which finishes the proof. □

The following theorem is the integral version of Jensen's inequality, which is termed as *Jensen's integral inequality*. Since we want to give a generalization of this inequality in Corollary 1.14 in a general framework, we will also give the Jensen integral inequality with the same generality.

Theorem 1.12 (Jensen's integral inequality). *Let μ be a positive measure on a σ-algebra $\mathscr{A}$ in a set Ω, such that $\mu(\Omega) = 1$. Let f be an integrable function on Ω, i.e.*

$$\int_\Omega f d\mu < \infty.$$

If $a < f(x) < b$ for all $x \in \Omega$ and φ is a convex function on (a,b), then

$$\varphi\left(\int_\Omega f d\mu\right) \leqslant \int_\Omega \varphi \circ f d\mu. \tag{1.17}$$

The equality is obtained if $f(x) = c$, for all $x \in \Omega$, where c is a real number.

Note: The cases $a = -\infty$ and $b = \infty$ are not excluded.

Proof. Let $t_0 = \int_\Omega f d\mu$. Since $a < f(x) < b$ for all $x \in \Omega$ and $\mu(\Omega) = 1$, we have that

$$a < t_0 < b,$$

and by Corollary 1.3 yields

$$\frac{\varphi(t_0) - \varphi(s)}{t_0 - s} \leqslant \frac{\varphi(u) - \varphi(t_0)}{u - t_0} \tag{1.18}$$

for all $a < s < t_0 < u < b$.

Let

$$\beta = \sup_{a<s<t_0}\left\{\frac{\varphi(t_0) - \varphi(s)}{t_0 - s}\right\}. \tag{1.19}$$

We affirm that

$$\varphi(y) \geqslant \varphi(t_0) + \beta(y - t_0)$$

for all $y \in (a,b)$. Indeed

(i) If $y = t_0$, there is nothing to prove.
(ii) If $t_0 < y < b$, then by (1.18) and (1.19) we have

$$\beta \leqslant \frac{\varphi(y) - \varphi(t_0)}{y - t_0}.$$

from which we get $\varphi(y) \geqslant \varphi(t_0) + \beta(y - t_0)$.
(iii) If $a < y < t_0$, then by (1.19) we get $\varphi(y) \geqslant \varphi(t_0) + \beta(y - t_0)$.

From (i), (ii), and (iii) we concluded that

$$\varphi(y) \geqslant \varphi(t_0) + \beta(y - t_0)$$

for all $y \in (a,b)$. Now taking $y = f(x)$ then

$$\varphi(f(x)) \geqslant \varphi(t_0) + \beta(f(x) - t_0)$$

integrating over Ω

$$\int_\Omega \varphi(f(x))\,\mathrm{d}\mu \geqslant \varphi(t_0) + \beta\left(\int_\Omega f(x)\,\mathrm{d}\mu - t_0\right).$$

It follows that

$$\varphi\left(\int_\Omega f\mathrm{d}\mu\right) \leqslant \int_\Omega \varphi \circ f\mathrm{d}\mu.$$

□

Remark 1.13. The condition $\mu(\Omega) = 1$ is necessary to the validity of Jensen's inequality, as the next example shows. Let $\Omega = [1,16]$, $f(x) = x^{-3/4}$ and $\varphi(x) = x^2$ be defined in Ω. We have

$$\varphi\left(\int_1^{16} x^{-3/4}\,\mathrm{d}x\right) = 16 > \int_1^{16} \varphi\left(x^{-3/4}\right)\mathrm{d}x = 3/2.$$

The formulation of Theorem 1.12 can be presented in various ways, depending on the type of proof. One alternative route of demonstration is to show the result for simple functions and use a density argument to obtain the result. For example, taking $(X, \mathscr{A}, \mu)$ as a probability space, f a simple function with $f(X) = \{r_1, \ldots, r_n\}$ and the fact that φ is convex we have

$$\varphi\left(\int_X f\mathrm{d}\mu\right) = \varphi\left(\sum r_i\mu(f^{-1}\{r_i\})\right) \leq \sum \varphi(r_i)\mu(f^{-1}\{r_i\}) = \int_X \varphi \circ f\mathrm{d}\mu,$$

where the inequality follows from Problem 1.24. Now the result is a consequence of the fact that the set of simple functions is dense in $L^1(X, \mathscr{A}, \mu)$ and the Lebesgue monotone convergence theorem.

The following corollary generalizes Jensen's integral inequality.

Corollary 1.14 *Under the assumptions of Theorem 1.12 it holds that*

$$\varphi\left(\frac{\int_\Omega f g d\mu}{\int_\Omega g d\mu}\right) \leqslant \frac{\int_\Omega \varphi(f) g d\mu}{\int_\Omega g d\mu}$$

where g is a positive and integrable function on Ω.

Proof. In the proof of Theorem 1.12 it was shown that if φ is convex, there is β such that

$$\varphi(y) \geqslant \varphi(t_0) + \beta(y - t_0) \tag{1.20}$$

for all $y \in (a,b)$. Now write $y = f(x)$ and multiplying (1.20) by a positive g function and integrating we obtain

$$\int_\Omega \varphi(f) g d\mu \geqslant \int_\Omega \varphi(t_0) g d\mu + \beta \int_\Omega f g d\mu - \beta t_0 \int_\Omega g d\mu. \tag{1.21}$$

Defining

$$t_0 = \frac{\int_\Omega f g d\mu}{\int_\Omega g d\mu}$$

and substituting in (1.21) we have

$$\int_\Omega \varphi(f) g d\mu \geqslant \varphi\left(\frac{\int_\Omega f g d\mu}{\int_\Omega g d\mu}\right) \int_\Omega g d\mu + \beta \int_\Omega f g d\mu - \beta \left(\frac{\int_\Omega f g d\mu}{\int_\Omega g d\mu}\right) \int_\Omega g d\mu$$

where

$$\varphi\left(\frac{\int_\Omega f g d\mu}{\int_\Omega g d\mu}\right) \leq \frac{\int_\Omega \varphi(f) g d\mu}{\int_\Omega g d\mu},$$

and we finish the proof. □

1.3 Problems

1.15. Show that Definition 1.1 is equivalent to the definition of convexity via convex epigraph as given in Figure 1.2.

1.16. An increasing function f is called superadditive if $f(x) - f(y) \leq f(x+y)$. If f is a convex function with $f(0) = 0$ then f is superadditive.

1.17. Prove that Definition 1.1 is equivalent to the following:
For $x, y \in (a,b)$, $p, q \geq 0$, $p+q > 0$ we have

$$f\left(\frac{px+qy}{p+q}\right) \leq \frac{p}{p+q} f(x) + \frac{q}{p+q} f(y).$$

1.18. Prove that Definition 1.1 is equivalent to the following:
If $x_1, x_2, x_3 \in (a,b)$ such that $x_1 < x_2 < x_3$ we have

$$\begin{vmatrix} x_1 & f(x_1) & 1 \\ x_2 & f(x_2) & 1 \\ x_3 & f(x_3) & 1 \end{vmatrix} = (x_3 - x_2) f(x_1) + (x_1 - x_3) f(x_2) + (x_2 - x_1) f(x_3) \geq 0.$$

1.19. Let $\varphi : [0, +\infty) \to [0, +\infty)$ be a function such that: (i) φ is convex, and (ii) $\varphi(x) = 0$ if and only if $x = 0$. Prove that:

(a) φ is strictly increasing on $[0, +\infty)$.
(b) If $0 < x < y$, then $\frac{\varphi(x)}{x} \leqslant \frac{\varphi(y)}{y}$.
(c) If $0 \leqslant x \leqslant 1$, then $\varphi(x) \leqslant x\varphi(1)$.

1.20. Show that the supremum of any collection of convex functions in (a,b) is convex and that the specific limits of sequences of convex functions are too. What can be said of the upper and lower limits of sequences of convex functions?

1.21. Suppose that φ is convex in (a,b) and ψ is convex nondecreasing in the path of φ. Prove that $\psi \circ \varphi$ is convex in (a,b). For $\varphi > 0$, show that the convexity of $\log \varphi$ implies that of φ, but not the other way.
Note: When $\log \circ \varphi$ is a convex function, we say that φ is *logarithmically convex*, log-*convex* or *superconvex*.

1.22. Prove that the composition of convex functions may not be convex.

1.23. Let $f : (a,b) \to \mathbb{R}$ be differentiable from the right at every point. If the derivative is increasing, show that f is convex.

1.24. If f is convex in (a,b), prove that

$$f\left(\sum_{j=1}^{n} \lambda_j x_j\right) \leqslant \sum_{j=1}^{n} \lambda_j f(x_j). \tag{1.22}$$

when $\sum_{j=1}^{n} \lambda_j = 1$.

1.25. Suppose that $\mu(\Omega) = 1$ and $h : \Omega \to [0, +\infty)$ is measurable. If $A = \int\limits_\Omega h \, d\mu$, show that

$$\sqrt{1+A^2} \leqslant \int\limits_\Omega \sqrt{1+h^2} \, d\mu \leq 1+A.$$

1.26. Suppose that φ is a real-valued function such that

$$\varphi\left(\int\limits_0^1 f(x)\,dx\right) \leqslant \int\limits_0^1 \varphi(f(x))\,dx$$

for every f bounded and measurable. Show that φ is convex.

1.27. Suppose that φ is strictly convex and $\mu(\Omega) = 1$. Show that the Jensen inequality

$$\varphi\left(\int\limits_\Omega f \, d\mu\right) \leqslant \int\limits_\Omega \varphi \circ f \, d\mu$$

is an equality if and only if f is almost everywhere constant.

1.28. Suppose that $\mu(\Omega) = 1$ and f, g are nonnegative measurable functions in Ω with $fg \geq 1$. Prove that

$$\int\limits_\Omega f \, d\mu \int\limits_\Omega g \, d\mu \geq 1.$$

1.29. Let μ be a positive measure on X and suppose that $f : X \to (0, +\infty)$ satisfies $\int\limits_X f \, d\mu = 1$. Show that the inequalities

(a) $\displaystyle\int\limits_E \log(f) \, d\mu \leqslant \mu(E) \log \frac{1}{\mu(E)}$,

(b) $\displaystyle\int\limits_E f^p \, d\mu \leqslant (\mu(E))^{1-p}, \qquad 0 < p < 1,$

are valid for all measurable set E with $0 < \mu(E) < \infty$.

1.30. Let g be a nonnegative measurable function on $[0, 1]$. Show that

$$\log \int\limits_0^1 g(t)\,dt \geqslant \int\limits_0^1 \log g(t)\,dt.$$

1.31. Let f be a positive measurable function on $[0, 1]$. Which of the two quantities

$$\int\limits_0^1 f(x) \log f(x)\,dx \qquad \text{or} \qquad \int\limits_0^1 f(s)\,ds \int\limits_0^1 \log f(t)\,dt$$

is greater?

1.32. Let $\{\alpha_n\}_{n\in\mathbb{N}}$ be a sequence of nonnegative numbers such that $\sum_{n=1}^{N}\alpha_n = 1$ and $\{\xi_n\}_{n\in\mathbb{N}}$ be a sequence of positive numbers. Show that

$$\prod_{n=1}^{N}\xi_n^{\alpha_n} \leqslant \sum_{n=1}^{N}\alpha_n\xi_n. \tag{1.23}$$

Use inequality (1.23) to show that the *geometric mean* is always less or equal to the *arithmetic mean*, namely

$$\sqrt[n]{\xi_1\cdot\xi_2\cdot\ldots\cdot\xi_n} \leqslant \frac{\xi_1+\xi_2+\ldots+\xi_n}{n}. \tag{1.24}$$

1.33. Show that the sequence $x_n = \left(1+\frac{1}{n}\right)^n$ is increasing using solely the arithmetic-geometric mean inequality (1.24).

1.34. Show that the *harmonic mean* is always less or equal to the geometric mean, namely

$$\frac{n}{\left(\frac{1}{x_1}+\frac{1}{x_2}+\ldots+\frac{1}{x_n}\right)} \leqslant \sqrt[n]{x_1\cdot x_2\cdot\ldots\cdot x_n}$$

for nonnegative x_i, $i = 1,2,\ldots,n$.

(a) Using (1.24) and reciprocals;
(b) Using the convexity of the function $f(x) = x\log x$.
Hint: Obtain the inequality

$$\frac{\sum p_ix_i}{\sum pi}\log\left(\frac{\sum p_ix_i}{\sum pi}\right) \leq \frac{\sum p_ix_i\log(x_i)}{\sum p_i},$$

use properties of the log function and specify $p_i = 1/x_i$.

1.35. Show that

$$a^\alpha b^\beta \leq \alpha a+\beta b$$

when α,β,a,b are nonnegative and $\alpha+\beta = 1$.
Hint: Study the convexity of the function $f(x) = e^x$.

1.36. Let $(\Omega,\mathscr{A},\mu)$ be a measure space such that $\mu(\Omega) = 1$ and $\int_\Omega |f|\,d\mu < \infty$ with $a < f(x) < b$ for all $x\in\Omega$. Show that if ψ is concave in (a,b) then

$$\psi\left(\int_\Omega f\,d\mu\right) \geqslant \int_\Omega \psi\circ f\,d\mu.$$

1.37. Let p and q be two real numbers with $0 < p < 1$ and $-\infty < q < 0$ such that $\frac{1}{p}+\frac{1}{q} = 1$. Prove that

$$ab \geqslant \frac{a^p}{p} + \frac{b^q}{q}$$

for a and b positive real numbers.

1.38. Use the functions

$$u(x) = \begin{cases} p\log a & \text{if } 0 \leqslant x \leqslant \frac{1}{p} \\ q\log b & \text{if } \frac{1}{p} \leqslant x \leq 1 \end{cases}$$

$(a, b > 0)$ and $f(x) = \mathrm{e}^x$ to demonstrate that

$$ab \leqslant \frac{a^p}{p} + \frac{b^q}{q} \qquad \text{with} \qquad \frac{1}{p} + \frac{1}{q} = 1.$$

1.39. Given $a, b > 0$ and $\varepsilon > 0$. Prove that

$$ab \leqslant \varepsilon a^p + c(\varepsilon) b^q \tag{1.25}$$

where $c(\varepsilon) = \frac{(\varepsilon p)^{-q/p}}{q}$.

Note: Inequality (1.25) is sometimes called the *Peter-Paul inequality* and examples of its usefulness can be found in ODEs and PDEs.

1.40. Show the Young inequality (1.9) by purely geometric insight.
Hint: See Fig. 1.3.

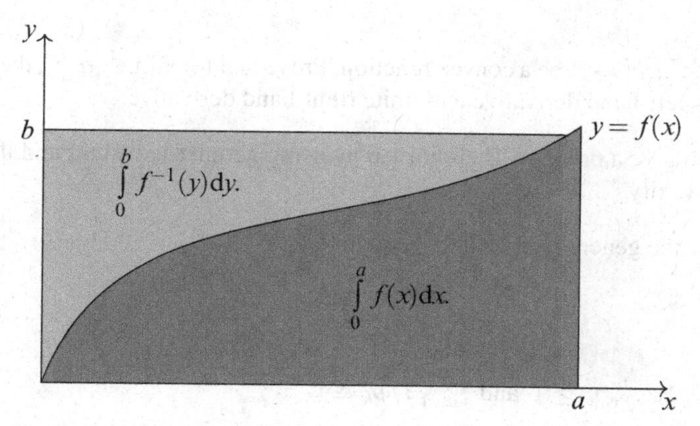

Fig. 1.3 Young's inequality via geometric inspection.

1.41. Prove that, for $x > 0$ and $p_i > 0$, we have

$$\left(\Pi_{i=1}^{n} x_i^{p_i}\right)^{\frac{1}{\sum_{i=1}^{n} p_i}} \le \frac{\sum_{i=1}^{n} p_i x_i}{\sum_{i=1}^{n} p_i},$$

studying the concavity of the function $f(x) = \log(x)$.

1.42. The *Legendre transform*, also known as *Legendre-Fenchel transform*, is defined in the following way. Take $f : \mathbb{R} \to \mathbb{R}$ a function and $f^* : \mathbb{R} \to \mathbb{R} \cup \{+\infty\}$ is its Legendre transform given by

$$f^*(y) = \sup_{x \in \mathbb{R}} \{xy - f(y)\}$$

for $y \in \mathbb{R}$. Prove:

(a) $xy \le f(x) + f^*(y)$, for all $x, y \in \mathbb{R}$;
(b) Using (a) show that $xy \le \frac{x^p}{p} + \frac{y^q}{q}$, for $1/p + 1/q = 1$.
Hint: Take $f(x) = \frac{x^p}{p}\chi_{[0,\infty)}$ and calculate its Legendre transform.
(c) Using (a) show that $xy \le e^x + y\log(y) - y$, for $x, y \ge 0$.
Hint: Take $f(x) = e^x$, $x \in \mathbb{R}$ and calculate its Legendre transform.

1.43. Let $1 < p < \infty$. Minimize the function $\phi(t) = \frac{t^p}{p} - t$ and show that $t \le \frac{t^p}{p} + \frac{1}{q}$, where $1/p + 1/q = 1$. Taking $t = \frac{r}{s^{q/p}}$, show the validity of Young's inequality $rs \le \frac{r^p}{p} + \frac{s^q}{q}$, for all $r, s > 0$.

1.44. 1. Show that $\varphi(x) = x\log x$ defined in $(0, \infty)$ is a convex function.
2. If $\int_X f(x)\mathrm{d}\mu = 1$, and f is nonnegative with $\int_X |f|\,\mathrm{d}\mu < \infty$, show that

$$\int_X f(x)\log(f(x))\mathrm{d}\mu \ge 0.$$

1.45. Let $f : [a,b] \to \mathbb{R}$ be a convex function. Prove that for all $x \in (a,b)$, the function f has finite left-hand derivative and finite right-hand derivative.

1.46. Try to give a proof of Theorem 1.6 by using geometric insight and the definition of convexity.

1.47. Show the generalized Young inequality

$$a_1 \cdots a_m \le \frac{a_1^{p_1}}{p_1} + \ldots + \frac{a_m^{p_m}}{p_m}.$$

where $p_1 \ge 1, \ldots, p_m \ge 1$ and $\sum_{k=1}^{m} 1/p_k \equiv 1$.

1.48. Prove the inequality

$$(AB)^r \le \frac{r}{p}A^p + \frac{r}{q}B^q$$

where $A > 0$, $B > 0$, $p > 0$, $q > 0$, and $\frac{1}{p} + \frac{1}{q} = \frac{1}{r}$.

1.4 Notes and Bibliographic References

The subject of convexity is a vast field, we will give only some small historical tidbits.

In 1889 Hölder [32] considered the concept of convexity connected with real functions having nonnegative second derivative. In 1893 Stolz [75] in his *Grundzüge der Differential- un Integralrechnung* showed already that if a continuous real-valued function is continuous and is mid-convex then it has lateral derivatives. On the other hand Hadamard [25] obtained an inequality between integrals for functions having increasing first derivative. It can be said that the father of convexity was Jensen [37, 38] which gave a detailed study where the Hölder and Minkowski inequalities are derived from Jensen's inequality.

A thrust in the study of convex functions was influenced by the classical book of Hardy, Littlewood, and Pólya [30]. For an encyclopedic monograph on inequalities see Mitrinović, Pečarić, and Fink [52] and for a thorough introduction to contemporary convex function theory see Niculescu and Persson [54].

Part I
Function Spaces

Chapter 2
Lebesgue Sequence Spaces

Abstract In this chapter, we will introduce the so-called Lebesgue sequence spaces, in the finite and also in the infinite dimensional case. We study some properties of the spaces, e.g., completeness, separability, duality, and embedding. We also examine the validity of Hölder, Minkowski, Hardy, and Hilbert inequality which are related to the aforementioned spaces. Although Lebesgue sequence spaces can be obtained from Lebesgue spaces using a discrete measure, we will not follow that approach and will prove the results in a direct manner. This will highlight some techniques that will be used in the subsequent chapters.

2.1 Hölder and Minkowski Inequalities

In this section we study the Hölder and Minkowski inequality for sums. Due to their importance in all its forms, they are sometimes called the *workhorses of analysis*.

Definition 2.1. The space ℓ_p^n, with $1 \le p < \infty$, denotes the n-dimensional vector space $\mathbb{R}^n$ for which the functional

$$\|\mathbf{x}\|_{\ell_p^n} = \left(\sum_{i=1}^{n} |x_i|^p \right)^{\frac{1}{p}} \tag{2.1}$$

is finite, where $\mathbf{x} = (x_1, \ldots, x_n)$. In the case of $p = \infty$, we define ℓ_∞^n as

$$\|\mathbf{x}\|_{\ell_\infty^n} = \sup_{i \in \{1,\ldots,n\}} |x_i|.$$

⊘

From Lemma 2.4 we obtain in fact that $\|\cdot\|_{\ell_p^n}$ defines a norm in $\mathbb{R}^n$.

R.E. Castillo, H. Rafeiro, *An Introductory Course in Lebesgue Spaces*, CMS Books in Mathematics, DOI 10.1007/978-3-319-30034-4_2

Example 2.2. Let us draw the unit ball for particular values of p for $n = 2$, as in Figs. 2.1, 2.2, and 2.3.

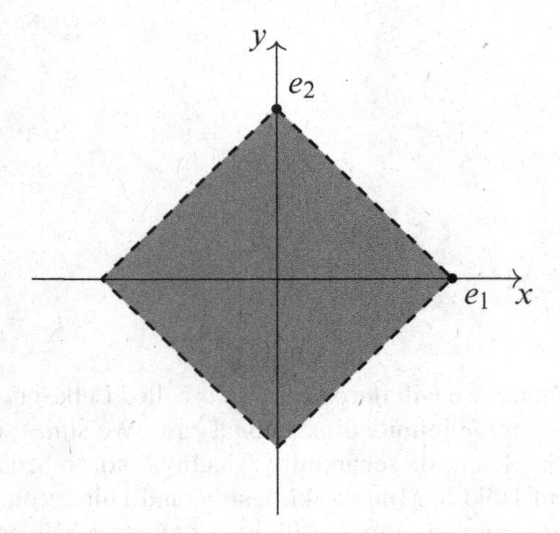

Fig. 2.1 Unit ball for ℓ_1^2

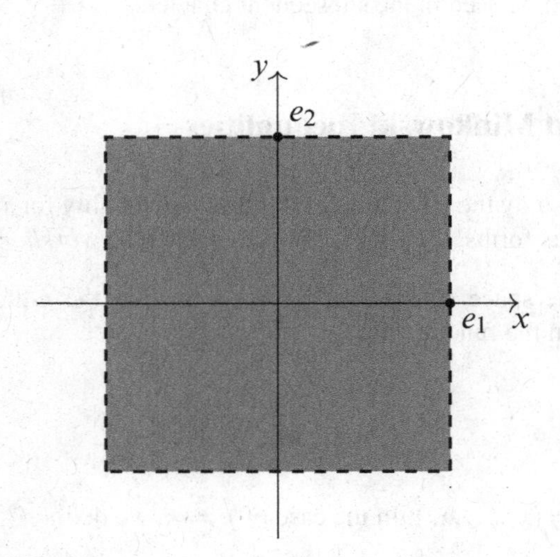

Fig. 2.2 Unit ball for ℓ_∞^2

Lemma 2.3 (Hölder's inequality). *Let p and q be real numbers with $1 < p < \infty$ such that $\frac{1}{p} + \frac{1}{q} = 1$. Then*

$$\sum_{k=1}^{n} |x_k y_k| \leq \left(\sum_{k=1}^{n} |x_k|^p \right)^{1/p} \left(\sum_{k=1}^{n} |y_k|^q \right)^{1/q} . \tag{2.2}$$

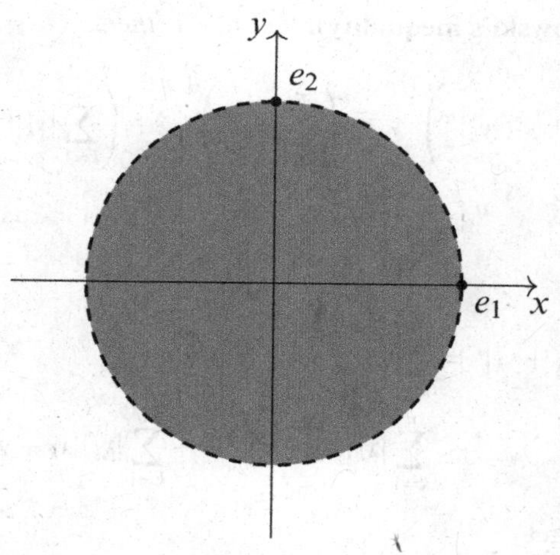

Fig. 2.3 Unit ball for ℓ_2^2

for $x_k, y_k \in \mathbb{R}$.

Proof. Let us take

$$\alpha = \frac{|x_k|}{\left(\sum_{k=1}^n |x_k|^p\right)^{1/p}}, \quad \beta = \frac{|y_k|}{\left(\sum_{k=1}^n |y_k|^q\right)^{1/q}}.$$

By Young's inequality (1.15) we get

$$\frac{|x_k||y_k|}{\left(\sum_{k=1}^n |x_k|^p\right)^{1/p}\left(\sum_{k=1}^n |y_k|^q\right)^{1/q}} \le \frac{1}{p}\frac{|x_k|^p}{\sum_{k=1}^n |x_k|^p} + \frac{1}{q}\frac{|y_k|^q}{\sum_{k=1}^n |y_k|^q}.$$

Termwise summation gives

$$\frac{\sum_{k=1}^n |x_k||y_k|}{\left(\sum_{k=1}^n |x_k|^p\right)^{1/p}\left(\sum_{k=1}^n |y_k|^q\right)^{1/q}} \le \frac{1}{p} + \frac{1}{q}$$

and from this we get

$$\sum_{k=1}^n |x_k y_k| \le \left(\sum_{k=1}^n |x_k|^p\right)^{1/p}\left(\sum_{k=1}^n |y_k|^q\right)^{1/q}.$$

□

We can interpret the inequality (2.2) in the following way: If $\mathbf{x} \in \ell_p^n$ and $\mathbf{y} \in \ell_q^n$ then $\mathbf{x} \odot \mathbf{y} \in \ell_1^n$ where $\odot$ stands for component-wise multiplication and moreover

$$\|\mathbf{x} \odot \mathbf{y}\|_{\ell_1^n} \le \|\mathbf{x}\|_{\ell_p^n} \|\mathbf{y}\|_{\ell_q^n}.$$

Lemma 2.4 (Minkowski's inequality). *Let $p \geq 1$, then*

$$\left(\sum_{k=1}^{n} |x_k + y_k|^p\right)^{1/p} \leq \left(\sum_{k=1}^{n} |x_k|^p\right)^{1/p} + \left(\sum_{k=1}^{n} |y_k|^p\right)^{1/p} \tag{2.3}$$

for $x_k, y_k \in \mathbb{R}$.

Proof. We have

$$\begin{aligned}\sum_{k=1}^{n} |x_k + y_k|^p &= \sum_{k=1}^{n} |x_k + y_k|^{p-1} |x_k + y_k| \\ &\leq \sum_{k=1}^{n} |x_k||x_k + y_k|^{p-1} + \sum_{k=1}^{n} |y_k||x_k + y_k|^{p-1}\end{aligned}$$

By Lemma 2.3 we get

$$\sum_{k=1}^{n} |x_k + y_k|^p \leq \left[\left(\sum_{k=1}^{n} |x_k|^p\right)^{1/p} + \left(\sum_{k=1}^{n} |y_k|^p\right)^{1/p}\right] \left(\sum_{k=1}^{n} |x_k + y_k|^{(p-1)q}\right)^{1/q}.$$

Since $\dfrac{1}{p} + \dfrac{1}{q} = 1$, then $p = (p-1)q$, from which

$$\sum_{k=1}^{n} |x_k + y_k|^p \leq \left[\left(\sum_{k=1}^{n} |x_k|^p\right)^{1/p} + \left(\sum_{k=1}^{n} |y_k|^p\right)^{1/p}\right] \left(\sum_{k=1}^{n} |x_k + y_k|^p\right)^{1/q},$$

then

$$\left(\sum_{k=1}^{n} |x_k + y_k|^p\right)^{1-\frac{1}{q}} \leq \left(\sum_{k=1}^{n} |x_k|^p\right)^{1/p} + \left(\sum_{k=1}^{n} |y_k|^p\right)^{1/p},$$

which entails (2.3). □

2.2 Lebesgue Sequence Spaces

We now want to extend the n-dimensional ℓ_p^n space into an infinite dimensional sequence space in a natural way.

Definition 2.5. The *Lebesgue sequence space* (also known as *discrete Lebesgue space*) with $1 \leq p < \infty$, denoted by ℓ_p or sometimes also by $\ell_p(\mathbb{N})$, stands for the set of all sequences of real numbers $\mathbf{x} = \{x_n\}_{n \in \mathbb{N}}$ such that $\sum_{k=1}^{\infty} |x_k|^p < \infty$. We endow the Lebesgue sequence space with the norm,

$$\|\mathbf{x}\|_{\ell_p} = \|\{x_n\}_{n \in \mathbb{N}}\|_{\ell_p} = \left(\sum_{k=1}^{\infty} |x_k|^p\right)^{1/p}, \tag{2.4}$$

where $\mathbf{x} \in \ell_p$. ⊘

We leave as Problem 2.24 to show that this is indeed a norm in ℓ_p, therefore $(\ell_p, \|\cdot\|_{\ell_p})$ is a normed space.

We will denote by $\mathbb{R}^\infty$ the set of all sequences of real numbers $\mathbf{x} = \{x_n\}_{n\in\mathbb{N}}$.

Example 2.6. The *Hilbert cube* $\mathfrak{H}$ is defined as the set of all real sequences $\{x_n\}_{n\in\mathbb{N}}$ such that $0 \le x_n \le 1/n$, i.e.

$$\mathfrak{H} := \{\mathbf{x} \in \mathbb{R}^\infty : 0 \le x_n \le 1/n\}.$$

By the hyper-harmonic series we have that the Hilbert cube is not contained in ℓ_1 but is contained in all ℓ_p with $p > 1$. ⊘

Let us show that ℓ_p is a subspace of the space $\mathbb{R}^\infty$. Let $\mathbf{x}$ and $\mathbf{y}$ be elements of ℓ_p and α, β be real numbers. By Lemma 2.4 we have that

$$\left(\sum_{k=1}^{n} |\alpha x_k + \beta y_k|^p\right)^{1/p} \le |\alpha| \left(\sum_{k=1}^{n} |x_k|^p\right)^{1/p} + |\beta| \left(\sum_{k=1}^{n} |y_k|^p\right)^{1/p}. \tag{2.5}$$

Taking limits in (2.5), first to the right-hand side and after to the left-hand side, we arrive at

$$\left(\sum_{k=1}^{\infty} |\alpha x_k + \beta y_k|^p\right)^{1/p} \le |\alpha| \left(\sum_{k=1}^{\infty} |x_k|^p\right)^{1/p} + |\beta| \left(\sum_{k=1}^{\infty} |y_k|^p\right)^{1/p}, \tag{2.6}$$

and this shows that $\alpha\mathbf{x} + \beta\mathbf{y}$ is an element of ℓ_p and therefore ℓ_p is a subspace of $\mathbb{R}^\infty$.

The Lebesgue sequence space ℓ_p is a complete normed space for all $1 \le p \le \infty$. We first prove for the case of finite exponent and for the case of $p = \infty$ it will be shown in Theorem 2.11.

Theorem 2.7. *The space $\ell_p(\mathbb{N})$ is a Banach space when $1 \le p < \infty$.*

Proof. Let $\{\mathbf{x}_n\}_{n\in\mathbb{N}}$ be a Cauchy sequence in $\ell_p(\mathbb{N})$, where we take the sequence $\mathbf{x}_n$ as $\mathbf{x}_n = (x_1^{(n)}, x_2^{(n)}, \ldots)$. Then for any $\varepsilon > 0$ there exists an $n_0 \in \mathbb{N}$ such that if $n, m \ge n_0$, then $\|\mathbf{x}_n - \mathbf{x}_m\|_{\ell_p} < \varepsilon$, i.e.

$$\left(\sum_{j=1}^{\infty} |x_j^{(n)} - x_j^{(m)}|^p\right)^{1/p} < \varepsilon, \tag{2.7}$$

whenever $n, m \ge n_0$. From (2.7) it is immediate that for all $j = 1, 2, 3, \ldots$

$$|x_j^{(n)} - x_j^{(m)}| < \varepsilon, \tag{2.8}$$

whenever $n, m \ge n_0$. Taking a fixed j from (2.8) we see that $(x_j^{(1)}, x_j^{(2)}, \ldots)$ is a Cauchy sequence in $\mathbb{R}$, therefore there exists $x_j \in \mathbb{R}$ such that $\lim_{m\to\infty} x_j^{(m)} = x_j$.

Let us define $\mathbf{x} = (x_1, x_2, \ldots)$ and show that $\mathbf{x}$ is in ℓ_p and $\lim_{n\to\infty} \mathbf{x}_n = \mathbf{x}$. From (2.7) we have that for all $n, m \geq n_0$

$$\sum_{j=1}^{k} |x_j^{(m)} - x_j^{(n)}|^p < \varepsilon^p, \quad k = 1, 2, 3, \ldots$$

from which

$$\sum_{j=1}^{k} |x_j - x_j^{(n)}|^p = \sum_{j=1}^{k} |\lim_{m\to\infty} x_j^{(m)} - x_j^{(n)}|^p \leq \varepsilon^p,$$

whenever $n \geq n_0$, This shows that $\mathbf{x} - \mathbf{x}_n \in \ell_p$ and we also deduce that $\lim_{n\to\infty} \mathbf{x}_n = \mathbf{x}$. Finally in virtue of the Minkowski inequality we have

$$\begin{aligned}\left(\sum_{j=1}^{\infty} |x_j|^p\right)^{1/p} &= \left(\sum_{j=1}^{\infty} |x_j^{(n)} + x_j - x_j^{(n)}|^p\right)^{1/p} \\ &\leq \left(\sum_{j=1}^{\infty} |x_j^{(n)}|^p\right)^{1/p} + \left(\sum_{j=1}^{\infty} |x_j - x_j^{(n)}|^p\right)^{1/p},\end{aligned}$$

which shows that $\mathbf{x}$ is in $\ell_p(\mathbb{N})$ and this completes the proof. □

The next result shows that the Lebesgue sequence spaces are separable when the exponent p is finite, i.e., the space ℓ_p admits an enumerable dense subset.

Theorem 2.8. *The space $\ell_p(\mathbb{N})$ is separable whenever $1 \leq p < \infty$.*

Proof. Let M be the set of all sequences of the form $\mathbf{q} = (q_1, q_2, \ldots, q_n, 0, 0, \ldots)$ where $n \in \mathbb{N}$ and $q_k \in \mathbb{Q}$. We will show that M is dense in ℓ_p. Let $\mathbf{x} = \{x_k\}_{k\in\mathbb{N}}$ be an arbitrary element of ℓ_p, then for $\varepsilon > 0$ there exists n which depends on ε such that

$$\sum_{k=n+1}^{\infty} |x_k|^p < \varepsilon^p/2.$$

Now, since $\overline{\mathbb{Q}} = \mathbb{R}$, we have that for each x_k there exists a rational q_k such that

$$|x_k - q_k| < \frac{\varepsilon}{\sqrt[p]{2n}},$$

then

$$\sum_{k=1}^{n} |x_k - q_k|^p < \varepsilon^p/2,$$

which entails

$$\|\mathbf{x} - \mathbf{q}\|_{\ell_p}^p = \sum_{k=1}^{n} |x_k - q_k|^p + \sum_{k=n+1}^{\infty} |x_k|^p < \varepsilon^p,$$

and we arrive at $\|\mathbf{x} - \mathbf{q}\|_{\ell_p} < \varepsilon$. This shows that M is dense in ℓ_p, implying that ℓ_p is separable since M is enumerable. □

With the notion of Schauder basis (recall the definition of Schauder basis in Definition B.3), we now study the problem of duality for the Lebesgue sequence space.

Theorem 2.9. *Let $1 < p < \infty$. The dual space of $\ell_p(\mathbb{N})$ is $\ell_q(\mathbb{N})$ where $\frac{1}{p} + \frac{1}{q} = 1$.*

Proof. A Schauder basis of ℓ_p is $e_k = \{\delta_{kj}\}_{j\in\mathbb{N}}$ where $k \in \mathbb{N}$ and δ_{kj} stands for the Kronecker delta, i.e., $\delta_{kj} = 1$ if $k = j$ and 0 otherwise. If $f \in (\ell_p)^*$, then $f(\mathbf{x}) = \sum_{k\in\mathbb{N}} \alpha_k f(e_k)$, $\mathbf{x} = \{\alpha_k\}_{k\in\mathbb{N}}$. We define $T(f) = \{f(e_k)\}_{k\in\mathbb{N}}$. We want to show that the image of T is in ℓ_q, for that we define for each n, the sequence $\mathbf{x}^n = (\xi_k^{(n)})_{k=1}^{\infty}$ with

$$\xi_k^{(n)} = \begin{cases} \frac{|f(e_k)|^q}{f(e_k)} & \text{if} \quad k \le n \text{ and } f(e_k) \ne 0, \\ 0 & \text{if} \quad k > n \text{ or } f(e_k) = 0. \end{cases}$$

Then

$$f(\mathbf{x}^n) = \sum_{k\in\mathbb{N}} \xi_k^{(n)} f(e_k) = \sum_{k=1}^{n} |f(e_k)|^q.$$

Moreover

$$\begin{aligned} f(\mathbf{x}^n) &\le \|f\| \|\mathbf{x}^n\|_p \\ &= \|f\| \left(\sum_{k=1}^{n} |\xi_k^{(n)}|^p \right)^{\frac{1}{p}} \\ &= \|f\| \left(\sum_{k=1}^{n} |f(e_k)|^{qp-p} \right)^{\frac{1}{p}} \\ &= \|f\| \left(\sum_{k=1}^{n} |f(e_k)|^q \right)^{\frac{1}{p}}, \end{aligned}$$

from which

$$\begin{aligned} \left(\sum_{k=1}^{n} |f(e_k)|^q \right)^{1-\frac{1}{p}} &= \left(\sum_{k=1}^{n} |f(e_k)|^q \right)^{\frac{1}{q}} \\ &\le \|f\|. \end{aligned}$$

Taking $n \to \infty$, we obtain

$$\left(\sum_{k=1}^{\infty} |f(e_k)|^q \right)^{\frac{1}{q}} \le \|f\|$$

where $\{f(e_k)\}_{k\in\mathbb{N}} \in \ell_q$.

Now, we affirm that:

(i) T is onto. In effect given $b = (\beta_k)_{k\in\mathbb{N}} \in \ell_q$, we can associate a bounded linear functional $g \in (\ell_p)^*$, given by $g(\mathbf{x}) = \sum_{k=1}^{\infty} \alpha_k \beta_k$ with $\mathbf{x} = (\alpha_k)_{k\in\mathbb{N}} \in \ell_p$ (the boundedness is deduced by Hölder's inequality). Then $g \in (\ell_p)^*$.

(ii) T is 1-1. This is almost straightforward to check.

(iii) T is an isometry. We see that the norm of f is the ℓ_q norm of Tf

$$
\begin{aligned}
|f(\mathbf{x})| &= \left| \sum_{k\in\mathbb{N}} \alpha_k f(e_k) \right| \\
&\leq \left(\sum_{k\in\mathbb{N}} |\alpha_k|^p \right)^{\frac{1}{p}} \left(\sum_{k\in\mathbb{N}} |f(e_k)|^q \right)^{\frac{1}{q}} \\
&= \|x\| \left(\sum_{k\in\mathbb{N}} |f(e_k)|^q \right)^{\frac{1}{q}}.
\end{aligned}
$$

Taking the supremum over all x of norm 1, we have that

$$
\|f\| \leq \left(\sum_{k\in\mathbb{N}} |f(e_k)|^q \right)^{\frac{1}{q}}.
$$

Since the other inequality is also true, we can deduce the equality

$$
\|f\| = \left(\sum_{k\in\mathbb{N}} |f(e_k)|^q \right)^{\frac{1}{q}},
$$

with which we establish the desired isomorphism $f \to \{f(e_k)\}_{k\in\mathbb{N}}$.

□

The ℓ_p spaces satisfy an embedding property, forming a nested sequence of Lebesgue sequences spaces.

Theorem 2.10. *If* $0 < p < q < \infty$, *then* $\ell_p(\mathbb{N}) \subsetneq \ell_q(\mathbb{N})$.

Proof. Let $\mathbf{x} \in \ell_p$, then $\sum_{n=1}^{\infty} |x_n|^p < \infty$. Therefore there exists $n_0 \in \mathbb{N}$ such that if $n \geq n_0$, then $|x_n| < 1$. Now, since $0 < p < q$, then $0 < q - p$ and $|x_n|^{q-p} < 1$ if $n > n_0$, by which $|x_n|^q < |x_n|^p$ if $n > n_0$. Let $M = \max\{|x_1|^{q-p}, |x_2|^{q-p}, \ldots, |x_{n_0}|^{q-p}, 1\}$, then

$$
\sum_{n=1}^{\infty} |x_n|^q = \sum_{n=1}^{\infty} |x_n|^p |x_n|^{q-p} < M \sum_{n=1}^{\infty} |x_n|^p < +\infty,
$$

implying that $\mathbf{x} \in \ell_q$.

To show that $\ell_p(\mathbb{N}) \neq \ell_q(\mathbb{N})$, we take the following sequence $x_n = n^{-1/p}$ for all $n \in \mathbb{N}$ with $1 \leq p < q \leq \infty$, and since $p < q$, then $\frac{q}{p} > 1$. Now we have

$$\sum_{n=1}^{\infty} |x_n|^q = \sum_{n=1}^{\infty} \frac{1}{n^{q/p}} < \infty.$$

The last series is convergent since it is a hyper-harmonic series with exponent bigger than 1, therefore $\mathbf{x} \in \ell_q(\mathbb{N})$. On the other hand

$$\sum_{n=1}^{\infty} |x_n|^p = \sum_{n=1}^{\infty} \frac{1}{n}$$

and we get the harmonic series, which entails that $\mathbf{x} \notin \ell_p(\mathbb{N})$. □

2.3 Space of Bounded Sequences

The *space of bounded sequences*, denoted by ℓ_∞ or sometimes $\ell_\infty(\mathbb{N})$, is the set of all real bounded sequences $\{x_n\}_{n\in\mathbb{N}}$ (it is clear that ℓ_∞ is a vector space). We will take the norm in this space as

$$\|\mathbf{x}\|_\infty = \|\mathbf{x}\|_{\ell_\infty} = \sup_{n\in\mathbb{N}} |x_n|, \tag{2.9}$$

where $\mathbf{x} = (x_1, x_2, \ldots, x_n, \ldots)$. The verification that (2.9) is indeed a norm is left to the reader.

An almost immediate property of the ℓ_∞-space is its completeness, inheriting this property from the completeness of the real line.

Theorem 2.11. *The space ℓ_∞ is a Banach space.*

Proof. Let $\{\mathbf{x}_n\}_{n\in\mathbb{N}}$ be a Cauchy sequence in ℓ_∞, where $\mathbf{x}_n = (x_1^{(n)}, x_2^{(n)}, \ldots)$. Then for any $\varepsilon > 0$ there exists $n_0 > 0$ such that if $m, n \geq n_0$ then

$$\|\mathbf{x}_m - \mathbf{x}_n\|_\infty < \varepsilon.$$

Therefore for fixed j we have that if $m, n \geq n_0$, then

$$|x_j^{(m)} - x_j^{(n)}| < \varepsilon \tag{2.10}$$

resulting that for all fixed j the sequence $(x_j^{(1)}, x_j^{(2)}, \ldots)$ is a Cauchy sequence in $\mathbb{R}$, and this implies that there exists $x_j \in \mathbb{R}$ such that $\lim_{m\to\infty} x_j^{(m)} = x_j$.

Let us define $\mathbf{x} = (x_1, x_2, \ldots)$. Now we want to show that $\mathbf{x} \in \ell_\infty$ and $\lim_{n\to\infty} \mathbf{x}_n = \mathbf{x}$.

From (2.10) we have that for $n \geq n_0$, then

$$\left|x_j - x_j^{(n)}\right| = \left|\lim_{n\to\infty} x_j^{(m)} - x_j^{(n)}\right| \leq \varepsilon, \tag{2.11}$$

since $\mathbf{x}_n = \{x_j^{(n)}\}_{j\in\mathbb{N}} \in \ell_\infty$, there exists a real number M_n such that $\left|x_j^{(n)}\right| \leq M_n$ for all j.

By the triangle inequality, we have

$$|x_j| \leq \left|x_j - x_j^{(n)}\right| + \left|x_j^{(n)}\right| \leq \varepsilon + M_n$$

whenever $n \geq n_0$, this inequality being true for any j. Moreover, since the right-hand side does not depend on j, therefore $\{x_j\}_{j\in\mathbb{N}}$ is a sequence of bounded real numbers, this implies that $\mathbf{x} = \{x_j\}_{j\in\mathbb{N}} \in \ell_\infty$.

From (2.11) we also obtain

$$\|\mathbf{x}_n - \mathbf{x}\|_{\ell_\infty} = \sup_{j\in\mathbb{N}} \left|x_j^{(n)} - x_j\right| < \varepsilon.$$

whenever $n \geq n_0$. From this we conclude that $\lim_{n\to\infty} \mathbf{x}_n = \mathbf{x}$ and therefore ℓ_∞ is complete. □

The following result shows a "natural" way to introduce the norm in the ℓ_∞ space via a limiting process.

Theorem 2.12. *Taking the norm of Lebesgue sequence space as in* (2.4) *we have that* $\lim_{p\to\infty} \|\mathbf{x}\|_{\ell_p} = \|\mathbf{x}\|_{\ell_\infty}$.

Proof. Observe that $|x_k| \leq \left(\sum_{k=1}^n |x_k|^p\right)^{\frac{1}{p}}$, therefore $|x_k| \leq \|\mathbf{x}\|_{\ell_p}$ for $k = 1,2,3,\ldots,n$, from which

$$\sup_{1\leq k\leq n} |x_k| \leq \|\mathbf{x}\|_{\ell_p},$$

whence

$$\|\mathbf{x}\|_{\ell_\infty} \leq \liminf_{p\to\infty} \|\mathbf{x}\|_{\ell_p}. \tag{2.12}$$

On the other hand, note that

$$\left(\sum_{k=1}^n |x_k|^p\right)^{\frac{1}{p}} \leq \left(\sum_{k=1}^n \left(\sup_{1\leq k\leq n} |x_k|\right)^p\right)^{\frac{1}{p}} \leq n^{\frac{1}{p}} \|\mathbf{x}\|_{\ell_\infty},$$

then for all $\varepsilon > 0$, there exists N such that

$$\|\mathbf{x}\|_{\ell_p} \leq \left(\sum_{k=1}^N |x_k|^p + \varepsilon\right)^{\frac{1}{p}} \leq \left(\|\mathbf{x}\|_{\ell_\infty}^p N + \varepsilon\right)^{\frac{1}{p}} \leq \|\mathbf{x}\|_{\ell_\infty} \left(N + \frac{\varepsilon}{\|\mathbf{x}\|_{\ell_\infty}^p}\right)^{\frac{1}{p}},$$

therefore

$$\limsup_{p\to\infty} \|\mathbf{x}\|_{\ell_p} \leq \|\mathbf{x}\|_{\ell_\infty}. \tag{2.13}$$

Combining (2.12) and (2.13) results

$$\|\mathbf{x}\|_{\ell_\infty} \leq \liminf_{p\to\infty} \|\mathbf{x}\|_{\ell_p} \leq \limsup_{p\to\infty} \|\mathbf{x}\|_{\ell_p} \leq \|\mathbf{x}\|_{\ell_\infty},$$

and from this we conclude that $\lim_{p\to\infty} \|\mathbf{x}\|_{\ell_p} = \|\mathbf{x}\|_{\ell_\infty}$. □

Now we study the dual space of ℓ_1 which is ℓ_∞.

Theorem 2.13. *The dual space of ℓ_1 is ℓ_∞.*

Proof. For all $\mathbf{x} \in \ell_1$, we can write $\mathbf{x} = \sum_{k=1}^{\infty} \alpha_k e_k$, where $e_k = (\delta_{kj})_{j=1}^{\infty}$ forms a Schauder basis in ℓ_1, since

$$\mathbf{x} - \sum_{k=1}^{n} \alpha_k e_k = (\underbrace{0,\ldots,0}_{n}, \alpha_{n+1},\ldots)$$

and

$$\left\| \mathbf{x} - \sum_{k=1}^{n} \alpha_k e_k \right\|_{\ell_1} = \left\| \sum_{k=n+1}^{\infty} \alpha_k e_k \right\|_{\ell_1} \to 0$$

since the series $\sum_{k=1}^{\infty} \alpha_k e_k$ is convergent.

Let us define $T(f) = \{f(e_k)\}_{k\in\mathbb{N}}$, for all $f \in (\ell_1)^*$. Since $f(\mathbf{x}) = \sum_{k\in\mathbb{N}} \alpha_k f(e_k)$, then $|f(e_k) \leq \|f\|$, since $\|e_k\|_{\ell_1} = 1$. In consequence, $\sup_{k\in\mathbb{N}} |f(e_k)| \leq \|f\|$, therefore $\{f(e_k)\}_{k\in\mathbb{N}} \in \ell_\infty$.

We affirm that:

(i) T is onto. In fact, for all $b = \{\beta_k\}_{k\in\mathbb{N}} \in \ell_\infty$, let us define $q : \ell_1 \to \mathbb{R}$ as $g(\mathbf{x}) = \sum_{k\in\mathbb{N}} \alpha_k \beta_k$ if $\mathbf{x} = \{\alpha_k\}_{k\in\mathbb{N}} \in \ell_\infty$. The functional g is bounded and linear since

$$|g(\mathbf{x})| \leq \sum_{k\in\mathbb{N}} |\alpha_k \beta_k| \leq \sup_{k\in\mathbb{N}} |\beta_k| \sum_{k\in\mathbb{N}} |\alpha_k| = \|\mathbf{x}\|_{\ell_1} \cdot \sup_{k\in\mathbb{N}} |\beta_k|,$$

then $g \in (\ell_1)^*$. Moreover, since $g(e_k) = \sum_{j\in\mathbb{N}} \delta_{kj} \beta_j$,

$$T(g) = \{g(e_k)\}_{k\in\mathbb{N}} = \{\beta_k\}_{k\in\mathbb{N}} = b.$$

(ii) T is 1-1. If $Tf_1 = Tf_2$, then $f_1(e_k) = f_2(e_k)$, for all k. Since we have $f_1(\mathbf{x}) = \sum_{k\in\mathbb{N}} \alpha_k f_1(e_k)$ and $f_2(\mathbf{x}) = \sum_{k\in\mathbb{N}} \alpha_k f_2(e_k)$, then $f_1 = f_2$.

(iii) T is an isometry. In fact,

$$\|Tf\|_\infty = \sup_{k\in\mathbb{N}} |f(e_k)| \leq \|f\| \tag{2.14}$$

and

$$|f(\mathbf{x})| = \left|\sum_{k\in\mathbb{N}} \alpha_k f(e_k)\right| \leq \sup_{j\in\mathbb{N}} |f(e_k)| \sum_{k\in\mathbb{N}} |\alpha_k| = \|\mathbf{x}\|_{\ell_1} \sup_{k\in\mathbb{N}} |f(e_k)|.$$

Then

$$\|f\| \leq \sup_{k\in\mathbb{N}} |f(e_k)| = \|Tf\|_\infty. \tag{2.15}$$

Combining (2.14) and (2.15) we get that $\|Tf\|_\infty = \|f\|$. We thus showed that the spaces $(\ell_1)^*$ and ℓ_∞ are isometric.

□

One of the main difference between ℓ_p and ℓ_∞ spaces is the separability issue. The space of bounded sequence ℓ_∞ is not separable, contrasting with the separability of the ℓ_p spaces whenever $1 \leq p < \infty$, see Theorem 2.8.

Theorem 2.14. *The space ℓ_∞ is not separable.*

Proof. Let us take any enumerable sequence of elements of ℓ_∞, namely $\{\mathbf{x}_n\}_{n\in\mathbb{N}}$, where we take the sequences in the form

$$\mathbf{x}_1 = \left(x_1^{(1)}, x_2^{(1)}, x_3^{(1)}, \ldots, x_k^{(1)}, \ldots\right)$$
$$\mathbf{x}_2 = \left(x_1^{(2)}, x_2^{(2)}, x_3^{(2)}, \ldots, x_k^{(2)}, \ldots\right)$$
$$\mathbf{x}_3 = \left(x_1^{(3)}, x_2^{(3)}, x_3^{(3)}, \ldots, x_k^{(3)}, \ldots\right)$$
$$\cdots\cdots\cdots\cdots\cdots\cdots\cdots\cdots\cdots$$
$$\mathbf{x}_k = \left(x_1^{(k)}, x_2^{(k)}, x_3^{(k)}, \ldots, x_k^{(k)}, \ldots\right)$$
$$\cdots\cdots\cdots\cdots\cdots\cdots\cdots\cdots\cdots$$

We now show that there exists an element in ℓ_∞ which is at a distance bigger than 1 for all elements of $\{\mathbf{x}_n\}_{n\in\mathbb{N}}$, showing the non-separability nature of the ℓ_∞ space.
Let us take $\mathbf{x} = \{x_n\}_{n\in\mathbb{N}}$ as

$$x_n = \begin{cases} 0, & \text{if } |x_n^{(n)}| \geq 1; \\ x_n = x_n^{(n)} + 1, & \text{if } |x_n^{(n)}| < 1. \end{cases}$$

It is clear that $\mathbf{x} \in \ell_\infty$ and $\|\mathbf{x} - \mathbf{x}_n\|_{\ell_\infty} > 1$ for all $n \in \mathbb{N}$, which entails that ℓ_∞ is not separable. □

We now define some subspaces of ℓ_∞, which are widely used in functional analysis, for example, to construct counter-examples.

Definition 2.15. Let $\mathbf{x} = (x_1, x_1, \ldots)$.
By c we denote the subspace of ℓ_∞ such that $\lim_{n\to\infty} x_n$ exists and is finite.
By c_0 we denote the subspace of ℓ_∞ such that $\lim_{n\to\infty} x_n = 0$.
By c_{00} we denote the subspace of ℓ_∞ such that $\mathrm{supp}(\mathbf{x})$ is finite. ⊘

These newly introduced spaces enjoy some interesting properties, e.g., c_0 is the closure of c_{00} in ℓ_∞. For more properties, see Problem 2.20.

2.4 Hardy and Hilbert Inequalities

We now deal with the discrete version of the well-known Hardy inequality.

Theorem 2.16 (Hardy's inequality). *Let $\{a_n\}_{n\in\mathbb{N}}$ be a sequence of real positive numbers such that $\sum_{n=1}^{\infty} a_n^p < \infty$. Then*

$$\sum_{n=1}^{\infty}\left(\frac{1}{n}\sum_{k=1}^{n} a_k\right)^p \le \left(\frac{p}{p-1}\right)^p \sum_{n=1}^{\infty} a_n^p.$$

Proof. Let $\alpha_n = \frac{A_n}{n}$ where $A_n = a_1 + a_2 + \cdots + a_n$, i.e., $A_n = n\alpha_n$, then

$$a_1 + a_2 + \cdots + a_n = n\alpha_n, \tag{2.16}$$

from which we get that $a_n = n\alpha_n - (n-1)\alpha_{n-1}$. Let us consider now

$$\begin{aligned}\alpha_n^p - \frac{p}{p-1}\alpha_n^{p-1}a_n &= \alpha_n^p - \frac{p}{p-1}\left[n\alpha_n - (n-1)\alpha_{n-1}\right]\alpha_n^{p-1}\\ &= \alpha_n^p - \frac{pn}{p-1}\alpha_n\alpha_n^{p-1} + \frac{p(n-1)}{p-1}\alpha_{n-1}\alpha_n^{p-1}.\end{aligned}$$

In virtue of Corollary 1.10 we have

$$\begin{aligned}\frac{p(n-1)}{p-1}\alpha_{n-1}\alpha_n^{p-1} &\le \frac{p(n-1)}{p-1}\frac{\alpha_{n-1}^p}{p} + \frac{p(n-1)}{p-1}\frac{\alpha_n^{q(p-1)}}{q}\\ &= \frac{n-1}{p-1}\alpha_{n-1}^p + \frac{p(n-1)}{p-1}\left(1-\frac{1}{p}\right)\alpha_n^p\\ &= \frac{n-1}{p-1}\alpha_{n-1}^p + (n-1)\alpha_n^p,\end{aligned}$$

therefore

$$\begin{aligned}\alpha_n^p - \frac{p}{p-1}\alpha_n^{p-1}a_n &\le \alpha_n^p - \frac{pn}{p-1}\alpha_n^p + \frac{n-1}{p-1}\alpha_{n-1}^p + (n-1)\alpha_n^p\\ &= \frac{p\alpha_n^p - \alpha_n^p - pn\alpha_n^p}{p-1} + \frac{(n-1)\alpha_{n-1}^p + (p-1)(n-1)\alpha_n^p}{p-1}\\ &= \frac{p\alpha_n^p - \alpha_n^p - pn\alpha_n^p + (n-1)\alpha_{n-1}^p + (pn-p-n+1)\alpha_n^p}{p-1}\\ &= \frac{1}{p-1}\left[(n-1)\alpha_{n-1}^p - n\alpha_n^p\right],\end{aligned}$$

from which

$$\begin{aligned}\sum_{n=1}^{N}\alpha_n^p-\frac{p}{p-1}\sum_{n=1}^{N}\alpha_n^{p-1}a_n&\leq\frac{1}{p-1}\sum_{n=1}^{N}\left[(n-1)\alpha_{n-1}^p-n\alpha_n^p\right]\\&=\frac{1}{p-1}\left[-\alpha_1^p+\alpha_1^p-2\alpha_2^p+\cdots-N\alpha_N^p\right]\\&=-\frac{N\alpha_N^p}{p-1}\leq 0.\end{aligned}$$

Then

$$\sum_{n=1}^{N}\alpha_n^p\leq\frac{p}{p-1}\sum_{n=1}^{N}\alpha_n^{p-1}a_n.$$

By Hölder's inequality we have that

$$\begin{aligned}\sum_{n=1}^{\infty}\alpha_n^p&\leq\frac{p}{p-1}\left(\sum_{n=1}^{\infty}a_n^p\right)^{\frac{1}{p}}\left(\sum_{n=1}^{\infty}\alpha_n^{q(p-1)}\right)^{\frac{1}{q}}\\&=\frac{p}{p-1}\left(\sum_{n=1}^{\infty}a_n^p\right)^{\frac{1}{p}}\left(\sum_{n=1}^{\infty}\alpha_n^p\right)^{\frac{1}{q}},\end{aligned}$$

then

$$\left(\sum_{n=1}^{\infty}\alpha_n^p\right)^{1-\frac{1}{q}}\leq\frac{p}{p-1}\left(\sum_{n=1}^{\infty}a_n^p\right)^{\frac{1}{p}}$$

and this implies

$$\sum_{n=1}^{\infty}\left(\frac{1}{n}\sum_{k=1}^{\infty}a_k\right)^p\leq\left(\frac{p}{p-1}\right)^p\sum_{n=1}^{\infty}a_n^p.$$

□

We now want to study the so-called Hilbert inequality. We need to remember some basic facts about complex analysis, namely

$$\frac{\pi}{\sin(\pi z)}=\frac{1}{z}+\sum_{n=1}^{\infty}(-1)^n\left(\frac{1}{z+n}+\frac{1}{z-n}\right). \tag{2.17}$$

Let us consider the function

$$f(z)=\frac{1}{\sqrt[p]{z}(z+1)}\qquad(p>1)$$

defined in the region $D_1=\{z\in\mathbb{C}:0<|z|<1\}$. We want to obtain the Laurent expansion. In fact, if $|z|<1$, then

$$\frac{1}{1+z} = \frac{1}{1-(-z)} = \sum_{n=0}^{\infty}(-z)^n = \sum_{n=0}^{\infty}(-1)^n z^n,$$

therefore

$$f(z) = \sum_{n=0}^{\infty}(-1)^n z^{n-\frac{1}{p}}. \tag{2.18}$$

By the same reasoning, let us consider

$$g(z) = \frac{1}{z^{1+\frac{1}{p}}\left(1+\frac{1}{z}\right)}$$

defined in the region $D_2 = \{z \in \mathbb{C} : |z| > 1\}$. Since $\left|\frac{1}{z}\right| < 1$, then

$$\frac{1}{1+\frac{1}{z}} = \frac{1}{1-\left(-\frac{1}{z}\right)} = \sum_{n=0}^{\infty}\left(-\frac{1}{z}\right)^n = \sum_{n=0}^{\infty}(-1)^n z^{-n}.$$

Therefore

$$g(z) = \sum_{n=0}^{\infty}(-1)^n z^{-n-1-\frac{1}{p}}. \tag{2.19}$$

We now obtain some auxiliary inequality before showing the validity of the Hilbert inequality (2.20).

Theorem 2.17 *For each positive number m and for all real $p > 1$ we have*

$$\sum_{n=1}^{\infty}\frac{m^{\frac{1}{p}}}{n^{\frac{1}{p}}(m+n)} \leq \frac{\pi}{\sin\left(\frac{\pi}{p}\right)}.$$

Proof. Note that

$$\begin{aligned}
\sum_{n=1}^{\infty}\frac{m^{\frac{1}{p}}}{n^{\frac{1}{p}}(m+n)} &\leq \int_0^{\infty}\frac{m^{\frac{1}{p}}}{x^{\frac{1}{p}}(m+x)}\,dx \\
&= \int_0^{\infty}\frac{dz}{z^{\frac{1}{p}}(1+z)} \\
&= \int_0^{1}\frac{dz}{z^{\frac{1}{p}}(1+z)} + \int_1^{\infty}\frac{dz}{z^{1+\frac{1}{p}}\left(1+\frac{1}{z}\right)}.
\end{aligned}$$

By (2.18) and (2.19) we deduce that

$$\sum_{n=1}^{\infty}\frac{m^{\frac{1}{p}}}{n^{\frac{1}{p}}(m+n)} \leq \int_0^{1}\left(\sum_{n=0}^{\infty}(-1)^n z^{n-\frac{1}{p}}\right)dz + \int_1^{\infty}\left(\sum_{n=0}^{\infty}(-1)^n z^{-n-1-\frac{1}{p}}\right)dz$$

$$= \sum_{n=0}^{\infty}(-1)^n \int_0^1 z^{n-\frac{1}{p}}\,\mathrm{d}z + \sum_{n=0}^{\infty}(-1)^n \int_1^{\infty} z^{-n-1-\frac{1}{p}}\,\mathrm{d}z$$

$$= \sum_{n=0}^{\infty} \frac{(-1)^n}{n-\frac{1}{p}+1} + \sum_{n=0}^{\infty} \frac{(-1)^n}{\frac{1}{p}+n}$$

$$= \sum_{n=1}^{\infty} \frac{(-1)^n}{\frac{1}{p}-n} + p + \sum_{n=1}^{\infty} \frac{(-1)^n}{\frac{1}{p}+n}$$

$$= p + \sum_{n=1}^{\infty}(-1)^n \left(\frac{1}{\frac{1}{p}-n} + \frac{1}{\frac{1}{p}+n} \right)$$

$$= \frac{\pi}{\sin\left(\frac{\pi}{p}\right)}.$$

This last one is obtained by (2.17) with $z = \frac{1}{p}$. □

Remark 2.18. In fact the proof of Theorem 2.17 is a two line proof if we remember that

$$\int_0^{\infty} \frac{x^{\alpha-1}}{(1+x)^{\alpha+\beta}}\,\mathrm{d}x = B(\alpha,\beta)$$

and the fact that $B(1-\alpha,\alpha) = \frac{\pi}{\sin \pi\alpha}$, $0 < \alpha < 1$, see Appendix C. ⊘

Before stating and proving the Hilbert inequality we need to digress into the concept of double series. Let $\{x_{k,j}\}_{j,k\in\mathbb{N}}$ be a double sequence, viz. a real-valued function $x : \mathbb{N}\times\mathbb{N} \to \mathbb{R}$. We say that a number L is the limit of the double sequence, denoted by

$$\lim_{k,j\to\infty} x_{k,j} = L,$$

if, for all $\varepsilon > 0$ there exists $n = n(\varepsilon)$ such that

$$|x_{k,j} - L| < \varepsilon$$

whenever $k > n$ and $j > n$. We can now introduce the notion of double series using the known construction for the series, namely

$$\sum_{k,j=1}^{\infty} x_{k,j} = \Sigma$$

if there exists the double limit

$$\lim_{k,j\to\infty} \Sigma_{k,j} = \Sigma$$

where $\Sigma_{k,j}$ is the rectangular partial sum given by

$$\Sigma_{k,j} = \sum_{m=1}^{k} \sum_{n=1}^{j} x_{m,n}.$$

A notion related to the double series is the notion of iterated series, given by

$$\sum_{k=1}^{\infty} \left(\sum_{j=1}^{\infty} x_{k,j} \right) \quad \text{and} \quad \sum_{j=1}^{\infty} \left(\sum_{k=1}^{\infty} x_{k,j} \right).$$

We can visualize the iterated series in the following way. We first represent the double sequence as numbers in an infinite rectangular array and then sum by lines and by columns in the following way:

$$\begin{array}{cccl}
x_{1,1} & x_{1,2} & x_{1,3} & \cdots \to \sum_{j=1}^{\infty} x_{1,j} =: L_1 \\
x_{2,1} & x_{2,2} & x_{2,3} & \cdots \to \sum_{j=1}^{\infty} x_{2,j} =: L_2 \\
x_{3,1} & x_{3,2} & x_{3,3} & \cdots \to \sum_{j=1}^{\infty} x_{2,j} =: L_3 \\
\vdots & \vdots & \vdots & \\
\downarrow & \downarrow & \downarrow & \\
C_1 := \sum_{k=1}^{\infty} x_{k,1} & C_2 := \sum_{k=1}^{\infty} x_{k,2} & C_3 := \sum_{k=1}^{\infty} x_{k,3} &
\end{array}$$

and now the iterated series are given by $\sum_{j=1}^{\infty} C_j$ and $\sum_{k=1}^{\infty} L_k$.

It is necessary some caution when dealing with iterated series since the equality $\sum_{j=1}^{\infty} C_j = \sum_{k=1}^{\infty} L_k$ is in general not true even if the series converges, as the following example shows

$$\begin{array}{cccccccl}
\frac{1}{2} & -\frac{1}{2} & 0 & 0 & 0 & \cdots & \to 0 \\
0 & \frac{3}{4} & -\frac{3}{4} & 0 & 0 & \cdots & \to 0 \\
0 & 0 & \frac{7}{8} & -\frac{7}{8} & 0 & \cdots & \to 0 \\
0 & 0 & 0 & \frac{15}{16} & -\frac{15}{16} & \cdots & \to 0 \\
\vdots & \vdots & \vdots & \vdots & \vdots & & \\
\downarrow & \downarrow & \downarrow & \downarrow & \downarrow & & \\
\frac{1}{2} & \frac{1}{4} & \frac{1}{8} & \frac{1}{16} & \frac{1}{32} & &
\end{array}$$

and clearly the obtained series are different. Fortunately we have a Fubini type theorem for series which states that when a double series is absolutely convergent then the double series and the iterated series are the same, i.e.

$$\sum_{k,j=1}^{\infty} x_{k,j} = \sum_{k=1}^{\infty} \left(\sum_{j=1}^{\infty} x_{k,j} \right) = \sum_{j=1}^{\infty} \left(\sum_{k=1}^{\infty} x_{k,j} \right).$$

Not only that, it is also possible to show a stronger result, that if the terms of an absolutely convergent double series are permuted in any order as a simple series, their sum tends to the same limit.

Theorem 2.19 (Hilbert's inequality). *Let $p,q>1$ be such that $\frac{1}{p}+\frac{1}{q}=1$ and $\{a_n\}_{n\in\mathbb{N}}$, $\{b_n\}_{n\in\mathbb{N}}$ be sequences of nonnegative numbers such that $\sum_{m=1}^{\infty} a_m^p$ and $\sum_{n=1}^{\infty} b_n^q$ are convergent. Then*

$$\sum_{m,n=1}^{\infty} \frac{a_m b_n}{m+n} \leq \frac{\pi}{\sin\left(\frac{\pi}{p}\right)} \left(\sum_{m=1}^{\infty} a_m^p\right)^{\frac{1}{p}} \left(\sum_{n=1}^{\infty} b_n^q\right)^{\frac{1}{q}}. \tag{2.20}$$

Proof. Using Hölder's inequality and Proposition 2.17 we get

$$\begin{aligned}
&\sum_{m,n=1}^{\infty} \frac{a_m b_n}{m+n} \\
&= \sum_{m,n=1}^{\infty} \frac{m^{\frac{1}{pq}}}{n^{\frac{1}{pq}}} \frac{a_m}{(m+n)^{\frac{1}{p}}} \frac{n^{\frac{1}{pq}}}{m^{\frac{1}{pq}}} \frac{b_n}{(m+n)^{\frac{1}{q}}} \\
&\leq \left(\sum_{m,n=1}^{\infty} \left(\frac{m^{\frac{1}{q}}}{n^{\frac{1}{q}}(m+n)}\right) a_m^p\right)^{\frac{1}{p}} \left(\sum_{m,n=1}^{\infty} \left(\frac{n^{\frac{1}{p}}}{m^{\frac{1}{p}}(m+n)}\right) b_n^q\right)^{\frac{1}{q}} \\
&= \left(\sum_{m=1}^{\infty} \left(\sum_{n=1}^{\infty} \frac{m^{\frac{1}{q}}}{n^{\frac{1}{q}}(m+n)}\right) a_m^p\right)^{\frac{1}{p}} \left(\sum_{n=1}^{\infty} \left(\sum_{m=1}^{\infty} \frac{n^{\frac{1}{p}}}{m^{\frac{1}{p}}(m+n)}\right) b_n^q\right)^{\frac{1}{q}} \\
&\leq \left(\sum_{m=1}^{\infty} \frac{\pi}{\sin\frac{\pi}{q}} a_m^p\right)^{\frac{1}{p}} \left(\sum_{n=1}^{\infty} \frac{\pi}{\sin\frac{\pi}{p}} b_n^q\right)^{\frac{1}{q}} \\
&\leq \left(\sum_{m=1}^{\infty} \frac{\pi}{\sin\frac{\pi}{p}} a_m^p\right)^{\frac{1}{p}} \left(\sum_{n=1}^{\infty} \frac{\pi}{\sin\frac{\pi}{p}} b_n^q\right)^{\frac{1}{q}} \\
&= \left(\frac{\pi}{\sin\frac{\pi}{p}}\right)^{\frac{1}{p}} \left(\frac{\pi}{\sin\frac{\pi}{p}}\right)^{\frac{1}{q}} \left(\sum_{m=1}^{\infty} a_m^p\right)^{\frac{1}{p}} \left(\sum_{n=1}^{\infty} b_n^q\right)^{\frac{1}{q}} \\
&= \frac{\pi}{\sin\frac{\pi}{p}} \left(\sum_{m=1}^{\infty} a_m^p\right)^{\frac{1}{p}} \left(\sum_{n=1}^{\infty} b_n^q\right)^{\frac{1}{q}},
\end{aligned}$$

which shows the result. □

2.5 Problems

2.20. Prove the following properties of the subspaces of ℓ_∞ introduced in Definition 2.15

(a) The space c_0 is the closure of c_{00} in ℓ_∞.
(b) The space c and c_0 are Banach spaces.
(c) The space c_{00} is not complete.

2.21. Show that (s,ρ) is a complete metric space, where s is the set of all sequences $\mathbf{x} = (x_1, x_2, \ldots)$ and ρ is given by

$$\rho(x,y) = \sum_{k=1}^{\infty} \frac{1}{2^k} \frac{|x_k - y_k|}{1 + |x_k - yk|}.$$

2.22. Let $\ell_p(\mathbf{w})$, $p \geq 1$ be the set of all real sequences $\mathbf{x} = (x_1, x_2, \ldots)$ such that

$$\sum_{k=1}^{\infty} |x_k|^p w_k < \infty$$

where $\mathbf{w} = (w_1, w_2, \ldots)$ and $w_k > 0$. Does $\mathcal{N} : \ell_p(\mathbf{w}) \longrightarrow \mathbb{R}$ given by

$$\mathcal{N}(\mathbf{x}) := \left(\sum_{k=1}^{\infty} |x_k|^p w_k \right)^{\frac{1}{p}}$$

defines a norm in $\ell_p(\mathbf{w})$?

2.23. As in the case of Example 2.2, draw the unit ball for ℓ_1^3, ℓ_∞^3, and ℓ_2^3.

2.24. Prove that (2.4) defines a norm in the space $\ell_p(\mathbb{N})$.

2.25. Prove the *Cauchy-Bunyakovsky-Schwarz inequality*

$$\left(\sum_{i=1}^{n} x_i y_i \right)^2 \leqslant \left(\sum_{i=1}^{n} x_i^2 \right) \left(\sum_{i=1}^{n} y_i^2 \right)$$

without using Jensen's inequality. This inequality is sometimes called *Cauchy*, *Cauchy-Schwarz* or *Cauchy-Bunyakovsky*.
Hint: Analyze the quadratic form $\sum_{i=1}^{n} (x_i u + y_i v)^2 = u^2 \sum_{i=1}^{n} x_i^2 + 2uv \sum_{i=1}^{n} x_i y_i + v^2 \sum_{i=1}^{n} y_i^2$.

2.26. Let $\{a_n\}_{n\in\mathbb{Z}}$ and $\{b_n\}_{n\in\mathbb{Z}}$ be sequences of real numbers such that

$$k = \sum_{n=-\infty}^{\infty} |a_n| < \infty \quad \text{and} \quad \sum_{m=-\infty}^{\infty} |b_m|^p < \infty$$

where $p > 1$. Let $C_n = \sum_{m=-\infty}^{\infty} a_{n-m} b_m$. Prove that

(a) $|C_n| \leq k^{1/q} \left(\sum_{m=-\infty}^{\infty} |a_{n-m}| |b_m|^p \right)^{1/p}$ where $\frac{1}{p} + \frac{1}{q} = 1$.
(b) $\left(\sum_{n=-\infty}^{\infty} |C_n|^p \right)^{1/p} \leq k \left(\sum_{n=-\infty}^{\infty} |b_n|^p \right)^{1/p}$.

2.27. If $a_n > 0$ for $n = 1, 2, 3, \ldots$ show that

$$\sum_{n=1}^{\infty} \sqrt[n]{a_1 a_2 \cdots a_n} \leq \mathrm{e} \sum_{n=1}^{\infty} a_n.$$

If $a_1 \geq a_2 \geq \cdots \geq a_k \geq \cdots \geq a_n \geq 0$ and $\alpha \geq \beta > 0$. Demonstrate that

$$\left(\sum_{k=1}^{n} a_k^{\alpha}\right)^{1/\alpha} \leq \left(\sum_{k=1}^{n} a_k^{\beta}\right)^{1/\beta}.$$

2.28. Use Theorem 10.5 to show the Theorem 2.16.
Hint: Choose a sequence $\{a_n\}_{n\in\mathbb{N}}$ of positive numbers such that $a_{n+1} \geq a_n$ $\forall n \in \mathbb{N}$. Consider $A_N = \sum_{n=1}^{N} a_n$ and define $f = \sum_{n=1}^{\infty} a_n \chi_{(n-1,n)}$.

2.29. Demonstrate that ℓ_1 is not the dual space of ℓ_∞.

2.30. Show that

$$\|\mathbf{x}\|_{\ell_q} \leq \|\mathbf{x}\|_{\ell_p} \tag{2.21}$$

whenever $1 \leq p < q < \infty$.
Hint: First, show the inequality (2.21) when $\|\mathbf{x}\|_{\ell_p} \leq 1$. Use that result and the homogeneity of the norm to get the general case.

2.6 Notes and Bibliographic References

The history of Hölder's inequality can be traced back to Hölder [32] but the paper of Rogers [61] preceded the one from Hölder just by one year, for the complete history see Maligranda [48].

The Minkowski inequality is due to Minkowski [51] but it seems that the classical approach to the Minkowski inequality via Hölder's inequality is due to Riesz [58].

The Hardy inequality (Theorem 2.16) appeared in Hardy [26] as a generalization of a tool to prove a certain theorem of Hilbert.

According to Hardy, Littlewood, and Pólya [30], the Hilbert inequality (Theorem 2.19) was included by Hilbert for $p = 2$ in his lectures, and it was published by Weyl [82], the general case $p > 1$ appeared in Hardy [27].

The Cauchy-Bunyakovsky-Schwarz inequality, which appears in Problem 2.25, was first proved by Cauchy [6].

Chapter 3
Lebesgue Spaces

There is much modern work, in real or complex function theory, in the theory of Fourier series, or in the general theory of orthogonal developments, in which the 'Lebesgue classes L^k' occupy the central position.

GODFREY HAROLD HARDY, JOHN EDENSOR LITTLEWOOD & GEORGE PÓLYA

Abstract Lebesgue spaces are without doubt the most important class of function spaces of measurable functions. In some sense they are the prototype of all such function spaces. In this chapter we will study these spaces and this study will be used in the subsequent chapters. After introducing the space as a normed space, we also obtain denseness results, embedding properties and study the Riesz representation theorem using two different proofs. Weak convergence, uniform convexity, and the continuity of the translation operator are also studied. We also deal with weighted Lebesgue spaces and Lebesgue spaces with the exponent between 0 and 1. We give alternative proofs for the Hölder inequality based on Minkowski inequality and also study the Markov, Chebyshev, and Minkowski integral inequality.

The reader will notice that some of the results are not given in full generality. We invite the reader to try to obtain such statements in an appropriate more general setting.

3.1 Essentially Bounded Functions

Definition 3.1. Let $(X, \mathscr{A}, \mu)$ be a measure space and f an $\mathscr{A}$-measurable function. For each $M > 0$ define $E_M = \{x \in X : |f(x)| > M\}$. We have $E_M \in \mathscr{A}$ since f is an $\mathscr{A}$-measurable function. Let

$$A = \{M > 0 : \mu(E_M) = 0\} = \{M > 0 : |f(x)| \leq M \ \mu\text{-a.e.}\}.$$

The essential supremum of f, denoted by $\operatorname{ess\,sup} f$ or $\|f\|_\infty$, is defined by

$$\|f\|_\infty = \|f\|_{L_\infty} = \operatorname{ess\,sup} f = \inf(A),$$

with the usual convention that $\inf(\emptyset) = +\infty$. ⊘

R.E. Castillo, H. Rafeiro, *An Introductory Course in Lebesgue Spaces*, CMS Books in Mathematics, DOI 10.1007/978-3-319-30034-4_3

Note that if $A \neq \emptyset$, then 0 is a lower bound on A, and thus $\inf(A) \in \mathbb{R}$. Let $\alpha = \|f\|_\infty < \infty$, we state that $\alpha \in A$. Notice that

$$E_\alpha = \{x \in X : |f(x)| > \alpha\} = \bigcup_{n=1}^{\infty} \{x \in X : |f(x)| > \alpha + 1/n\}$$

moreover, for each n the set $\{x \in X : |f(x)| > \alpha + 1/n\} \in A$. As $\alpha = \inf(A)$, then for every $n \in \mathbb{N}$ there exists $\alpha_n \in A$ such that $\alpha \leq \alpha_n < \alpha + 1/n$, hence

$$\{x \in X : |f(x)| > \alpha + 1/n\} \subset \{x \in X : |f(x)| > \alpha_n\}$$

then

$$\mu(\{x \in X : |f(x)| > \alpha + 1/n\}) \leq \mu(\{x \in X : |f(x)| > \alpha_n\}) = 0,$$

therefore $\mu(E_\alpha) = \mu(\{x \in X : |f(x)| > \alpha\}) = 0$, showing that $\alpha \in A$, then if $\alpha = \|f\|_\infty < \infty$ we get

$$|f(x)| \leq \|f\|_\infty \quad \mu\text{-a.e.}$$

Now we define

$$f^*(x) = \begin{cases} f(x) & \text{if } x \notin E_\alpha, \\ 0 & \text{if } x \in E_\alpha, \end{cases}$$

since $f^*(x) = f(x)$ μ-a.e. it follows

$$\|f\|_\infty = \|f^*\|_\infty = \sup_{x \in X} |f^*(x)| = \sup_{x \in X \setminus E_\alpha} |f(x)|.$$

Definition 3.2. We define $\mathbf{L}_\infty(X, \mathscr{A}, \mu)$, called the set of *essentially bounded functions*, by

$$\mathbf{L}_\infty(X, \mathscr{A}, \mu) = \left\{ f : X \to \mathbb{R} \text{ is an } \mathscr{A}\text{-measurable function and } \|f\|_\infty < \infty \right\}.$$

⊘

The set $\mathbf{L}_\infty(X, \mathscr{A}, \mu)$ is a very large set, since it includes all bounded functions in X.

Example 3.3. Let f be a Dirichlet-type function given by

$$f(x) = \begin{cases} 1 & \text{if } x \in (\mathbb{R} \setminus \mathbb{Q}) \cap [0,1], \\ \infty & \text{if } x \in \mathbb{Q} \cap [0,1]. \end{cases}$$

Since $\|f\|_\infty = \inf \left\{ M > 0 : \mu(\{x \in [0,1] : |f(x)| > M\}) = 0 \right\} = 1$ we have that $f \in \mathbf{L}_\infty(X, \mathscr{A}, \mu)$. ⊘

Example 3.4. Let $X = \mathbb{R}$, $\mathscr{A} = \mathscr{L}$, $\mu = m$, and $\mathbb{Q} = \{r_1, r_2, \ldots, r_n, \ldots\}$ be an enumeration of the rational numbers in $\mathbb{R}$. Define

$$f(x) = \begin{cases} n & \text{if } x = r_n \in \mathbb{Q}, \\ 1 & \text{if } x \in \mathbb{R} \setminus \mathbb{Q}. \end{cases}$$

We want to show that $A = \left\{ M > 0 : m(\{x \in X : |f(x)| > M\}) = 0 \right\} = [1, \infty)$. In fact, let $M \in [1, \infty)$, then

$$\{x \in X : |f(x)| > M\} \subset \mathbb{Q},$$

therefore

$$m(\{x \in X : |f(x)| > M\}) \leq m(\mathbb{Q}) = 0,$$

which gives $M \in A$, i.e.

$$[1, \infty) \subseteq A. \tag{3.1}$$

On the other hand, suppose that $y \notin [1, \infty)$, then $y < 1$, which implies

$$\mathbb{R} \setminus \mathbb{Q} \subset \{x \in X : |f(x)| > y\},$$

and we obtain that

$$m(\{x \in X : |f(x)| > y\}) \neq 0$$

which means that $y \notin A$, then

$$A \subset [1, \infty). \tag{3.2}$$

From (3.1) and (3.2) we have

$$A = \{M > 0 : m(\{x \in X : |f(x)| > M\}) = 0\} = [1, \infty).$$

Note that

$$\inf \left\{ M > 0 : m(\{x \in X : |f(x)| > M\}) = 0 \right\} = 1,$$

therefore $f \in \mathbf{L}_\infty(X, \mathscr{A}, \mu)$. ⊘

Example 3.5. Let $X = \mathbb{N}$, $\mathscr{A} = \mathscr{P}(\mathbb{N})$, $\mu = \#$ the counting measure and the function $f : \mathbb{N} \to \mathbb{N}$ given by $n \mapsto n$. We state that

$$A = \{M > 0 : \#(\{x \in X : |f(x)| > M\}) = 0\} = \emptyset.$$

In fact, let $M > 0$ be arbitrary, and choose $k > M$, $k \in \mathbb{N}$ then

$$\#(\{x \in X : |f(x)| > M\}) \geq \#(\{k\}) = 1,$$

which implies that $M \notin A$ and since M is arbitrary, we conclude that $A = \emptyset$, therefore $\|f\|_\infty = \infty$. ⊘

3.2 Lebesgue Spaces with $p \geq 1$

We now study the set of p-th integrable functions.

Definition 3.6. Let $(X, \mathscr{A}, \mu)$ be a measure space and p a positive real number. The function $f : X \to \mathbb{R}$ is said to belong to the *pre-Lebesgue space* $\mathbf{L}_p(X, \mathscr{A}, \mu)$ if $\int_X |f|^p \, d\mu < \infty$, that is,

$$\mathbf{L}_p(X,\mathscr{A},\mu) = \left\{ f : X \to \mathbb{R} \text{ is an } \mathscr{A}\text{-measurable and } \int_X |f|^p \, \mathrm{d}\mu < \infty \right\}.$$

Sometimes we use other notation, e.g., $\mathbf{L}_p(X)$ or $\mathbf{L}_p(\mu)$ if it will be clear from the context and we want to emphasize the underlying space or the measure. ⊘

We now give some examples.

Example 3.7. Let $X = [0,1/2]$ and $f : X \to \mathbb{R}$ be defined by $f(x) = \left[x\log^2\left(\frac{1}{x}\right)\right]^{-1}$, then $f \in \mathbf{L}_1(m)$. ⊘

Example 3.8. Let $X = (0,\infty)$ and $f : X \to \mathbb{R}$ be defined by $f(x) = (1+x)^{-1/2}$, then $f \in \mathbf{L}_p(X,\mathscr{A},\mu)$ for $2 < p < \infty$. ⊘

The following example show us that, in general, the spaces $\mathbf{L}_p$ are not comparable for different values of p.

Example 3.9. Let $X = [0,16]$ and $f : X \to \mathbb{R}$ be defined by $f(x) = x^{-1/4}$. We have that $f \in \mathbf{L}_1(m)$ but $f \notin \mathbf{L}_4(m)$, where m denotes the Lebesgue measure. ⊘

The next theorem tells us under what conditions it is possible to compare $\mathbf{L}_p$ spaces with different exponents.

Theorem 3.10. *Let $(X,\mathscr{A},\mu)$ be a measure space such that $\mu(X) < \infty$. Then*

$$\mathbf{L}_q(X,\mathscr{A},\mu) \subseteq \mathbf{L}_p(X,\mathscr{A},\mu)$$

for any $1 \le p \le q \le \infty$.

Proof. We first prove when $p = \infty$. Indeed, let $f \in \mathbf{L}_\infty(X,\mathscr{A},\mu)$, thus $|f| \le \|f\|_\infty$ μ-a.e., then

$$\int_X |f|^p \mathrm{d}\mu \le \|f\|_\infty^p \int_X \mathrm{d}\mu = \mu(X)\|f\|_\infty^p < \infty,$$

so $f \in \mathbf{L}_p(X,\mathscr{A},\mu)$.

For the remaining cases, let $f \in \mathbf{L}_q(X,\mathscr{A},\mu)$ if $A = \{x \in X : |f(x)| \le 1\}$, then $\chi_X = \chi_A + \chi_{X\setminus A}$ and $|f(x)|^p < |f(x)|^q$ for $x \in X \setminus A$ and $|f(x)| \le 1$, for $x \in A$, then

$$\begin{aligned}
\|f\|_p^p = \int_X |f|^p \, \mathrm{d}\mu &= \int_X \chi_A |f|^p \, \mathrm{d}\mu + \int_X \chi_{X\setminus A} |f|^p \, \mathrm{d}\mu \\
&\le \int_X \chi_A \mathrm{d}\mu + \int_X \chi_{X\setminus A} |f|^p \, \mathrm{d}\mu \le \mu(A) + \int_X \chi_{X\setminus A} |f|^q \, \mathrm{d}\mu \\
&\le \mu(X) + \|f\|_q^q < \infty,
\end{aligned}$$

therefore $f \in \mathbf{L}_p(X,\mathscr{A},\mu)$. □

The inclusion in Theorem 3.10 is strict. To see this, consider the following example.

Example 3.11. Let $X = [0,1]$ and $1 \leq p < \alpha < q \leq \infty$, where $\alpha = \frac{p+q}{2}$ then if $p < \alpha < q$ we have that $p/\alpha < 1$ and $q/\alpha > 1$. Choose $\beta = 1/\alpha$ and define

$$f(x) = \begin{cases} \frac{1}{x^\beta} & \text{if } x \neq 0, \\ 0 & \text{if } x = 0. \end{cases}$$

then consider

$$\int_0^1 |f(x)|^p \, dx = \int_0^1 \frac{dx}{x^{p\beta}} = \int_0^1 \frac{dx}{x^{p/\alpha}} < \infty,$$

since $p/\alpha < 1$, then $f \in \mathbf{L}_p(m)$, on the other hand,

$$\int_0^1 |f(x)|^q \, dx = \int_0^1 \frac{dx}{x^{q\beta}} = \int_0^1 \frac{dx}{x^{q/\alpha}},$$

and this last integral is divergent since $q/\alpha > 1$, which gives that $f \notin \mathbf{L}_q(m)$. Thus $\mathbf{L}_q(m) \subsetneq \mathbf{L}_p(m)$. ⊘

We cannot drop the condition $\mu(X) = \infty$ in Theorem 3.10 as the next example shows.

Example 3.12. Consider the constant function $f(x) = c$ with $c \neq 0$ in $(0, \infty)$, it is easy to see that $f \in \mathbf{L}_\infty(\mu)$ but $f \notin \mathbf{L}_p(X, \mathscr{A}, \mu)$ for $0 < p < \infty$. On the other hand, let $X = [1, \infty)$ and define $f : X \to \mathbb{R}$ with $f(x) = \frac{1}{x}$, then

$$\int_1^\infty \frac{1}{x^2} dx = 1,$$

implying that $f \in \mathbf{L}_2(m)$, but the integral

$$\int_1^\infty \frac{dx}{x}$$

is divergent, therefore $f \notin \mathbf{L}_1(m)$. ⊘

It is not difficult to verify that $\mathbf{L}_p$, with $1 \leq p < \infty$, is a vector space. In fact, note that if $f, g \in \mathbf{L}_p(X, \mathscr{A}, \mu)$, then by the inequality

$$|f+g|^p \leq (|f| + |g|)^p \leq (2\max\{|f|, |g|\})^p = 2^p \max\{|f|^p, |g|^p\} \leq 2^p(|f|^p + |g|^p),$$

we have that $f+g \in \mathbf{L}_p(X,\mathscr{A},\mu)$. Moreover, if $f \in \mathbf{L}_p(X,\mathscr{A},\mu)$ and $\alpha \in \mathbb{R}$, then $\alpha f \in \mathbf{L}_p(X,\mathscr{A},\mu)$. On the other hand, the inequalities $0 \leq f^+ \leq |f|$, $0 \leq f^- \leq |f|$ imply that f^+, f^- and $|f|$ are in $\mathbf{L}_p(X,\mathscr{A},\mu)$.

The following result permits to obtain ℓ^p by means of $\mathbf{L}_p(X,\mathscr{A},\mu)$ choosing an appropriate measure space.

Theorem 3.13. *Let X be a countable set and # be the counting measure over X, then*

$$\mathbf{L}_p(X,\mathscr{P}(X),\#) = \ell^p.$$

Proof. Let # be the counting measure over X, i.e.

$$\#(E) = \begin{cases} \text{number of elements of } E & \text{if } E \text{ is a finite set;} \\ \infty & \text{if } E \text{ is an infinite set.} \end{cases}$$

Without loss of generality, we suppose that $X = \mathbb{Z}^+$, since X, endowed with the counting measure, is isomorphic to $\mathbb{Z}^+$, then we can write $\mathbb{Z}^+ = \bigcup_{k=1}^{\infty} \{k\}$. Let $f \in \mathbf{L}_p(\mathbb{Z}^+,\mathscr{P}(\mathbb{Z}^+),\#)$ and

$$\varphi_n = \sum_{k=1}^{n} |f(k)|^p \chi_{\{k\}}$$

be a sequence of simple functions such that

$$\lim_{n\to\infty} \varphi_n(k) = |f(k)|^p \quad \text{for each} \quad k,$$

now

$$\int_{\mathbb{Z}^+} \varphi_n \, d\# = \sum_{k=1}^{n} |f(k)|^p \#\Big(\mathbb{Z}^+ \cap \{k\}\Big) = \sum_{k=1}^{n} |f(k)|^p \#\Big(\{k\}\Big) = \sum_{k=1}^{n} |f(k)|^p,$$

since $\#\Big(\{k\}\Big) = 1$.

It is clear that $\varphi_1 \leq \varphi_2 \leq \varphi_3 \leq \ldots$, and using the monotone convergence theorem we get

$$\int_{\mathbb{Z}^+} |f(k)|^p \, d\# = \int_{\mathbb{Z}^+} \lim_{n\to\infty} \varphi_n(k) \, d\# = \sum_{k=1}^{\infty} |f(k)|^p,$$

$$\sum_{k=1}^{\infty} |f(k)|^p = \int_{\mathbb{Z}^+} |f(k)|^p \, d\#.$$

This last result shows that $|f|^p$ is integrable if and only if

$$\sum_{k=1}^{\infty} |f(k)|^p < \infty.$$

In other words, to say that f belongs to $\mathbf{L}_p(X, \mathscr{P}(X), \#)$ endowed with the counting measure is equivalent to say that the sequence $\{f(k)\}_{k\in\mathbb{N}}$ is a member of ℓ^p, therefore

$$\mathbf{L}_p(X, \mathscr{P}(X), \#) = \ell^p$$

which ends the proof. □

Let $(X, \mathscr{A}, \mu)$ be a measure space, define the functional $\|\cdot\|_p : \mathbf{L}_p(X, \mathscr{A}, \mu) \to \mathbb{R}^+$ by

$$\|f\|_p = \|f\|_{L_p} = \left(\int_X |f|^p \, \mathrm{d}\mu \right)^{1/p}$$

with $1 \leq p < \infty$.

We now show that the functional $\|\cdot\|_p$ is not a norm since it does not hold the definite positive property of the norm.

Example 3.14. Let $X = [0,1]$ and consider the Dirichlet function

$$f(x) = \begin{cases} 1 \text{ if } x \in \mathbb{Q} \cap [0,1], \\ 0 \text{ if } x \in (\mathbb{R} \setminus \mathbb{Q}) \cap [0,1]. \end{cases}$$

For $p = 1$, we have

$$\|f\|_1 = \int_{[0,1]} f(x) \, \mathrm{d}m = \int_{\mathbb{Q}\cap[0,1]} 1 \, \mathrm{d}m = m(\mathbb{Q} \cap [0,1]) = 0.$$

On the other hand, if $p = \infty$, we obtain

$$\|f\|_\infty = \inf \Big\{ M > 0 : m(\{x \in [0,1] : |f(x)| > M\}) = 0 \Big\} = 0.$$

However $f(x) \neq 0$ for all $x \in X$. This shows that both $\|\cdot\|_1$ and $\|\cdot\|_\infty$ do not define a norm in $\mathbf{L}_1$ and $\mathbf{L}_\infty$ respectively. ⊘

To correct this nuisance we resort to the notion of quotient space, i.e., we will split all the elements of $\mathbf{L}_p$ into equivalence classes. In other words, two functions f and g in $\mathbf{L}_p$ are said to belong to the same equivalence class if and only if $f = g$ μ-almost everywhere, in symbols $f \sim g \Leftrightarrow f = g$ μ-almost everywhere. It is just a matter of routine calculations to verify that the relation $\sim$ defines an equivalence relation. Once this is verified, we denote the class generated by f as

$$[f] = \{g \in \mathbf{L}_p(X, \mathscr{A}, \mu) : g \sim f\} \tag{3.3}$$

and we define the norm of g as $\|g\|_p = \|[f]\|_p$ for $g \in [f]$. For arbitrary $g_1 \in [f]$ and $g_2 \in [f]$ we have that $g_1 = g_2$ μ-a.e. since $g_1 \sim f$ and $g_2 \sim f$. This tell us that $\|[f]\|_p = \|g\|_p$ is well defined being independent of the representative of the class $[f]$.

With the above taken into account, we now define a normed space based upon the pre-Lebesgue space.

Definition 3.15. We define the Lebesgue space $L_p(X, \mathscr{A}, \mu)$ as the set of equivalence classes

$$L_p(X, \mathscr{A}, \mu) = \{[f] : f \in \mathbf{L}_p(X, \mathscr{A}, \mu)\},$$

where $[\cdot]$ is defined in (3.3). ⊘

We went to a lot of work to define the Lebesgue space L_p space via quotient spaces just to have $\|f\|_p = 0$ if and only if $f = [0]$, but in practice we never think of L_p spaces as equivalence classes. With some patience, we can see that L_p is a vector space over $\mathbb{R}$.

We now show that the functional $\|\cdot\|_p$ satisfies the triangle inequality.

Theorem 3.16 (Minkowski's inequality). *Let $1 \le p \le \infty$ and $f, g \in L_p(X, \mathscr{A}, \mu)$. Then $f + g \in L_p(X, \mathscr{A}, \mu)$ and*

$$\|f+g\|_p \le \|f\|_p + \|g\|_p.$$

The equality holds if $A|f| = B|g|$ μ-a.e. for A and B of the same sign and not simultaneously zero.

Proof. Let us check equality. Let A and B be numbers of the same sign and not simultaneously zero such that $A|f| = B|g|$ μ-a.e., then $A\|f\|_p = B\|g\|_p$, i.e., $\|f\|_p = \frac{B}{A}\|g\|_p$. Moreover,

$$\|f+g\|_p = \left\|\frac{B}{A}g + g\right\|_p = \frac{B+A}{A}\|g\|_p = \frac{B}{A}\|g\|_p + \|g\|_p = \|f\|_p + \|g\|_p.$$

When $p = \infty$ and $p = 1$ the inequality is immediate, as well as when $\|f\|_p = \|g\|_p = 0$. Suppose that $1 < p < \infty$ and $\|f\|_p = \alpha \neq 0$ and $\|g\|_p = \beta \neq 0$, then there are functions f_0 and g_0 such that $|f| = \alpha f_0$ and $|g| = \beta g_0$ with $\|f_0\|_p = \|g_0\|_p = 1$.

Now, consider $\lambda = \frac{\alpha}{\alpha+\beta}$ and $1 - \lambda = \frac{\beta}{\alpha+\beta}$ note that $0 < \lambda < 1$, then

$$\begin{aligned}
|f(x)+g(x)|^p &\leq (|f(x)|+|g(x)|)^p \\
&= (\alpha f_0(x)+\beta g_0(x))^p \\
&= [(\alpha+\beta)\lambda f_0(x)+(\alpha+\beta)(1-\lambda)g_0(x)]^p \\
&= (\alpha+\beta)^p(\lambda f_0(x)+(1-\lambda)g_0(x))^p \\
&\leq (\alpha+\beta)^p[\lambda(f_0(x))^p+(1-\lambda)(g_0(x))^p].
\end{aligned} \tag{3.4}$$

Since $\varphi(t)=t^p$ is convex in $[0,\infty)$, integrating in (3.4) we have

$$\begin{aligned}
\int_X |f(x)+g(x)|^p \mathrm{d}\mu &\leq (\alpha+\beta)^p[\lambda\|f_0\|_p^p+(1-\lambda)\|g_0\|_p^p] \\
&= (\alpha+\beta)^p < \infty,
\end{aligned}$$

i.e., $f+g \in L_p(X,\mathscr{A},\mu)$. Finally,

$$\|f+g\|_p^p \leq (\|f\|_p+\|g\|_p)^p$$

thus

$$\|f+g\|_p \leq \|f\|_p+\|g\|_p,$$

which ends the proof. □

We are now in condition to introduce a norm in the Lebesgue space.

Definition 3.17. The *Lebesgue space* $\left(L_p(X,\mathscr{A},\mu),\|\cdot\|_{L_p}\right)$ is a normed space with the norm

$$\|f\|_p = \|f\|_{L_p} := \left(\int_X |f|^p \mathrm{d}\mu\right)^{\frac{1}{p}}, \tag{3.5}$$

whenever $1 \leq p < +\infty$. ⊘

To see that (3.5) does not define a norm when $p<1$, we can take $f=\chi_{[0,1/2]}$, $g=\chi_{[1/2,1]}$ and we see that we have a *reverse* triangle inequality in $L^{\frac{1}{2}}([0,1],\mathscr{L},m)$. We could hope to define another norm in $L^{\frac{1}{2}}$ that could turn the vector space into a normed space but this is not possible, see Theorem 3.79.

We now want to see if the product of two functions in some L_p is still in L_p. The following example shows us that this is not always true.

Example 3.18. Consider the function

$$f(x) = \begin{cases} |x|^{-1/2} & \text{if } |x|<1, \\ 0 & \text{if } |x| \geq 1. \end{cases}$$

note that

$$\int_{\mathbb{R}} f(x)\,\mathrm{d}x = \int_{[-1,1]} \frac{\mathrm{d}x}{\sqrt{|x|}} = 4,$$

therefore $f \in L_1(m)$, but

$$\int_{\mathbb{R}} f^2(x)\,\mathrm{d}x = \int_{[-1,1]} \frac{\mathrm{d}x}{|x|}$$

is a divergent integral, therefore $f^2 \notin L_1(m)$. ⊘

Now, we study under which conditions the product of two functions stays in $L_1(X,\mathscr{A},\mu)$. The following result says that if $f \in L_p(X,\mathscr{A},\mu)$ and $g \in L_q(X,\mathscr{A},\mu)$ for p and q *conjugated numbers*, i.e., $\frac{1}{p}+\frac{1}{q}=1$, we have that $fg \in L_1(X,\mathscr{A},\mu)$. Prior to the demonstration of this powerful result, we need the following lemma.

Lemma 3.19. *Let $1 \leq p < \infty$. Then for nonnegative numbers a, b, and t we have*

$$(a+tb)^p \geq a^p + ptba^{p-1}.$$

Proof. Let us define

$$\phi(t) = (a+tb)^p - a^p - ptba^{p-1}.$$

Note that $\varphi(0) = 0$ and $\phi'(t) = bp[(a+tb)^{p-1} - a^{p-1}] \geq 0$ since $p \geq 1$ and a, b, t are nonnegative numbers. Therefore φ is increasing in $[0,\infty)$ which gives that is nonnegative for $t \geq 0$. Thus, $\varphi(t) \geq \varphi(0)$ and $(a+tb)^p \geq a^p + ptba^{p-1}$. □

The next proof of the Hölder inequality is not the standard textbook proof. Traditionally, the Minkowski inequality is obtained using the Hölder inequality as was done in the case of Lebesgue sequence space in Lemma 2.4. Here we get the Hölder inequality from Minkowski's inequality, which highlights the fact that both inequalities are intertwined in some sense. If we carefully analyze the situation in question, we can see that the Young inequality (Corollary 1.10) provides us with a tool which allows us to prove Hölder's inequality without using Minkowski's inequality as was done in Lemma 2.3.

Theorem 3.20 (Hölder's inequality.). *Let p and q be extended nonnegative numbers such that $\frac{1}{p}+\frac{1}{q}=1$ and $f \in L_p(X,\mathscr{A},\mu)$, $g \in L_q(X,\mathscr{A},\mu)$. Then $fg \in L_1(X,\mathscr{A},\mu)$ and*

$$\int_X |fg|\,\mathrm{d}\mu \leq \|f\|_p \|g\|_q. \tag{3.6}$$

Equality holds if there are constants a and b, not simultaneously zero, such that $a|f|^p = b|g|^q$ μ-a.e.

Proof. First consider $p = 1$ and $q = \infty$, then clearly

$$|g| \leq \|g\|_\infty \quad \mu\text{-a.e.}$$

since $|f| \geq 0$, we have that $|fg| \leq |f| \|g\|_\infty$ μ-a.e. therefore

$$\int_X |fg| \, d\mu \leq \left(\int_X |f| \, d\mu \right) \|g\|_\infty,$$

thus

$$\int_X |fg| \, d\mu \leq \|f\|_1 \|g\|_\infty.$$

Now, suppose that $1 < p < \infty$, $1 < q < \infty$ and $f \geq 0$, $g \geq 0$. Define $h(x) = [g(x)]^{q/p}$, then

$$g(x) = [h(x)]^{p/q} = [h(x)]^{p-1}.$$

Using Lemma 3.19 we have

$$ptf(x)g(x) = ptf(x)[h(x)]^{p-1} \leq (h(x) + tf(x))^p - [h(x)]^p.$$

thus,

$$pt \int_X f(x)g(x) \, d\mu \leq \int_X (h(x) + tf(x))^p \, d\mu - \int_X [h(x)]^p \, d\mu = \|h + tf\|_p^p - \|h\|_p^p.$$

From Minkowski's inequality (Theorem 3.16) we have

$$p \int_X f(x)g(x) \, d\mu \leq \frac{(\|h\|_p + t\|f\|_p)^p - \|h\|_p^p}{t}.$$

Taking $f(t) = (\|h\|_p + t\|f\|_p)^p$, we get $f(0) = \|h\|_p^p$. Then

$$\begin{aligned} p \int_X fg \, d\mu &\leq \lim_{t \to 0} \frac{f(t) - f(0)}{t} = f'(0) \\ &= p(\|h\|_p)^{p-1} \|f\|_p. \end{aligned}$$

Note that

$$\begin{aligned} \left(\int_X [h(x)]^p \, d\mu \right)^{\frac{p-1}{p}} &= \left(\int_X [g(x)]^q \, d\mu \right)^{1-\frac{1}{p}} \\ &= \left(\int_X [g(x)]^q \, d\mu \right)^{\frac{1}{q}}, \end{aligned}$$

then, $\|h\|_p^{p-1} = \|g\|_q$. Thus

$$\int_X fg \, d\mu \leq \|f\|_p \|g\|_q.$$

Finally, choosing $a = \|g\|_q^q$ and $b = \|f\|_p^p$ such that $a|f|^p = b|g|^q$, then

$$|f| = \|f\|_p \frac{|g|^{q/p}}{\|g\|_q^{q/p}},$$

and integrating we get (3.6). □

We will give another proof of the Hölder inequality using Minkowski's inequality, but first an auxiliary lemma.

Lemma 3.21. *Let a, b, and θ be real numbers such that $0 < \theta < 1$ and $a, b \geq 0$. Then*

$$\lim_{n \to +\infty} \left[\theta a^{1/n} + (1-\theta) b^{1/n}\right]^n = a^\theta b^{(1-\theta)}.$$

Proof. Let $a, b > 0$. Taking $I(n) := \left[\theta a^{1/n} + (1-\theta) b^{1/n}\right]^n$, we have

$$\begin{aligned} I(n) &= \exp\left\{ n \log\left[\theta a^{1/n} + (1-\theta) b^{1/n}\right]\right\} \\ &= \exp\left\{ \frac{\varphi\left(\frac{1}{n}\right) - \varphi(0)}{\frac{1}{n}} \right\} \end{aligned}$$

where $\varphi(t) = \log\left[\theta a^t + (1-\theta) b^t\right]$. Passing now to the limit, we get

$$\begin{aligned} \lim_{n \to +\infty} I(n) &= \exp\left(\varphi'(0)\right) \\ &= \exp\left(\theta \log(a) + (1-\theta)\log(b)\right) \\ &= a^\theta b^{1-\theta}, \end{aligned}$$

which ends the proof. □

We are now in a position to provide one more alternative proof of the Hölder inequality.

Proof (Alternative proof of Theorem 3.20). Let $f \in L_p$ and $g \in L_q$. Define $F = |f|^p$ and $G = |g|^q$, which entails that $F^{1/p} \in L_p$ and $G^{1/p} \in L_q$. Now we get

$$\begin{aligned} \left\|\theta F^{1/p} + (1-\theta) G^{1/p}\right\|_{L_p} &\leq \left\|\theta F^{1/p}\right\|_{L_p} + \left\|(1-\theta) G^{1/p}\right\|_{L_p} \\ &= \theta \left\|F^{1/p}\right\|_{L_p} + (1-\theta) \left\|G^{1/p}\right\|_{L_p} \end{aligned}$$

or in integral terms

$$\int_X \left(\theta F^{1/p} + (1-\theta)G^{1/p}\right)^p \mathrm{d}\mu$$
$$\leq \left[\theta\left(\int_X F\mathrm{d}\mu\right)^{1/p} + (1-\theta)\left(\int_X G\mathrm{d}\mu\right)^{1/p}\right]^p. \quad (3.7)$$

Applying Lemma 3.21 and Lebesgue theorem in (3.7) we obtain

$$\int_X F^\theta G^{(1-\theta)}\mathrm{d}\mu \leq \left(\int_X F\mathrm{d}\mu\right)^\theta \cdot \left(\int_X G\mathrm{d}\mu\right)^{(1-\theta)}$$

which is exactly

$$\int_X |f|^{p\theta} \cdot |g|^{q(1-\theta)}\mathrm{d}\mu \leq \left(\int_X |f|^p\mathrm{d}\mu\right)^\theta \left(\int_X |g|^q\mathrm{d}\mu\right)^{(1-\theta)}.$$

Taking $\theta = 1/p$ we get Hölder's inequality (3.6). □

The Hölder inequality can be extended in the following way.

Corollary 3.22 *Let $p_k > 1$ be such that $\sum_{k=1}^n \frac{1}{p_k} = 1$. If $f_k \in L_{p_k}(X, \mathscr{A}, \mu)$, for all $k = 1, 2, \ldots, n$, than we have that $f_1 \times f_2 \cdots \times f_n \in L_1(X, \mathscr{A}, \mu)$ and*

$$\int_X \left|\prod_{k=1}^n f_k\right| \mathrm{d}\mu \leq \prod_{k=1}^n \|f\|_{p_k}.$$

Proof. We give the proof for $n = 3$. Let $\frac{1}{p} + \frac{1}{q} + \frac{1}{s} = 1$ and take $\frac{1}{p}$, then $\frac{s}{p} + \frac{s}{q} = 1$, implying that $\frac{1}{s} + \frac{1}{r} = 1$. We want to show that $fg \in L_s(X, \mathscr{A}, \mu)$. Indeed, by Theorem 3.20

$$\int_X |fg|^s \mathrm{d}\mu \leq \left(\int_X |f|^{sp/s} \mathrm{d}\mu\right)^{s/p} \left(\int_X |g|^{sq/s} \mathrm{d}\mu\right)^{s/q}$$

i.e.

$$\left(\int_X |fg|^s \mathrm{d}\mu\right)^{1/s} \leq \|f\|_p \|g\|_q,$$

therefore $fg \in L_s(X,\mathscr{A},\mu)$. Finally, once again invoking Theorem 3.20 we get

$$\int_X |fgh|\,\mathrm{d}\mu \leq \left(\int_X |fg|^s\,\mathrm{d}\mu\right)^{1/s}\left(\int_X |h|^r\,\mathrm{d}\mu\right)^{1/r} \leq \|f\|_p\|g\|_q\|h\|_r.$$

The general case follows by similar arguments. □

Example 3.23. As an application of Holdër's inequality, we show that the Gamma function (see Appendix C for more details) $\Gamma : (0,\infty) \to \mathbb{R}$ given by

$$\Gamma(p) = \int_0^\infty \mathrm{e}^{-t}t^{p-1}\mathrm{d}t$$

is a log-convex function, i.e., it satisfies $\varphi(\lambda x + (1-\lambda)y) \leq \varphi(x)^\lambda \varphi(y)^{1-\lambda}$ for $0 < \lambda < 1$ and x, y in the domain of φ. Let $x, y \in (0,\infty)$, $0 < \lambda < 1$, $p = 1/\lambda$ and $q = 1/(1-\lambda)$. Let us take

$$f(t) = t^{\frac{x-1}{p}}\mathrm{e}^{-\frac{t}{p}}, \quad g(t) = t^{\frac{y-1}{q}}\mathrm{e}^{-\frac{t}{q}},$$

and now by Holdër's inequality we get

$$\int_\varepsilon^N f(t)g(t)\mathrm{d}t \leq \left(\int_\varepsilon^N f(t)^p\mathrm{d}t\right)^{\frac{1}{p}}\left(\int_\varepsilon^N g(t)^q\mathrm{d}t\right)^{\frac{1}{q}}.$$

Now taking $\varepsilon \to 0$ and $N \to \infty$ we get

$$\Gamma\left(\frac{x}{p} + \frac{y}{q}\right) \leq \Gamma(x)^{\frac{1}{p}}\Gamma(y)^{\frac{1}{q}}$$

since $f(t)g(t) = t^{x/p+y/q-1}\mathrm{e}^{-t}$, $f(t)^p = t^{x-1}\mathrm{e}^{-t}$ and $g(t)^q = t^{y-1}\mathrm{e}^{-t}$. ⊘

There is a reverse Hölder type inequality where we can obtain information from one of the integrand functions knowing an a priori uniformly bound with respect to the other integrand function, see Lemma 3.40 and Lemma 3.41.

The following result provides us with another characterization of the norm $\|\cdot\|_p$.

Theorem 3.24. *Let $f \in L_p(X,\mathscr{A},\mu)$ with $1 \leq p < \infty$, then*

$$\|f\|_p = \|f\|_{L_p} = \sup_{g\in L_q(X,\mathscr{A},\mu)}\left\{\|fg\|_1\|g\|_q^{-1} : g \neq 0,\ \frac{1}{p} + \frac{1}{q} = 1\right\}. \tag{3.8}$$

Proof. Using Hölder's inequality we have

$$\|fg\|_1 = \int_X |fg| \, \mathrm{d}\mu \leq \|f\|_p \|g\|_q,$$

then

$$\|fg\|_1 \|g\|_q^{-1} \leq \|f\|_p$$

for $g \neq 0$, which implies

$$\sup_{g \in L_q(X,\mathscr{A},\mu)} \left\{ \|fg\|_1 \|g\|_q^{-1} : g \neq 0, \frac{1}{p} + \frac{1}{q} = 1 \right\} \leq \|f\|_p. \tag{3.9}$$

Moreover, suppose $f \neq 0$ and $g = c|f|^{p-1}$ (c constant), then

$$|fg| = c|f|^p,$$

thus

$$\|fg\|_1 = c\|f\|_p^p.$$

If we choose $c = \|f\|_p^{1-p}$ we obtain

$$\|fg\|_1 = \|f\|_p^{1-p} \|f\|_p^p = \|f\|_p. \tag{3.10}$$

Now

$$|g|^q = c^q |f|^{q(p-1)}$$

and integrating both sides give us

$$\|g\|_q = c \left(\int_X |f|^p \, \mathrm{d}\mu \right)^{1/q} = \|f\|_p^{1-p} \|f\|_p^{p/q} = \|f\|_p^{1-p} \|f\|_p^{p-1} = 1$$

since $f \neq 0$, then $\|g\|_q^{-1} = 1$.

Thus, we can write (3.10) as

$$\|f\|_p = \|fg\|_1 \|g\|_q^{-1} \leq \sup_{g \in L_q(X,\mathscr{A},\mu)} \left\{ \|fg\|_1 \|g\|_q^{-1} : g \neq 0, \frac{1}{p} + \frac{1}{q} = 1 \right\}. \tag{3.11}$$

combining (3.9) and (3.11) we obtain the result. □

We now give a result, sometimes called the *integral Minkowski inequality* or even *generalized Minkowski inequality*, which is a corollary of the characterization of the Lebesgue norm given in (3.8). Nonetheless this inequality is widely used, for example in the theory of integral equations, among many others.

Theorem 3.25 (Integral Minkowski inequality). *Let $(X, \mathscr{A}_1, \mu)$ and $(Y, \mathscr{A}_2, \mu)$ be σ-finite measure spaces. Suppose that f is a measurable $\mathscr{A}_1 \times \mathscr{A}_2$ function and $f(\cdot, y) \in L_p(\mu)$ for all $y \in Y$. Then for $1 \leq p < \infty$ we have*

$$\left\| \int_Y f(\cdot, y) \mathrm{d}y \right\|_{L_p(X)} \leq \int_Y \|f(\cdot, y)\|_{L_p(X)} \mathrm{d}y \tag{3.12}$$

where the dot means that the norm is taken with respect to the first variable.

Proof. Let us define $a(x) = \int_Y f(x,y)\mathrm{d}y$. We have

$$\begin{aligned}
\|a\|_{L_p(X)} &= \sup_{\substack{g\in L^q(X)\\ \|g\|_q=1}} \int_X |a(x)g(x)|\mathrm{d}x \\
&= \sup_{\substack{g\in L^q(X)\\ \|g\|_q=1}} \int_X \left| \int_Y f(x,y)g(x)\mathrm{d}y \right| \mathrm{d}x \\
&\leq \sup_{\substack{g\in L^q(X)\\ \|g\|_q=1}} \int_Y \int_X |f(x,y)g(x)|\mathrm{d}x\mathrm{d}y \\
&= \int_Y \|f(\cdot,y)\|_{L_p(X)} \mathrm{d}y
\end{aligned}$$

where the first and last equalities are just consequences of the characterization given in (3.8) for the norm of an L_p function whereas the inequality is a consequence of Fubini-Tonelli theorem and the inequality $|\int f| \leq \int |f|$. □

As an immediate consequence of the integral Minkowski inequality we get the so-called *Young's Theorem for convolution*, see Chapter 11 for definitions and in particular Theorem 11.10 for a different proof.

Theorem 3.26. *Let $k \in L^1(\mathbb{R}^n)$ and $f \in L_p(\mathbb{R}^n)$. Then*

$$\left\| \int_{\mathbb{R}^n} k(x-t)f(t)\mathrm{d}t \right\|_{L_p(\mathbb{R}^n)} \leq \|k\|_{L^1(\mathbb{R}^n)} \|f\|_{L_p(\mathbb{R}^n)}.$$

Proof. By a linear change of variables we have

$$\int_{\mathbb{R}^n} k(x-t)f(t)\mathrm{d}t = \int_{\mathbb{R}^n} k(t)f(x-t)\mathrm{d}t,$$

which gives

$$\begin{aligned}
\left\| \int_{\mathbb{R}^n} k(x-t)f(t)\mathrm{d}t \right\|_{L_p(\mathbb{R}^n)} &= \left\| \int_{\mathbb{R}^n} k(t)f(x-t)\mathrm{d}t \right\|_{L_p(\mathbb{R}^n)} \\
&\leq \int_{\mathbb{R}^n} \left\| k(t)f(\cdot - t) \right\|_{L_p(\mathbb{R}^n)} \mathrm{d}t \\
&\leq \int_{\mathbb{R}^n} |k(t)| \left\| f(\cdot - t) \right\|_{L_p(\mathbb{R}^n)} \mathrm{d}t \\
&\leq \int_{\mathbb{R}^n} |k(t)| \|f\|_{L_p(\mathbb{R}^n)} \mathrm{d}t \\
&= \|k\|_{L^1(\mathbb{R}^n)} \|f\|_{L_p(\mathbb{R}^n)} .
\end{aligned}$$

The first inequality is a consequence of the integral Minkowski inequality, the third inequality is due to the fact that $\left\| f(\cdot - t) \right\|_{L_p(\mathbb{R}^n)} = \|f\|_{L_p(\mathbb{R}^n)}$, see Problem 3.85. □

We now show that the L_∞-norm can be obtained from the L_p-norm by a limiting process.

Theorem 3.27. *Let $f \in L_1(X,\mathscr{A},\mu) \cap L_\infty(X,\mathscr{A},\mu)$. Then*

(a) $f \in L_p(X,\mathscr{A},\mu)$ for $1 < p < \infty$.
(b) $\lim\limits_{p\to\infty} \|f\|_p = \|f\|_\infty$.

Proof. (a) Let $f \in L_1(X,\mathscr{A},\mu) \cap L_\infty(X,\mathscr{A},\mu)$. Since $|f| \leq \|f\|_\infty$ μ-a.e., then we have $|f|^{p-1} \leq \|f\|_\infty^{p-1}$ therefore $|f|^p \leq \|f\|_\infty^{p-1}|f|$ whence

$$\|f\|_p \leq \|f\|_\infty^{1-\frac{1}{p}} \|f\|_1^{\frac{1}{p}}, \tag{3.13}$$

i.e., $f \in L_p(X,\mathscr{A},\mu)$.

(b) By (3.13) we have

$$\limsup_{p\to\infty} \|f\|_p \leq \|f\|_\infty. \tag{3.14}$$

On the other hand, let $0 < \varepsilon < \frac{1}{2}\|f\|_\infty$ and

$$A = \{x \in X : |f(x)| > \|f\|_\infty - \varepsilon\},$$

note that $\mu(A) > 0$, then

$$\int_X |f|^p \mathrm{d}\mu \geq \int_A |f|^p \mathrm{d}\mu \geq (\|f\|_\infty - \varepsilon)^p \mu(A),$$

then

$$\liminf_{p\to\infty} \|f\|_p \geq (\|f\|_\infty - \varepsilon) \liminf_{p\to\infty} [\mu(A)]^{\frac{1}{p}},$$

since ε is arbitrary, we get

$$\liminf_{p\to\infty} \geq \|f\|_\infty, \tag{3.15}$$

combining (3.14) and (3.15) we get

$$\|f\|_\infty \leq \liminf_{p\to\infty} \|f\|_p \leq \limsup_{p\to\infty} \|f\|_p \leq \|f\|_\infty.$$

So $\lim_{p\to\infty} \|f\|_p = \|f\|_\infty$. □

The following result gives an upper bound for the measure of a set that depends on the function f using an integral upper bound depending on the function f, namely:

Lemma 3.28 (Markov's inequality). *Let $f \in L_p(X, \mathscr{A}, \mu)$ and g be an increasing function in $[0, \infty)$. Then*

$$\mu(\{x \in X : |f(x)| > \lambda\}) \leq \frac{1}{g(\lambda)} \int_X g \circ |f| \mathrm{d}\mu, \tag{3.16}$$

where $g(x) \neq 0$ for all $x \in [0, \infty)$.

Proof. Let $A_\lambda = \{x \in X : |f(x)| > \lambda\}$ with $\lambda > 0$. Then, for all $x \in A_\lambda$, we have $\lambda < |f(x)|$, and thus

$$g(\lambda)\chi_{A_\lambda}(x) \leq g(|f(x)|)\chi_{A_\lambda}(x).$$

Now integrating both sides we obtain (3.16). □

In the case g is the identity function, the Markov inequality is widely known as *Chebyshev's inequality*.

The next results show that the Lebesgue spaces L_p, $1 \leq p \leq \infty$, are not only normed spaces, but are in fact Banach spaces.

Theorem 3.29. *Let $1 \leq p \leq \infty$. Then $(L_p(X, \mathscr{A}, \mu), \|\cdot\|_p)$ is a complete space.*

Proof. We will split the proof in two cases.

Case 1. When $1 \leq p < \infty$. Take $\{f_n\}_{n\in\mathbb{N}}$ a Cauchy sequence in $L_p(X, \mathscr{A}, \mu)$. Then, for all $\varepsilon > 0$ there exists $n_0 \in \mathbb{N}$ such that

$$\|f_n - f_m\|_p^p < \varepsilon^p$$

if $n, m \geq n_0$. By the Markov inequality with $g(\lambda) = \lambda^p$, we obtain

$$\varepsilon^p \mu(\{x : |f_n(x) - f_m(x)| \geq \varepsilon\}) \leq \|f_n - f_m\|_p^p$$

if $n, m \geq n_0$. The latter tells us that $\{f_n\}_{n\in\mathbb{N}}$ is a Cauchy sequence in measure, therefore there exists a subsequence $\{f_{n_k}\}_{k\in\mathbb{N}}$ of $\{f_n\}_{n\in\mathbb{N}}$ that converges μ-a.e. to a measurable function f (see Theorems 5.7 and 5.8). By Fatou's lemma we have

$$\|f\|_p^p = \int |f|^p \, d\mu \leq \liminf_{k\to\infty} \int |f_{n_k}|^p \, d\mu < \infty.$$

So $f \in L_p(X, \mathscr{A}, \mu)$. Invoking again Fatou's lemma we see that

$$\|f_n - f\|_p^p = \int |f_n - f|^p \, d\mu \leq \liminf_{k\to\infty} \int |f_n - f_{n_k}|^p \, d\mu < \varepsilon^p$$

whenever $n \geq n_0$. It means that f_n converges to f in $L_p(X, \mathscr{A}, \mu)$.
Case 2. When $p = \infty$. Let $\{f_n\}_{n\in\mathbb{N}}$ be a Cauchy sequence in $L_\infty(X, \mathscr{A}, \mu)$. For each $n \in \mathbb{N}$ define

$$A_k = \{x : |f_k(x)| > \|f_k\|_\infty\}$$

and for each $n, m \in \mathbb{N}$, let

$$B_{n,m} = \{x : |f_n(x) - f_m(x)| > \|f_n - f_m\|_\infty\}.$$

Note that each A_k and each $B_{n,m}$ have measure zero. Let

$$E = \left(\bigcup_{k=1}^{\infty} A_k\right) \cup \left(\bigcup_{n,m}^{\infty} B_{n,m}\right),$$

then $\mu(E) = 0$. Note that each $f_n(x)$ is a real function and also

$$|f_n(x) - f_m(x)| \leq \|f_n - f_m\|_\infty, \quad \forall x \in X \setminus E.$$

The latter tells us that $\{f_n\}_{n\in\mathbb{N}}$ is a uniform Cauchy sequence in $X \setminus E$. Now, let us define

$$f(x) = \begin{cases} \lim_{n\to\infty} f_n(x) & \text{if } x \in X \setminus E, \\ 0 & \text{if } x \in E. \end{cases}$$

Then f is measurable since $f = \lim_{n\to\infty} f_n \chi_{E^c}$, i.e., $f_n \to f$ uniformly in E^c. Finally, we show that

$$\lim_{n\to\infty} \|f_n - f\|_\infty = 0.$$

Indeed, since $\varepsilon > 0$, there $n_1 \in \mathbb{N}$ such that $|f_n(x) - f(x)| < \varepsilon/4 \; \forall x \in X \setminus E$ when $\geq n_1$. Thus,

$$\{x : |f_n(x) - f(x)| \geq \varepsilon/2\} \subset E \quad \text{for} \quad n \geq n_1.$$

Since $\mu(E) = 0$ we conclude that

$$\|f_n - f\|_\infty \leq \varepsilon/2 < \varepsilon \quad \text{if} \quad n \geq n_1,$$

i.e., $\lim_{n\to\infty} \|f_n - f\|_\infty = 0$, in particular $\|f_{n_1} - f\|_\infty < \varepsilon$ and $f_{n_1} - f \in L_\infty(X, \mathscr{A}, \mu)$. now, as $f_{n_1} \in L_\infty(X, \mathscr{A}, \mu)$ and $L_\infty(X, \mathscr{A}, \mu)$ is a vector space, then $f = f_{n_1} - (f_{n_1} - f) \in L_\infty(X, \mathscr{A}, \mu)$. □

We now characterize the sequence of μ almost everywhere convergent functions that converge in norm.

Theorem 3.30. *Let $\{f_n\}_{n\in\mathbb{N}}$ be a sequence of functions in $L_p(X,\mathscr{A},\mu)$ with $1\le p<\infty$, which converge μ-a.e. to a function $f\in L_p(X,\mathscr{A},\mu)$. Then*

$$\lim_{n\to\infty}\|f_n-f\|_p=0 \quad \textit{iff} \quad \lim_{n\to\infty}\|f_n\|_p=\|f\|_p.$$

Proof. Since $\varphi(t)=t^p$ is convex in $[0,\infty)$ when $1\le p<\infty$ we obtain

$$\left|\frac{a-b}{2}\right|^p\le\left(\frac{|a|+|b|}{2}\right)^p\le\frac{1}{2}(|a|^p+|b|^p),$$

which implies

$$|a-b|^p\le 2^{p-1}(|a|^p+|b|^p). \tag{3.17}$$

Taking $f_n\to f$ a.e., then $|f_n-f|^p\to 0$ μ-a.e. and by (3.17) we get

$$0\le 2^{p-1}\left(|f_n|^p+|f|^p\right)-|f_n-f|^p,$$

then

$$\lim_{n\to\infty}\left[2^{p-1}\left(|f_n|^p+|f|^p\right)-|f_n-f|^p\right]=2^p|f|.$$

By Fatou's lemma we have that

$$\begin{aligned}
2^p\int_X|f|^p\,\mathrm{d}\mu&=\int_X\liminf_{n\to\infty}\left[2^{p-1}\left(|f_n|^p+|f|^p\right)-|f_n-f|^p\right]\mathrm{d}\mu\\
&\le\liminf_{n\to\infty}\int_X\left[2^{p-1}\left(|f_n|^p+|f|^p\right)-|f_n-f|^p\right]\mathrm{d}\mu\\
&=2^p\int_X|f|^p\,\mathrm{d}\mu+\liminf_{n\to\infty}\left\{-\int_X|f_n-f|^p\,\mathrm{d}\mu\right\},
\end{aligned}$$

i.e.

$$2^p\int_X|f|^p\,\mathrm{d}\mu\le 2^p\int_X|f|^p-\limsup\int_X|f_n-f|^p\mathrm{d}\mu,$$

therefore

$$\limsup\int_X|f_n-f|^p\,\mathrm{d}\mu\le 0.$$

Since

$$0\le\int_X|f_n-f|^p\,\mathrm{d}\mu\le\limsup\int_X|f_n-f|^p\,\mathrm{d}\mu\le 0,$$

then

$$\lim_{n\to\infty} \|f_n - f\|_p = 0$$

if $\lim_{n\to\infty} \|f_n\|_p = \|f\|_p$.

Now, if $\lim_{n\to\infty} \|f_n - f\| = 0$, then we have the inequality

$$\left| \|f_n\|_p - \|f\|_p \right| \le \|f_n - f\|_p$$

and the result follows. □

It should be pointed out, that for $p = \infty$, the Theorem 3.30 is false. In fact, let $\{f_n\}_{n\in\mathbb{N}} \subset L_\infty([0,1])$ be defined by $f_n = \chi_{(1/n,1]}$ and note that $f_n \to 1$ μ-a.e. in $[0,1]$. Moreover,

$$\|f_n\|_\infty = \inf\left\{ M : \mu\left(\left\{ x \in [0,1] : \left|\chi_{(1/n,1]}(x)\right| > M \right\} \right) = 0 \right\} = 1$$

and $\|1\|_\infty = 1$ then $\lim_{n\to\infty} \|f_n\|_\infty = \|1\|_\infty$, but

$$\|f_n - 1\|_\infty = \sup_{x\in[0,1]} \left|\chi_{(1/n,1]}(x) - 1\right| = 1.$$

3.3 Approximations

Let $(X, \mathscr{A}, \mu)$ be a measure space. A simple function s vanishes outside a set of finite measure means that

$$\mu(\{x \in X : s(x) \neq 0\}) < \infty.$$

Now, suppose that $s = \sum_{k=1}^{n} \alpha_k \chi_{E_k}$ where $\alpha_k \neq 0$ for $1 \le k \le n$ and $E_k \in \mathscr{A}$. If s vanishes outside a set of finite measure, then $\mu(E_k) < \infty$ for $1 \le k \le n$, whence

$$\|s\|_p^p = \sum_{k=1}^{n} |\alpha_k|^p \mu(E_k).$$

So $s \in L_p(X, \mathscr{A}, \mu)$ if and only if s vanishes outside a set of finite measure.

Lemma 3.31. *For $1 \le p < \infty$, the set of simple $\mathscr{A}$-measurable functions which vanish outside a set of finite measure is dense in $L_p(X, \mathscr{A}, \mu)$.*

Proof. Let $1 \le p < \infty$ and $f \in L_p(X, \mathscr{A}, \mu)$, we show that given $\varepsilon > 0$ there exists an $\mathscr{A}$-measurable s simple function which vanishes outside a set of finite measure such that $\|f - s\|_p < \varepsilon$. To do this we consider two cases:

Case 1 $f \geq 0$. We know that there exists a sequence $\{s_n\}_{n\in\mathbb{N}}$ of simple nonnegative and $\mathscr{A}$-measurable function such that $s_n \to f$ pointwise at X, since $0 \leq s_n \leq f$ for all n and $f \in L_p(X,\mathscr{A},\mu)$ implies that $s_n \in L_p(X,\mathscr{A},\mu)$ and this means that every s_n vanishes outside a set of finite measure. Now, note that

$$\lim_{n\to\infty} |s_n - f|^p = 0 \quad \text{in} \quad X$$

and

$$|s_n - f|^p \leq (|s_n| + |f|)^p \leq (2|f|)^p = 2^p|f|^p,$$

since $f \in L_p(X,\mathscr{A},\mu)$, then $2^p|f|^p \in L_1(X,\mathscr{A},\mu)$. By the dominated convergence theorem we have that

$$\lim_{n\to\infty} \int_X |s_n - f|^p \,\mathrm{d}\mu = 0,$$

i.e.

$$\lim_{n\to\infty} \|s_n - f\|_p^p = 0,$$

then, since $\varepsilon > 0$ exists $n_0 \in \mathbb{N}$ such that

$$\|s_n - f\|_p^p < \varepsilon^p,$$

now, we choose $s = s_{n_0}$, then

$$\|s - f\|_p < \varepsilon.$$

Case 2 Let f be $\mathscr{A}$-measurable, then $f = f^+ - f^-$ where f^+ and f^- are nonnegative $\mathscr{A}$-measurable functions. Using the Case 1 exist nonnegative simple functions s_1 and s_2 which are $\mathscr{A}$-measurable and which vanish outside a set of finite measure and such that

$$\|f^+ - s_1\|_p < \varepsilon/2 \quad \text{and} \quad \|f^- - s_2\| < \varepsilon/2.$$

Let $s = s_1 - s_2$, note that s is a simple $\mathscr{A}$-measurable function which vanishes outside a set of finite measure. Finally, by the Minkowski inequality we have

$$\|f - s\|_p = \|(f^+ - s_1) - (f^- - s_2)\|_p \leq \|f^+ - s_1\|_p + \|f^- s_2\|_p < \varepsilon,$$

which entails the denseness. □

For the case of L_∞ we need to suppose a stronger condition on the dense set.

Lemma 3.32. *The set of simple functions is dense in $L_\infty(X,\mathscr{A},\mu)$.*

Proof. Let $\varepsilon > 0$ and $f \in L_\infty(X,\mathscr{A},\mu)$, then $|f| \leq \|f\|_\infty$ μ-a.e. in X, so there exists $E \in \mathscr{A}$ with $\mu(E) = 0$ such that $|f(x)| \leq \|f\|_\infty$ for all $x \in X \setminus E$. Define

$$\overline{f}(x) = \begin{cases} f(x) & \text{if } x \in X \setminus E, \\ 0 & \text{if } x \in E. \end{cases}$$

Then, $|\overline{f}(x)| \leq \|f\|_\infty$ for all $x \in X$, then there exists a sequence $\{t_n\}_{n\in\mathbb{N}}$ of simple functions such that $t_n \to \overline{f}$ uniformly in X so there exists $n_0 \in \mathbb{N}$ such that

$$|t_n(x) - \overline{f}(x)| < \varepsilon/2,$$

for all $x \in X$, then

$$|t_{n_0}(x) - f(x)| < \varepsilon/2$$

for all $x \in X \setminus E$. This means that with $s = t_{n_0}$ we get $\|s - f\| \leq \varepsilon/2 < \varepsilon$. □

Let (X,d) be a metric space and $E \subset X$. Define, as usual, the distance of an element x to the set E as

$$d(x,E) = \inf\{d(x,e) : e \in E\}.$$

It is almost immediate that

(a) $d(x,E) = 0$ if and only if $x \in \overline{E}$.
(b) $d(\cdot,E) : X \to \mathbb{R}^+$ is continuous on X.

The following lemma is a well-known result in the theory of metric spaces.

Lemma 3.33 (Urysohn lemma). *Let (X,d) be a metric space. Let F be a closed set in X and V an open set in X such that $F \subset V$. Then there exists a function $g : X \to [0,1]$ such that g is continuous in X, $g(x) = 1$ for all $x \in F$ and $g(x) = 0$ for all $x \in X \setminus V$.*

Proof. For $x \in X$, define

$$g(x) = \frac{d(x,X \setminus V)}{d(x,F) + d(x,X \setminus V)}.$$

By (b) it is clear that g is continuous in X.

If $x \in F$, then $g(x) = 1$. If $x \in X \setminus V$, then $d(x,X \setminus V) = 0$, and $g(x) = 0$ for all $x \in X \setminus V$. Now, since $d(x,X \setminus V) \geq 0$ and $d(X,F) \geq 0$ we can see that $0 \leq g(x) \leq 1$, which completes the proof. □

The next theorem is reminiscent of Luzin's theorem.

Theorem 3.34. *Let $f \in L_p(\mathbb{R},\mathscr{L},m)$ with $1 \leq p < \infty$. For all $\varepsilon > 0$, there is a continuous function $g \in L_p(\mathbb{R},\mathscr{L},m)$ such that $\|f - g\|_p < \varepsilon$.*

Proof. We proceed by cases.

Case 1 Let $f = \chi_E$, then $E \in \mathscr{L}$ and $m(E) < \infty$, so we can find a closed set F and open set V such that $F \subset E \subset V$ and $m(V \setminus F) < \left(\frac{\varepsilon}{2}\right)^p$. Consider now g as defined in the Uryshon lemma, then g is continuous in $\mathbb{R}$, $g = 1$ in F and $g = 0$ in $X \setminus V$. Moreover,

$$\{x : g(x) \neq f(x)\} \subset V \setminus F,$$

indeed, if $x_0 \notin V \setminus F$, then $x_0 \in (V \setminus F)^c$ and $x_0 \in V^c \cup F$, then if $x \in V^c$, then $f(x_0) = g(x_0) = 0$, which means that $x_0 \notin \{x : g(x) \neq f(x)\}$, so we have shown that

$$\{x : g(x) \neq f(x)\} \subset V \setminus F.$$

Then

$$\int_{\mathbb{R}} |f-g|^p \, \mathrm{d}m \leq \int_{V \setminus F} |f-g|^p \, \mathrm{d}m.$$

But $|f-g| \leq 2$, then

$$\begin{aligned}\int_{\mathbb{R}} |f-g|^p \, \mathrm{d}m &\leq 2^p \int_{V\setminus F} \mathrm{d}m \\ &= 2^p m(V \setminus F) \\ &< 2^p \frac{\varepsilon^p}{2^p},\end{aligned}$$

i.e.

$$\|f-g\|_p < \varepsilon,$$

and thus showing the Case 1.

Case 2 Let F be a simple $\mathscr{A}$-measurable function, which vanishes outside a set of finiteness measure, is

$$f = \sum_{k=1}^{n} \alpha_k \chi_{E_k}, \quad \text{where} \quad \alpha_k \neq 0 \quad \text{for} \quad 1 \leq k \leq n$$

and $m(E_k) < \infty$, $E_k \in \mathscr{A}$.

Using the Case 1, for all $k \in \mathbb{N}$ exists a continuous function g_k such that

$$\|\chi_{E_k} - g_k\|_p < \frac{\varepsilon}{n|\alpha_k|}.$$

Note that $g = \sum_{k=1}^{n} \alpha_k g_k$ is a continuous function on $\mathbb{R}$ therefore by Minkowski inequality

$$\begin{aligned}\|g-f\|_p &= \left\| \sum_{k=1}^{n} \alpha_k (g_k - \chi_{E_k}) \right\|_p \\ &\leq \sum_{k=1}^{n} |\alpha_k| \|g_k - \chi_{E_k}\|_p \\ &< \sum_{k=1}^{n} |\alpha_k| \frac{\varepsilon}{n|\alpha_k|} \\ &= \varepsilon.\end{aligned}$$

proving the Case 2.

Case 3 Let f be an arbitrary function. By Lemma 3.31 exists a simple function s which vanishes outside a set of finite measure such that

$$\|f-s\|_p < \varepsilon/2,$$

by the Case 2 there is a continuous function g such that

$$\|s-g\|_p < \varepsilon/2.$$

Then

$$\|f-g\|_p \leq \|f-s\|_p + \|s-g\|_p < \varepsilon,$$

which completes the proof. □

Theorem 3.34 is not true for $p=\infty$. To see this, consider the following example. Let $0<\varepsilon<1/2$ and $f=\chi_{(a,b)}$ with $a<b$, $a,b\in\mathbb{R}$. Suppose there is a continuous function g such that $\|f-g\|_\infty<\varepsilon$, hence we obtain that $|\chi_{(a,b)}(x)-g(x)|<\varepsilon$ μ-a.e. Now, for each $\delta>0$ we can find $x_0\in(a,a+\delta)$ and $x_1\in(a-\delta,a)$ such that

$$|\chi_{(a,b)}(x_0)-g(x_0)|<\varepsilon/2$$

and

$$|\chi_{(a,b)}(x_1)-g(x_1)|<\varepsilon/2,$$

i.e.

$$|1-g(x_0)|<\varepsilon/2 \quad \text{and} \quad |g(x_1)|<\varepsilon/2. \tag{3.18}$$

As g is continuous in a, then

$$g(a+)=g(a)=g(a-). \tag{3.19}$$

By the definitions of $g(a+)$ and $g(a-)$ there exists $\delta>0$ such that

$$\begin{aligned} |g(x)-g(a+)|<\varepsilon/2 \quad &\text{if} \quad a<x<a+\delta \\ |g(x)-g(a-)|<\varepsilon/2 \quad &\text{if} \quad a-\delta<x<a. \end{aligned} \tag{3.20}$$

For this $\delta>0$, by (3.18) we have

$$1-\varepsilon/2<g(x_0)<1+\varepsilon/2$$

and

$$-\varepsilon/2<g(x_1)<\varepsilon/2. \tag{3.21}$$

By (3.20)

$$|g(x_0)-g(a+)|<\varepsilon/2$$

and

$$|g(x_1)-g(a-)|<\varepsilon/2.$$

By (3.21)

$$\begin{aligned} g(a+) &> g(x_0)-\varepsilon/2 \\ &> 1-\varepsilon/2-\varepsilon/2 \\ &= 1-\varepsilon \\ &> 1/2 \end{aligned}$$

and

$$\begin{aligned} g(a-) &< g(x_1) + \varepsilon/2 \\ &< \varepsilon/2 + \varepsilon/2 \\ &= \varepsilon \\ &< 1/2, \end{aligned}$$

this means that $g(a+) \neq g(a-)$, which contradicts (3.19), therefore such g does not exist.

Definition 3.35. A step function is a function of the form

$$\sum_{k=1}^{n} \alpha_k \chi_{I_k}$$

where $\alpha_k \neq 0$, $1 \leq k \leq n$ and each I_k is a bounded interval. ⊘

Remark 3.36. Note that each step function vanishes outside a set of finite measure. Thus, every step function is a member of $L_p(\mathbb{R}, \mathscr{L}, m)$. Moreover, the set of all step functions forms a vector subspace of $L_p(\mathbb{R}, \mathscr{A}, m)$.

Theorem 3.37. *Let* $1 \leq p < \infty$. *Then the set of all step functions is dense in* $L_p(\mathbb{R}, \mathscr{L}, m)$.

Proof. Let $f \in L_p(\mathbb{R}, \mathscr{L}, m)$ and $\varepsilon > 0$.

Case 1 If $f = \chi_E$, then $E \in \mathscr{L}$, as $f \in L_p(\mathbb{R}, \mathscr{L}, m)$, we have $m(E) < \infty$, we can conclude that there is a finite union of disjoint open intervals, say I such that $m(E \triangle I) < \varepsilon^p$, is $I = \bigcup_{k=1}^n I_k$, choose $\phi = \sum_{k=1}^n \chi_{I_k} = \chi_I$ then

$$\begin{aligned} \int_{\mathbb{R}} |f - \phi|^p \, dm &= \int_{\mathbb{R}} |\chi_E - \chi_i|^p \, dm \\ &= \int_{\mathbb{R}} |\chi_{E \triangle I}|^p \, dm \\ &= m(E \triangle I) \\ &< \varepsilon, \end{aligned}$$

thus

$$\|f - \varphi\|_p < \varepsilon.$$

which shows the case 1.

Case 2 $f = \sum_{k=1}^n \alpha_k \chi_{E_k}$ where $\alpha_k \neq 0$ for all k and $m(E_k) < \infty$. Using the case 1 there is a step function φ_k such that

$$\|\chi_{E_k} - \varphi_k\|_p < \frac{\varepsilon}{n|\alpha_k|}.$$

Note that

$$\varphi = \sum_{k=1}^{n} \alpha_k \varphi_k$$

is a step function

$$\|f-\varphi\|_p = \|\sum_{k=1}^{n} \alpha_k(\chi_{E_k} - \varphi_k)\|_p \leq \sum_{k=1}^{n} |\alpha_k| \|\chi_{E_k} - \varphi_k\|_p < \varepsilon,$$

thus demonstrating case 2.

Case 3 Let f an arbitrary function, by virtue of Lemma 3.32 there exists a simple function s which vanishes outside a set of finite measure such that $\|f-s\|_p < \varepsilon/2$, for case 2 we can find a step function φ such that $\|s-\varphi\|_p < \varepsilon/2$. Gathering everything we get that $\|f-\varphi\|_p < \varepsilon$. □

Another interesting property of the Lebesgue space is that it is a separable space, i.e., it has an enumerable dense set.

Theorem 3.38. *The space $L_p(\mathbb{R}, \mathscr{L}, m)$ is separable for $1 \leq p < \infty$.*

Proof. Let us define

$$S = \left\{ \sum_{k=1}^{n} b_k \chi_{J_k} : n \in \mathbb{N}, b_k \in \mathbb{Q} \right\},$$

where J_k is a finite interval with rational endpoints $1 \leq k \leq n$. Let $\varepsilon > 0$ and $f \in L_p(\mathbb{R}, \mathscr{L}, m)$, then there exists a step function (Theorem 3.37)

$$\sum_{k=1}^{n} a_k \chi_{I_k}$$

such that

$$\left\| \sum_{k=1}^{n} a_k \chi_{I_k} - f \right\|_p < \varepsilon/2.$$

Now, let $\delta > 0$. For all $1 \leq k \leq n$, choose $b_k \in \mathbb{Q}$ such that $|b_k - a_k| < \delta/2$ and J_k an interval with rational endpoints to $I_k \subset J_k$ with $m(J_k \setminus I_k) < \delta/n$. Using the Minkowski inequality, we have

$$\begin{aligned}
\left\| \sum_{k=1}^{n} a_k \chi_{I_k} - \sum_{k=1}^{n} b_k \chi_{J_k} \right\|_p &= \left\| \sum_{k=1}^{n} (a_k - b_k) \chi_{I_k} + \sum_{k=1}^{n} b_k (\chi_{I_k} - \chi_{J_k}) \right\|_p \\
&\leq \left\| \sum_{k=1}^{n} (a_k - b_k) \chi_{I_k} \right\|_p + \left\| \sum_{k=1}^{n} b_k (\chi_{I_k} - \chi_{J_k}) \right\|_p \\
&\leq \sum_{k=1}^{n} |a_k - b_k| \left\| \chi_{I_k} \right\|_p + \sum_{k=1}^{n} |b_k| \left\| \chi_{I_k} - \chi_{J_k} \right\|_p
\end{aligned}$$

$$= \sum_{k=1}^{n} |a_k - b_k| \left(m(I_k)\right)^{1/p} + \sum_{k=1}^{n} |b_k| \left(m(J_k \setminus I_k)\right)^{1/p}$$
$$< \frac{\delta}{2} \sum_{k=1}^{n} \left(m(I_k)\right)^{1/p} + \left(\frac{\delta}{2} + \max_{1\le k\le n} |a_k|\right) \delta^{1/p}$$
$$= \delta_0,$$

since $|b_k| \le |a_k - b_k| + |a_k| \le \frac{\delta}{2} + \max_{1\le k\le n} |a_k|$. Note that $\delta_0 \to 0$ if $\delta \to 0$, so we can choose δ such that $\delta_0 < \varepsilon/2$. Again invoking the Minkowski inequality we get

$$\left\| f - \sum_{k=1}^{n} b_k \chi_{j_k} \right\|_p \le \left\| f - \sum_{k=1}^{n} a_k \chi_{i_k} \right\|_p + \left\| \sum_{k=1}^{n} a_k \chi_{i_k} - \sum_{k=1}^{n} b_k \chi_{j_k} \right\|_p$$
$$< \varepsilon/2 + \varepsilon/2$$
$$= \varepsilon.$$

Finally, note that S is a countable set since

$$\mathbb{Q} \times \mathbb{Q} \times \mathbb{Q} = \bigcup_{q\in\mathbb{Q}} \mathbb{Q} \times \mathbb{Q} \times \{q\}$$

is countable. Thus we have shown that the set of all step functions is dense in $L_p(\mathbb{R}, \mathscr{L}, m)$ with $1 \le p < \infty$. □

3.4 Duality

Let g be a fixed function in $L_q(X, \mathscr{A}, \mu)$, we will show that F given by

$$F(f) = \int_X fg \, \mathrm{d}\mu$$

defines a linear functional in $L_p(X, \mathscr{A}, \mu)$. In effect, let α and β be real numbers and f and h elements of X, then

$$F(\alpha f + \beta h) = \int_X (\alpha f + \beta h) g \, \mathrm{d}\mu$$
$$= \alpha \int_X fg \, \mathrm{d}\mu + \beta \int_X hg \, \mathrm{d}\mu$$
$$= \alpha F(f) + \beta F(h).$$

On the other hand

$$|F(f)| = \left|\int_X fg\,\mathrm{d}\mu\right| \leq \int_X |fg|\,\mathrm{d}\mu \leq \|f\|_p\|g\|_q$$

from this it follows that

$$\frac{|F(f)|}{\|f\|_p} \leq \|g\|_q$$

meaning

$$\|F\| \leq \|g\|_q,$$

and this shows that F is bounded. Moreover, for $1 < p < \infty$, let us define

$$f = |g^{q-1}|\,\mathrm{sgn}(g),$$

then

$$fg = |g|^{q-1}\,\mathrm{sgn}(g)g = |g|^q.$$

Furthermore,

$$|f| = |g|^{q-1}|\,\mathrm{sgn}(g)|,$$

from which $|f| = |g|^{q-1}$, then $|f|^p = |g|^{p(q-1)}$ and $|f|^p = |g|^q$. We now have

$$F(f) = \int_X fg\,\mathrm{d}\mu = \int_X |g|^q\,\mathrm{d}\mu = \|g\|_q^q,$$

then

$$\int_X |g|^q\,\mathrm{d}\mu = \|g\|_q^q = \|g\|_q^{p(q-1)} = \frac{\|g\|_q^{pq}}{\|g\|_q^p},$$

where

$$\|g\|_q^p \int_X |g|^q\,\mathrm{d}\mu = \|g\|_q^{pq},$$

therefore

$$\|g\|_q^p \int_X |f|^p\,\mathrm{d}\mu = \|g\|_q^{pq},$$

from which we get

$$\|g\|_q^q = \|f\|_p\|g\|_q,$$

which entails

$$F(f) = \|f\|_p\|g\|_q,$$

so

$$\frac{|F(f)|}{\|f\|_p} \geq \|g\|_q,$$

then there is $f = |g|^{q-1}\,\mathrm{sgn}(g)$ for which

$$\frac{|F(f)|}{\|f\|_p} \geq \|g\|_q,$$

therefore, $\|f\| \geq \|g\|_q$. Consequently the norm attains the supremum and

$$\|F\| = \|g\|_q.$$

Now consider the case $p = 1$ and $p = \infty$. Let $g \in L_1(X, \mathscr{A}, \mu)$ and $f = \text{sign}(g)$, then $\|f\|_\infty = 1$ and

$$\int_X fg\,\mathrm{d}\mu = \int_X g\,\text{sgn}(g)\,\mathrm{d}\mu = \int_X |g|\,\mathrm{d}\mu = \|g\|_1,$$

then

$$F(f) = \|f\|_\infty \|g\|_1$$

thus

$$\frac{F(f)}{\|f\|_\infty} = \|g\|_1,$$

therefore

$$\|F\| \geq \|g\|_1.$$

The another inequality is obtained using the Hölder inequality, and we get

$$\|F\| = \|g\|_1.$$

Now, if $g \in L_\infty$ given $\varepsilon > 0$, let $E = \{x \in X : g(x) > \|g\|_\infty - \varepsilon\}$ and let $f = \chi_E$. Then

$$\begin{aligned}\int_X fg\,\mathrm{d}\mu &= \int_E g\,\mathrm{d}\mu \\ &\geq (\|g\|_\infty - \varepsilon) \int_X \mathrm{d}\mu \\ &= (\|g\|_\infty - \varepsilon)\|f\|_1,\end{aligned}$$

thus

$$\frac{F(f)}{\|f\|_1} \geq (\|g\|_\infty - \varepsilon).$$

By the arbitrariness of $\varepsilon > 0$ we have

$$\frac{F(f)}{\|f\|_1} \geq \|g\|_\infty, \tag{3.22}$$

from which it follows that $\|F\| \geq \|f\|_\infty$.

On the other hand, $|g| \leq \|g\|_\infty$ μ-a.e. Then $|fg| \leq |f|\|g\|_\infty$, therefore

$$\int_X |fg|\,\mathrm{d}\mu \le \left(\int_X |f|\,\mathrm{d}\mu\right)\|g\|_\infty$$

but

$$\left|\int_X fg\,\mathrm{d}\mu\right| \le \int_X |fg|\mathrm{d}\mu,$$

therefore

$$|F(f)| \le \left(\int_X |f|\,\mathrm{d}\mu\right)\|g\|_\infty,$$

so

$$\frac{|F(f)|}{\|f\|_1} \le \|g\|_\infty,$$

from here we obtain

$$\|F\| \le \|g\|_\infty, \tag{3.23}$$

in view of (3.22) and (3.23) is

$$\|F\| = \|g\|_\infty.$$

Thus, we have proved the following theorem.

Theorem 3.39. *Each function $g \in L_q(X, \mathscr{A}, \mu)$ defines a linear functional F bounded in $L_p(X, \mathscr{A}, \mu)$ given by*

$$F(f) = \int_X fg\,\mathrm{d}\mu,$$

and $\|F\| = \|g\|_q$.

We now investigate a reverse Hölder type inequality.

Lemma 3.40. *Let g be an integrable function in $[0,1]$ and suppose that there exists a constant M such that*

$$\left|\int_0^1 fg\,\mathrm{d}\mu\right| \le M\|f\|_p$$

for every bounded measurable function f. Then $g \in L_q([0,1], \mathscr{L}, m)$ and $\|g\|_q \le M$, where q is the conjugate exponent.

Proof. First suppose that $1 < p < \infty$ and define the sequence of measurable and bounded functions by

$$g_n(x) = \begin{cases} g(x) & \text{if } |g(x)| \le n, \\ 0 & \text{if } |g(x)| > n. \end{cases}$$

and define

$$f_n = |g_n|^{q-1} \operatorname{sgn}(g_n).$$

Since

$$|f_n|^p = |g_n|^{p(q-1)},$$

then

$$\|f_n\|_p^p = \|g_n\|_q^q$$

and further, $|f_n|^p = |g_n|^q$, moreover,

$$f_n g_n = g_n |g_n|^{q-1} \operatorname{sgn}(g_n) = |g_n|^q$$

but $f_n g_n = f_n g$, then

$$\|g_n\|_q^q = \int_0^1 f_n g \, \mathrm{d}\mu \leq M \|f_n\|_p = M \|g_n\|_q^{q/p},$$

here

$$\|g_n\|_q^q \leq M \|g_n\|_q^{q/p},$$

therefore

$$\|g_n\|_q^{q-\frac{q}{p}} \leq M$$

but

$$\frac{qp - q}{p} = \frac{q(p-1)}{p} = \frac{p}{p} = 1,$$

then

$$\|g_n\|_q \leq M$$

and

$$\int_0^1 |g_n|^q \, \mathrm{d}\mu \leq M^q$$

since

$$\lim_{n \to \infty} |g_n|^q = |g|^q$$

almost everywhere $[0,1]$, then by Fatou's lemma

$$\int_0^1 |g|^q \, \mathrm{d}\mu \leq \lim_{n \to \infty} \int_0^1 |g_n|^q \, \mathrm{d}\mu \leq M^q,$$

this means that $g \in L_q[0,1]$ and also

$$\|g\|_q \leq M.$$

For the case $p = 1$, let $\varepsilon > 0$ and consider the set

$$E = \{x : |g(x)| \geq M + \varepsilon\}$$

and function $f = \operatorname{sgn}(g)\chi_E$, then

$$\|f\|_1 = \mu(E),$$

where

$$M\mu(E) = M\|f\|_1 \geq \left| \int_0^1 fg\,\mathrm{d}\mu \right| \geq (M+\varepsilon)\mu(E),$$

from this it follows that $0 \leq \varepsilon\mu(E) \leq 0$ therefore $\mu(E) = 0$ consequently $\|f\|_1 = 0$ and $\|g\|_\infty \leq M$. □

The previous lemma can be extended to any finite measure space in the following sense.

Lemma 3.41. *Let $(X, \mathscr{A}, \mu)$ be a finite measure space. Let $g \in L_1(X, \mathscr{A}, \mu)$ be such that for any $M > 0$ and for every simple function s it holds that*

$$\left| \int_X sg\,\mathrm{d}\mu \right| \leq M\|s\|_p$$

$1 \leq p < \infty$. Then $g \in L_q(X, \mathscr{A}, \mu)$ and $\|g\|_q \leq M$, where q is the conjugate exponent p.

Proof. **Case** $p = 1$. Let $A = \{x : g(x) > M\}$ and $B = \{x : g(x) < -M\}$. Note that A and B are in $\mathscr{A}$. If we choose $s = \chi_A$, then by hypothesis we have

$$\left| \int_X \chi_A g\,\mathrm{d}\mu \right| \leq M\|\chi_A\|_1,$$

namely

$$\left| \int_A g\,\mathrm{d}\mu \right| \leq M\mu(A) = \int_A M\,\mathrm{d}\mu,$$

where

$$\int_A (g - M)\,\mathrm{d}\mu \leq 0$$

as $g > M$, we conclude that $\mu(A) = 0$.

Similarly, choose $s = -\chi_B$, then we can show that

$$\mu(B) = 0.$$

and $\mu(A \cup B) = 0$, which means that $|g(x)| \leq M$ μ-a.e., then $\|g\|_\infty \leq M$, and thus the lemma is proved for the case $p = 1$.

Case $1 < p < \infty$**.** Since $|g|^q > 0$ we can find $\{s_n\}_{n\in\mathbb{N}}$ a sequence of nonnegative simple functions such that $s_n \to |g|^q$ pointwise.

Let us define $t_n = s_n^{1/p}(\operatorname{sgn}(g))$, $n \in \mathbb{N}$, note that each t_n is a simple function and

$$\|t_n\|_p = \left(\int_X |t_n|^p \mathrm{d}\mu \right)^{1/p} = \left(\int_X s_n \,\mathrm{d}\mu \right)^{1/p}.$$

Since

$$\begin{aligned} g t_n &= s_n^{1/p} g \operatorname{sgn}(g) \\ &= s_n^{1/p} |g| \\ &\geq s_n^{1/p} s_n^{1/q} \\ &= s_n, \end{aligned}$$

then

$$0 \leq \int_X s_n \,\mathrm{d}\mu \leq \int_X g t_n \,\mathrm{d}\mu \leq M \|t_n\|_p,$$

from which we get

$$\int_X s_n \,\mathrm{d}\mu \leq M^q,$$

and by the monotone convergence theorem we conclude that

$$\int_X |g|^q \,\mathrm{d}\mu \leq M^q,$$

where $g \in L_q(X, \mathscr{A}, \mu)$ and $\|g\|_q \leq M$. □

The Riesz representation theorem is an important theorem in functional analysis since it characterizes the dual of Lebesgue spaces in a very easy way. We will give different proofs of this key result. We start with the simple case of $X = [0,1]$.

Theorem 3.42 (Riesz Representation Theorem). *Let F be a bounded linear functional on $L_p[0,1]$, $1 \leq p < \infty$, then there exists a function $g \in L_q[0,1]$ such that*

$$F(f) = \int_0^1 fg\,\mathrm{d}\mu$$

for any $f \in L_p[0,1]$ *and also*

$$\|F\| = \|g\|_q.$$

Proof. Let χ_S be the characteristic function of the interval $[0,s]$. Let us define the function $\phi : [0,1] \to \mathbb{R}$ such that $\phi(s) = F(\chi_s)$. We will show that ϕ is absolutely continuous. Let $\{s_k, \widehat{s}_k\}_{k=1}^n$ be any collection of disjoint subintervals of $[0,1]$ such that

$$\sum_{k=1}^n |\widehat{s}_k - s_k| < \delta,$$

then if

$$\alpha_k = \operatorname{sign}\left(\phi(\widehat{s}_k) - \phi(s_k)\right))$$

we have

$$\begin{aligned}
\sum_{k=1}^n |\phi(\widehat{S}_k) - \phi(S_k)| &= \sum_{k=1}^n \left(\phi(\widehat{S}_k) - \phi(S_k)\right) \operatorname{sgn}\left(\phi(\widehat{S}_k) - \phi(S_k)\right) \\
&= \sum_{k=1}^n \left(F(\chi_{\widehat{S}_k}) - F(\chi_{S_k})\right) \alpha_k \\
&= \sum_{k=1}^n F(\alpha_k(\chi_{\widehat{S}_k} - \chi_{S_k})) \\
&= F\left(\sum_{k=1}^n \alpha_k(\chi_{\widehat{S}_k} - \chi_{S_k})\right),
\end{aligned}$$

then

$$\sum_{k=1}^n \left|\phi(\widehat{s}_k) - \phi(s_k)\right| = F(f),$$

where

$$f = \sum_{k=1}^n \alpha_k(\chi_{\widehat{s}_k} - \chi_{s_k}).$$

On the other hand consider

$$\begin{aligned}
\|f\|_p^p &= \int_0^1 |f|^p\,\mathrm{d}\mu \\
&= \int_0^1 \left|\sum_{k=1}^n \alpha_k\left(\chi_{\widehat{s}_k} - \chi_{s_k}\right)\right|^p \mathrm{d}\mu
\end{aligned}$$

$$\leq \int_0^1 \left(\sum_{k=1}^{n} |\alpha_k (\chi_{\widehat{s_k}} - \chi_{s_k})| \right)^p \mathrm{d}\mu$$

$$= \int_0^1 \left(\sum_{k=1}^{n} |\chi_{\widehat{s_k}} - \chi_{s_k}| \right)^p \mathrm{d}\mu$$

$$= \int_0^1 \left(\sum_{k=1}^{n} \chi_{[s_k, \widehat{s_k}]} \right)^p \mathrm{d}\mu$$

$$= \int_0^1 \left(\chi_{\bigcup_{k=1}^{n} [s_k, \widehat{s_k}]} \right)^p \mathrm{d}\mu$$

since $[s_k, \widehat{s_k}]$ are disjoint, continuing

$$\begin{aligned} \int_0^1 \left(\chi_{\bigcup_{k=1}^{n} [s_k, \widehat{s_k}]} \right)^p \mathrm{d}\mu &= \int_0^1 \chi_{\bigcup_{k=1}^{n} [s_k, \widehat{s_k}]} \mathrm{d}\mu \\ &= \mu \left(\bigcup_{k=1}^{n} [s_k, \widehat{s_k}] \right) \\ &= \sum_{k=1}^{n} \mu([s_k, \widehat{s_k}]) \\ &= \sum_{k=1}^{n} |\widehat{s_k} - s_k| \\ &< \delta \end{aligned}$$

implying that

$$\|f\|_p^p < \delta.$$

Now

$$\sum_{k=1}^{n} |\phi(\widehat{s_k}) - \phi(s_k)| = F(f) \leq \|F\| \|f\|_p < \|F\| \delta^{1/p},$$

if

$$\delta = \frac{\varepsilon^p}{\|f\|^p},$$

then

$$\sum_{k=1}^{n} |\phi(\widehat{s_k}) - \phi(s_k)| < \varepsilon$$

if $\|f\|_p^p < \delta$, which shows that ϕ is absolutely continuous in $[0,1]$. Since all absolutely continuous function is integrable, then there exists $g \in [0,1]$ such that

$$\phi(s) = \int_0^s g(t)\mathrm{d}\mu,$$

hence

$$F(\chi_s) = \int_0^1 g(t)\chi_s(t)\mathrm{d}\mu.$$

On the other hand, since any step function ψ of $[0,1]$ can be written as

$$\psi = \sum_{k=1}^{n} c_k \chi_s,$$

then we have in particular

$$F(\chi_{s_k}) = \int_0^1 g\chi_{s_k}\,\mathrm{d}\mu$$

$$c_k F(\chi_{s_k}) = c_k \int_0^1 g\chi_{s_k}\,\mathrm{d}\mu$$

$$F(c_k\chi_{s_k}) = \int_0^1 g c_k\chi_{s_k}\,\mathrm{d}\mu$$

$$\sum_{k=1}^{n} F(c_k\chi_{s_k}) = \sum_{k=1}^{n} \int_0^1 g c_k\chi_{s_k}\,\mathrm{d}\mu$$

$$F\left(\sum_{k=1}^{n} c_k\chi_{s_k}\right) = \int_0^1 g\sum_{k=1}^{n} c_k\chi_{s_k}\,\mathrm{d}\mu$$

$$F(\psi) = \int_0^1 g\psi\mathrm{d}\mu.$$

Now, consider a measurable and bounded function f in $[0,1]$, then by a known theorem in measure theory, there exists a sequence $\{\psi_n\}_{n\in\mathbb{N}}$ of step functions such that $\psi \to f$ a.e., with the result that the sequence $\{|\psi - f|^p\}_{n\in\mathbb{N}}$ is uniformly bounded and tends to zero in almost all $[0,1]$, then by the dominated convergence theorem

$$\lim_{n\to\infty} \|\psi_n - f\|_p = 0$$

and since f is bounded, then

$$|F(f) - F(\psi_n)| = |F(f - \psi_n)| \leq \|F\|\|f - \psi_n\|_p,$$

which entails

$$\lim_{n\to\infty} F(\psi_n) = F(f).$$

On the other hand, there is $M > 0$ such that

$$|\psi_n| \leq M$$

since $\{\psi_n\}_{n\in\mathbb{N}}$ is convergent, whence

$$-gM \leq g\psi_n \leq gM$$

then

$$|g\psi_n| \leq gM \leq M|g|,$$

therefore

$$\lim_{n\to\infty} \int_0^1 g\psi_n \, \mathrm{d}\mu = \int_0^1 fg \, \mathrm{d}\mu$$

$$\lim_{n\to\infty} F(\psi_n) = \int_0^1 fg \, \mathrm{d}\mu$$

$$F(f) = \int_0^1 fg \, \mathrm{d}\mu$$

for each measurable and bounded function f since

$$|F(f)| \leq \|F\| \|f\|_p$$

i.e.

$$\left| \int_0^1 fg \, \mathrm{d}\mu \right| \leq \|F\| \|f\|_p,$$

then by Lemma 3.40 $g \in L_q[0,1]$ and

$$\|g\|_q \leq \|F\|.$$

Now we only have to show that

$$F(f) = \int_0^1 fg \, \mathrm{d}\mu$$

for each $f \in L_p[0,1]$. Let f be an arbitrary function in $L_p[0,1]$. By virtue of Theorem 3.37 for each $\varepsilon > 0$ there exists a step function φ such that

$$\|f-\psi\|_p < \varepsilon.$$

Since ψ is bounded, then we have

$$F(\psi) = \int_0^1 \psi g \mathrm{d}\mu,$$

then

$$\begin{aligned}
\left| F(f) - \int_0^1 fg\mathrm{d}\mu \right| &= \left| F(f) - F(\psi) + F(\psi) - \int_0^1 fg\mathrm{d}\mu \right| \\
&\leq |F(f) - F(\psi)| + \left| F(\psi) - \int_0^1 fg\mathrm{d}\mu \right| \\
&= |F(f-\psi)| + \left| \int_0^1 (\psi - f) g\mathrm{d}\mu \right| \\
&\leq \|F\| \|f-\psi\|_p + \|g\|_q \|f-\psi\|_p \\
&< (\|f\| + \|g\|_q)\, \varepsilon,
\end{aligned}$$

and now by the arbitrariness of ε we get

$$F(f) = \int_0^1 fg\mathrm{d}\mu.$$

Equality $\|F\| = \|g\|_q$ follows from Theorem 3.39. □

Theorem 3.42 can be extended to a σ-finite measure space using a standard approach from passing to finite to σ-finite measure spaces.

Theorem 3.43 (Riesz Representation Theorem). *Let $(X, \mathscr{A}, \mu)$ be a σ-finite space and T a linear functional in $L_p(X, \mathscr{A}, \mu)(1 \leq p < \infty)$. Then there exists a unique $g \in L_q(X, \mathscr{A}, \mu)$ such that*

$$T(f) = \int_X fg\mathrm{d}\mu \tag{3.24}$$

for all $f \in L_p(X, \mathscr{A}, \mu)$ and

$$\|T\| = \|g\|_q. \tag{3.25}$$

where q is the conjugate exponent of p.

Proof. We first show the uniqueness of g. Suppose that there exists g_1 and g_2 in $L_q(X, \mathscr{A}, \mu)$ such that satisfy (3.24) namely

$$\int_E g_1 \, \mathrm{d}\mu = \int_E g_2 \, \mathrm{d}\mu$$

for all $E \in \mathscr{A}$ with $\mu(E) < \infty$. Since μ is σ-finite, we can find a sequence of disjoints sets $\{X_n\}_{n\in\mathbb{N}}$ in $\mathscr{A}$ such that $\mu(X_n) < \infty$ for all n and

$$X = \bigcup_{n=1}^{\infty} X_n.$$

Let $A := \{x \in X : g_1(x) > g_2(x)\}$ and $B := \{x \in X : g_1(x) < g_2(x)\}$.

Then

$$\int_{X_n \cap A} g_1 \, \mathrm{d}\mu = \int_{X_n \cap A} g_2 \mathrm{d}\mu,$$

so

$$\int_{X_n \cap A} (g_1 - g_2) \, \mathrm{d}\mu = 0,$$

but $g_1 > g_2$ in $X_n \cap A$, which means that $\mu(X_n \cap A) = 0$ for all $n \in \mathbb{N}$, then

$$\mu(A) = \sum_{n=1}^{\infty} \mu(A \cap X_n) = 0.$$

Similarly $\mu(B) = 0$ therefore $g_1 = g_2$ μ-a.e. and this proves the uniqueness.

We now prove the existence of g by cases.

Case 1. $\mu(X) < \infty$ then for each $E \in \mathscr{A}$ define $\nu(E) = T(\chi_E)$. Note that $\mu(X) < \infty$ implies $\mu(E) < \infty$, so

$$\chi_E \in L_p(X, \mathscr{A}, \mu).$$

We now show that ν is a signed measure on $\mathscr{A}$. Clearly $\chi_\emptyset$ is the zero function $L_p(X, \mathscr{A}, \mu)$, then

$$\nu(\emptyset) = T(\chi_\emptyset)$$

Note that T is a real function, then ν also is a real function. In the same way, choose $\{E_n\}_{n\in\mathbb{N}}$ a sequence of sets in $\mathscr{A}$ disjoint. Now, let us define

$$E = \bigcup_{n=1}^{\infty} E_n \quad \text{and} \quad A_n = \bigcup_{i=1}^{n} E_i,$$

then

$$\bigcup_{n=1}^{\infty} A_n = E,$$

note that $\{A_n\}_{n\in\mathbb{N}}$ is an increasing sequence, by induction it is easy to show that

$$\chi_{A_n} = \sum_{k=1}^{n} \chi_{E_k},$$

by linearity of T, it must be

$$T(\chi_{A_n}) = \sum_{k=1}^{n} T(\chi_{E_k}),$$

namely

$$T(\chi_{A_n}) = \sum_{k=1}^{n} \nu(E_k).$$

To show that $\chi_{A_n} \to \chi_E$ in $L_p(X,\mathscr{A},\mu)$ we consider

$$\begin{aligned}
\|\chi_{A_n} - \chi_E\|_p^p &= \int_X |\chi_{A_n} - \chi_E|^p \, \mathrm{d}\mu \\
&= \int_X \chi_{E\setminus A_n} \, \mathrm{d}\mu \\
&= \mu(E\setminus A_n) \\
&= \mu(E) - \mu(A_n).
\end{aligned}$$

Since $\{A_n\}_{n\in\mathbb{N}}$ is an increasing sequence of sets such that $E = \bigcup_{n=1}^{\infty} A_n$, then

$$\mu(E) = \lim_{n\to\infty} \mu(A_n),$$

namely $\lim_{n\to\infty}[\mu(E) - \mu(A_n)] = 0$ so $\lim_{n\to\infty} \|\chi_{A_n} - \chi_E\|_p^p = 0$, which shows that

$$\lim_{n\to\infty} \|\chi_{A_n} - \chi_E\|_p = 0.$$

Under the continuity of T in $L_q(X,\mathscr{A},\mu)$, it follows that

$$\lim_{n\to\infty} T(\chi_{A_n}) = T(\chi_E),$$

then

$$\begin{aligned}
\nu(E) &= T(\chi_E) \\
&= \lim_{n\to\infty} T(\chi_{A_n}) \\
&= \lim_{n\to\infty} \sum_{k=1}^{n} \nu(E_k)
\end{aligned}$$

and the latter tells us that ν is a signed measure. Now we want to prove that $\nu \ll \mu$. Suppose that $E \in \mathscr{A}$ with $\mu(E) = 0$, then

$$\|\chi_E\|_p = \left(\int_X \chi_E \, \mathrm{d}\mu \right)^{1/p}$$
$$= [\mu(E)]^{1/p}$$
$$= 0.$$

This shows that χ_E is the zero function in $L_p(X, \mathscr{A}, \mu)$ and $T(\chi_E) = 0$, i.e., $\nu(E) = 0$ therefore $\nu \ll \mu$. Using the Radon-Nikodym theorem for measures (signed) finite, there is a measurable function g such that

$$\nu(E) = \int_E g \, \mathrm{d}\mu$$

for all $E \in \mathscr{A}$. Then

$$\int_X g \, \mathrm{d}\mu = \nu(X)$$
$$= T(\chi_X)$$
$$= T(1)$$
$$< \infty,$$

and $g \in L_1(X, \mathscr{A}, \mu)$.

Let us verify that g satisfies the hypotheses of Lemma 3.41. Let $s \in L_p(X, \mathscr{A}, \mu)$ a simple $\mathscr{A}$-measurable function with canonical representation

$$s = \sum_{k=1}^{n} \alpha_k \chi_{E_k},$$

then

$$T(s) = T\left(\sum_{k=1}^{n} \alpha_k \chi_{E_k} \right)$$
$$= \sum_{k=1}^{n} \alpha_k \nu(E_k)$$
$$= \sum_{k=1}^{n} \alpha_k \int_{E_k} g \, \mathrm{d}\mu$$
$$= \int_X g \left(\sum_{k=1}^{n} \alpha_k \chi_{E_k} \right) \mathrm{d}\mu$$
$$= \int_X g s \, \mathrm{d}\mu.$$

therefore

$$T(s) = \int_X sg\,\mathrm{d}\mu$$

for every simple function $s \in L_p(X, \mathscr{A}, \mu)$, hence it follows that

$$\left| \int_X sg\,\mathrm{d}\mu \right| = |T(s)|.$$

If $M = \|T\|$ then $0 < M < \infty$ which shows that g satisfies the conditions of Lemma 3.41, therefore we can conclude that $g \in L_q(X, \mathscr{A}, \mu)$ and

$$\|g\|_q \leq M = \|T\|. \tag{3.26}$$

We will prove that

$$T(f) = \int_X fg\,\mathrm{d}\mu$$

for all $f \in L_p(X, \mathscr{A}, \mu)$.

Let $f \in L_p(X, \mathscr{A}, \mu)$ and for arbitrary $\varepsilon > 0$, there is a simple function $s \in L_p(X, \mathscr{A}, \mu)$ such that

$$\|f - s\|_p < \frac{\varepsilon}{\|g\|_q + \|T\| + 1}$$

by Lemma 3.31. Then

$$\begin{aligned}
\left| T(f) - \int_X fg\,\mathrm{d}\mu \right| &= \left| T(f) - T(s) + T(s) - \int_X fg\,\mathrm{d}\mu \right| \\
&\leq |T(f-s)| + \left| \int_X sg\mathrm{d}\mu - \int_X fg\,\mathrm{d}\mu \right| \\
&\leq |T(f-s)| + \int_X |s - f||g|\,\mathrm{d}\mu \\
&\leq \|T\|\|f-s\|_p + \|s-f\|_p\|g\|_q \\
&= (\|T\| + \|g\|_q)\,\|f-s\|_p \\
&< \frac{(\|T\| + \|g\|_q)\,\varepsilon}{\|T\| + \|g\|_q + 1} \\
&< \varepsilon.
\end{aligned}$$

Since ε is arbitrary, we conclude that $T(f) = \int_X fg \, d\mu$ for all $f \in L_p(X, \mathscr{A}, \mu)$.

Finally by the Hölder inequality (Theorem 3.20) we have $|T(f)| \le \|g\|_q \|f\|_p$ where

$$\|T\| \le \|g\|_q. \tag{3.27}$$

Now from (3.26) and (3.27) we have $\|T\| = \|g\|_q$, thus showing the case 1.

Case 2. $\mu(X) = \infty$. Under the σ-finiteness of μ exists $\{X_n\}_{n\in\mathbb{N}}$ such that $X = \bigcup_{n=1}^{\infty} X_n$ with $X_n \subset X_{n+1}$ and $\mu(X_n) < \infty$ for all n. We apply the case 1 to the measure space $(X_n, \mathscr{A} \cap X_n, \mu_n)$ where $\mu_n = \mu|_{\mathscr{A} \cap X_n}$.

Let $T_n = T|_{L_p(\mu_n)}$, for case 1 for all $n \in \mathbb{N}$ there exists $g_n \in L_q(\mu_n)$ such that

$$T_n(h) = \int_{X_n} h g_n \, d\mu_n \tag{3.28}$$

for all $h \in L_p(X, \mathscr{A}, \mu)$ which vanishes outside X_n, and

$$\|g_n\|_q = \|T_n\| \le \|T\|. \tag{3.29}$$

Define

$$\overline{g_n}(x) = \begin{cases} g_n(x) & \text{if } x \in X_n, \\ 0 & \text{if } x \notin X_n. \end{cases}$$

then we can write (3.28) as

$$T(h) = \int_X h \overline{g_n} \, d\mu \tag{3.30}$$

for all $h \in L_p(X, \mathscr{A}, \mu)$ which vanishes outside X_n.

Now, since $\overline{g_{n+1}}$ restricted to X_n have the same properties as $\overline{g_n}$ under the uniqueness we have $\overline{g_n} = \overline{g_{n+1}}$ in X_n. Now we define

$$g(x) = g_n(x) \text{ if } x \in X_n.$$

Since

$$|\overline{g_n}(x)| \le |\overline{g_{n+1}}(x)|$$

for all $x \in X$ and

$$\lim_{n\to\infty} \overline{g_n}(x) = g(x),$$

then by the monotone convergence theorem

$$\int_X |g|^q \, d\mu = \lim_{n\to\infty} \int_X |\overline{g_n}|^q \, d\mu$$

$$\leq \|T\|^q,$$

which implies that $g \in L_q(X, \mathscr{A}, \mu)$ and

$$\|g\|_q \leq \|T\|. \tag{3.31}$$

Let $f \in L_p(X, \mathscr{A}, \mu)$ and $f_n = f\chi_{X_n}$ and note that f_n vanishes outside X_n and $f_n \to f$ (pointwise) in X. Clearly

$$|f_n - f| \leq |f|,$$

so

$$|f_n - f|^p \leq |f|^p,$$

and by the dominated convergence theorem we have

$$\lim_{n\to\infty} \int_X |f_n - f|^p \, \mathrm{d}\mu = 0.$$

By the continuity of T it follows that $T(f_n) \to T(f)$ when $n \to \infty$. Moreover, note that $|f_n g| \leq |fg|$, $fg \in L_1(X, \mathscr{A}, \mu)$ and $\lim_{n\to\infty} f_n g = fg$. Now, invoking the dominated convergence theorem to obtain

$$\begin{aligned}
\int_X fg \, \mathrm{d}\mu &= \lim_{n\to\infty} \int_X f_n g \, \mathrm{d}\mu \\
&= \lim_{n\to\infty} \int_X f\chi_{X_n} g \, \mathrm{d}\mu \\
&= \lim_{n\to\infty} \int_X (f\chi_{X_n})\,(g\chi_{X_n}) \, \mathrm{d}\mu \\
&= \lim_{n\to\infty} \int_X f_n \overline{g_n} \, \mathrm{d}\mu \\
&= \lim_{n\to\infty} T(f_n) \\
&= T(f).
\end{aligned}$$

Thus, we have demonstrated (3.24) and once again invoking the Hölder inequality we get

$$|T(f)| \leq \|f\|_p \|g\|_q,$$

from which $\|T\| \leq \|g\|_q$, and by (3.31) we get $\|T\| = \|g\|_q$, which ends the proof.

□

3.5 Reflexivity

In this section we will show that the Lebesgue spaces are reflexive whenever $1 \leq p < \infty$. We recall the following result, which is a consequence of the Hahn-Banach norm-version theorem.

Theorem 3.44. *Let $(X, \|\cdot\|)$ be a normed space, Y a subspace of X and $x_0 \in X$ such that*

$$\delta = \inf_{y \in Y} \|x_0 - y\| > 0$$

i.e., the distance from x_0 to Y is strictly positive. Then, there exists a bounded and linear functional f in X such that $f(y) = 0$ for all $y \in Y$, $f(x_0) = 1$ and $\|f\| = 1/\delta$.

Suppose that X is a normed spaces. For each $x \in X$ let $\phi(x)$ be a linear functional in X^* defined by

$$\phi(x)(f) = f(x)$$

for each $f \in X^*$. Since

$$|\phi(x)(f)| \leq \|f\| \|x\|,$$

the functional $\phi(x)$ is bounded, indeed $\|\phi(x)\| \leq \|x\|$. Then $\phi(x) \in X^{**}$.

Theorem 3.45 *Let $(X, \|\cdot\|)$ be a normed space. Then for each $x \in X$*

$$\|x\| = \sup\{|f(x)| : f \in X^*; \|f\| = 1\}.$$

Proof. Let us fix $x \in X$. If $f \in X^*$ with $\|f\| = 1$, then

$$|f(x)| \leq \|f\| \|x\| \leq \|x\|.$$

On the other hand, if $x \neq 0$, then

$$\delta = \text{dist}(x, \{0\}) = \inf_{y \in \{0\}} \|x - y\| = \|x\| > 0,$$

by the Theorem 3.44 there exists $g \in X^*$ such that

$$g(x) = 1 \quad \text{and} \quad \|g\| = \frac{1}{\|x\|}.$$

Let $f = \|x\| g$, and note that

$$\|f\| = \|x\| \|g\| = \|x\| \frac{1}{\|x\|} = 1$$

and

$$f(x) = \|x\| g(x) = \|x\|,$$

therefore

$$\|x\| \leq \sup\{|f(x)| : f \in X^*, \|f\| = 1\} \leq \|x\|,$$

which gives

$$\|x\| = \sup\{|f(x)| : f \in X^*, \|f\| = 1\}.$$

□

Remark 3.46. By the previous result it is not hard to show that

$$\|\phi(x)\| = \|x\|,$$

which states that ϕ is an isometric isomorphism from X to $\phi(X)$. To the functional ϕ we will call it the *natural immersion* from X into X^{**}. ⊘

We denote by $X^{**} := (X^*)^*$ the dual space of X^*, which is called the *bidual space of* X.

Definition 3.47. A normed space $(X, \|\cdot\|)$ is called *reflexive* if

$$\phi(X) = X^{**},$$

where ϕ is the canonical isomorphism. In this case X is isometrically isomorphic to X^{**}. ⊘

We now show that Lebesgue spaces are reflexive, whenever $1 \leq p < \infty$. Sometimes the following reasoning is given for justifying the reflexivity of the L_p spaces:

$$\left((L_p)^*\right)^* = (L_q)^* = L_p$$

where the equalities follow from the Riesz representation theorem. In fact we need to show the following:

$$\phi\left(L_p\right) = ((L_p)^*)^*,$$

which means that we have an isometric isomorphism between L_p and its bidual space $((L_p)^*)^*$ by the natural immersion ϕ.

Theorem 3.48. *The space $L_p(\mu)$ with $1 \leq p < \infty$ is reflexive.*

Proof. If $1 < p < \infty$ and $\frac{1}{p} + \frac{1}{q} = 1$, let us consider

$$\psi : L_q(\mu) \longrightarrow \left(L_p(\mu)\right)^*,$$

defined by

$$\psi(g) = \phi(F_g)$$

for $g \in L_q(\mu)$ where $F_g(f) = \int_X gf\,d\mu$.

Observe that ψ is a linear and bounded functional, the boundedness is obtained by

$$|\psi(g)| = |\phi(F_g)| \leq \|\phi\|\|F_g\| = \|\phi\|\|g\|_q,$$

where the last inequality is a consequence of Theorem 3.43. Again by Theorem 3.43 there exists some $f \in L_p(\mu)$ such that $\psi(g) = \int_X gf\,d\mu$ for all $g \in L_q(\mu)$.

On the other hand, for $w \in (L_p(\mu))^{**}$ we have

$$w(F_g) = F_g(f) = \int_X fgd\mu = \psi(g) = \phi(F_g)$$

for all $g \in L_q(\mu)$. Theorem 3.43 guarantees that $w = \phi$. It means that the natural immersion

$$\phi : L_p(\mu) \to (L_p(\mu))^{**}$$

is onto. This shows that $L_p(\mu)$ is reflexive. □

3.6 Weak Convergence

Consider the function $x \mapsto \cos(nx)$ for $n = 1, 2, \ldots$. Note that

$$\int_0^{2\pi} \cos^2(nx)\mathrm{d}x = \pi,$$

for all $n \in \mathbb{N}$.

Thus the sequence $\{\cos nx\}_{n\in\mathbb{N}}$ of functions in $L_2([0,2\pi],\mathscr{L},m)$ does not converge to zero in the $L_2([0,2\pi],\mathscr{L},m)$ sense. However, such a sequence converges to zero in the following sense. Let $g = \chi_{[a,b]}$ where $[a,b] \subset [0,2\pi]$. A direct calculation shows that

$$\int_0^{2\pi} \chi_{[a,b]} \cos(nx)\mathrm{d}x = \frac{1}{n}[\sin(nb) - \sin(na)] \to 0$$

when $n \to \infty$. Now consider $\{(a_j,b_j)\}_{j=1}^m$ a finite collection of disjoint subintervals of $[0,2\pi]$ and simple function φ of the form

$$\varphi = \sum_{j=1}^m \alpha_j \chi_{[a_j,b_j]}.$$

Observe that

$$\lim_{n\to\infty} \int_0^{2\pi} \varphi(x)\cos(nx)\mathrm{d}x = 0.$$

The above considerations motivate the following definition.

Definition 3.49. Let $(X,\mathscr{A},\mu)$ be a measure space and $1 \le p < \infty$ and $q = p/(p-1)$ with the convention $q = +\infty$ if $p = 1$. A sequence of functions $\{f_n\}_{n\in\mathbb{N}}$ in $L_p(X,\mathscr{A},\mu)$, $1 \le p < \infty$ is said to *converge weakly* to $f \in L_p(X,\mathscr{A},\mu)$ if

$$\lim_{n\to\infty} \int_X f_n g d\mu = \int_X fg d\mu$$

for all $g \in L_q(X, \mathscr{A}, \mu)$. We denote this convergences as $f_n \rightharpoonup f$. ⊘

By the Riesz representation theorem $f_n \rightharpoonup f$, $f_n, f \in X = L_p$ if $F(f_n) \to F(f)$ for all $F \in X^* = L_q$.

The next theorem characterizes weak convergence in Lebesgue spaces.

Theorem 3.50. *Let $(X, \mathscr{A}, \mu)$ be a finite measure space and $1 < p < \infty$. Let $\{f_n\}_{n\in\mathbb{N}}$ be a sequence of functions in $L_p(X, \mathscr{A}, \mu)$ such that $f_n \to f$ μ-a.e. Then*

$$\lim_{n\to\infty} \varphi(f_n) = \varphi(f)$$

for all $\varphi \in L_p^*(X, \mathscr{A}, \mu)$ if and only if $\{\|f_n\|_p\}_{n\in\mathbb{N}}$ is bounded.

Proof. ($\Rightarrow$). Suppose that $\lim_{n\to\infty} \varphi(f_n) = \varphi(f)$ for all $\varphi \in L_p^*(X, \mathscr{A}, \mu)$. Define $\psi_n : L_q \to \mathbb{R}$ by $\int_X g f_n \, d\mu$ and note that $\|\psi_n\| = \|f_n\|_p$. By hypothesis $\sup_{n\in\mathbb{N}} |\psi_n(g)| < \infty$. By the uniform boundedness principle we get that $\sup_{n\in\mathbb{N}} \|\psi_n\| < C$, from which we get $\sup_{n\in\mathbb{N}} \|f_n\|_p < C$.

($\Leftarrow$). Now, if $\|f_n\|_p \le M$, then by Fatou's lemma we have

$$\begin{aligned} \int_X |f|^p d\mu &= \int_X \liminf_{n\to\infty} |f_n|^p d\mu \\ &\le \liminf_{n\to\infty} \int_X |f_n|^p d\mu \\ &\le M. \end{aligned}$$

On the other hand, by the Riesz representation Theorem 3.42 for all $\varphi \in L_p^*(X, \mathscr{A}, \mu)$ there exists $g \in L_q(X, \mathscr{A}, \mu)$ such that

$$\varphi(h) = \int_X hg d\mu$$

for all $h \in L_p(X, \mathscr{A}, \mu)$. Since $g \in L_q(X, \mathscr{A}, \mu)$, therefore $|g|^q \in L_1(X, \mathscr{A}, \mu)$, then there exists $\delta > 0$ that for all $E \in \mathscr{A}$ such that $\mu(E) < \delta$ implies

$$\int_E |g|^q d\mu < \left(\frac{\varepsilon}{4M}\right)^q,$$

since $\mu(X) < \infty$ we can use Egorov's theorem to guarantee that there exists $n_0 \in \mathbb{N}$ and $A \in \mathscr{A}$ such that $\mu(A) < \delta$ and

$$|f_n(x) - f(x)| < \frac{\varepsilon}{2(\|g\|_1 + 1)}$$

for all $x \in A^c$ and all $n \geq n_0$.

Finally, by Hölder's inequality we have

$$\begin{aligned}
|\varphi(f_n) - \varphi(f)| &= \left| \int_X (f_n - f) g \mathrm{d}\mu \right| \\
&= \left| \int_A (f_n - f) g \mathrm{d}\mu + \int_{A^c} (f_n - f) g \mathrm{d}\mu \right| \\
&\leq \|f_n - f\|_p \left(\int_A |g|^q \mathrm{d}\mu \right)^{1/q} + \int_{A^c} |f_n - f| |g| \mathrm{d}\mu \\
&\leq 2M \frac{\varepsilon}{4M} + \frac{\varepsilon \|g\|_1}{2(\|g\|_1 + 1)} \\
&= \varepsilon
\end{aligned}$$

if $n \geq n_0$, which ends the proof. □

Corollary 3.51 *Let $\{f_n\}_{n\in\mathbb{N}}$ be a sequence of functions in L_p $1 < p < \infty$ which converges μ-a.e. to a function $f \in L_p$ and moreover, suppose that exists a constant M such that $\|f_n\|_p \leq M$, for all $n \in \mathbb{N}$. Then for all $g \in L_q$ we have*

$$\lim_{n\to\infty} \int f_n g \mathrm{d}\mu = \int f g \mathrm{d}\mu.$$

Remark 3.52. The Corollary 3.51 is not true for $p = 1$. Let $X = [0,1]$ and $n \in \mathbb{N}$.

Consider $f_n(x) = n\chi_{[0,1/n]}(x)$ and take $g(x) = 1$. Since $f_n \to f = 0$ we have

$$\int_0^1 f_n g dx = \int_0^{1/n} n \, \mathrm{d}x = 1$$

but

$$\int_0^1 f g \, \mathrm{d}x = 0.$$

Nevertheless, the Corollary 3.51 is true for $p = \infty$. Let $\{f_n\}_{n\in\mathbb{N}}$ be a sequence of functions in L_∞ such that $\|f_n\|_\infty \leq M$ for all $n \in \mathbb{N}$ and $f_n \to f$ μ-almost everywhere. Then we get $|f_n| \leq M$ almost everywhere. Let $g \in L_1$, then $|f_n g| \leq M|g| \in L_1$. From the Lebesgue dominated convergence theorem we obtain

$$\lim_{n\to\infty}\int f_n g\,\mathrm{d}\mu = \int fg\,\mathrm{d}\mu.$$

⊘

The next result is a type of Lebesgue dominated convergence theorem tailored for the p-th integrable functions.

Theorem 3.53. *Let $1 < p < \infty$ and $1/p + 1/q = 1$. Suppose that $\{f_n\}_{n\in\mathbb{N}} \subset L_p(\mathbb{R})$ with $M = \sup_n \|f_n\|_p < \infty$. Then, there exists $f \in L_p(\mathbb{R}, \mathscr{L}, m)$ such that $\|f\|_p \leq M$ and a subsequence $\{f_{n_k}\}_{k\in\mathbb{N}} \subset \{f_n\}_{n\in\mathbb{N}}$ such that*

$$\lim_{n\to\infty}\int_{\mathbb{R}} f_{n_k} g\,\mathrm{d}m = \int_{\mathbb{R}} fg\,\mathrm{d}m,$$

with $g \in L_q(\mathbb{R}, \mathscr{L}, m)$.

Proof. Let us first suppose that g belongs to the family $\{g_n\}_{n\in\mathbb{N}} \subset L_q(\mathbb{R})$. We want to show that there exists

$$\lim_{k\to\infty}\int_{\mathbb{R}} f_{n_k} g\,\mathrm{d}m.$$

Let us define

$$C_{k,n} = \int_{\mathbb{R}} f_k g_n\,\mathrm{d}m.$$

By Hölder's inequality we have that

$$\begin{aligned}\left|C_{k,1}\right| &= \left|\int_{\mathbb{R}} f_k g_1\,\mathrm{d}m\right| \\ &\leq \|f_k\|_p \|g_1\|_q,\end{aligned}$$

and, since $f_k \in L_p(\mathbb{R})$ and the hypothesis, we get

$$\left|C_{k,1}\right| \leq M\|g_1\|_q. \tag{3.32}$$

Since $\{C_{k,1}\}_{k\in\mathbb{N}}$ is a sequence of real numbers and is also bounded by (3.32), we can invoke Bolzano-Weierstrass Theorem and obtain a subsequence $\{C_{k_1,1}\} \subset \{C_{k,1}\}$ such that there exists

$$\lim_{k_1\to\infty} C_{k_1,1}$$

We can repeat the argument with $\{C_{k_1,1}\}$ to obtain a new subsequence $\{C_{k_2,h}\}$ with $h = 1,2$ such that there exists

$$\lim_{k_2\to\infty} C_{k_2,h}.$$

We can now proceed inductively and using Cantor's diagonal argument and we obtain a subsequence $\{C_{k_m,h}\}$ such that $\lim_{k_m\to\infty} C_{k_m,h}$ exists.

On the other hand, we can choose from $\{g_n\}_{n\in\mathbb{N}}$ a dense family in $L_q(\mathbb{R})$, let us denote it by $\mathscr{G}$ and define

$$T(g) = \lim_{m\to\infty} \int_{\mathbb{R}} f_{k_m} g \,\mathrm{d}m$$

for all $g \in \mathscr{G}$. Note that

$$T\left(\alpha g_{n_1} + \beta g_{n_2}\right) = \alpha T\left(g_{n_1}\right) + T\left(g_{n_2}\right)$$

for all $g_{n_1}, g_{n_2} \in \mathscr{G}$. Again, by Hölder's inequality we obtain

$$\begin{aligned} |T(g)| &\leq \|f_{k_m}\|_p \|g\|_q \\ &\leq M\|g\|_q, \end{aligned}$$

and we showed that T is a linear bounded functional, i.e., T is continuous. We now want to extend T to all $L_q(\mathbb{R})$. For all $g \in L_q(\mathbb{R})$ there exists a sequence $\{g_n\}_{n\in\mathbb{N}}$ in $\mathscr{G}$ such that

$$\lim_{t\to\infty} \|g - g_n\|_q = 0$$

and

$$\lim_{n\to\infty} \|g_n\|_q = \|g\|_q.$$

Observe that

$$|T(g_n) - T(g_L)| \leq M\|g_n - g_L\|_q \to 0$$

when $n \to \infty$, this means that $\{T(g_n)\}_{n\in\mathbb{N}}$ is a Cauchy sequence in $\mathbb{R}$, therefore it converges in $\mathbb{R}$, let us say that the limit is $T(g)$, i.e.,

$$T(g) = \lim_{n\to\infty} T(g_n),$$

which is well defined in $L_q(\mathbb{R})$. Moreover, note that

$$\begin{aligned} \left|T(g)\right| &\leq \limsup_{n\to\infty} \left|T(g) - T(g_n)\right| + \limsup_{n\to\infty} \left|T(g_n)\right| \\ &\leq \limsup_{n\to\infty} \left|T(g_n)\right| \\ &\leq M \limsup_{n\to\infty} \|g_n\|_q \\ &= M\|g\|_q \end{aligned}$$

for all $g \in L_q(\mathbb{R})$. By the Riesz representation theorem, there exists $f \in L_p(\mathbb{R})$ such that

$$\|T\| = \|f\|_p,$$

but $\|T\| \leq M$, then $\|f\|_p \leq M$ and

$$T(g) = \int_{\mathbb{R}} f g \, \mathrm{d}m,$$

therefore

$$\lim_{n\to\infty} \int_{\mathbb{R}} f_{k_m} g \, \mathrm{d}m = \int_{\mathbb{R}} f g \, \mathrm{d}m,$$

which ends the proof. □

The next theorem is quite important since it permits to calculate an integral in a general space via a one-dimensional integral. Formula (3.33) is sometimes denoted as *Cavalieri's principle* or even by the fancy designation of *layer cake representation*.

Theorem 3.54. *Let $(X, \mathscr{A}, \mu)$ be a σ-finite measure space and f an $\mathscr{A}$-measurable function. Let $\varphi : [0,\infty) \longrightarrow [0,\infty)$ be a C^1 class function such that $\varphi(0) = 0$. Then,*

$$\int_X \varphi(|f|) \mathrm{d}\mu = \int_0^\infty \varphi'(\lambda) \mu(\{x \in X : |f(x)| > \lambda\}) \mathrm{d}\lambda. \tag{3.33}$$

Proof. If $\mu(\{x \in X : |f(x)| > \lambda\}) = \infty$ there is nothing to prove.

Therefore, let us suppose that $\mu(\{x \in X : |f(x)| > \lambda\}) < \infty$. We want to show that $\{x \in X : |f(x)| > \lambda\}$ is measurable over $[0,\infty)$. Let us consider the set

$$E = \{(x,\lambda) \in X \times [0,\infty) : 0 < \lambda < |f(x)|\}.$$

We now show that the E set is measurable in $X \times [0,\infty)$. Since $|f| \geq 0$, therefore, there exists a sequence $\{s_n\}_{n\in\mathbb{N}}$ of simple functions such that $s_n \uparrow f$. We can write

$$s_n = \sum_{j=1}^{\infty} a_j^n \chi_{A_j^n},$$

with $A_j^n \in \mathscr{A}$ and $j = 1, 2, \ldots, n \in \mathbb{N}$. Therefore

$$\begin{aligned} E_n &= \{(x,\lambda) \in X \times [0,\infty) : 0 < \lambda < s_n(x)\} \\ &= \bigcup_{j=1}^{\infty} A_j^n \times (0, a_j^n) \end{aligned}$$

is measurable in $X \times [0,\infty)$. Observe now that

$$\lim_{n\to\infty} \chi_{E_n}(x) = \chi_E(x),$$

since χ_{E_n} is a measurable function in $X \times [0,\infty)$ because E_n is measurable in $X \times [0,\infty)$, therefore χ_E is measurable since it is the limit of a sequence of measurable functions. As a consequence

$$E = \{(x,\lambda) \in X \times [0,\infty) : 0 < \lambda < |f(x)|\}$$

is measurable, therefore

$$E^\lambda = \{x \in X : |f(x)| > \lambda\}$$

is measurable in $[0,\infty)$.

On the other hand, since φ is C^1, therefore

$$(x,\lambda) \longrightarrow \varphi'(\lambda)\chi_{E^\lambda}(x)$$

is measurable, moreover, since $\varphi(0) = 0$ we get

$$\varphi(t) = \int_0^t \varphi'(\lambda)\mathrm{d}\lambda.$$

We now obtain

$$\begin{aligned}
\int_X \varphi(|f|)\mathrm{d}\mu &= \int_X \int_0^{|f|} \varphi'(\lambda)\mathrm{d}\lambda\mathrm{d}\mu \\
&= \int_X \int_0^{\infty} \chi_{[0,|f|]}(\lambda)\varphi(\lambda)\mathrm{d}\lambda\mathrm{d}\mu \\
&= \int_0^{\infty} \int_X \chi_{\{x\in X:|f(x)|>\lambda\}}(x)\varphi'(\lambda)\mathrm{d}\mu\mathrm{d}\lambda \\
&= \int_0^{\infty} \varphi'(\lambda)\mu(\{x \in X : |f(x)| > \lambda\})\mathrm{d}\lambda,
\end{aligned}$$

ending the proof. □

In the particular case $\varphi(t) = t^p$ we get the following result.

Corollary 3.55 *Let $(X,\mathscr{A},\mu)$ be a σ-finite measure space and f where $0 < p < \infty$. Therefore,*

$$\int_X |f|^p\mathrm{d}\mu = p\int_0^{\infty} \lambda^{p-1}\mu(\{x \in X : |f(x)| > \lambda\})\mathrm{d}\lambda. \tag{3.34}$$

The following theorem gives us information regarding the size, at the origin and at the infinite, of the distribution function $D_f(t) := \mu(\{x \in X : |f(x)| > t\})$, see Chapter 4 for a comprehensive study of the distribution function.

Theorem 3.56. *Let $f \in L_p(X, \mathscr{A}, \mu)$ with $1 \le p < \infty$. We have*

$$\lim_{t\to 0} t^p \mu(\{x \in X : |f(x)| > t\}) = \lim_{t\to\infty} t^p \mu(\{x \in X : |f(x)| > t\}) = 0.$$

Proof. Let $f \in L_p(X, \mathscr{A}, \mu)$, $1 \le p < \infty$, and now using the Markov inequality (Lemma 3.28) with $g(t) = t^p$, we obtain

$$\mu(\{x \in X : |f(x)| > t\}) \le \frac{1}{t^p} \int_X |f|^p \mathrm{d}\mu < \infty,$$

then by (3.34) we have

$$\lim_{t\to 0} t^p \mu(\{x \in X : |f(x)| > t\}) = 0.$$

On the other hand, let us define

$$f_t = \begin{cases} |f|, & \text{if } |f| > t; \\ 0, & \text{if } |f| \le t, \end{cases} \tag{3.35}$$

with $f \in L_p(X, \mathscr{A}, \mu)$, then $|f| < \infty$ μ-a.e. To see that, note

$$\{x : |f(x)| = \infty\} = \bigcap_{n=1}^{\infty} \{x : |f(x)| > n\}.$$

Therefore

$$\mu(\{x : |f(x)| > n\}) \le \frac{1}{n^p} \int_X |f|^p \mathrm{d}\mu,$$

from which

$$\lim_{n\to\infty} \mu(\{x : |f(x)| > n\}) = 0,$$

and we get

$$\mu(\{x \in X : |f(x)| = \infty\}) \le \lim_{n\to\infty} \mu(\{x \in X : |f(x)| > n\}) = 0,$$

resulting in $|f| < \infty$ μ-a.e. Going back to (3.35), observe that $\lim_{t\to\infty} f_t = 0$ μ-a.e. implies that we have $\lim_{t\to\infty} f_t^p = 0$ μ-a.e., and now by the dominated convergence theorem we get

$$\lim_{t\to\infty} \int_X f_t^p \mathrm{d}\mu = 0.$$

Then

$$\lim_{t\to\infty} \int_{\{x:|f(x)|>t\}} |f|^p \mathrm{d}\mu = \lim_{t\to\infty} \int_X f_t^p \mathrm{d}\mu = 0,$$

and finally

$$\lim_{t\to\infty} t^p\mu(\{x\in X : |f(x)|>t\}) \le \lim_{t\to\infty} \int_{\{x:|f(x)|>t\}} |f|^p \mathrm{d}\mu = 0.$$

The proof is complete. □

Now we can define a Borel measure in $(0,\infty)$ given by

$$\nu((a,b]) = D_f(b) - D_f(a)$$

for all $a,b>0$, where D_f stands for the *distribution function* $D_f : [0,\infty) \longrightarrow [0,\infty)$, given by $D_f(\lambda) = \mu(\{x\in X : |f(x)|>\lambda\})$. For further properties of the distribution function see Proposition 4.3.

In this sense we can consider the Lebesgue-Stieltjes integral

$$\int_0^\infty \varphi \mathrm{d}D_f = \int_0^\infty \varphi \mathrm{d}\nu,$$

where φ is a nonnegative and Borel measure function in $(0,\infty)$.

The following results show us that the integral over X of the function $|f|$ can be expressed as a Lebesgue-Stieltjes integral.

Theorem 3.57. *If $D_f(\lambda)<\infty$ for all $\lambda>0$ and φ is a nonnegative Borel measurable function in $(0,\infty)$. Then*

$$\int_X \varphi\circ|f|\mathrm{d}\mu = -\int_0^\infty \varphi(\lambda)\mathrm{d}D_f(\lambda).$$

Proof. Let ν be a Borel measure in $(0,\infty)$ given by

$$\nu((a,b]) = D_f(b) - D_f(a) \tag{3.36}$$

for all $a,b>0$. We affirm that

$$D_f(a) - D_f(b) = \mu(\{x : a<|f(x)|\le b\}).$$

Indeed, since $b>a$, then

$$\{x\in X : |f(x)|>b\} \subset \{x\in X : |f(x)|>a\}.$$

By the fact that $D_f(a)<\infty$ we have

$$\mu(\{x\in X : |f(x)|>a\}) < \infty,$$

then

$$\{x\in X : a<|f(x)|\le b\} = \{x\in X : |f(x)|>a\} \cap \{x\in X : |f(x)|\le b\}$$

$$= \{x \in X : |f(x)| > a\} \cap \{x \in X : |f(x)| > b\}^c$$
$$= \{x \in X : |f(x)| > a\} \setminus \{x \in X : |f(x)| > b\}.$$

Therefore

$$\begin{aligned}
&\mu(\{x \in X : a < |f(x)| \leq b\}) \\
&\qquad = \mu(\{x \in X : |f(x)| > a\}) - \mu(\{x \in X : |f(x)| > b\}) \\
&\qquad = D_f(a) - D_f(b).
\end{aligned}$$

By (3.36) we can write

$$\begin{aligned}
\nu((a,b]) &= -[D_f(a) - D_f(b)] \\
&= -\mu(\{x \in X : a < |f(x)| \leq b\}) \\
&= -\mu(|f|^{-1}(a,b]).
\end{aligned}$$

Using the unique extension theorem for measure, we obtain

$$\nu(E) = -\mu(|f|^{-1}(E))$$

for all Borel set $E \subset [0,\infty)$.

Now, let us consider

(a) $\varphi = \chi_E$ and observe that

$$\begin{aligned}
\varphi \circ |f|(x) = \chi_E \circ |f|(x) &= \begin{cases} 1, & \text{if } |f(x)| \in E; \\ 0, & \text{if } |f(x)| \notin E, \end{cases} \\
&= \begin{cases} 1, & \text{if } x \in |f|^{-1}(E); \\ 0, & \text{if } x \notin |f|^{-1}(E), \end{cases} \\
&= \chi_{|f|^{-1}(E)}(x).
\end{aligned}$$

Thus

$$\begin{aligned}
\int_X \varphi \circ |f| \mathrm{d}\mu &= \int_X \chi_{|f|^{-1}(E)}(x) \mathrm{d}\mu \\
&= \mu(|f|^{-1}(E)) \\
&= -\nu(E) \\
&= -\int_0^\infty \chi_E \mathrm{d}\nu
\end{aligned}$$

$$= -\int_0^\infty \varphi \mathrm{d}\nu$$
$$= -\int_0^\infty \varphi \mathrm{d}D_f.$$

(b) If $\varphi = \sum_{k=1}^n a_k \chi_{E_k}$, then

$$\int_X \varphi \circ |f| \mathrm{d}\mu = \sum_{k=1}^n a_k \int_X \chi_{|f|^{-1}(E_k)}(x) \mathrm{d}\mu$$
$$= \sum_{k=1}^n a_k \mu(|f|^{-1}(E_k))$$
$$= -\sum_{k=1}^n a_k \nu(E_k)$$
$$= -\sum_{k=1}^n a_k \int_0^\infty \chi_{E_k} \mathrm{d}\nu$$
$$= -\int_0^\infty \varphi \mathrm{d}\nu$$
$$= -\int_0^\infty \varphi \mathrm{d}D_f.$$

(c) If φ is any nonnegative Borel measurable function, then there exists a sequence $\{s_n\}_{n\in\mathbb{N}}$ of simple functions such that $s_n \uparrow \varphi$, therefore $s_n \circ |f| \uparrow \varphi \circ |f|$, and using (b)

$$\int_X s_n \circ |f| \mathrm{d}\mu = -\int_0^\infty s_n \mathrm{d}D_f,$$

now by the monotone convergence theorem we have

$$\lim_{n\to\infty} \int_X s_n \circ |f| \mathrm{d}\mu = -\lim_{n\to\infty} \int_X s_n \mathrm{d}D_f,$$

$$\int_X \varphi \circ |f| \mathrm{d}\mu = -\int_0^\infty \varphi \mathrm{d}D_f.$$

The proof is complete. □

3.7 Continuity of the Translation

Let $\Omega \subset \mathbb{R}^n$ be a measurable set. For $f \in L_p(\Omega)$ and $h \in \mathbb{R}^n$, define the *translation of* f as

$$T_h f(x) = \begin{cases} f(x+h) & \text{if } x+h \in \Omega \\ 0 & \text{if } x+h \in \mathbb{R}^n \setminus \Omega. \end{cases}$$

Sometimes the translation operator is introduced in a slightly different way, namely $\tau_h f(x) = T_{-h} f(x)$, where τ_h is the other definition of the translation operator.

The following result shows that the translation operation is continuous in the topology generated by the $L_p(\Omega)$ norm for $p \in [1, +\infty)$.

Theorem 3.58 *Let $\Omega \subset \mathbb{R}^n$ be a measurable set and $f \in L_p(\Omega)$ for $1 \leq p < \infty$. For any $\varepsilon > 0$ there exists $\delta = \delta(\varepsilon)$ such that*

$$\sup_{|h| \leq \delta} \|T_h f - f\|_p \leq \varepsilon.$$

Proof. First suppose that Ω is a bounded subset of $\mathbb{R}^n$, i.e., it is contained in a ball $B_R(0)$ centered at the origin and radius R, for any $R > 0$ sufficiently large. Without loss of generality we can assume that $\Omega = B_R(0)$ and define $f = 0$ in $\mathbb{R}^n \setminus \Omega$. For $E \subset \Omega$ and $\lambda \in \mathbb{R}$ define

$$E - \lambda = \{x \in \mathbb{R}^n : x + \lambda \in E\}.$$

By the absolute continuity of the integral, for $\varepsilon > 0$ there exists $\delta > 0$ such that for all $E \subset \Omega$ measurable with $m(E) < \delta$ then

$$\int_E |f|^p \, \mathrm{d}m < \frac{\varepsilon^p}{2^{p+1}}.$$

Since the Lebesgue measure is translation invariant, we have that

$$m\big((E - \lambda) \cap \Omega\big) < \delta.$$

Therefore

$$\int_E |T_h f - f|^p \mathrm{d}m \leq 2^{p-1} \left[\int_E |f|^p \mathrm{d}m + \int_{(E-h)\cap\Omega} |f|^p \mathrm{d}m \right] \leq \frac{\varepsilon^p}{2}.$$

Since f is measurable, by Lusin's theorem, f is quasi-continuous. Therefore, we set a positive number

$$\sigma = \frac{\delta}{2 + m(\Omega)},$$

then there exists a closed set $\Omega_\sigma \subset \Omega$ such that $m(\Omega \setminus \Omega_\sigma) \le \sigma$ and f is uniformly continuous on Ω_σ. In particular, there exists $\delta_\sigma > 0$ such that

$$|T_h f(x) - f(x)| \le \frac{\varepsilon^p}{2m(\Omega_\sigma)},$$

whenever $|h| < \delta_\sigma$, and $x, x+h$ belong to Ω_σ.

For any h, we have

$$\int\limits_{\Omega_\sigma \cap (\Omega_\sigma - h)} |T_h f - f|^p \, dm \le \frac{1}{2}\varepsilon^p.$$

For any $\eta \in \mathbb{R}^n$ such that $|\eta| < \sigma$, we get

$$\begin{aligned} m\left(\Omega \setminus (\Omega - \eta)\right) &= m\left((\Omega + \eta) \setminus \Omega_\sigma\right) \\ &\le m(\Omega \setminus \Omega_\sigma) + m\left((\Omega + \eta) \setminus \Omega\right) \\ &\le \sigma + \sigma m(\Omega). \end{aligned}$$

We now obtain

$$m(\Omega \setminus [\Omega_\sigma \cap (\Omega_\sigma - \eta)]) \le m(\Omega \setminus \Omega_\sigma) + m\left(\Omega \setminus (\Omega_\sigma - \eta)\right) \le \sigma\left(2 + m(\Omega)\right) = \delta.$$

We now consider the case when Ω is an unbounded subset of $\mathbb{R}^n$. Then since $\varepsilon > 0$ exists $R > 0$ sufficiently large such that for all $|h| < 1$

$$\int\limits_{\Omega \cap \{|x| > 2R\}} |T_h f - f|^p dm \le 2^p \int\limits_{\Omega \cap \{|x| > R\}} |f|^p dm \le \frac{1}{2^p}\varepsilon^p.$$

For such R, there is $\delta_0 = \delta_0(\varepsilon)$ such that

$$\sup_{|h| < \delta_0} \|T_h f - f\|_{p, \Omega \cap \{|x| < 2r\}} \le \frac{1}{2}\varepsilon.$$

Therefore

$$\sup_{|h| < \delta_0} \|T_h f - f\|_{p,\Omega} \le \|T_h f - f\|_{p, \Omega \cap \{|x| < 2r\}} + 2\|f\|_{p, \Omega \cap \{|x| > r\}} < \varepsilon.$$

The proof is complete. □

It should be mentioned that Theorem 3.58 is false when $p = \infty$.

3.8 Weighted Lebesgue Spaces

A *weight* is a nonnegative locally integrable function on $\mathbb{R}^n$ that takes values in $(0,\infty)$ almost everywhere. Therefore, weights are allowed to be zero or infinite only on a set of Lebesgue measure zero. Hence, if w is a weight and $1/w$ is locally integrable, then $1/w$ is also a weight.

Given a weight w and a measurable set E, we use the notation $w(E) = \int\limits_E w(x)\mathrm{d}x$ to denote the w-measure of the set E. Since weights are locally integrable functions, $w(E) < \infty$ for all sets E contained in some finite ball.

From now on in this section we will restrict to work on $\Omega = \mathbb{R}_+$.

Definition 3.59. The *weighted L_p spaces* are denoted by $L_p(w)$, for $1 \leq p < \infty$ it consists of all measurable functions f such that

$$\|f\|_{L_p((0,\infty),w)} = \|f\|_{L_p(w)} = \left(\int_0^\infty |f(x)|^p w(x)\mathrm{d}x \right)^{1/p} < \infty,$$

i.e.

$$L_p(w) = \left\{ f : \|f\|_{L_p(w)} = \left(\int_0^\infty |f(x)|^p w(x)\mathrm{d}x \right)^{1/p} < \infty \right\}.$$

We also denote

$$(L_p(w))^d = \{ f \in L_p(w) : f \text{ is decreasing} \}.$$

⊘

Some authors define the weighted Lebesgue spaces with the weight as a multiplier, i.e., $f \in L_p(w) \Leftrightarrow wf \in L_p$.

The next result gives another characterization of the norm of weighted Lebesgue spaces based on the so-called Minkowski functional. This type of norm will be fully exploited in Chapter 7.

Theorem 3.60 *Let $f \in L_p(w)$ where $1 \leq p < \infty$. Then*

$$\|f\|_{L_p(w)} = \inf \left\{ \lambda > 0 : \int_0^\infty \left| \frac{f(x)}{\lambda} \right|^p w(x)\mathrm{d}x \leq 1 \right\}. \tag{3.37}$$

Proof. On the one hand, let us take $\lambda = \|f\|_{L_p(w)}$, then

$$\inf\left\{\lambda>0:\int\limits_0^\infty\left|\frac{f(x)}{\lambda}\right|^p w(x)\mathrm{d}x\le 1\right\}\le\|f\|_{L_p(w)}. \tag{3.38}$$

On the other hand, if $\int\limits_0^\infty |f(x)/\lambda|^p w(x)\mathrm{d}x \le 1$, then $\int\limits_0^\infty |f(x)|^p w(x)\mathrm{d}x \le \lambda^p$ and thus $(\int\limits_0^\infty |f(x)|^p w(x)\mathrm{d}x)^{1/p} \le \lambda$. Therefore

$$\|f\|_{L_p(w)}\le\inf\left\{\lambda>0:\int\limits_0^\infty\left|\frac{f(x)}{\lambda}\right|^p w(x)\mathrm{d}x\le 1\right\}. \tag{3.39}$$

Combining (3.38) and (3.39) we have (3.37). □

We define the duality on the weighted Lebesgue space $L_p(w)$ with $1<p<\infty$ by the inner product

$$\langle f,g\rangle=\int\limits_0^\infty f(x)g(x)\mathrm{d}x,\quad f\in L_p(w).$$

Theorem 3.61 *Let $\frac{1}{p}+\frac{1}{q}=1$ and $g\in L_q(w^{1-q})$. Then*

$$\sup_{\|f\|_{L_p(w)}=1}|\langle f,g\rangle|=\|g\|_{L^q(w^{1-q})}$$

and $\left(L_p(w)\right)^*=L^q\left(w^{1-q}\right)$.

Proof. By Hölder's inequality we have

$$\begin{aligned}|\langle f,g\rangle| &= \int\limits_0^\infty |f(x)|(w(x))^{\frac{1}{p}}|g(x)|(w(x))^{-\frac{1}{p}}\mathrm{d}x\\ &\le\left(\int\limits_0^\infty |f(x)|w(x)\mathrm{d}x\right)^{\frac{1}{p}}\left(\int\limits_0^\infty |g(x)|^q w(x)^{-\frac{q}{p}}\mathrm{d}x\right)^{\frac{1}{q}}.\end{aligned}$$

Since $q/p=q-1$ and if

$$f=\frac{|g|^{q-1}(\operatorname{sgn}g)w^{1-q}}{\|g\|^{q-1}_{L_q\left(w^{1-q}\right)}}$$

then $\|f\|_{L_p(w)}=1$ and $\langle f,g\rangle=\|g\|_{L^q(w^{1-q})}$. Hence

$$\|g\|_{L^q(w^{1-q})}\le\sup_{\|f\|_{L_p(w)}=1}|\langle f,g\rangle|\le\|g\|_{L^q(w^{1-q})}$$

and the result follows. □

We now obtain a Cavalieri's type principle for the weighted Lebesgue decreasing functions.

Theorem 3.62 *If $f \in \left(L_p(w)\right)^d$ then for any measurable weight function w we have*

$$\int_0^\infty |f(x)|^p w(x)\mathrm{d}x = p\int_0^\infty \lambda^{p-1}\left(\int_0^{D_f(\lambda)} w(x)\mathrm{d}x\right)\mathrm{d}\lambda, \quad 0<p<\infty,$$

where $D_f(\lambda) = m\left(\{x : |f(x)| > \lambda\}\right)$.

Proof. First of all, notice that since $f \in \left(L_p(w)\right)^d$, then

$$\{t \in (0,\infty) : f(t) > \lambda\} = (0, D_f(\lambda)).$$

Let us denote $E_f(\lambda) = \{t \in (0,\infty) : f(t) > \lambda\}$, then $D_f(\lambda) = m\left(E_f(\lambda)\right)$. Now, by Fubini's theorem, we obtain

$$\begin{aligned}
\int_0^\infty |f(x)|^p w(x)\mathrm{d}x &= \int_0^\infty \left(p\int_0^{|f(x)|} \lambda^{p-1}\mathrm{d}\lambda\right) w(x)\mathrm{d}x \\
&= p\int_0^\infty \left(\int_0^\infty \lambda^{p-1}\chi_{E_f(x)}(x)\mathrm{d}\lambda\right) w(x)\mathrm{d}x \\
&= p\int_0^\infty \lambda^{p-1}\left(\int_0^\infty \chi_{E_f(\lambda)}(x)w(x)\mathrm{d}x\right)\mathrm{d}\lambda \\
&= p\int_0^\infty \lambda^{p-1}\left(\int_{E_f(\lambda)} w(x)\mathrm{d}x\right)\mathrm{d}\lambda \\
&= p\int_0^\infty \lambda^{p-1}\left(\int_{(0,D_f(\lambda))} w(x)\mathrm{d}x\right)\mathrm{d}\lambda \\
&= p\int_0^\infty \lambda^{p-1}\int_0^{D_f(\lambda)} w(x)\mathrm{d}x\mathrm{d}\lambda,
\end{aligned}$$

which ends the proof. □

The weighted Lebesgue spaces enjoy the same embedding type result than the Lebesgue spaces.

Theorem 3.63 *Let $1 \leq p < q \leq \infty$. Then $L_q(E,w) \hookrightarrow L_p(E,w)$ for any finite measurable set $E \subset (0,\infty)$.*

Proof. Let $f \in L^q(w)$ and $E \subset (0,\infty)$ a measurable set. Let $r = q/p$ and $s = r/(r-1)$, then $1/r + 1/s = 1$. Observe that

$$\int_0^\infty |f(x)|^{pr} w(x)\mathrm{d}x = \int_0^\infty |f(x)|^q w(x)\mathrm{d}x < \infty.$$

By the Hölder inequality we have

$$\begin{aligned}\int_0^\infty |f(x)|^p \chi_E(x) w(x)\mathrm{d}x &= \int_0^\infty |f(x)|^p \chi_E(x)(w(x))^{1/r}(w(x))^{1/s}\mathrm{d}x \\ &\leq \left(\int_0^\infty |f(x)|^q w(x)\mathrm{d}x\right)^{1/r} \left(\int_E w(x)\mathrm{d}x\right)^{1/s} \\ &= w(E)^{1/s}\left(\int_o^\infty |f(x)|^q w(x)\mathrm{d}x\right)^{1/r},\end{aligned}$$

which shows the embedding. □

The Hardy inequality is valid in the framework of weighted Lebesgue spaces in the following form.

Theorem 3.64 *Let $f \in L_p(w)$ with $1 < p < \infty$. If w is a nondecreasing weight, then*

$$\left(\int_0^\infty \left[\frac{1}{x}\int_0^x f(t)\mathrm{d}t\right]^p w(x)\mathrm{d}x\right)^{1/p} \leq \frac{p}{p-1}\left(\int_0^\infty f(x)^p w(x)\mathrm{d}x\right)^{1/p}. \tag{3.40}$$

Proof. Making a convenient change of variable twice and using the Minkowski integral inequality we have

$$\left(\int_0^\infty \left[\frac{1}{x}\int_0^x f(t)\mathrm{d}t\right]^p w(x)\mathrm{d}x\right)^{1/p} = \left(\int_0^\infty \left[\frac{1}{x}\int_0^1 f(xu)x\mathrm{d}u\right]^p w(x)\mathrm{d}x\right)^{1/p}$$

$$= \left(\int_0^\infty \left[\int_0^1 f(xu)\mathrm{d}u \right]^p w(x)\mathrm{d}x \right)^{1/p}$$

$$\leq \int_0^1 \left(\int_0^\infty (f(xu))^p w(x)\mathrm{d}x \right)^{1/p} \mathrm{d}u$$

$$= \int_0^1 \left(\int_0^\infty (f(v))^p w\left(\frac{v}{u}\right) \frac{\mathrm{d}v}{u} \right)^{1/p} \mathrm{d}u.$$

Since $0 < u < 1$, then $1/u > 1$ and so $w(v/u) \leq w(v)$ since w is nonincreasing, therefore we have

$$\int_0^1 \left(\int_0^\infty (f(v))^p w\left(\frac{v}{u}\right) \frac{\mathrm{d}v}{u} \right)^{1/p} du \leq \int_0^1 \left[\int_0^\infty (f(v))^p w(v) dv \right]^{1/p} \frac{du}{u^{1/p}}$$

$$= \left(\int_0^1 u^{-1/p}\mathrm{d}u \right) \left(\int_0^\infty (f(v))^p w(v)\mathrm{d}v \right)^{1/p}$$

$$= \frac{p}{p-1} \left(\int_0^\infty (f(v))^p w(v)\mathrm{d}v \right)^{1/p}.$$

Finally, gathering all of these inequalities we obtain (3.40). □

Theorem 3.65 *Let $f \in L_p(w)$ with $1 < p < \infty$. If w is a nondecreasing weight such that*

$$\left(\int_0^\infty \left[\frac{1}{x} \int_0^x f(t)\mathrm{d}t \right]^p w(x)\mathrm{d}x \right)^{1/p} \leq \frac{p}{p-1} \left(\int_0^\infty [f(x)]^p w(x)\mathrm{d}x \right)^{1/p}$$

holds, then

$$r^p \int_r^\infty \frac{w(x)}{x^p}\mathrm{d}x \leq B \int_0^r w(x)\mathrm{d}x,$$

when $r > 0$ where $B = C - 1$, $C = p/(p-1)$.

Proof. Let $C = (p/(p-1))$, then

$$\int_0^\infty \left[\frac{1}{x}\int_0^x f(t)\mathrm{d}t\right]^p w(x)\mathrm{d}x \leq C\int_0^\infty [f(x)]^p w(x)\mathrm{d}x.$$

Now, let us take $f = \chi_{[0,r]}$ for $r > 0$, then

$$\int_0^\infty \left(\frac{1}{x}\int_0^x \chi_{[0,r]}(t)\mathrm{d}t\right)^p w(x)\mathrm{d}x \leq C\int_0^\infty \left[\chi_{[0,r]}(x)\right]^p w(x)\mathrm{d}x$$

and this finishes the proof. □

3.9 Uniform Convexity

Definition 3.66. A Banach space $(\mathbb{R}, V, +, \cdot, \|\cdot\|)$ is said to the *uniformly convex* if for all $\varepsilon > 0$ there exists a number $\delta > 0$ such that for all $x, y \in V$ the conditions

$$\|x\| = \|y\| = 1, \quad \|x - y\| \geq \varepsilon$$

imply

$$\left\|\frac{x+y}{2}\right\| \leq 1 - \delta.$$

The number

$$\delta(\varepsilon) = \inf\left\{1 - \left\|\frac{x+y}{2}\right\| : \|x\| = \|y\| = 1, \|x - y\| \geq \varepsilon\right\}$$

is called the *modulus of convexity*. Note that if $\varepsilon_1 < \varepsilon_2$, then $\delta(\varepsilon_1) < \delta(\varepsilon_2)$ and $\delta(0) = 0$ since $x = y$ if $\varepsilon = 0$. ⊘

The immediate example of an uniformly convex space is the Hilbert space since its norm satisfies the parallelogram law,

$$\left\|\frac{x+y}{2}\right\|^2 = \frac{1}{2}(\|x\|^2 + \|y\|^2) - \left\|\frac{x-y}{2}\right\|^2,$$

in this sense, if $\|x\| = \|y\| = 1$ and $\|x - y\| \geq \varepsilon$, then

$$\left\|\frac{x+y}{2}\right\|^2 \leq 1 - \frac{\varepsilon^2}{4},$$

from which we obtain

$$1-\sqrt{1-\frac{\varepsilon^2}{4}} \leq 1-\left\|\frac{x+y}{2}\right\|,$$

which entails

$$\delta(\varepsilon) \geq 1-\sqrt{1-\frac{\varepsilon^2}{4}}.$$

In order to prove the main result in this section, namely that the L_p spaces are uniformly convex whenever $1 < p < \infty$, we need some auxiliary lemmas dealing with inequalities.

Lemma 3.67. *If $p \geq 2$ then*

$$\left(|a+b|^p+|a-b|^p\right)^{1/p} \leq \left(|a+b|^2+|a-b|^2\right)^{1/2} \tag{3.41}$$

for all $a, b \in \mathbb{R}$.

Proof. Let us consider the function

$$f(t) = (1+t^p)^{1/p}(1+t^2)^{-1/2}$$

for all $t \in [-1,1]$. It is clear that f is differentiable in $[-1,1]$ and calculating the first derivative of f we obtain

$$f'(t) = (1+t^p)^{\frac{1}{p}-1}(1+t^2)^{-3/2}t(t^{p-2}-1),$$

which entails that $f'(1)=0$ and $f'(0)=0$. We also observe that $f'(t)<0$ if $t \in (0,1)$ and $f'(t)>0$ if $t \in (-1,0)$, which means that the function f attains its maximum at $t=0$. Let $t \neq 0$, then

$$f(t) \leq f(0),$$

i.e., $f(t) \leq 1$ since $f(0)=1$, from which we get

$$(1+t^p)^{1/p}(1+t^2)^{-1/2} \leq 1,$$

therefore

$$(1+t^p)^{1/p} \leq (1+t^2)^{1/2}.$$

Choosing

$$t = \frac{|a-b|}{|a+b|},$$

we get (3.41). □

We will also need the following inequality.

Lemma 3.68. *If $p \geq 2$ for all $a, b \in \mathbb{R}$. Then*

$$|a+b|^p+|a-b|^p \leq 2^{p-1}(|a|^p+|b|^p).$$

Proof. In virtue of Lemma 3.67 and the parallelogram law we can write

$$\begin{aligned}(|a+b|^p+|a-b|^p)^{1/p} &\le (|a+b|^2+|a-b|^2)^{1/2}\\ &= [2(|a|^2+|b|^2)]^{1/2}\\ &\le \sqrt{2}(|a|^2+|b|^2)^{1/2}. \end{aligned} \tag{3.42}$$

By the Hölder inequality for

$$\frac{2}{p}+\frac{p-2}{p}=1,$$

we get

$$\begin{aligned}|a|^2+|b|^2 &\le (|a|^p+|b|^p)^{2/p}(1+1)^{\frac{p-2}{p}}\\ &\le 2^{\frac{p-2}{p}}(|a|^p+|b|^p)^{2/p},\end{aligned}$$

from which we obtain

$$\begin{aligned}\sqrt{2}(|a|^2+|b|^2)^{1/2} &\le \sqrt{2}\cdot 2^{\frac{p-2}{2p}}(|a|^p+|b|^p)^{1/p}\\ &= 2^{\frac{p-1}{p}}(|a|^p+|b|^p)^{1/p}.\end{aligned}$$

By (3.42) it follows that

$$(|a+b|^p+|a-b|^p)^{1/p} \le 2^{\frac{p-1}{p}}(|a|^p+|b|^p)^{1/p},$$

which ends the proof. □

Lemma 3.69. *Let $a \ge 0$, $b \ge 0$, and $s \ge 2$, then*

$$a^s+b^s \le \left(a^2+b^2\right)^{s/2}$$

and

$$\left(\frac{a}{2}+\frac{b}{2}\right)^s \le \frac{a^s}{2}+\frac{b^s}{2}.$$

Proof. Consider the functions

$$\varphi(t)=t^s+1$$

and

$$\psi(t)=\left(t^2+1\right)^{s/2}$$

for $t \ge 0$ and $s > 2$. Observe that

$$\varphi'(t)=st^{s-1}$$

and

$$\psi'(t) = \frac{s}{2}\left(t^2+1\right)^{s/2-1}(2t) = st(t^2+1)^{(s-2)/2}$$

for all $t \geq 0$. For $t > 0$, note that $t^2 < t^2+1$ from which

$$\begin{aligned}(t^2)^{s-2} &< (t^2+1)^{s-2}\\ t^{2s-4} &< (t^2+1)^{s-2}\\ t^{2(s-1)} &< t^2(t^2+1)^{s-2}\\ t^{s-1} &< t(t^2+1)^{\frac{s-2}{2}}\\ st^{s-1} &< st(t^2+1)^{\frac{s-2}{2}}.\end{aligned}$$

This tell us that $\varphi'(t) < \psi'(t)$ for all $t > 0$, moreover since $\varphi'(0) = 0 = \psi'(0)$. We have

$$\varphi'(t) \leq \psi'(t),$$

for all $t \geq 0$. Thus, for every $x > 0$,

$$\varphi(x) - \varphi(0) = \int_0^x \varphi'(t)\,\mathrm{d}t \leq \int_0^x \psi'(t)\,\mathrm{d}t = \psi(x) - \psi(0),$$

but $\varphi(0) = 1 = \psi(0)$. Therefore

$$x^s + 1 \leq (x^2+1)^{s/2}.$$

In particular, for $x = a/b$ with $a \geq 0$ and $b \geq 0$, we obtain

$$\left(\frac{a}{b}\right)^s + 1 \geq \left[(\frac{a}{b})^2 + 1\right]^{s/2}$$

$$\begin{aligned}a^s + b^s &\leq b^s\left(\frac{a^2+b^2}{b^2}\right)^{s/2}\\ &= b^s\left(\frac{a^2+b^2}{(b^s)^{s/2}}\right)^{s/2}\\ &= (a^2+b^2)^{s/2}\end{aligned}$$

Finally $a^s + b^s \leq (a^2+b^2)^{s/2}$. Note that this last inequality is valid for $b = 0$, thus

$$a^s + b^s \leq \left(a^2+b^2\right)^{s/2}$$

for all $a \geq 0$, $b \geq 0$ and $s > 2$.

On the other hand, the function $p(t) = t^s$ is convex on $[0,\infty)$ for $s > 2$, then

$$p''(t) = s(s-1)t^{s-2} > 0$$

for all $t > 0$ without lost of generality we might suppose that $a < b$ since $a, b > 0$, we also have

$$p(ta + (1-t)b) \leq tp(a) + (1-t)p(b),$$

for $0 \leq t \leq 1$. By the definition of p and for $t = 1/2$ it follows that

$$\left(\frac{a}{2} + \frac{b}{2}\right)^s \leq \frac{a^s}{2} + \frac{b^s}{2}$$

for all $a \geq 0$, $b \geq 0$ and $s > 2$. □

We now show the uniform convexity of Lebesgue spaces.

Theorem 3.70. *If $1 < p < \infty$, the space L_p is uniformly convex.*

Proof. Let us distinguish two cases:

First case $p \geq 2$. In virtue of Lemma 3.68 we have that

$$|a+b|^p + |a-b|^p \leq 2^{p-1}(|a|^p + |b|^p).$$

for $a, b \in \mathbb{R}$. If f and g are L_p functions, then

$$|f+g|^p + |f-g|^p \leq 2^{p-1}(|f|^p + |g|^p),$$

and integrating, we get

$$\|f+g\|_p^p + \|f-g\|_p^p \leq 2^{p-1}(\|f\|_p^p + \|g\|_p^p).$$

Now, if $\|f\|_p = \|g\|_p = 1$ and $\|f-g\|_p \geq \varepsilon$, then

$$\|f+g\|_p^p \leq 2^{p-1}(\|f\|_p^p + \|g\|_p^p) - \|f-g\|_p^p$$

which implies that

$$\begin{aligned}\|f+g\|_p^p &\leq 2^{p-1}2 - \varepsilon^p \\ &= 2^p - \varepsilon^p,\end{aligned}$$

therefore

$$\left\|\frac{f+g}{2}\right\|_p^p \leq 1 - \left(\frac{\varepsilon}{2}\right),$$

from which

$$\left\|\frac{f+g}{2}\right\|_p \leq \left(1 - \left(\frac{\varepsilon}{2}\right)\right)^{1/p} \leq 1 - \left(\frac{\varepsilon}{2}\right)^p,$$

and from this we have

$$\left(\frac{\varepsilon}{2}\right)^p \leq 1 - \left\|\frac{f+g}{2}\right\|_p,$$

therefore

$$\delta_{L_p}(\varepsilon) \geq \left(\frac{\varepsilon}{2}\right)^p.$$

Second case $1 < p < 2$. From Lemma 3.69 and if $1/p + 1/q = 1$ then $q = p/(p-1) > 2$ and we have

$$\begin{aligned}
\left\|\frac{f+g}{2}\right\|_p^q + \left\|\frac{f-g}{2}\right\|_p^q &\leq \left(\left\|\frac{f}{2}\right\|_p + \left\|\frac{g}{2}\right\|_p\right)^q + \left(\left\|\frac{f}{2}\right\|_p + \left\|\frac{g}{2}\right\|_p\right)^q \\
&\leq 2\left(\left\|\frac{f}{2}\right\|_p + \left\|\frac{g}{2}\right\|_p\right)^q \\
&\leq 2\left(\left\|\frac{f}{2}\right\|_p + \left\|\frac{g}{2}\right\|_p\right)^{\frac{p}{p-1}} \\
&\leq 2\left(\left\|\frac{f}{2}\right\|_p^p + \left\|\frac{g}{2}\right\|_p^p\right)^{\frac{1}{p-1}} \\
&\leq 2\left[\frac{1}{2^p}\left(\|f\|_p^p + \|g\|_p^p\right)\right]^{\frac{1}{p-1}} \\
&\leq 2\frac{1}{2^{p/(p-1)}}\left(\|f\|_p^p + \|g\|_p^p\right)^{\frac{1}{p-1}} \\
&= 2^{1-q}\left(\|f\|_p^p + \|g\|_p^p\right)^{\frac{1}{p-1}} \\
&= 2^{\frac{-1}{p-1}}\left(\|f\|_p^p + \|g\|_p^p\right)^{\frac{1}{p-1}} \\
&= \left(\frac{1}{2}\right)^{\frac{1}{p-1}}\left(\|f\|_p^p + \|g\|_p^p\right)^{\frac{1}{p-1}} \\
&= \left(\frac{\|f\|_p^p + \|g\|_p^p}{2}\right)^{\frac{1}{p-1}}.
\end{aligned}$$

Therefore we conclude that

$$\left\|\frac{f+g}{2}\right\|_p^q + \left\|\frac{f-g}{2}\right\|_p^q \leq \left(\frac{\|f\|_p^p + \|g\|_p^p}{2}\right)^{\frac{1}{p-1}}.$$

from which the result follows. □

Remark 3.71. The uniform convexity is a geometric property of the norm, an equivalent norm is not necessarily uniformly convex. On the other hand, the reflexivity is a topological property, i.e., a reflexive space continues to be reflexive for its equivalent norms. The Theorem of Milman-Pettis (see, e.g., Brezis [3]) give us an unusual tool, it states that a geometric property imply a topologies property. In other words, all uniformly convex Banach space is reflexive. Before talking about the reciprocal, let us analyze the following example. Consider in $\mathbb{R}^2$ the Euclidean norm $\|(x,y)\|_2 = (|x|^2+|y|^2)^{1/2}$ and the taxicab norm $\|(x,y)\|_1 = |x|+|y|$, it is not difficult to show that

$$\frac{1}{2}\|(x,y)\|_1 \leq \|(x,y)\|_2 \leq \sqrt{2}\|(x,y)\|_1$$

in other words, the norms $\|(x,y)\|_1$ and $\|(x,y)\|_2$ are equivalent, moreover $\mathbb{R}^2$ with the Euclidean norm is uniformly convex space, but with the norm ℓ_1 it is not, nonetheless $\mathbb{R}^2$ is reflexive with respect to both norms. The following result, proved by Day [14], affirms that the reciprocal of the Milman-Pettis theorem is not valid.

Theorem 3.72 (M.M. Day). *There exist Banach spaces which are separable, reflexive, and strictly convex, but are not isomorphic to any uniformly convex space.*

$\oslash$

3.10 Isometries

Definition 3.73. The linear operator $T : L_p(\mu) \to L_p(\mu)$ is said to be an *isometry* if

$$\|T(f)\|_p = \|f\|_p,$$

for all $f \in L_p(\mu)$. $\oslash$

Lemma 3.74. *Let ξ and ν be real numbers. Then if $2 \leq p < \infty$*

$$|\xi+\nu|^p + |\xi-\nu|^p \leq 2(|\xi|^p + |\nu|^p),$$

and

$$|\xi+\nu|^p + |\xi-\nu|^p \geq 2(|\xi|^p + |\nu|^p).$$

for $p < 2$. If $p \neq 2$, the inequality occurs if ξ or ν is zero.

Proof. If $p = 2$, we have the equality for all ξ and ν. If $2 < p < \infty$, then $1 \leq p/2$, then with the conjugate exponent $p/2$ and $p/(p-2)$ we apply the Hölder inequality to $\alpha^2 + \beta^2$ (see Lemma 2.3), therefore

$$\alpha^2 + \beta^2 \leq (\alpha^p + \beta^r)^{\frac{2}{p}}(1+1)^{\frac{p-2}{p}}$$

from which we obtain

$$\alpha^p + \beta^p \geq 2^{\frac{2-p}{2}} (\alpha^2 + \beta^2)^{\frac{p}{2}}. \tag{3.43}$$

If $0 < p < 2$, we replace p by $4/p$ in (3.43), we get

$$\alpha^{\frac{4}{p}} + \beta^{\frac{4}{p}} \geq 2^{\frac{p-2}{p}} (\alpha^2 + \beta^2)^{\frac{2}{p}}.$$

If we replace α by $\alpha^{\frac{p}{2}}$ and β by $\beta^{\frac{p}{2}}$, the last inequality is transformed into

$$\alpha^2 + \beta^2 \geq 2^{\frac{p-2}{p}} (\alpha^p + \beta^p)^{\frac{2}{p}}$$

or

$$\alpha^p + \beta^p \leq 2^{\frac{2-p}{2}} (\alpha^2 + \beta^2)^{\frac{p}{2}}. \tag{3.44}$$

Since

$$0 \leq \frac{\xi^2}{\xi^2 + \nu^2} \leq 1$$

and

$$0 \leq \frac{\nu^2}{\xi^2 + \nu^2} \leq 1$$

for $2 < p$, results

$$0 \leq \frac{|\xi|^{2(p-2)}}{(\xi^2 + \nu^2)^{p-2}} \leq 1$$

and

$$0 \leq \frac{|\nu|^{2(p-2)}}{(\xi^2 + \nu^2)^{p-2}} \leq 1$$

which is equivalent to

$$0 \leq \frac{|\xi|^{p-2}}{(\xi^2 + \nu^2)^{\frac{p-2}{2}}} \leq 1$$

and

$$0 \leq \frac{|\nu|^{p-2}}{(\xi^2 + \nu^2)^{\frac{p-2}{2}}} \leq 1.$$

Therefore

$$0 \leq \frac{|\xi|^p}{(\xi^2 + \nu^2)^{\frac{p}{2}}} \leq \frac{|\xi|^1}{\xi^2 + \nu^2}$$

and

$$0 \leq \frac{|\nu|^p}{(\xi^2 + \nu^2)^{\frac{p}{2}}} \leq \frac{|\nu|^1}{\xi^2 + \nu^2}.$$

Summing these last inequalities we obtain

$$|\xi|^p + |\nu|^p \leq (\xi^2 + \nu^2)^{\frac{p}{2}} \quad \text{if } 2 < p. \tag{3.45}$$

In a similar fashion, we get

$$|\xi|^p + |\nu|^p \geq (\xi^2 + \nu^2)^{\frac{p}{2}} \quad \text{if } p < 2. \tag{3.46}$$

Note that the equality in (3.45) and (3.46) happens if and only if $\xi = 0$ and $\nu = 0$. Finally if $p > 2$, we replace α and β in (3.43) by $|\xi + \nu|$ and $|\xi - \nu|$. Therefore

$$\begin{aligned} |\xi+\nu|^p + |\xi-\nu|^p &\geq 2^{\frac{2-p}{p}}(|\xi+\nu|^2 + |\xi-\nu|^2)^{\frac{p}{2}} \\ &= 2^{\frac{2-p}{p}}(2[|\xi|^2 + |\nu|^2])^{\frac{p}{2}} \\ &= 2(\xi^2+\nu^2)^{\frac{p}{2}} \\ &\geq 2(|\xi|^p + |\nu|^p). \end{aligned}$$

Therefore we obtain the first inequality. The second inequality is obtained in a similar way using $\xi = 0$ and $\nu = 0$. □

As a consequence of the previous lemma we obtain the following integral version.

Lemma 3.75. *Let $1 \leq p < \infty$, $p \neq 2$ and suppose that $f, g \in L_p(\mu)$. Then*

$$\|f+g\|_p^p + \|f-g\|_p^p = 2(\|f\|_p^p + \|g\|_p^p)$$

if and only if $f \cdot g = 0$ μ-a.e.

Proof. If $f \cdot g = 0$ μ-a.e., then $\mu(\operatorname{supp} f \cap \operatorname{supp} g) = 0$. Then

$$\begin{aligned} \|f+g\|_p^p &= \int\limits_{\operatorname{supp}(g)} |f+g|^p \mathrm{d}\mu + \int\limits_{\operatorname{supp} g} |f+g|^p \mathrm{d}\mu + \int\limits_{X\setminus(\operatorname{supp} f \cap \operatorname{supp} g)} |f+g|^p \mathrm{d}\mu \\ &= \|f\|_p^p + \|g\|_p^p \end{aligned} \tag{3.47}$$

In a similar way we obtain

$$\|f-g\|_p^p = \|f\|_p^p + \|g\|_p^p \tag{3.48}$$

Summing (3.47) and (3.48) we get

$$\|f+g\|_p^p + \|f-g\|_p^p = 2(\|f\|_p^p + \|g\|_p^p).$$

($\Rightarrow$) Now, if $\|f+g\|_p^p + \|f-g\|_p^p = 2(\|f\|_p^p + \|g\|_p^p)$, then

$$\int\limits_X |f+g|^p \mathrm{d}\mu + \int\limits_X |f-g|^p \mathrm{d}\mu - 2\left(\int\limits_X |f|^p \mathrm{d}\mu + \int\limits_X |g|^p \mathrm{d}\mu\right) = 0$$

i.e.

$$\int\limits_X \left[|f+g|^p + |f-g|^p - 2(|f|^p + |g|^p)\right] \mathrm{d}\mu = 0.$$

Now by Lemma 3.74 we get

$$|f+g|^p + |f-g|^p - 2(|f|^p + |g|^p) \geq 0$$

or

$$|f+g|^p + |f-g|^p - 2(|f|^p + |g|^p) \leq 0.$$

In both cases, by a known result in the Lebesgue integration theory we arrive at

$$|f+g|^p + |f-g|^p - 2(|f|^p + |g|^p) = 0$$

μ-a.e. This clearly implies that for almost all $x \in X$, $f(x) = 0$ when $g(x) \neq 0$ and $g(x) = 0$ when $f(x) \neq 0$, or in alternative $\mu(\text{supp}(f) \cap \text{supp}(g)) = 0$. □

The next result gives a characterization of linear operators that are also isometries between $L_p(m)$ spaces.

Theorem 3.76 (Lamperti). *Let the linear operator $T : L_p(m) \to L_p(m)$ be an isometry in $L_p(m)$ where $1 \leq p < \infty$, $p \neq 2$. Then there exists a Lebesgue measurable function $\phi : \mathbb{R} \to \mathbb{R}$ and a unique Lebesgue measurable function h defined m-a.e. such that*

$$T(f)(x) = h(x)f(\phi(x)).$$

The function ϕ is defined m-a.e. on supp(h) *and for each Lebesgue measurable set E with $m(E) < \infty$, we get*

$$m(E) = \int_{\phi^{-1}(E)} |h(t)|^p \mathrm{d}t.$$

Proof. For each Lebesgue measurable set $A \subset \mathbb{R}$ such that $m(A) < \infty$, let us define $\varphi(A)$ by

$$\varphi(A) = \text{supp}(T(\chi_A)).$$

If $A \cap B = \emptyset$, then by Lemma 3.75

$$\|\chi_A + \chi_B\|_p^p + \|\chi_A - \chi_B\|_p^p = 2(\|\chi_A\|_p^p + \|\chi_B\|_p^p).$$

Since T is an isometry in $L_p(m)$ and by Lemma 3.75 it follows that

$$T(\chi_A) \cdot T(\chi_B) = 0$$

m-a.e. in en $\mathbb{R}$. Moreover $\chi_{A \cup B} = \chi_A + \chi_B$ and $T(\chi_{A\cup B}) = T(\chi_A) + T(\chi_B)$, then

$$\varphi(A \cup B) = \varphi(A) \cup \varphi(B), \quad \text{if } A \cap B = \emptyset.$$

Since for all set A and B we have that

$$\varphi(A) = \varphi(A \setminus (A \cap B)) \cup \varphi(A \cap B),$$

and

$$\varphi(A \cup B) = \varphi(A \setminus (A \cap B)) \cup \varphi(B).$$

It is clear that for Lebesgue measurable sets A and B such that $m(A) < \infty$ and $m(B) < \infty$ we get

$$\varphi(A \cup B) \cup \varphi(A \cap B) = \varphi(A) \cap \varphi(B).$$

On the other hand, let $\{X_j\}_{j \in \mathbb{N}}$ be an increasing sequence of subsets of $\mathbb{R}$ such that $m(X_j) < \infty$ for $j = 1, 2, \ldots$ and $\mathbb{R} = \bigcup_{j=1}^{\infty} X_j$. Moreover let $E = \bigcup_{j=1}^{\infty} \varphi(X_j)$. Then φ sends measurable subsets in $\mathbb{R}$ in measurable subsets of E and $\varphi(\mathbb{R} \setminus A) = E \setminus \varphi(A)$.

Also, if $\{A_j\}_{j \in \mathbb{N}}$ is a sequence of Lebesgue measurable subsets of $\mathbb{R}$ with $m(A_j) < \infty$, $j = 1, 2, \ldots$ and if $A = \bigcup_{j=1}^{\infty} A_j$ then $\chi_A = \lim_{k \to \infty} \sum_{j=1}^{k} \chi_{A_j}$, from the continuity and boundedness of T, we obtain that

$$T(\chi_A) = \lim_{k \to \infty} \sum_{j=1}^{k} T(\chi_{A_j}).$$

Nonetheless, the sequence $\{\operatorname{supp} T(\chi_{A_j})\}_{j \in \mathbb{N}}$ is a disjoint class, therefore

$$\varphi\left(\bigcup_{j=1}^{\infty}\right) = \bigcup_{j=1}^{\infty} \varphi(A_j).$$

This shows that φ is a σ-homomorphism, now invoking the Theorem A.20, we can find a Lebesgue measurable function $\phi : E \to \mathbb{R}$ such that

$$\varphi(A) = \phi^{-1}(A)$$

for any Borel set A. Then, since the set $\varphi(A)$ is unique m-a.e., then the function ϕ is unique m-a.e. Given a Lebesgue measurable set C with $m(C) < \infty$, let us define

$$h_C = T(\chi_X),$$

therefore

$$\|T(\chi_c)\|_p = \|h_C\|_p = (m(C))^{\frac{1}{p}},$$

then, for each measurable set A, we get

$$\chi_C = \chi_{C \cap A} + \chi_{C \cap A^c},$$

and from where we obtain

$$h_C = T(\chi_{C \cap A}) + T(\chi_{C \cap A^c}).$$

Nonetheless, above we proved that

$$(\operatorname{supp}(T(\chi_{A \cap C})) \cap (\operatorname{supp} T(\chi_{C \cap A^c}))) = \emptyset.$$

Then

$$T(\chi_{C \cap A}) = h_C \chi_{\varphi(C \cap A)} = h_C \chi_{C \cap A}(\phi(\cdot)).$$

For the sequence $\{X_j\}_{i\in\mathbb{N}}$ considered above, let us define a Lebesgue measurable function h given by

$$h = \lim_{j\to\infty} h_{X_j}.$$

Then, given any measurable subset A of $\mathbb{R}$ with $m(A) < \infty$, it follows from the boundedness and continuity of T

$$T(\chi_A) = \lim_{j\to\infty} T(\chi_{X_j\cap A}) = h\chi_A(\phi(\cdot)).$$

Then for each simple and measurable function f we have

$$T(f) = hf(\phi(\cdot)).$$

Let $f \in L_p(m)$, then invoking Lemma 3.31 there exists a sequence $\{s_n\}_{n\in\mathbb{N}}$ of simple functions such that

$$\|f - s_n\|_p \to 0,$$

when $n \to \infty$. Now

$$\|T(f) - hS_n(\phi(\cdot))\|_p = \|T(f - S_n)\|_p = \|f - S_n\|_p.$$

Then by Lemma 3.28 with $g(\varepsilon) = \varepsilon^p$ we get that

$$\begin{aligned} m(\{x \in \mathbb{R} : |T(f)(x) - hS_n(\phi(x))| > \varepsilon\}) &\le \varepsilon^{-p}\|T(f) - hS_n(\phi(\cdot))\|_p^p \\ &= \varepsilon^{-p}\|f - S_n\|_p^p. \end{aligned}$$

Then, for $\varepsilon > 0$, we have that $m(\{x \in \mathbb{R} : |T(f)(x) - hS_n(\phi(x))| > \varepsilon\}) \to 0$ when $n \to \infty$, and from this it follows that there exists a subsequence $\{S_{n_k}\}_{j\in\mathbb{N}}$ such that for almost all $x \in \mathbb{R}$

$$\begin{aligned} T(f)(x) &= \lim_{j\to\infty} h(x)S_{n_k}(\phi(x)) \\ &= h(x)f(\phi(x)). \end{aligned}$$

Finally, for each Lebesgue measurable set A with $m(A) < \infty$, we have

$$\begin{aligned} m(A) &= \int_{\mathbb{R}} (\chi_A(x))^p \mathrm{d}m = \int_{\mathbb{R}} |T(\chi_A)(x)|^p \mathrm{d}m \\ &= \int_{\mathbb{R}} |h(x)|^p \chi_A(\phi(x)) \mathrm{d}m \\ &= \int_{\phi^{-1}(A)} |h(x)|^p \mathrm{d}m. \end{aligned}$$

The function ϕ is unique m-a.e., in fact let h^0 be another function such that $T(f) = h^0 f(\phi(\cdot))$ given that f is arbitrary, we get equality m-a.e., i.e., $h = h^0$. □

3.11 Lebesgue Spaces with $0 < p < 1$

We now study the Lebesgue space whenever the exponent is strictly less than one. Since many of the properties of the L_p spaces with $p \geq 1$ rely heavily on the convexity property of the function $x \mapsto x^p$, it is expected that many of those properties will not be valid when the exponent p is between $0 < p < 1$.

Theorem 3.77 *Let*

$$L_p = \left\{ f : [0,1] \to \mathbb{R} \text{ such that } f \text{ is } \mathscr{L}\text{-measurable and } \int_0^1 |f(t)|^p \, dt < \infty \right\}.$$

Then,

(a) $d(f,g) = \int_0^1 |f(t) - g(t)|^p \, dt$ *is a metric in* L_p.
(b) d *is translation invariant.*
(c) $(L_p)^* = \{0\}$.

Proof. (a) Since $|f - g|^p \geq 0$ for all $f, g \in L_p$ then

$$\int_0^1 |f - g|^p \, dt \geq 0,$$

therefore $d(f,g) \geq 0$.
If $d(f,g) = 0$, then

$$\int_0^1 |f - g|^p \, dt = 0,$$

therefore $|f - g|^p = 0$ a.e., which implies that $|f - g| = 0$ a.e., from which $f = g$ a.e.
On the other hand, if $f = g$ for $f, g \in L_p$, then $f - g = 0$ from which $|f - g|^p = 0$, which entails

$$\int_0^1 |f - g|^p \, dt = 0,$$

i.e., $d(f,g) = 0$.
For f and g in L_p note that

$$d(f,g) = \int_0^1 |f - g|^p \, dt = \int_0^1 |g - f|^p \, dt = d(g,f)$$

i.e., $d(f,g) = d(g,f)$.

Finally, if a and b are real positive, then $a+b \geq b$ and $a+b \geq a$, since $0 < p < 1$, then $p-1 < 0$, therefore

$$(a+b)^{p-1} \leq b^{p-1}$$

and

$$(a+b)^{p-1} \leq a^{p-1}$$

from which

$$b(a+b)^{p-1} \leq b^p$$

and

$$a(a+b)^{p-1} \leq a^p,$$

and summing

$$\begin{aligned} a(a+b)^{p-1} + b(a+b)^{p-1} &\leq a^p + b^p \\ (a+b)(a+b)^{p-1} &\leq a^p + b^p \\ (a+b)^p &\leq a^p + b^p \end{aligned}$$

Now, using this last inequality, we can show the triangle inequality. Let f, g and h in L_p, then

$$|f-g|^p = |f-h+h-g|^p \leq |f-h|^p + |h-g|^p$$

for $0 < p < 1$, therefore

$$\begin{aligned} d(f,g) &= \int_0^1 |f-g| \, dt \\ &\leq \int_0^1 |f-h|^p \, dt + \int_0^1 |h-g|^p \, dt \\ &= d(f,h) + d(h,g), \end{aligned}$$

in other words, $d(f,g) \leq d(f,h) + d(h,g)$.

(b) Now we can show that d is translation invariant. In fact

$$\begin{aligned} d(f+h, g+h) &= \int_0^1 |f+h-(g+h)|^p \, dt \\ &= \int_0^1 |f-g| \, dt \\ &= d(f,g). \end{aligned}$$

(c) Let us suppose that $(L_p)^* \neq \{0\}$. Let us take a non-null linear continuous functional $T \in (L^p[0,1])^*$. There exists a function $f \in L_p[0,1]$ such that $T(f) = 1$. We now take the function

$$\begin{aligned} \varphi : [0,1] &\longrightarrow \mathbb{R} \\ x &\longmapsto T(f\chi_{[0,x]}) \end{aligned}$$

which is a continuous real-valued function with $\varphi(0) = 0$ and $\varphi(1)$. By the intermediate value property, there exists a point $0 < x_0 < 1$ such that $\varphi(x_0) = 1/2$. We now define the functions $\psi_1 = f\chi_{[0,x_0]}$ and $\psi_2 = f\chi_{[x_0,1]}$. We observe that since $T(\psi_1) = 1/2$ it implies, taking into account that $T(\psi_1 + \psi_2) = T(\psi_1) + T(\psi_2)$, that $T(\psi_2) = 1/2$. On the other hand, since

$$\int_0^1 |\psi_1(x)|^p + |\psi_2(x)|^p \,\mathrm{d}x = \int_0^1 |f(x)|^p \,\mathrm{d}x$$

we have that at least one of the norms $\|\psi_j\|_p$, $j = 1,2$ should be less or equal to $\|f\|_p/2$. From this we have that one of the norms ($j = 1$ or $j = 2$) satisfy

$$\|2\psi_j\|_p^p \leq 2^{p-1}\|f\|_p^p. \tag{3.49}$$

From the above construction, we obtain a function $f_1 \in L^p$, which is defined as $f_1 = 2\psi_j$ where ψ_j satisfy (3.49), such that:

(i) $T(f_1) = 1$, and
(ii) $\|f_1\|_p^p \leq 2^{p-1}\|f\|_p^p$.

By repeating the above argument, we construct a sequence of functions $(f_n)_{n\in\mathbb{N}}$ belonging to $L_p[0,1]$ such that

(a) $T(f_n) = 1$, and
(b) $\|f_n\|_p^p \leq 2^{n(p-1)}\|f\|_p^p$.

But the conditions (a) and (b) will imply a contradiction, since $f_n \to 0$ in $L_p[0,1]$ but $T(f_n) = 1$.

With the metric given in Theorem 3.77(a), the L_p space is a complete space.

Theorem 3.78. *The space $L_p(\mu)$ with $0 < p < 1$ is a complete space.*

Proof. Let $\{f_n\}_{n\in\mathbb{N}}$ be a Cauchy sequence in $L_p(\mu)$ with $0 < p < 1$, then

$$d(f_n, f_m) = \lim_{\substack{n\to\infty \\ m\to\infty}} \int_X |f_n(x) - f_m(x)|^p \,\mathrm{d}\mu = 0,$$

from here it follows that for each natural number k there exists a small natural number n_k such that

$$\int_X |f_n(x) - f_m(x)|^p \, d\mu < \frac{1}{3^k} \quad \text{for } m \geq n_k, n \geq n_k,$$

in particular, we can take $n = n_{k+1}$ and $m = n_k$ such that

$$\int_X |f_{n_k+1}(x) - f_{n_k}(x)|^p \, d\mu < \frac{1}{3^k} \quad \text{for } k = 1, 2, 3, \ldots \tag{3.50}$$

Let us define

$$E_k = \left\{ x \in X : |f_{n_k+1} - f_{n_k}(x)| > \frac{1}{2^{k/p}} \right\},$$

then,

$$\int_{E_k} |f_{n_k+1}(x) - f_{n_k}(x)|^p \, d\mu \geq \int_{E_k} \left(\frac{1}{2^{k/p}} \right)^p d\mu = \frac{1}{2^k} \mu(E_k)$$

therefore for (3.50) we have that

$$\mu(E_k) < \left(\frac{2}{3} \right)^k.$$

Let us consider now

$$\bigcup_{k=N}^{\infty} E_k = \bigcup_{k=n}^{\infty} \left\{ x \in X : |f_{n_k+1} - f_{n_k}(x)| > \frac{1}{2^{k/p}} \right\}. \tag{3.51}$$

If x does not belong to this set, it is clear that

$$|f_{n_N+1} - f_{n_N}(x)| \leq \frac{1}{2^{N/p}},$$

$$|f_{n_N+2} - f_{n_N+1}(x)| \leq \frac{1}{2^{(N+1)/p}},$$

$$\vdots$$

in such a way that the series

$$\sum_{k=1}^{\infty} |f_{n_k+1}(x) - f_{n_k}(x)| \tag{3.52}$$

converges, but the measure of (3.51) is

$$\sum_{k=N}^{\infty} \mu(E_k) < \sum_{k=N}^{\infty} \left(\frac{2}{3} \right)^k = 3 \left(\frac{2}{3} \right)^N$$

which tends to zero when N tends to infinity. This shows that the set of all x for which (3.52) does not converge has zero measure, it means, the series (3.52) converges almost everywhere, therefore the series

$$\sum_{k=1}^{\infty}(f_{n_k+1}(x)-f_{n_k}(x))$$

also converges, since an absolutely convergent series is convergent. Writing

$$f_{n_k}(x)=f_{n_1}(x)+\sum_{j=1}^{k-1}(f_{n_j+1}(x)-f_{n_j}(x))$$

it follows that

$$\lim_{k\to\infty}f_{n_k}(x)=f_{n_1}(x)+\sum_{j=1}^{\infty}(f_{n_j+1}(x)-f_{n_j}(x))$$

the limit exists, we denote this limit by $f(x)$, we see that $\lim_{k\to\infty}f_{n_k}=f(x)$ almost everywhere.

We need to prove that $\{f_n\}_{n\in\mathbb{N}}$ converges to f. To do that, we observe that by Fatou's lemma

$$\begin{aligned}\int_X |f_{n_k}(x)-f(x)|^p\mathrm{d}\mu &= \int_X \liminf_{j\to\infty}|f_{n_k}(x)-f_{n_j}(x)|^p\mathrm{d}\mu\\ &\leq \liminf_{j\to\infty}\int_X |f_{n_k}(x)-f_{n_j}(x)|^p\mathrm{d}\mu\\ &< \varepsilon^p/2.\end{aligned}$$

Finally

$$\begin{aligned}\int_X |f_{n_k}(x)-f(x)|^p\mathrm{d}\mu &\leq \int_X |f_n(x)-f_{n_k}(x)|^p\mathrm{d}\mu+\int_X |f_{n_k}(x)-f(x)|^p\mathrm{d}\mu\\ &<\varepsilon^p\end{aligned}$$

therefore, $f_n-f\in L_p(\mu)$ and $f=(f-f_n)+f_n\in L_p(\mu)$. □

We already know from the observation given after Definition 3.17 that the functional given by $f\mapsto\sqrt[p]{\int|f|^p\mathrm{d}m}$ do not define a norm. The question remains: Is the space $L_p([0,1],\mathscr{L},m)$ normalizable? The following results give a negative answer.

Theorem 3.79. *Let $\{f_n\}_{n\in\mathbb{N}}$ be a sequence in $L_p([0,1],\mathscr{L},m)$. Then there is not a norm $\|\cdot\|$ in $L_p(m)$ such that if $f_n\to 0$ in $L_p(m)$ implies that $\|f_n\|\to 0$ when $n\to\infty$, for any sequence $\{f_n\}_{n\in\mathbb{N}}\subset L_p(m)$.*

Proof. Let us suppose that the norm $\|\cdot\|$ exists. We state that there exists a positive constant $C<\infty$ such that $\|f\|\leq C\|f\|_p$ for all $f\in L_p(m)$. Since the application $f\mapsto\|f\|$ is continuous, we can find a $\delta>0$ such that if $\|f\|_p<\delta$, then $\|f\|\leq 1$, therefore, for all $f\in L_p(m)$ we have that $\frac{\alpha\delta f}{\|f\|_p}\in B(0,\delta)$ with $0<|\alpha|<1$, where

$B(0,\delta)$ is the ball of center 0 and radius δ. Hence

$$\left\| \frac{\alpha\delta f}{\|f\|_p} \right\| \leq 1$$

which implies that

$$\|f\| \leq \frac{1}{\alpha\delta}\|f\|_p. \tag{3.53}$$

If $\alpha \to 1$, then (3.53) is true for $C = \frac{1}{\delta}$. Let us choose

$$C = \inf\{k : \|f\| \leq K\|f\|_p, \quad \text{for all } f \in L_p(m)\}$$

(Note that we do not exclude the possibility that $C = 0$). By the intermediate value theorem, there exists $c \in (0,1)$ such that

$$\int_0^c |f|^p \mathrm{d}m = \int_c^1 |f|^p \mathrm{d}m = \frac{1}{2}\int_0^1 |f|^p \mathrm{d}m.$$

Now, for $g = f\chi_{[0,c)}$ and $h = f\chi_{(c,1]}$, we have that $f = g + h$ and $\|g\|_p = \|h\|_p = 2^{\frac{-1}{p}}\|f\|_p$, by the triangle inequality

$$\|f\| \leq \|g\| + \|h\| \leq C(\|g\|_p + \|h\|_p) = \frac{C}{2^{1/p-1}}.$$

Since $p \in (0,1)$, then

$$\frac{C}{2^{1/p-1}} \leq C,$$

from which $C = 0$, therefore $\|f\| = 0$ for all $f \in L_p(m)$ which contradicts the fact that $\|\cdot\|$ is a norm. □

Another way to see that $L_p(m)$ cannot be normalizable is to notice that if this was the case, the Hahn-Banach theorem would be true in $L_p(m)$, but $(L_p(m))^* = \{0\}$, which is a contradiction.

The following result shows that, although the space is not normalizable, we can nevertheless obtain a *quasi-triangle inequality* in the framework of L_p spaces with $0 < p < 1$.

Theorem 3.80. *Let $(X, \mathscr{A}, \mu)$ be a measure space and p a real number such that $0 < p < 1$. Let f and g functions in $L_p(X)$, such that $g \neq 0$, then*

$$\|f+g\|_p \leq 2^{\frac{1}{p}-1}(\|f\|_p + \|g\|_p). \tag{3.54}$$

Proof. Let $0 \leq t < \infty$ and $0 < p < 1$, then $p - 1 < 0$ and is clear that $1 + t \geq t$ and $1 + t \geq 1$, by which

$$(1+t)^{p-1} \leq t^{p-1}$$

and

$$(1+t)^{p-1} \leq 1. \tag{3.55}$$

From (3.55) we obtain that

$$t(1+t)^{p-1} \leq t^p, \tag{3.56}$$

adding (3.55) and (3.56) we obtain that

$$(1+t^p) \leq 1+t^p. \tag{3.57}$$

Let us define $t = \frac{|f|}{|g|}$, substituting in (3.57) it results that

$$(|f|+|g|)^p \leq |f|^p + |g|^p,$$

but $|f+g| \leq |f|+|g|$ and $0 < p < 1$, then

$$|f+g|^p \leq |f|^p + |g|^p. \tag{3.58}$$

Now, since the function $f(t) = t^{\frac{1}{p}}$ is convex in $[0,\infty]$, therefore in virtue of (3.58) we obtain

$$\frac{\|f+g\|_p^p}{2} = \frac{\int |f+g|^p \mathrm{d}\mu}{2} \leq \frac{\int |f|^p \mathrm{d}\mu + \int |g|^p \mathrm{d}\mu}{2},$$

therefore

$$\frac{\|f+g\|_p}{2^{\frac{1}{p}}} = \left(\frac{\int |f+g|^p \mathrm{d}\mu}{2} \right)^{\frac{1}{p}} \leq \left(\frac{\int |f|^p \mathrm{d}\mu + \int |g|^p \mathrm{d}\mu}{2} \right)^{\frac{1}{p}}$$
$$\leq 2^{\frac{1}{p}-1} \left[\|f\|_p + \|g\|_p \right]$$

and this shows (3.54). □

As already mentioned before, Hölder's inequality is closely related to convexity. We now show that an inversion in the inequality sign in the Hölder inequality holds in the case of L_p spaces with $0 < p < 1$, since we do not have convexity in $x \mapsto x^p$.

Theorem 3.81. *Let p and q be real numbers with* $0 < p < 1$ *and* $-\infty < q < 0$ *such that*

$$\frac{1}{p} + \frac{1}{q} = 1.$$

Let f and g be positive functions such that f^p *and* g^q *are integrable and, moreover, suppose that fg is integrable, therefore*

$$\|f\|_p \|g\|_q \leq \int_X fg \, \mathrm{d}\mu.$$

Proof. Let us define $r = 1/p$ and $s = -q/p$, from which r and s are conjugate exponents since

$$\frac{1}{r} + \frac{1}{s} = p - \frac{p}{q} = p\left(1 - \frac{1}{q}\right) = 1.$$

Moreover, observe that $s > 1$ and write

$$f^p = f^p g^p g^{-p} = (fg)^p g^{-p},$$

in order to prove that $(fg)^p \in L_r(X)$ and $g^{-p} \in L_s(X)$. In fact

$$\int_X [(fg)^p]^r \mathrm{d}\mu = \int_X [(fg)^p]^{1/p} \mathrm{d}\mu = \int_X fg \mathrm{d}\mu < \infty,$$

which shows that $(fg)^p \in L_r(X)$. On the other hand,

$$\int_X (g^{-p})^s \mathrm{d}\mu = \int_X (g^{-p})^{-q/p} \mathrm{d}\mu = \int_X g^q \mathrm{d}\mu < \infty,$$

then $g^{-p} \in L_s(X)$, therefore in virtue of the Hölder inequality we obtain that

$$\begin{aligned}\int_X f^p \mathrm{d}\mu &\leq \left(\int_X [(fg)^p]^r \mathrm{d}\mu\right)^{1/r} \left(\int_X (g^{-p})^s \mathrm{d}\mu\right)^{1/s} \\ &= \left(\int_X [(fg)^p]^{1/p} \mathrm{d}\mu\right)^{p} \left(\int_X (g^{-p})^{-q/p} \mathrm{d}\mu\right)^{-p/q} \\ &= \left(\int_X fg \mathrm{d}\mu\right)^{p} \left(\int_X g^q \mathrm{d}\mu\right)^{-p/q},\end{aligned}$$

from which

$$\left(\int_X |f|^p \mathrm{d}\mu\right)^{1/p} \leq \left(\int_X fg \mathrm{d}\mu\right) \left(\int_X |g|^q \mathrm{d}\mu\right)^{-1/q},$$

hence $\|f\|_p \|g\|_q \leq \int_X fg \mathrm{d}\mu$. □

A similar phenomenon to Hölder's inequality occurs with the Minkowski inequality in the case of L_p with $0 < p < 1$, namely, the sign of the inequality is reversed.

Corollary 3.82 *Let $0 < p < 1$, f and g be positive functions in $L_p(X)$, then $f + g \in L_p(X)$ and*

$$\|f + g\|_p \geq \|f\|_p + \|g\|_p.$$

Proof. In virtue of Theorem 3.80 we have that $f+g \in L_p(X)$. On the other hand

$$(f+g)^p = f(f+g)^{p-1} + g(f+g)^{p-1}, \tag{3.59}$$

hence

$$\begin{aligned}
\left\|(f+g)^{p-1}\right\|_q &= \left(\int_X |f+g|^{q(p-1)} \mathrm{d}\mu\right)^{1/q} \\
&= \left(\int_X |f+g|^p \mathrm{d}\mu\right)^{1/q} \\
&= \|f+g\|_p^{p/q} \\
&\leq \left[2^{\frac{1}{p}-1}(\|f\|_p + \|g\|_p)\right]^{p/q},
\end{aligned}$$

which shows that $(f+g)^{p-1} \in L_q(X)$. Again by (3.59) and using the Theorem 3.81, it follows

$$\begin{aligned}
\int_X (f+g)^p \mathrm{d}\mu &\geq \left[\left(\int_X f^p \mathrm{d}\mu\right)^{1/p} + \left(\int_X g^p \mathrm{d}\mu\right)^{1/p}\right]\left(\int_X (f+g)^{q(p-1)} \mathrm{d}\mu\right)^{1/q} \\
&= \left[\left(\int_X f^p \mathrm{d}\mu\right)^{1/p} + \left(\int_X g^p \mathrm{d}\mu\right)^{1/p}\right]\left(\int_X (f+g)^p \mathrm{d}\mu\right)^{1/q},
\end{aligned}$$

from which

$$\left(\int_X (f+g)^p \mathrm{d}\mu\right)^{1-\frac{1}{q}} \geq \left(\int_x f^p \mathrm{d}\mu\right)^{1/p} + \left(\int_X g^p \mathrm{d}\mu\right)^{1/p},$$

therefore

$$\left(\int_X (f+g)^p \mathrm{d}\mu\right)^{1/p} \geq \left(\int_x f^p \mathrm{d}\mu\right)^{1/p} + \left(\int_X g^p \mathrm{d}\mu\right)^{1/p},$$

obtaining $\|f+g\|_p \geq \|f\|_p + \|g\|_p$. □

3.12 Problems

3.83. Let us define the weighted Lebesgue space $\widetilde{L}_p(w)$ as the set of all measurable functions such that $wf \in L_p$. Find the dual space of $\widetilde{L}_p(w)$.

3.84. Prove that $L_p(w)$ equipped with the norm defined by (3.37) is a Banach space.

3.85. Let $f \in L_p(\mathbb{R}^n)$ and t be a fixed constant in $\mathbb{R}^n$. Show that $\|f(\cdot - t)\|_{L_p(\mathbb{R}^n)} = \|f\|_{L_p(\mathbb{R}^n)}$.

3.86. If $f \in L_\infty(X, \mathscr{A}, \mu)$ and $\mu(X) < \infty$, show that $\lim_{p\to\infty} \|f\|_p = \|f\|_\infty$.

3.87. Show that the result of item (1) is false if $\mu(X) = +\infty$.

3.88. Show that the following functions are essentially bounded and find its essential norm.

(a) $f(x) = e^x$ for all $x \in [0,1]$.
(b) $f(x) = \frac{1}{x^2}$ for all $x \in [1, +\infty)$.
(c) $f(x) = e^{-\lambda x^2}$ for all $x \in \mathbb{R}$ and $\lambda > 0$.
(d) $f(x) = x^n$ for all $x \in [0,a]$, $n \in \mathbb{N}$ and $a \in \mathbb{R}^+$.
(e) $f(x) = 1 - e^{-x}$ for all $x \in [0,1]$.

3.89. In all the cases of Problem 3.88 show that $\lim_{p\to\infty} \|f\|_p = \|f\|_\infty$.

3.90. Let $f(x) = -\log(1-x)$. Show that $f \notin L_\infty([0,1), \mathscr{L}, m)$ but $f \in L_p([0,1], \mathscr{L}, m)$ with $1 \leq p < \infty$.

3.91. Let $f \in L_\infty(X, \mathscr{A}, \mu)$ be such that $\|f\|_\infty > 0$. If $0 < \mu(X) < \infty$, show that

$$\lim_{n\to\infty} \frac{\alpha_{n+1}}{\alpha_n} = \|f\|_\infty$$

where

$$\alpha_n = \int_X |f|^n d\mu, \qquad n = 1, 2, 3, \ldots$$

3.92. Let $I = [0,1]$, $f \in L_1(I, \mathscr{L}, m)$ and $S = \{x \in r : f(x) \in \mathbb{Z}\}$. Prove that

$$\lim_{n\to\infty} \int_I |\cos \pi f(x)|^n \, dm = m(S).$$

3.93. Let $f \in L_p(X, \mathscr{A}, \mu)$ with $\mu(X) = 1$ and $p > 0$. Prove that

$$\lim_{p\to 0} \|f\|_p = \exp\left(\int_X (f) d\mu \right).$$

3.94. Let $(X, \mathscr{A}, \mu)$ a measure space and $f \in L_p(X, \mathscr{A}, \mu), 1 \leq p < \infty$. Suppose that $\{E_n\}_{n \in \mathbb{N}}$ is a sequence of measurable sets such that $\mu(E_n) = \frac{1}{n}$, $n \in \mathbb{N}$. Prove that

$$\lim_{n \to \infty} n^{\frac{p-1}{p}} \int_{E_n} |f| \, \mathrm{d}\mu = 0.$$

3.95. Let $(X, \mathscr{A}, \mu)$ be a measure space such that $\mu(X) < \infty$ and f a positive $\mathscr{A}$-measurable function. If $\lim_{n \to \infty} \int_X f^n \, \mathrm{d}\mu < \infty$, show that

$$\lim_{n \to \infty} \int_X f^n \, \mathrm{d}\mu = \mu\left(\{x \in X : f(x) = 1\}\right).$$

3.96. Let $f \in L_{p_0}(X, \mathscr{A}, \mu)$ for some $0 < p_0 < \infty$. Prove that

$$\lim_{p \to 0} \int_X |f|^p \, \mathrm{d}\mu = \mu\left(\{x \in X : f(x) \neq 0\}\right).$$

3.97. Let $u, v \in L_4(X, \mathscr{A}, \mu)$ and $w \in L_2(X, \mathscr{A}, \mu)$. Prove that

$$\left| \int_X uvw \mathrm{d}\mu \right| \leq \left(\int_X |u|^4 \mathrm{d}\mu \right)^{1/4} \left(\int_X |v|^4 \mathrm{d}\mu \right)^{1/4} \left(\int_X |w|^2 \mathrm{d}\mu \right)^{1/2}.$$

3.98. Let $p > 1$ and $f \in L_p\left([1, +\infty), \mathscr{L}, m\right)$. Define $g(x) = \int_1^\infty f(t) e^{-tx} \, \mathrm{d}t$. Prove

(a) $g \in L_1\left([1, +\infty), \mathscr{L}, m\right)$.

(b) $\|g\|_1 \leq \left(1 - \frac{1}{p}\right)^{1/q} \|f\|_p$ where $\frac{1}{p} + \frac{1}{q} = 1$.

3.99. For $1 \leq p < \infty$ and $0 < \mu(X) < \infty$ let us define

$$N_p(f) = \left(\frac{1}{\mu(X)} \int_X |f|^p \, \mathrm{d}\mu \right)^{1/p}.$$

Prove that:

(a) If $p_1 < p_2$ then $N_{p_1}(f) \leq N_{p_2}(f)$.

(b) $N_p(f + g) \leq N_p(f) + N_p(g)$.

(c) $\frac{1}{\mu(X)} \int_X |fg| \, \mathrm{d}\mu \leq N_p(f) N_q(g)$ with $\frac{1}{p} + \frac{1}{q} = 1$.

(d) $\lim_{p \to \infty} N_p(f) = \|f\|_\infty$.

3.100. Let $(X, \mathscr{A}, \mu)$ be a measure space, f and g be positive $\mathscr{A}$-measurable functions in X. Let $0 < t < r < m < \infty$. If $\int_X fg^t \,\mathrm{d}\mu < \infty$ and $\int_X fg^m \,\mathrm{d}\mu < \infty$, show that

$$\left(\int_X fg^r \mathrm{d}\mu \right)^{m-t} \leq \left(\int_X fg^t \mathrm{d}\mu \right)^{m-r} \left(\int_X fg^m \mathrm{d}\mu \right)^{r-t}. \tag{3.60}$$

Inequality (3.60) is known as *Roger's inequality.*

3.101. If $\int_X f \,\mathrm{d}\mu < \infty$ and $\int_X fg^m \,\mathrm{d}\mu < \infty$ for $M > 1$. Prove that

$$\left(\int_X fg\mathrm{d}\mu \right)^{m} \leq \left(\int_X f\mathrm{d}\mu \right)^{m-1} \left(\int_X fg^m \mathrm{d}\mu \right).$$

3.102. Use the inequality given in (1.23) to give an alternative proof of the result:
Let p, q, and r be real numbers such that $\frac{1}{p} + \frac{1}{q} + \frac{1}{r} = 1$. Let $f \in L_p(X, \mathscr{A}, \mu)$, $g \in L_q(X, \mathscr{A}, \mu)$ and $h \in L_r(X, \mathscr{A}, \mu)$. Then

$$\int_X |fgh| \,\mathrm{d}\mu \leq \|f\|_p \|g\|_q \|h\|_r.$$

3.103. Prove Corollary 3.22.

3.104. Let $I = [0, \pi]$ and $f \in L_2([0.\pi], \mathscr{L}, m)$. Is it possible to have

$$\int_I (f(x) - \sin x)^2 \,\mathrm{d}x \leq 4 \quad \text{and} \quad \int_I (f(x) - \cos x)^2 \mathrm{d}x \leq \frac{1}{9}$$

simultaneously?

3.105. Let h be an increasing function in $(0, +\infty)$. If $0 < \alpha \leq 1$ and $\beta \geq 0$ show that

$$\left(\int_0^\infty t^{\beta-1} h(t) \mathrm{d}t \right)^{\alpha} \leq \alpha \beta^{1-\alpha} \int_0^\infty t^{\alpha-1} [h(t)]^\alpha \mathrm{d}t.$$

3.106. Let f be a nonnegative and decreasing function in $(0, +\infty)$ for $p \geq 1$. Prove that

$$\int_0^\infty f^p \,\mathrm{d}x^p \leq \left(\int_0^\infty f \mathrm{d}x \right)^p.$$

3.107. Let $f \in L_1(X, \mathscr{A}, \mu)$. Prove that

$$\mu\left(\{x \in X : |f(x)| > \lambda\}\right) \leq 2\mathrm{e}^{-\lambda^2/2}\lambda \int_X \cosh\left(\frac{\lambda}{2} f\right) \mathrm{d}\mu.$$

3.108. Let $0 < \alpha < 1$, $b > 1$ and $M > 1$. Show that

(a) $\left|a^{-m} - b^{-m}\right| \leq m|b-a|^{\alpha} \max\{a^{-\alpha-m}, b^{-\alpha-m}\}$.

(b) $|x-y|^p \leq |x^p - y^p|$ to $x, y \in \mathbb{R}^+$ and $1 \leq p < \infty$.

(c) $|\log b - \log a| \leq \frac{1}{\alpha}|b-a|^{\alpha} \max\{a^{-\alpha}, b^{-\alpha}\}$.

(d) $\left||b|^p - |a|^p\right| \leq p|b-a| \max\{|b|^{p-1}, |a|^{p-1}\}(1 \leq p < \infty)$.

3.109. Let $I = [0,1]$. Prove that $f \in L_2(I, \mathscr{L}, m)$ if and only if $f \in L_1(I, \mathscr{L}, m)$ such that there exists an increasing g function, such that for all closed interval $[a,b] \subset [0,1]$ we have

$$\left|\int_A^b f(x)\mathrm{d}x\right|^2 \leq (g(b) - g(a))\,|b-a|.$$

3.110. Let $1 \leq p < \infty$. Prove that

$$\|f\|_p = \inf\left\{\lambda \geq 0 : \int_X \left|\frac{|f|}{\lambda}\right|^p \mathrm{d}m \leq 1\right\}$$

coincides with the standard L_p norm.

3.111. Let $E = \{p \in (0,\infty) : \|f\|_p < \infty\}$. Show that E is an interval.

3.112. Let $\{f_n\}_{n \in \mathbb{N}}$ be a sequence of real functions belonging to $L_{4/3}((0,1), \mathscr{L}, m)$ such that $f_n \to 0$ in measure as $n \to \infty$ and $\int_{(0,1)} \left|f_n(x)\right|^{4/3} \mathrm{d}x \leq 1$. Prove that

$$\lim_{n\to\infty} \int_{(0,1)} \left|f_n(x)\right| \mathrm{d}x = 0.$$

3.113. Let $(X, \mathscr{A}, \mu)$ be a measure space with $\mu(X) = 1$. If $f \in L_p(X, \mathscr{A}, \mu)$ for $0 < p < \infty$, prove that

$$\exp\left(\int_X \log|f|\mathrm{d}\mu\right) \leq \left(\int_X |f|^p \mathrm{d}\mu\right)^{1/p}.$$

3.114. Let $(X,\mathscr{A},\mu)$ be a measure space and $\{f_n\}_{n\in\mathbb{N}}$ be a sequence of measurable functions such that $f_n \in L_p(X,\mathscr{A},\mu)$ for all $n\in\mathbb{N}$. If $\sum_{n=1}^{\infty}\|f_n\|_p<\infty$, prove that

$$\left\|\sum_{n=1}^{\infty} f_n\right\|_p \leq \sum_{n=1}^{\infty}\|f_n\|_p.$$

3.115. Prove that if $\|f+g\|_p=\|f\|_p+\|g\|_p$, then

$$\frac{f}{\|f\|_p}=\frac{g}{\|g\|_p} \quad \text{a.e.}$$

3.116. Let $f_n \to f$ in L_p, $1\leq p<\infty$ and let $\{g_n\}_{n\in\mathbb{N}}$ be a sequence of measurable functions such that $|g_n|\leq M$, for all $\mathbb{N}$ and $g_n\to g$ μ-a.e. Prove that $g_nf_n\to gf$ in L_p.

3.117. Use the Corollary 1.10 to deduce the Hölder inequality.

3.118. Use Problem 29 to derive the Minkowski inequality.

3.119. Let $(X,\mathscr{A},\mu)$ be a measure space such that $\mu(X)=1$. Find all functions ϕ in $(0,+\infty)$ such that

$$\phi\left(\lim_{p\to 0}\|f\|_p\right)=\int_X \phi(f)\mathrm{d}\mu.$$

3.120. Let $p,q,r\in\mathbb{R}^+$ such that $\frac{1}{p}+\frac{1}{q}=\frac{1}{r}$. If $f\in L_p(X,\mathscr{A},\mu)$ and $g\in L_q(X,\mathscr{A},\mu)$, show that $fg\in L_r(X,\mathscr{A},\mu)$ and

$$\|fg\|_r\leq\|f\|_p\|g\|_q.$$

3.121. If $0<p<q$ and $\mu(X)=1$, show that

$$\left(\int_X |f|^p\,\mathrm{d}\mu\right)^{1/p}\leq\left(\int_X |f|^q\,\mathrm{d}\mu\right)^{1/q}.$$

3.122. Use Corollary 1.10 to prove Theorem 3.20 (Hölder's inequality).

3.123. The function $f:[a,b]\to\mathbb{R}$ is said to have p-bounded variation if

$$V_p(f,[a,b])=\sup_{\Pi}\sum_{k=1}^{n}\frac{|f(x_k)-f(x_{k-1})|^p}{|x_k-x_{k-1}|^{p-1}}<+\infty,$$

with $1<p<\infty$, where the supremum is taken over all partitions Π of the interval $[a,b]$. The set of all functions with p-variation is denoted by $BV_p([a,b])$. If $f\in BV_p([a,b])$ show that $f'\in L_p([a,b],\mathscr{L},m)$ and also that

$$V_p(f,[a,b])=\|f'\|_p.$$

3.124. Let $\Delta = \{\xi_0, \xi_1, \ldots, \xi_m\}$ be a finite partition of the interval $[a,b]$ and let $f \in L_p\big([a,b], \mathscr{L}, m\big)$ where $p \geq 1$. The function T_Δ defined by

$$T_\Delta f(\xi_k) = \frac{1}{\xi_k - \xi_{k-1}} \int_{\xi_{k-1}}^{\xi_k} f(t)\,\mathrm{d}t$$

is called the Δ-*approximation of* f *in average*. Prove that

$$\|T_\Delta f\|_p \leq \|f\|_p.$$

3.125. Let $f \in L_p\big([a,b], \mathscr{L}, m\big)$. Show that $\|T_\Delta f - f\|_p \to 0$ when the length Δ of the largest subinterval of Δ tends to zero.

3.126. Prove that $T_\Delta f \xrightarrow{\mu} f$ when $\Delta \to 0$.

3.127. Let $f \in L_p\big([a,b], \mathscr{L}, m\big)$, $1 \leq p < \infty$. Show that given $\varepsilon > 0$ there exists a measurable function f_M such that $|f_M| \leq M$ and $\|f - f_M\|_p < \varepsilon$.

3.128. Let μ be a positive measure and assume that $f, g \in L_p(X, \mathscr{A}, \mu)$. Demonstrate:

(a) If $0 < p < 1$, then

$$\int_X \big(|f|^p - |g|^p\big)\,\mathrm{d}\mu \leq \int_X |fg|^p \mathrm{d}\mu$$

(b) If $1 \leq p < \infty$ and $\|f\|_p \leq M$, $\|g\|_p \leq M$, then

$$\int_X \big||f|^p - |g|^p\big|\,\mathrm{d}\mu \leq 2pM^{p-1}\|fg\|_p.$$

3.129. If $0 < p < q < r \leq \infty$, show that

$$L_p(X, \mathscr{A}, \mu) \cap L_r(X, \mathscr{A}, \mu) \subset L_q(X, \mathscr{A}, \mu)$$

where $\frac{1}{q} = \frac{\lambda}{p} + \frac{1-\lambda}{r}$, $\lambda > 0$. Prove also that

$$\|f\|_q \leq \|f\|_p^{\lambda} \|f\|_r^{1-\lambda}.$$

3.130. If $0 < p < q < r \leq \infty$, prove that

$$L_q(X, \mathscr{A}, \mu) \subset L_p(X, \mathscr{A}, \mu) + L_r(X, \mathscr{A}, \mu)$$

i.e., every $f \in L_q(X, \mathscr{A}, \mu)$ is the sum of a function in $L_p(X, \mathscr{A}, \mu)$ and a function in $L_r(X, \mathscr{A}, \mu)$.

3.131. Let $1 \leq p < \infty$ and $g \in L_p(X, \mathscr{A}, \mu)$ be such that $|f_n| \leq g$ μ-a.e. for all $\mathbb{N}$. If $f_n \to f$ μ-a.e. show that $f_n \to f$ in $L_p(X, \mathscr{A}, \mu)$.

3.132. Let $(X,\mathscr{A},\mu)$ be a positive measure space. Let f be an $\mathscr{A}$-measurable function. Show that

$$\|f\|_\infty \leq \liminf_{p\to\infty} \|f\|_p.$$

3.133. Let $1 \leq p < q < \infty$. Suppose that for $g \in L_q(X,\mathscr{A},\mu)$ we have

$$\int_X fg\mathrm{d}\mu = 0$$

for all $f \in L_p(X,\mathscr{A},\mu)$. Prove that $g = 0$ μ-a.e. in X.

3.134. Let $(X,\mathscr{A},\mu)$ be a measure space and $f \in L_1(X,\mathscr{A},\mu)$. Let us define

$$L_f(h) = \int_X fh\mathrm{d}\mu$$

for $h \in L_\infty(X,\mathscr{A},\mu)$. Show that L_f is a linear and bounded operator in $L_\infty(X,\mathscr{A},\mu)$ and moreover

$$\|L_f\| = \|f\|_1.$$

3.135. Let $(X,\mathscr{A},\mu)$ be a σ-finite measure space. Given $g \in L_\infty(X,\mathscr{A},\mu)$ and $\varepsilon > 0$, show that there exists $f \in L_1(X,\mathscr{A},\mu)$ such that $\|f\|_1 = 1$ and

$$\|g\|_\infty \geq \int_X fg\mathrm{d}\mu > \|g\|_\infty - \varepsilon.$$

3.136. Let $(X,\mathscr{A},\mu)$ be a measure space with $\mu(X) = 1$ and $1 \leq p < \infty$. Suppose that

(a) S is a closed subspace of $L_p(X,\mathscr{A},\mu)$,
(b) $S \subset L_\infty(X,\mathscr{A},\mu)$.

Prove that S has finite dimension.

3.137. Let $X = C[0,1]$ be the Banach space of continuous functions in $[0,1]$ endowed with the supremum norm. Let S be a subspace of X which is closed as a subspace of $L_2([0,1],\mathscr{L},m)$. Show that

(a) S is a closed subspace of X.
(b) There exists a constant M such that

$$\|f\|_\infty \leq M\|f\|_2$$

for all $f \in S$.
(c) For all $y \in [0,1]$ there exists $k_y \in L_2([0,1],\mathscr{L},m)$ such that

$$f(y) = \int_0^1 k_y(x)f(x)\,\mathrm{d}x$$

for all $f \in S$.

3.138. Prove that $f \in L_p([0,\infty),\mathscr{L},m)$, $p>1$, if and only if

$$\sum_{n=1}^{\infty} \frac{1}{n^{p+1}} m\left(\left\{|f| > \frac{1}{n}\right\}\right) + \sum_{n=1}^{\infty} n^{p-1} m\Big(\{|f|>n\}\Big) < \infty.$$

3.139. Let $\|x\|_p = \left(\sum_{k=1}^n |x_k|^p\right)^{1/p}$ and $\ell_p^n = \{x \in \mathbb{R}^n : \|x\|_p < \infty\}$ for $1 \le p < \infty$. Let us define $B_{\ell_p^n}(0,1) = \{x : \|x\|_p \le 1\}$ which stands for the unit ball of ℓ_p^n(observe that $\mathbb{R}^n \cong \ell_p^n$). Let us denote

$$v_n(p) = \mathrm{vol}(B_{\ell_p^n}) = \mathrm{vol}\{x \in \ell_p^n : \|x\|_p \le 1\}.$$

Calculate $v_n(p)$.
Hint: Calculate the following integral

$$I_p = \int_{\mathbb{R}^n} \mathrm{e}^{-\|x\|_p^p} \,\mathrm{d}x.$$

3.140. Let $(X,\mathscr{A},\mu)$ be a measure space and f an $\mathscr{A}$-measurable functions. Prove that

$$\int_X \sin|f(x)|\mathrm{d}\mu(x) = \int_0^\infty \cos\lambda\, \mu\Big(\{x \in X : |f(x)| > \lambda\}\Big)\,\mathrm{d}\lambda.$$

3.141. For $g \in L_q(X,\mathscr{A},\mu)$, let Φ be a linear function in $L_p(X,\mathscr{A},\mu)$ defined by $\Phi(f) = \int fg\mathrm{d}\mu$. Without using the Riesz representation theorem, show that $\|\Phi\| = \|g\|_q$.

3.142. Let $(X,\mathscr{A},\mu)$ be a finite measure spaces. Demonstrate that the dual space of $L_1(X,\mathscr{A},\mu)$ is $L_\infty(X,\mathscr{A},\mu)$.

3.143. Let $(X,\mathscr{A},\mu)$ be a measure spaces and $f : X \to \mathbb{R}$ a measurable function. If $\mu\Big(\{x \in X : |f(x)| > \lambda\}\Big) \le \mathrm{e}^{-\lambda}$ for all $\lambda \ge 0$, then show that $f \in L_p(X,\mathscr{A},\mu)$ for all $1 \le p < \infty$.

3.144. Consider in $\mathbb{R}^2$ the Euclidean norm $\|(x,y)\|_2 = (|x|^2+|y|^2)^{1/2}$ and the taxicab norm $\|(x,y)\|_1 = |x|+|y|$. Prove that these norms are equivalent

$$\frac{1}{2}\|(x,y)\|_1 \le \|(x,y)\|_2 \le \sqrt{2}\|(x,y)\|_1.$$

3.145. Show that for any $\alpha_k > 0$, $k = 1,2,\ldots,n$, we have

$$\lim_{p\to+\infty} \left[\sum_{k=1}^n \theta_k(\alpha_k)^{1/p}\right]^p = \prod_{k=1}^n (\alpha_k)^{\theta_k},$$

where $\theta_k > 0$ and $\sum_{k=1}^{n} \theta_k = 1$.
Hint: Check Lemma 3.21.

3.13 Notes and Bibliographic References

In 1910 Riesz [57] published the foundational paper on $L_p([a,b])$ $(1 < p < \infty)$ where he showed many important properties of the Lebesgue spaces.

For bibliographic references to Hölder and Minkowski inequality see §2.6. The alternative proof of Theorem 3.20 (Hölder's inequality) given in p. 54 was taken from Maligranda [49].

Uniform convexity results from §3.9 were first obtained by Clarkson [8].

A more advanced study of Lebesgue spaces with exponent $0 < p < 1$ is given by Day [13], see also Köthe [40].

Chapter 4
Distribution Function and Nonincreasing Rearrangement

A mathematician is a person who can find analogies between theorems; a better mathematician is one who can see analogies between proofs and the best mathematician can notice analogies between theories.
STEFAN BANACH

Abstract In this chapter we study the distribution function which is a tool that provides information about the size of a function but not about its pointwise behavior or locality; for example, a function f and its translation are the same in terms of their distributions. Based on the distribution function we study the nonincreasing rearrangement and establish its basic properties. We obtain sub-additive and sub-multiplicative type inequalities for the decreasing rearrangement. The maximal function associated with the decreasing rearrangement is introduced and some important relations are obtained, e.g., Hardy's inequality. In the last section of this chapter we deal with the rearrangement of the Fourier transform.

4.1 Distribution Function

Suppose that $(X, \mathscr{A}, \mu)$ is a measure space and let $\mathfrak{F}(X, \mathscr{A})$ denote the set of all $\mathscr{A}$-measurable functions on X.

Definition 4.1. The *distribution function* D_f of a function f in $\mathfrak{F}(X, \mathscr{A})$ is given by

$$D_f(\lambda) := \mu\left(X(|f| > \lambda)\right) \tag{4.1}$$

where

$$X(|f| > \lambda) = \{x \in X : |f(x)| > \lambda\}$$

for $\lambda \geq 0$. In case we need to emphasize the underlying measure, we can write $D_f^{\mu}(\lambda)$.

⊘

Observe that the distribution function D_f depends only on the absolute value of the function f and its global behavior. Moreover, notice that D_f may even assume the value $+\infty$.

R.E. Castillo, H. Rafeiro, *An Introductory Course in Lebesgue Spaces*, CMS Books in Mathematics, DOI 10.1007/978-3-319-30034-4_4

It should be pointed out that the notation for the distribution function (4.1) is not standardized, other used notations include f_*, μ_f, d_f, λ_f, among others.

Definition 4.2. Let $(X, \mathscr{A}, \mu)$ and $(Y, \mathscr{M}, \nu)$ be two measure spaces. Two functions $f \in \mathfrak{F}(X, \mathscr{A})$ and $g \in \mathfrak{F}(Y, \mathscr{M})$ are said to be *equimeasurable* if they have the same distribution function, that is, if

$$D_f^{\mu}(\lambda) = D_g^{\nu}(\lambda) \tag{4.2}$$

for all $\lambda \geq 0$. ⊘

In what follows, we gather some useful properties of the distribution function.

Theorem 4.3. *Let f and g be two functions in $\mathfrak{F}(X, \mathscr{A})$. Then for all $\lambda, \lambda_1, \lambda_2 \geq 0$ we have:*

(a) D_f is decreasing and continuous from the right;

(b) $|g| \leq |f|$ μ-a.e. implies that $D_g(\lambda) \leq D_f(\lambda)$;

(c) $D_{cf}(\lambda_1) = D_f\left(\frac{\lambda_1}{|c|}\right)$ for all $c \in \mathbb{C} \setminus \{0\}$;

(d) $D_{f+g}(\lambda_1 + \lambda_2) \leq D_f(\lambda_1) + D_g(\lambda_2)$;

(e) $D_{fg}(\lambda_1 \lambda_2) \leq D_f(\lambda_1) + D_g(\lambda_2)$;

(f) If $|f| \leq \liminf |f_n|$ μ-a.e., then $D_f(\lambda) \leq \liminf D_{f_n}(\lambda)$;

(g) If $|f_n| \uparrow |f|$, then $\lim_{n\to\infty} D_{f_n}(\lambda) = D_f(\lambda)$.

Proof. (a) Let $0 \leq \lambda_1 \leq \lambda_2$ be arbitrary. Then

$$\left\{x \in X : |f(x)| > \lambda_2\right\} \subseteq \left\{x \in X : |f(x)| > \lambda_1\right\}.$$

Hence by the monotonicity of the measure we have that $D_f(\lambda_2) \leq D_f(\lambda_1)$, that is, D_f is decreasing. To prove that D_f is continuous from the right, let $\lambda_0 \geq 0$ and define

$$E_f(\lambda) = \left\{x \in X : |f(x)| > \lambda\right\},$$

then by the monotone convergence theorem, we have

$$\begin{aligned}
\lim_{n\to\infty} D_f(\lambda_0 + \tfrac{1}{n}) &= \lim_{n\to\infty} \mu\left(E_f(\lambda_0 + \tfrac{1}{n})\right) \\
&= \mu\left(\bigcup_{n=1}^{\infty} E_f(\lambda_0 + \tfrac{1}{n})\right)
\end{aligned}$$

$$= \mu\Big(E_f(\lambda_0)\Big)$$
$$= D_f(\lambda_0)$$

since $E_f(\lambda_1) \subseteq E_f(\lambda_2) \subseteq E_f(\lambda_3) \subseteq \ldots$ when $\lambda_1 \geq \lambda_2 \geq \lambda_3 \geq \ldots$, and this establishes the right-continuity.

(b) Let f and g be two functions in $\mathfrak{F}(X, \mathscr{A})$ such that $|g| \leq |f|$ μ-a.e. then

$$\Big\{x \in X : |g(x)| > \lambda\Big\} \subseteq \Big\{x \in X : |f(x)| > \lambda\Big\},$$

by the monotonicity of the measure we have that

$$\mu\Big(\{x \in X : |g(x)| > \lambda\}\Big) \leq \mu\Big(\{x \in X : |f(x)| > \lambda\}\Big),$$

and thus $D_g(\lambda) \leq D_f(\lambda)$.

(c) Let $f \in \mathfrak{F}(X, \mathscr{A})$ and $c \in \mathbb{C} \setminus \{0\}$. Then

$$\Big\{x \in X : |cf(x)| > \lambda\Big\} = \Big\{x \in X : |c||f(x)| > \lambda\Big\} = \Big\{x \in X : |f(x)| > \frac{\lambda}{|c|}\Big\},$$

and thus $\mu\Big(\{x \in X : |cf(x)| > \lambda\}\Big) = \mu\Big(\{x \in X : |f(x)| > \lambda/|c|\}\Big)$ which is simply

$$D_{cf}(\lambda) = D_f\left(\frac{\lambda}{|c|}\right).$$

(d) Let $f, g \in \mathfrak{F}(X, \mathscr{A})$ and $\lambda_1, \lambda_2 \geq 0$, then

$$\Big\{x \in X : |f(x) + g(x)| > \lambda_1 + \lambda_2\Big\} \subseteq \Big\{x \in X : |f(x)| + |g(x)| > \lambda_1 + \lambda_2\Big\}$$
$$\subseteq \Big\{x \in X : |f(x)| > \lambda_1\Big\} \cup \Big\{x \in X : |g(x)| > \lambda_2\Big\}.$$

Then

$$\mu\Big(\{x \in X : |f(x) + g(x)| > \lambda_1 + \lambda_2\}\Big) \leq \mu\Big(\{x \in X : |f(x)| > \lambda_1\}\Big) + \mu\Big(\{x \in X : |g(x)| > \lambda_2\}\Big),$$

which is $D_{f+g}(\lambda_1 + \lambda_2) \leq D_f(\lambda_1) + D_g(\lambda_2)$.

(e) Let $f, g \in \mathfrak{F}(X, \mathscr{A})$ and $\lambda_1, \lambda_2 \geq 0$, then

$$\Big\{x \in X : |f(x)g(x)| > \lambda_1\lambda_2\Big\} = \Big\{x \in X : |f(x)||g(x)| > \lambda_1\lambda_2\Big\}$$
$$\subseteq \Big\{x \in X : |f(x)| > \lambda_1\Big\} \cup \Big\{x \in X : |g(x)| > \lambda_2\Big\},$$

and thus $D_{fg}(\lambda_1\lambda_2) \leq D_f(\lambda_1) + D_g(\lambda_2)$.

(f) Fix $\lambda \geq 0$ and let $E = \{x \in X : |f(x)| > \lambda\}$ and $E_n = \{x \in X : |f_n(x)| > \lambda\}$ where $n = 1, 2, \ldots$. Clearly, $E \subset \cup_{m=1}^{\infty} \cap_{n>m} E_n$, therefore

$$\mu\left(\bigcap_{n>m} E_n\right) \leq \inf_{n>m} \mu(E_n) \leq \sup_m \inf_{n>m} \mu(E_n) = \liminf_{n\to\infty} \mu(E_n)$$

for each $m = 1, 2, \ldots$. And thus, an appeal to the monotone convergence theorem and the fact that $\cap_{n>m} E_n \subset \cap_{n>m+1} E_n$ we get

$$\mu(E) \leq \mu\left(\bigcup_{m=1}^{\infty} \bigcap_{n>m} E_n\right) = \lim_{m\to\infty} \mu\left(\bigcap_{n>m} E_n\right) \leq \liminf_{n\to\infty} \mu(E_n),$$

which gives $D_f(\lambda) \leq \liminf_{n\to\infty} D_{f_n}(\lambda)$.

(g) If $|f_n| \uparrow |f|$, then $E_{f_1}(\lambda) \subseteq E_{f_2}(\lambda) \subseteq E_{f_3}(\lambda) \subseteq \ldots$. Hence

$$E_f(\lambda) = \bigcup_{n=1}^{\infty} E_{f_n}(\lambda),$$

and thus

$$D_f(\lambda) = \mu(E_f(\lambda)) = \mu\left(\bigcup_{n=1}^{\infty} E_{f_n}(\lambda)\right) = \lim_{n\to\infty} \mu(E_{f_n}(\lambda)) = \lim_{n\to\infty} D_{f_n}(\lambda).$$

□

4.2 Decreasing Rearrangement

With the notion of distribution function we are ready to introduce the decreasing rearrangement and its important properties (Fig. 4.1).

Definition 4.4. Let $f \in \mathfrak{F}(X, \mathscr{A})$. The *decreasing rearrangement* of f is the function $f^* : [0, \infty) \longrightarrow [0, \infty]$ defined by

$$f^*(t) = \inf\left\{\lambda \geq 0 : D_f(\lambda) \leq t\right\},$$

taking the usual convention that $\inf(\emptyset) = \infty$. ⊘

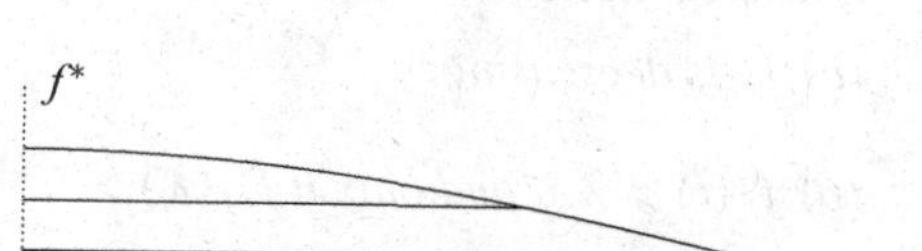

Fig. 4.1 The graph of a function f and its decreasing rearrangement f^*.

Notice that

$$f^*(0) = \inf\left\{\lambda \geq 0 : D_f(\lambda) = 0\right\} = \|f\|_\infty,$$

since

$$\|f\|_\infty = \inf\left\{\alpha \geq 0 : \mu\Big(\{x \in X : |f(x)| > \alpha\}\Big) = 0\right\}.$$

Also observe that if D_f is strictly decreasing, then

$$f^*\Big(D_f(t)\Big) = \inf\left\{\lambda \geq 0 : D_f(\lambda) \leq D_f(t)\right\} = t,$$

which shows that f^* is the left inverse function of the distribution function D_f.

In general we have the following result

$$f^*\Big(D_f(\lambda)\Big) \leq \lambda. \tag{4.3}$$

To see this fix $\lambda \geq 0$ and suppose $t = D_f(\lambda) < \infty$. Then Definition 4.4 gives

$$f^*\Big(D_f(\lambda)\Big) = f^*(t) = \inf\{\tilde{\lambda} \geq 0 : D_f(\tilde{\lambda}) \leq t = D_f(\lambda)\} \leq \lambda,$$

which establishes (4.3).

On the other hand, we also have

$$D_f\Big(f^*(t)\Big) \leq t. \tag{4.4}$$

In order to prove (4.4) let us assume $\lambda = f^*(t) < \infty$, by Definition 4.4 there exists a sequence $\{\lambda_n\}_{n\in\mathbb{N}}$ such that $\lambda_n \downarrow \lambda$ and $D_f(\lambda_n) \leq t$, thus the right continuity of D_f gives

$$D_f\Big(f^*(t)\Big) = D_f(\lambda) = \lim_{n\to\infty} D_f(\lambda_n) \leq t,$$

which establish (4.4).

The next theorem establishes some basic properties of the decreasing rearrangement.

Theorem 4.5. *The decreasing rearrangement has the following properties:*

(a) f^* is decreasing;

(b) $f^*(t) > \lambda$ if and only if $D_f(\lambda) > t$;

(c) f and f^* are equimeasurables, that is $D_f(\lambda) = D_{f^*}(\lambda)$ for all $\lambda \geq 0$;

(d) If $f \in \mathfrak{F}(X, \mathscr{A})$, then $f^*(t) = D_{D_f}(t)$ for all $t \geq 0$ (this result tell us that f^* is right-continuous);

(e) $(\alpha f)^*(t) = |\alpha| f^*(t)$; $\alpha \in \mathbb{R}$;

(f) If $|f_n| \uparrow |f|$, then $f_n^* \uparrow f^*$;

(g) If $|f| \leq \liminf_{n\to\infty} |f_n|$, then $f^* \leq \liminf_{n\to\infty} f_n^*$;

(h) For $0 < p < \infty$, $\left(|f|^p\right)^*(t) = [f^*(t)]^p$;

(i) If $|f| \leq |g|$, then $f^*(t) \leq g^*(t)$;

(j) If $E \in \mathscr{A}$, then $\left(\chi_E\right)^*(t) = \chi_{[0,\mu(E)]}(t)$;

(k) If $E \in \mathscr{A}$, then $\left(f\chi_E\right)^*(t) \leq f^*(t)\chi_{[0,\mu(E)]}(t)$;

(l) If f belong to $\mathfrak{F}(X, \mathscr{A})$, $\lambda > 0$ and $F = \chi_{E_f(\lambda)}$ we get $F^*(t) = \chi_{E_{f^*}(\lambda)}(t)$.

Proof. (a) Let $0 \leq t \leq u$, then

$$\left\{\lambda \geq 0 : D_f(\lambda) \leq t\right\} \subset \left\{\lambda \geq 0 : D_f(\lambda) \leq u\right\},$$

then

$$\inf\left\{\lambda \geq 0 : D_f(\lambda) \leq u\right\} \leq \inf\left\{\lambda \geq 0 : D_f(\lambda) \leq t\right\},$$

and thus $f^*(u) \leq f^*(t)$.

(b) If $s < f^*(t) = \inf\left\{\alpha > 0 : D_f(\alpha) \leq t\right\}$ then $s \notin \left\{\alpha > 0 : D_f(\alpha) \leq t\right\}$ which gives $D_f(s) > t$. Conversely, if for some $t < D_f(s)$ we have $f^*(t) \leq s$, then $D_f(s) \leq D_f\left(f^*(t)\right) \leq t$ which is a contradiction.

(c) By (b) we have

$$\left\{t \geq 0 : f^*(t) > \lambda\right\} = \left\{\lambda \geq 0 : D_f(\lambda) > t\right\} = \Big(0, D_f(\lambda)\Big),$$

then

$$D_{f^*}(\lambda) = m\Big(\{t \geq 0 : f^*(t) > \lambda\}\Big) = m\Big(0, D_f(\lambda)\Big) = D_f(\lambda)$$

for all $\lambda \geq 0$.

(d) Again by (b) we have

$$\left\{\lambda \geq 0 : D_f(\lambda) > t\right\} = \Big(0, f^*(t)\Big),$$

then

$$f^*(t) = m\Big(\{\lambda \geq 0 : D_f(\lambda) > t\}\Big) = D_{D_f}(t).$$

Theorem 4.3 (a) shows us that f^* is right-continuous.

(e) Let $f \in \mathfrak{F}(X, \mathscr{A})$ and $\alpha \in \mathbb{R}$, then

$$\begin{aligned}
\Big(\alpha f\Big)^*(t) &= \inf\left\{\lambda \geq 0 : D_{\alpha f}(\lambda) \leq t\right\} \\
&= \inf\left\{\lambda \geq 0 : D_f\left(\frac{\lambda}{|\alpha|}\right) \leq t\right\} \\
&= \inf\left\{|\alpha|\gamma \geq 0 : D_f(\gamma) \leq t\right\} \\
&= |\alpha| \inf\left\{\gamma \geq 0 : D_f(\gamma) \leq t\right\} \\
&= |\alpha| f^*(t),
\end{aligned}$$

where $\gamma = \frac{\lambda}{|\alpha|}$.

(f) We already know by Proposition 4.3 (g) that if $|f_n| \uparrow |f|$, then $\lim_{n\to\infty} D_{f_n}(\lambda) = D_f(\lambda)$. Let $F_n(t) = D_{f_n}(t)$, then

$$f_n^*(t) = m\Big(\{\lambda > 0 : D_{f_n}(\lambda) > t\}\Big) = D_{F_n}(t),$$

since $D_{f_n}(t) \leq D_{f_{n+1}}(t)$ we have $F_n(t) \leq F_{n+1}(t)$ and thus

$$E_{F_1}(t) \subseteq E_{F_2}(t) \subseteq \dots \quad \text{and} \quad E_F(t) = \bigcup_{n=1}^{\infty} E_{F_n}(t)$$

therefore $\lim_{n\to\infty} D_{F_n}(t) = D_F(t)$. That is $\lim_{n\to\infty} f_n^*(t) = f^*(t)$.

(g) Let $F_n(t) = \inf_{m>n} |f_m(t)|$ and observe that

$$F_n(t) \leq F_{n+1}(t)$$

for all $n \in \mathbb{N}$ and all $t \in X$. Taking $h(t) = \liminf_{n\to\infty} |f_n(t)| = \sup_{n\geq 1} F_n(t)$, we get that $F_n^* \uparrow h^*$ as $n \to \infty$ by the fact that $F_n \uparrow h$ and item (f).

By hypothesis we have $|f| \leq h$, hence

$$f^*(t) \leq h^*(t) = \sup_{n\geq 1} F_n^*(t).$$

Since $F_n \leq |f_m|$ for $m \geq n$, it follows that $F_n^* \leq \inf_{m\geq n} f_n^*(t)$. Putting these facts together we have

$$f^*(t) \leq h^*(t) = \sup_{n\geq 1} F_n^*(t) \leq \sup_{n\geq 1} \inf_{m>n} f_m^*(t) = \liminf f_n^*(t).$$

(h) Let $0 < p < \infty$, then

$$D_{|f|^p}(\lambda) = \mu\Big(\{x \in X : |f(x)|^p > \lambda\}\Big) = \mu\Big(\{x \in X : |f(x)| > \lambda^{1/p}\}\Big) = D_f(\lambda^{1/p}),$$

and

$$\begin{aligned}
(|f|^p)^*(t) &= \inf\Big\{\lambda > 0 : D_{|f|^p}(\lambda) \leq t\Big\} \\
&= \inf\Big\{\lambda > 0 : D_f(\lambda^{1/p}) \leq t\Big\} \\
&= \inf\Big\{u^p > 0 : D_f(u) \leq t\Big\} \\
&= \left(\inf\Big\{u > 0 : D_f(u) \leq t\Big\}\right)^p \\
&= [f^*(t)]^p,
\end{aligned}$$

where $u = \lambda^{1/p}$.

(i) Let $f, g \in \mathfrak{F}(X, \mathscr{A})$ such that $|f(x)| \leq |g(x)|$ for all $x \in X$. Then $D_f(\lambda) \leq D_g(\lambda)$ for all $\lambda \geq 0$, which yields

$$\left\{\lambda > 0 : D_g(\lambda) \le t\right\} \subseteq \left\{\lambda > 0 : D_f(\lambda) \le t\right\},$$

therefore

$$\inf\left\{\lambda > 0 : D_f(\lambda) \le t\right\} \le \inf\left\{\lambda > 0 : D_g(\lambda) \le t\right\},$$

and thus $f^*(t) \le g^*(t)$.

(j) Let $E \in \mathscr{A}$, if $f = \chi_E$ then

$$\left\{x \in X : \chi_E(x) > \lambda\right\} = \begin{cases} E \text{ if } 0 \le \lambda < 1 \\ \emptyset \text{ if } \lambda \ge 1. \end{cases}$$

Hence

$$D_f(\lambda) = \mu\Big(\{x \in X : \chi_E(x) > \lambda\}\Big) = \begin{cases} \mu(E) \text{ if } 0 \le \lambda < 1 \\ 0 \quad\;\; \text{if } \lambda \ge 1. \end{cases}$$

and thus

$$f^*(t) = \inf\left\{\lambda \ge 0 : D_f(\lambda) \le t\right\} = \begin{cases} 1 \text{ if } t < \mu(E) \\ 0 \text{ if } t \ge \mu(E), \end{cases}$$

which prove that

$$f^*(t) = \chi_{(0,\mu(E))}(t).$$

(k) Since $\Big(f\chi_E\Big)(x) \le f(x)$ for all $x \in X$, we have

$$D_{f\chi_E}(\lambda) \le D_f(\lambda),$$

then

$$\left\{\lambda \ge 0 : D_{f\chi_E}(\lambda) > t\right\} \subseteq \left\{\lambda \ge 0 : D_f(\lambda) > t\right\}.$$

Hence

$$D_{D_{f\chi_E}}(t) \le D_{D_f}(t),$$

which is equivalent to

$$\Big(f\chi_E\Big)^*(t) \le f^*(t), \tag{4.5}$$

for all $t \ge 0$.

(l) Let $F = \chi_{E_f(\lambda)}$, then

$$F^*(t) = \chi_{\left(0,\mu(E_f(\lambda))\right)}(t).$$

By Theorem 4.5 (b) we have $E_{f^*}(\lambda) = \left(0, \mu(E_f(\lambda))\right)$.

Thus

$$F^*(t) = \chi_{E_{f^*}(\lambda)}(t),$$

which ends the proof. □

Observe that property (c) of Theorem 4.5 does not hold if we remove the strict inequalities, that is

$$\mu\Big(\{x \in X : |f(x)| \geq \lambda\}\Big) = m\Big(\{t \geq 0 : |f^*(t)| \geq \lambda\}\Big),$$

does not hold in the general case. This can be seen by taking

$$f(x) = \frac{x}{x+1}$$

for all $x \in [0,\infty)$. Then $f^*(t) = 1$ which means that

$$\mu\left(\left\{x \in X : \frac{x}{x+1} \geq 1\right\}\right) = 0 \neq \infty = m\Big(\{t \geq 0 : 1 \geq 1\}\Big).$$

From Theorem 4.5 (c) we know that f and f^* are equimeasurables functions and this is a very important property of the decreasing rearrangement, since it permits to replace the function f by its decreasing rearrangement whenever we are working with *something* that only requires global information of the function, cf. Theorem 4.13. We now show that the decreasing rearrangement is the unique right-continuous decreasing function equimeasurable with f.

Theorem 4.6. *There exists only one right-continuous decreasing function f^* equimeasurable with f.*

Proof. Let f_1^* and f_2^* be two different right-continuous functions equimeasurable with f. Then there exists a t_0 such that $f_1^*(t_0) \neq f_2^*(t_0)$, we may assume without loss of generality that $f_1^*(t_0) > f_2^*(t_0)$. Choose $\varepsilon > 0$ such that

$$f_1^*(t_0) > f_2^*(t_0) + \varepsilon.$$

And then by the right continuity of f_1^* there exists an interval $[t_0, t_1]$ such that

$$f_1^*(t) > f_1^*(t_0) + \varepsilon$$

for all $t \in [t_0, t_1]$. On the one hand, observe that if $s \in (0, t_1]$, then $f_1^*(t_1) \leq f_1^*(s)$ and thus $f_1^*(s) \geq f_1^*(t_1) > f_2^*(t_1) + \varepsilon$, which means that $s \in \{t \geq 0 : f_1^*(t) > f_2^*(t_1) + \varepsilon\}$

that is

$$(0,t_1] \subset \{t \geq 0 : f_1^*(t) > f_2^*(t_1) + \varepsilon.\}$$

On the other hand, clearly

$$\left\{ t \geq 0 : f_2^*(t) > f_1^*(t_1) + \varepsilon \right\} \subset (0,t_1].$$

Then

$$m\Big(\{t \geq 0 : f_2^*(t) > f_1^*(t_1) + \varepsilon\} \Big) \leq t_1 \leq m\Big(\{t \geq 0 : f_1^*(t) > f_2^*(t_1) + \varepsilon\} \Big),$$

which is a contradiction to the equimeasurability with f. Hence $f_1^*(t) = f_2^*(t)$. □

Next, we take a look at some examples of distribution function and decreasing rearrangement.

Example 4.7. Let $f(x) = 1 - e^{-x}$ for $0 < x < \infty$. Then

$$D_f(\lambda) = m\Big(\{x \in (0,\infty) : |f(x)| > \lambda\} \Big) = \begin{cases} \infty \text{ if } 0 \leq \lambda < 1; \\ 0 \ \text{ if } \lambda \geq 1. \end{cases}$$

Since

$$f^*(t) = \inf\left\{ \lambda > 0 : D_f(\lambda) \leq t \right\},$$

we have $f^*(t) = 1 \quad$ for all $\quad t \geq 0$. ⊘

This example shows that a considerable amount of information may be lost in passing to the decreasing rearrangement. Such information, however, is irrelevant regarding *rearrangement-invariant norms*, such as L_p-norms, cf. Theorem 4.13. Thus, the L_p-norm of f and f^* are both infinite whenever $1 \leq p < \infty$, and the L_∞-norm are both equal to 1.

Example 4.8 (Decreasing rearrangement of a simple function). Let s be a simple function of the following form

$$s(x) = \sum_{j=1}^{n} \alpha_j \chi_{A_j}(x),$$

where $\alpha_1 > \alpha_2 > \ldots > \alpha_n > 0$, $A_j = \{x : s(x) = \alpha_j\}$ and χ_{A_j} is the characteristic function of the set A_j. Then if $\lambda \geq \alpha_1$ clearly $D_s(\lambda) = 0$. If $\alpha_2 \leq \lambda < \alpha_1$, then

$$D_s(\lambda) = \mu(A_1 \cup A_2) = \mu(A_1).$$

In general, for $\lambda \geq 0$ we have

$$D_s(\lambda) = \sum_{j=1}^{n} B_j \chi_{[\alpha_{j-1},\alpha_j)}(\lambda)$$

where

$$B_j = \sum_{i=1}^{j} \mu(A_i).$$

This shows that the distribution function of simple function is a simple function itself. Now

$$\begin{aligned} s^*(t) &= \inf\left\{\lambda \geq 0 : D_s(\lambda) \leq t\right\} \\ &= \inf\left\{\lambda \geq 0 : \sum_{j=1}^{n} B_j \chi_{[\alpha_{j-1},\alpha_j)}(\lambda) \leq t\right\} \\ &= \sum_{j=1}^{n} \alpha_j \chi_{[B_{j-1},B_j)(t)} \end{aligned}$$

which is also a simple function. ⊘

Example 4.9. Let $X = [0,\infty)$; $\mathscr{A} = \{\text{all Lebesgue measurable subsets of } X\}$ and $\mu = m$, where m denotes the Lebesgue measure on X.

Define $f : [0,\infty) \longrightarrow [0,\infty)$ as

$$f(x) = \begin{cases} 1-(x-1)^2 & \text{if } 0 \leq x \leq 2, \\ 0 & \text{if } x > 2. \end{cases}$$

After some routine calculations we get

$$\begin{aligned} m\Big(\{x \in \mathbb{R}^+ : |f(x)| > \lambda\}\Big) &= m\Big(\{x \in [0,2] : 1-(x-1)^2 > \lambda\}\Big) \\ &= m\Big(\{x \in [0,2] : 1-\sqrt{1-\lambda} < x < 1+\sqrt{1-\lambda}\}\Big) \\ &= 2\sqrt{1-\lambda} \quad \text{for} \quad \lambda \in [0,1]. \end{aligned}$$

Therefore

$$D_f(\lambda) = \begin{cases} 2\sqrt{1-\lambda} & \text{if } 0 \leq \lambda \leq 1, \\ 0 & \text{if } \lambda > 1. \end{cases}$$

Next, the decreasing rearrangement becomes

$$f^*(t) = \begin{cases} 1 - \dfrac{t^2}{4} \text{ if } 0 \le t \le 2, \\ 0 \text{ if } t > 2. \end{cases}$$

Finally, observe that

$$\int_0^\infty f(x)\,\mathrm{d}x = \int_0^2 (1-(x-1)^2)\,\mathrm{d}x = \int_0^1 2\sqrt{1-\lambda}\,\mathrm{d}\lambda = \int_0^2 \left(1 - \frac{t^2}{4}\right) \mathrm{d}t = \frac{4}{3}.$$

⊘

The operation $f \mapsto f^*$ is neither *subadditive*, in the sense that the inequality

$$(f+g)^*(t) \le f^*(t) + g^*(t)$$

is not true in general, nor *submultiplicative*, that is

$$(fg)^*(t) \le f^*(t) g^*(t)$$

does not hold for any t. The next example shows this.

Example 4.10. Let A and B be measurable sets such that $A \cap B \neq \emptyset$ and $0 < \mu(A) < \mu(B)$. Put $f(x) = \chi_A(x)$ and $g(x) = \chi_B(x)$. The decreasing rearrangements are

$$f^*(t) = \chi_{[0,\mu(A))}(t)$$

and

$$g^*(t) = \chi_{[0,\mu(B))}(t),$$

which means that

$$(f+g)^*(t) = \begin{cases} 2 \text{ if } 0 \le t < \mu(A), \\ 1 \text{ if } \mu(A) \le t < \mu(B), \\ 0 \text{ if } t \ge \mu(B). \end{cases}$$

Moreover, since $(f+g)(x) = \chi_A(x) + \chi_B(x)$ it follows that

$$f^*(t) + g^*(t) = \begin{cases} 2 \text{ if } 0 \le t < \mu(A \cap B), \\ 1 \text{ if } \mu(A \cap B) \le t \le \mu(A \cup B), \\ 0 \text{ if } t \ge \mu(A \cup B). \end{cases}$$

Hence, for any t such that $\mu\left(A \cap B\right) < t < \mu(A \cup B)$, we have

$$(f+g)^*(t) = 1 > 0 = f^*(t) + g^*(t).$$

Similar calculations show that $f \mapsto f^*$ is not submultiplicative. ⊘

Despite the previous comments and examples, we have a subadditive and submultiplicative type inequality regarding the decreasing rearrangement of a function, as shown in the next theorem.

Theorem 4.11. *Let $(X, \mathscr{A}, \mu)$ be a measure space and f and g be two functions in $\mathfrak{F}(X, \mathscr{A})$. Then the inequalities*

$$(f+g)^*(t+u) \leq f^*(t) + g^*(u) \tag{4.6}$$

and

$$(fg)^*(t+u) \leq f^*(t)g^*(u) \tag{4.7}$$

hold for all $t, u \geq 0$. In particular

$$(f+g)^*(t) \leq f^*(t/2) + g^*(t/2)$$

and

$$(fg)^*(t) \leq f^*(t/2)g^*(t/2)$$

for all $t \geq 0$.

Proof. If a and b are two numbers satisfying $a = f^*(t_0)$ and $b = g^*(u)$, then

$$\begin{aligned} D_f(a) &= \mu\Big(\{x \in X : |f(x)| > a\} \Big) \\ &= \mu\Big(\{x \in X : |f(x)| > f^*(t_0)\} \Big) \\ &= D_f(f^*(t_0)) \leq t_0. \end{aligned}$$

Similar argument shows that $D_f(b) \leq u$, and thus

$$\Big\{x : |f(x) + g(x)| > a + b\Big\} \subseteq \Big\{x : |f(x)| > a\Big\} \cup \Big\{x : |g(x)| > b\Big\},$$

which entails

$$D_{f+g}(a+b) \leq D_f(a) + D_g(b) \leq t_0 + u.$$

This shows that $(f+g)^*(t_0+u) \leq a+b = f^*(t_0) + g^*(u)$ for any such numbers a and b.

The proof of the second inequality is similar.

By Theorem 4.3 (e) we have

$$D_{fg}(ab) \leq D_f(a) + D_g(b) \leq t_0 + u.$$

Again, using the definition of the decreasing rearrangement, we get

$$(fg)^*(t_0+u) \le ab = f^*(t_0)g^*(u),$$

which is our desired inequality. The rest of the theorem now follows by taking $t_0 = u = t/2$. □

We now show that the integral of a decreasing rearrangement function satisfies an integration by parts type formula, namely:

Theorem 4.12. *We have the following equality*

$$\int_0^t f^*(s)\,\mathrm{d}s = tf^*(t) + \int_{f^*(t)}^{\infty} D_{f^*(t)}(\lambda)\,\mathrm{d}\lambda. \tag{4.8}$$

Proof. Let us start with

$$\int_{f^*(t)}^{\infty} D_{f^*(t)}(\lambda)\,\mathrm{d}\lambda = \int_{f^*(t)}^{\infty} m\Big(\{s : f^*(s) > \lambda\}\Big)\,\mathrm{d}\lambda.$$

Next, we use Fubini's theorem and the fact that f^* is a decreasing function, thus

$$\begin{aligned}
\int_{f^*(t)}^{\infty} D_{f^*(t)}(\lambda)\,\mathrm{d}\lambda &= \int_{f^*(t)}^{\infty}\int_0^t \chi_{\{s:f^*(s)>\lambda\}}(s)\,\mathrm{d}m(s)\,\mathrm{d}\lambda \\
&= \int_0^{\infty} \chi_{(f^*(t),\infty)}(\lambda)\left(\int_0^t \chi_{(0,f^*(s))}(\lambda)\,\mathrm{d}m(s)\right)d\lambda \\
&= \int_0^t\int_0^{\infty} \chi_{(f^*(t),\infty)}(\lambda)\chi_{(0,f^*(s))}(\lambda)\,\mathrm{d}\lambda\,\mathrm{d}s \\
&= \int_0^t\int_0^{\infty} \chi_{(f^*(t),f^*(s))}(\lambda)\,\mathrm{d}\lambda\,\mathrm{d}s \\
&= \int_0^t \Big[f^*(s) - f^*(t)\Big]\,\mathrm{d}s,
\end{aligned}$$

which entails (4.8). □

The next theorem is quite important. It shows, in particular, that the integral of a function and the decreasing rearrangement have the same value.

Theorem 4.13. *Let $(X,\mathscr{A},\mu)$ be a σ-finite measure space and φ be a differentiable and increasing function with $\varphi(0)=0$. Then*

$$\int_X \varphi(|f|)\,\mathrm{d}\mu = \int_0^\infty \varphi(f^*(t))\,\mathrm{d}t.$$

Proof. By Theorem 3.54 and Theorem 4.5(c) we have

$$\begin{aligned}\int_X \varphi(|f|)\,\mathrm{d}\mu &= \int_0^\infty \varphi'(\lambda)\mu\left(\{x\in X : |f(x)|>\lambda\}\right)\mathrm{d}\lambda \\ &= \int_0^\infty \varphi'(\lambda) m\left(\{t\in[0,\infty] : f^*(t)>\lambda\}\right)\mathrm{d}\lambda \\ &= \int_0^\infty \varphi'(\lambda)\left(\int_0^\infty \chi_{(0,f^*(t))}(\lambda)\,\mathrm{d}t\right)\mathrm{d}\lambda.\end{aligned}$$

Next, applying the Fubini theorem

$$\begin{aligned}\int_0^\infty \varphi'(\lambda)\left(\int_0^\infty \chi_{(0,f^*(t))}(\lambda)\,\mathrm{d}t\right)\mathrm{d}\lambda &= \int_0^\infty\int_0^\infty \varphi'(\lambda)\chi_{(0,f^*(t))}(\lambda)\,\mathrm{d}\lambda\,\mathrm{d}t \\ &= \int_0^\infty\int_0^{f^*(t)} \varphi'(\lambda)\,\mathrm{d}\lambda\,\mathrm{d}t \\ &= \int_0^\infty \varphi(f^*(t))\,\mathrm{d}t,\end{aligned}$$

which ends the proof. □

Remark 4.14. We now get some particular cases of Theorem 4.13, namely:

If $\varphi(t)=t^p$ for $1\le p<\infty$, then

$$\int_{\mathbb{R}} |f|^p\,\mathrm{d}\mu = \int_0^\infty \left(f^*(t)\right)^p\mathrm{d}t.$$

The above shows that we can calculate the L^p norm of a function via its decreasing rearrangement, i.e., $\|f \mid L^p(\mathbb{R},\mu)\| = \|f^* \mid L^p(\mathbb{R}_+,m)\|$. ⊘

The following inequality is due to Hardy and Littlewood.

Theorem 4.15. *Let $(X, \mathscr{A}, \mu)$ be a σ-finite measure space and f, g be two functions in $\mathfrak{F}(X, \mathscr{A})$. Then*

$$\int_X |fg| \, d\mu \leq \int_0^\infty f^*(t) g^*(t) \, dt. \tag{4.9}$$

Proof. Assume first that $f = \chi_A$ and $g = \chi_B$ where A and B are sets in $\mathscr{A}$. We suppose without loss of generality that $\mu(A)$ and $\mu(B)$ are finite. Then it follows from Theorem 4.5 (j) that

$$\begin{aligned}
\int_X |fg| \, d\mu &= \int_X \chi_{A \cap B} \, d\mu \\
&= \mu(A \cap B) \\
&\leq \int_0^{\min\{\mu(A), \mu(B)\}} dt \\
&= \int_0^{\mu(A)} \chi_{(0, \mu(B))}(t) \, dt \\
&= \int_0^{\mu(A)} g^*(t) \, dt \\
&= \int_0^\infty \chi_{(0, \mu(A))}(t) g^*(t) \, dt \\
&= \int_0^\infty f^*(t) g^*(t) \, dt.
\end{aligned}$$

In general let f and g be two functions belonging to $\mathfrak{F}(X, \mathscr{A})$. Then

$$\begin{aligned}
\int_X |fg| \, d\mu &= \int_X \left(\int_0^{|f|} d\alpha \right) \left(\int_0^{|g|} d\beta \right) d\mu \\
&= \int_X \left(\int_0^\infty \chi_{E_f(\alpha)} d\alpha \right) \left(\int_0^\infty \chi_{E_g(\beta)} d\beta \right) d\mu.
\end{aligned}$$

It follows from Fubini's Theorem and Theorem 4.5 (1) that

$$
\begin{aligned}
\int_X |fg|\,\mathrm{d}\mu &= \int_0^\infty \int_0^\infty \int_X \chi_{E_f(\alpha)}\chi_{E_g(\beta)}\,\mathrm{d}\mu\, d\alpha\,\mathrm{d}\beta \\
&\le \int_0^\infty \int_0^\infty \left\{ \int_0^\infty \left(\chi_{E_f(\alpha)}\right)^*(t) \left(\chi_{E_g(\beta)}\right)^*(t)\,\mathrm{d}t \right\} d\alpha\,\mathrm{d}\beta \\
&= \int_0^\infty \int_0^\infty \left\{ \int_0^\infty \chi_{E_{f^*}(\alpha)}(t)\chi_{E_{g^*}(\beta)}(t)\,\mathrm{d}t \right\} d\alpha\,\mathrm{d}\beta \\
&= \int_0^\infty \left(\int_0^\infty \chi_{E_{f^*}(\alpha)}(t)\,\mathrm{d}\alpha \right) \left(\int_0^\infty \chi_{E_{g^*}(\beta)}(t)\,\mathrm{d}\beta \right) \mathrm{d}t \\
&= \int_0^\infty \left(\int_0^{f^*(t)} \mathrm{d}\alpha \right) \left(\int_0^{g^*(t)} \mathrm{d}\beta \right) \mathrm{d}t \\
&= \int_0^\infty f^*(t) g^*(t)\,\mathrm{d}t,
\end{aligned}
$$

and thus we obtain (4.9). □

Corollary 4.16 *Let $(X, \mathscr{A}, \mu)$ be a measure space and $f \in \mathfrak{F}(X, \mathscr{A})$. Then*

$$
\int_E |f|\,\mathrm{d}\mu \le \int_0^{\mu(E)} f^*(t)\,\mathrm{d}t,
$$

for all $E \in \mathscr{A}$.

Proof. By applying Theorem 4.15, we have

$$
\int_E |f|\,\mathrm{d}\mu = \int_X |f|\chi_E\,\mathrm{d}\mu \le \int_0^\infty f^*(t)\left(\chi_E\right)^*(t)\,\mathrm{d}t = \int_0^\infty f^*(t)\chi_{(0,\mu(E))}(t)\,\mathrm{d}t,
$$

which ends the proof. □

At this point, we will use some facts about atoms, see Section A.6 for more details in atoms. In fact whenever invoking the necessity of nonatomic measure space we want to exploit the "continuous of values fact," i.e., if μ is a nonatomic measure and A is a measurable set with $\mu(A) > 0$, then for any real number b satisfying

$$\mu(A) \geq b \geq 0,$$

there exists a measurable subset B of A such that $\mu(B) = b$.

Theorem 4.17. *Let $(X, \mathscr{A}, m)$ be a nonatomic measure space. Then*

$$\sup\left\{\int_E |f|\,\mathrm{d}m : m(E) = t\right\} = \int_0^t f^*(s)\,\mathrm{d}s.$$

Proof. Given the real number $t > 0$, we have

$$\begin{aligned}\int_E |f|\,\mathrm{d}m &= \int_0^\infty m\Big(E \cap E_f(\lambda)\Big)\,\mathrm{d}\lambda \\ &= \int_{\{\lambda : D_f(\lambda) \leq t\}} m\Big(E \cap E_f(\lambda)\Big)\,\mathrm{d}\lambda + \int_{\{\lambda : D_f(\lambda) > t\}} m\Big(E \cap E_f(\lambda)\Big)\,\mathrm{d}\lambda.\end{aligned}$$

Since $(X, \mathscr{A}, \mu)$ is a nonatomic measure space, if $m\Big(E_f(\lambda)\Big) \leq t$, then there exists a set $E \in \mathscr{A}$ such that $E \subseteq E_f(\lambda)$ and $m(E) = t$.

Hence

$$\begin{aligned}\sup\left\{\int_E |f|\,\mathrm{d}m : m(E) = t\right\} &= \int_{f^*(t)}^\infty m\Big(E_f(\lambda)\Big)\,\mathrm{d}\lambda + \int_0^{f^*(t)} t\,\mathrm{d}\lambda \\ &= \int_{f^*(t)}^\infty D_f(\lambda)\,\mathrm{d}\lambda + \int_0^{f^*(t)} t\,\mathrm{d}\lambda \\ &= \int_{f^*(t)}^\infty D_{f^*}(\lambda)\,\mathrm{d}\lambda + t f^*(t) \\ &= \int_0^t \int_{f^*(t)}^{f^*(s)} \mathrm{d}\lambda\,\mathrm{d}s + t f^*(t) \\ &= \int_0^t f^*(s)\,\mathrm{d}s\end{aligned}$$

which ends the proof. □

Remark 4.18. Observe that, if f and g are two $\mathscr{A}$-measurable functions then as in the proof of Theorem 4.17, we have for each set $E \in \mathscr{A}$ with $m(E) = t$

$$\int_E |f+g|\,\mathrm{d}m \leq \int_E |f|\,\mathrm{d}m + \int_E |g|\,\mathrm{d}m$$
$$\leq \int_0^t f^*(s)\,\mathrm{d}s + \int_0^t g^*(s)\,\mathrm{d}s \tag{4.10}$$

since $(X, \mathscr{A}, m)$ is nonatomic, it follows from Theorem 4.17 that

$$\int_0^t (f+g)^*(s)\,\mathrm{d}s \leq \int_0^t f^*(s)\,\mathrm{d}s + \int_0^t g^*(s)\,\mathrm{d}s. \tag{4.11}$$

Theorem 4.19 (Hardy). *Let f and g be nonnegative measurable function on $(0,\infty)$ and suppose*

$$\int_0^t f(s)\,\mathrm{d}s \leq \int_0^t g(s)\,\mathrm{d}s \tag{4.12}$$

for all $t > 0$. Let k be any nonnegative decreasing function on $(0,\infty)$. Then

$$\int_0^\infty f(s)k(s)\,\mathrm{d}s \leq \int_0^\infty g(s)k(s)\,\mathrm{d}s. \tag{4.13}$$

Proof. Let k be a simple function given by

$$k(s) = \sum_{j=1}^n a_j \chi_{(0,t_j)}(s),$$

where the coefficients a_j are positive and $0 < t_1 < t_2 < \ldots < t_n$.

Using (4.12), we obtain

$$\int_0^\infty f(s)k(s)\,\mathrm{d}s = \sum_{j=1}^n a_j \int_0^{t_j} f(s)\,\mathrm{d}s \leq \sum_{j=1}^n a_j \int_0^{t_j} g(s)\,\mathrm{d}s = \int_0^\infty g(s)k(s)\,\mathrm{d}s,$$

which establishes (4.13). Next, let k be any nonnegative measurable function then there exists a sequence of increasing measurable simple functions $\{\phi_n\}_{n\in\mathbb{N}}$ such that $\lim_{n\to\infty} \phi_n = k$. Since ϕ_n are increasing, for all $n \in \mathbb{N}$ we have $\phi_n \leq k$ and thus $\phi_n f \leq fk$. By the monotone convergence theorem we obtain

$$\int_0^\infty f(s)k(s)\,\mathrm{d}s = \lim_{n\to\infty} \int_0^\infty f(s)\phi_n(s)\,\mathrm{d}s.$$

But ϕ_n is a simple function, then

$$\int_0^\infty f(s)\phi_n(s)\,\mathrm{d}s \le \int_0^\infty g(s)\phi_n(s)\,\mathrm{d}s$$

for all $n \in \mathbb{N}$. Hence

$$\lim_{n\to\infty}\int_0^\infty f(s)\phi_n(s)\,\mathrm{d}s \le \lim_{n\to\infty}\int_0^\infty g(s)\phi_n(s)\,\mathrm{d}s.$$

Finally, we have

$$\int_0^\infty f(s)k(s)\,\mathrm{d}s \le \int_0^\infty g(s)k(s)\,\mathrm{d}s.$$

□

Corollary 4.20 *Let $(X,\mathscr{A},\mu)$ be a nonatomic measure space and let f, g and h be three functions in $\mathfrak{F}(X,\mathscr{A})$. Then*

$$\int_X |fgh|\,\mathrm{d}\mu \le \int_0^\infty f^*(t)g^*(t)h^*(t)\,\mathrm{d}t.$$

Proof. By Remark 4.14 and Theorem 4.15 we have

$$\int_0^\infty (fg)^*(t)\,\mathrm{d}t = \int_X |fg|\,\mathrm{d}\mu \le \int_0^\infty f^*(t)g^*(t)\,\mathrm{d}t.$$

Next, we invoke Theorem 4.19 with $h(t) = f^*(t)$ to obtain

$$\int_X |fgh|\,\mathrm{d}\mu \le \int_0^\infty f^*(t)g^*(t)h^*(t)\,\mathrm{d}t.$$

□

Corollary 4.21 *Let $(X,\mathscr{A},\mu)$ be a nonatomic measure space and let $f_1, f_2, \cdots, f_n$ be functions in $\mathfrak{F}(X,\mathscr{A})$. Then*

$$\int_X \prod_{k=1}^n |f_k|\,\mathrm{d}\mu \le \int_0^\infty \prod_{k=1}^n f_k^*(t)\,\mathrm{d}t$$

for $n \in \mathbb{N}$.

Proof. This follows from Corollary 4.20 and the principle of induction. □

We will take a closer look on the question if there is a function h in $\mathfrak{F}(X,\mu)$ which is equimeasurable with g such that

$$\int_X |fh|\,\mathrm{d}\mu = \int_0^\infty f^*(t)g^*(t)\,\mathrm{d}t.$$

The answer is in general no and it depends if the measure space is nonatomic or completely atomic where all atoms are of equal measure and if the measure space is finite. To make the terminology less cumbersome, we first make a definition.

Definition 4.22. A measure space or a measure is said to be *resonant* if it is $\sigma-$finite and either nonatomic or a countable union of atoms of equal measure. ⊘

The next theorem shed light in the posed question.

Theorem 4.23. *Let $(X,\mathscr{A},\mu)$ be a resonant measure space. Then*

$$\sup \int_X |f\tilde{g}|\,\mathrm{d}\mu = \int_0^\infty f^*(t)g^*(t)\,\mathrm{d}t, \tag{4.14}$$

where the supremum is taken over all $\mathscr{A}-$measurable functions $\tilde{g}$ equimeasurable with g, that is, $D_{\tilde{g}}(\lambda) = D_g(\lambda)$, $\lambda \geq 0$.

Moreover, if $\mu(X) < \infty$, then there is an $\mathscr{A}-$measurable function $\tilde{g}$ such that

$$\int_X |f\tilde{g}|\,\mathrm{d}\mu = \int_0^\infty f^*(t)g^*(t)\,\mathrm{d}t. \tag{4.15}$$

Proof. We can clearly assume that both f and g are positive. Let $\mu(X) < \infty$. If μ is completely atomic where all atoms are of equal measure, then $\widetilde{g}$ can be constructed by permuting the atoms. More specific, this permutation is the composition of the permutation that takes f to f^* and the permutation that takes g to g^*. That is, $\widetilde{g}$ is g, but where all atoms are permutation so that the atom where $\widetilde{g}$ has its largest value is the same as the atom where f has its largest value, the atom where $\widetilde{g}$ has its second largest value is the same as the atom where f has its second largest value and so on. Clearly, $\widetilde{g}$ and g are equimeasurable which means that we have proved (4.15) for the case when μ is completely atomic where all atoms are of equal measure.

Now, let μ be nonatomic. Since g is positive we can find a sequence of simple positive functions $\{g_n\}_{n\in\mathbb{N}}$ such that $g_n \uparrow g$. Then, for any arbitrary fixed integer $n \geq 1$, we can represent g_n as

$$g_n(x) = \sum_{j=1}^{k} \delta_j \chi_{D_j}(x)$$

where $D_1 \subset D_2 \subset \ldots D_k$ and $\delta_j > 0$, $j = 1,2,\ldots,k$ since $\mu(X) < \infty$ we can apply Problem 4.44 with $\varphi(t) = t$ for each $\mu(D_j)$ and get a sequence of sets $E_1, E_2, \ldots, E_k$

such that $\mu(E_j) = \mu(D_j)$, $j = 1,2,\ldots,k$ and

$$\int_{E_j} |f|\,\mathrm{d}\mu = \int_0^{\mu(D_j)} f^*(t)\,\mathrm{d}t. \tag{4.16}$$

Define the simple function $\widetilde{g}_n : X \to [0,\infty)$ by

$$\widetilde{g}_n(x) = \sum_{j=1}^{k} \delta_j \chi_{E_j}(x).$$

Since the measure of E_j and D_j is equal for all $j = 1,2,\ldots,k$ the decreasing rearrangement of g_n and $\widetilde{g}_n$ are equal, that is

$$g_n^*(t) = \widetilde{g}_n^*(t) = \sum_{j=1}^{k} \delta_j \chi_{[0,\mu(E_j))}(t).$$

Hence, we get a sequence of positive simple functions $\{\widetilde{g}_n\}_{n\in\mathbb{N}}$ such that $\widetilde{g}_n$ and g_n are equimeasurable with $n = 1,2,\ldots$. Using Proposition 4.3 (f) it also follows that

$$\widetilde{g} = \lim_{n\to\infty} \widetilde{g}_n,$$

is equimeasurable with g. Moreover, by (4.16)

$$\begin{aligned}
\int_X f\widetilde{g}_n\,\mathrm{d}\mu &= \sum_{j=1}^{k} \delta_j \int_{E_j} f\,\mathrm{d}\mu = \sum_{j=1}^{k} \delta_j \int_0^{\mu(D_j)} f^*(t)\,\mathrm{d}t \\
&= \int_0^{\infty} \sum_{j=1}^{k} f^*(t)\delta_j \chi_{[0,\mu(D_j))}(t)\,\mathrm{d}t \\
&= \int_0^{\infty} f^*(t) g_n^*(t)\,\mathrm{d}t.
\end{aligned}$$

Thus, we have proved (4.15) for g_n, $n = 1,2,\ldots$ and by Theorem 4.3 (f), Theorem 4.5 (d), the monotone convergence theorem and the fact that $\widetilde{g}$ and g are equimeasurable, the general case follows.

Now, let $\mu(X) = \infty$, μ be nonatomic or completely atomic where all atoms are of equal measure and $\alpha > 0$ be a real number such that

$$\alpha < \int_0^{\infty} f^*(t) g^*(t)\,\mathrm{d}t.$$

Since f and g are assumed to be positive it will suffice to show that there exists a positive function $\widetilde{g}$ equimeasurable with g such that

$$\alpha < \int_X f\widetilde{g}\,\mathrm{d}\mu.$$

Because μ is a $\sigma-$finite measure we can find a sequence of pairwise disjoint sets $X_1, X_2, \ldots$, such that $\mu(X_n) < \infty$, $n = 1, 2, \ldots$, and $X = \bigcup_{n=1}^{\infty} X_n$.

We can therefore find two sequences of positive functions $\{f_n\}_{n\in\mathbb{N}}$ and $\{g_n\}_{n\in\mathbb{N}}$ such that $f_n \uparrow f$, $g_n \uparrow g$ and both f_n and g_n have support in X_n, $n = 1, 2, \ldots$. Since

$$\int_0^\infty f_n^*(t)g_n^*(t)\,\mathrm{d}t \to \int_0^\infty f^*(t)g^*(t)\,\mathrm{d}t,$$

by Theorem 4.3 (f), Theorem 4.5 (d) and the monotone convergence theorem it follows that there exists an integer $N \geq 1$ such that

$$0 < \int_0^\infty f_N^*(t)g_N^*(t)\,\mathrm{d}t.$$

Then $(X_N, \mathscr{A}, \mu)$ is a finite measure space which is either nonatomic or completely atomic with all atoms of equal measure. We can therefore apply the first part of the proof and find a positive function $\overline{h}$ on X_N equimeasurable with $g\chi_{X_N}$ such that

$$\int_{X_N} f\overline{h}\,\mathrm{d}\mu = \int_0^\infty (f\chi_{X_N})^*(t)(g\chi_{X_N})^*(t)\,\mathrm{d}t.$$

Since $f_N \leq f\chi_{X_N}$ and $g_N \leq g\chi_{X_N}$ by the construction of $\{f_n\}$ and $\{g_n\}$ it follows that

$$\begin{aligned}\alpha < \int_0^\infty f_N^*(t)g_N^*(t)\,\mathrm{d}t &\leq \int_0^\infty (f\chi_{X_N})^*(t)(g\chi_{X_N})^*(t)\,\mathrm{d}t \\ &= \int_{X_N} f\overline{h}\,\mathrm{d}\mu = \int_X fh\,\mathrm{d}\mu,\end{aligned}$$

where

$$h(x) = \begin{cases} \overline{h}(x) & \text{if } x \in X_N; \\ 0 & \text{if } x \notin X_N. \end{cases}$$

Thus, if we take

$$\widetilde{g}(x) = h(x)\chi_{X_N}(x) + g(x)\chi_{X\setminus X_N}(x), \quad x \in X,$$

then $\widetilde{g}$ is equimeasurable with g and $h \leq \widetilde{g}$.

Hence

$$\alpha < \int_X fh \, d\mu \leq \int_X f\widetilde{g} \, d\mu,$$

which complete the proof. □

We now introduce the notion of *maximal function of* f^*.

Definition 4.24. Let $f \in \mathfrak{F}(X,\mathscr{A})$. By f^{**} we denote the *maximal function of* f^* defined by

$$f^{**}(t) = \frac{1}{t}\int_0^t f^*(s)\,ds, \tag{4.17}$$

for $t > 0$. ⊘

Some elementary attributes of the maximal function are listed below.

Theorem 4.25. *We have the following properties:*

(a) $f^* \leq f^{**}$;
(b) f^{**} *is decreasing;*
(c) $(f+g)^{**} \leq f^{**} + g^{**}$.

Proof. Property (a) follows directly from the fact that f^* is decreasing, thus

$$f^{**}(t) = \frac{1}{t}\int_0^t f^*(s)\,ds \geq f^*(t)\frac{1}{t}\int_0^t ds = f^*(t).$$

(b) Since f^* is decreasing so $f^*(v) \leq f^*\left(\frac{tv}{s}\right)$ if $0 < t \leq s$, hence

$$f^{**}(s) = \frac{1}{s}\int_0^s f^*(v)\,dv \leq \frac{1}{s}\int_0^s f^*\left(\frac{tv}{s}\right)dv = \frac{1}{t}\int_0^t f^*(u)\,du = f^{**}(t)$$

and so f^{**} is decreasing.

(c) Follows directly from (4.11). □

Theorem 4.26. *Let* $(X,\mathscr{A},\mu)$ *be a totally* σ*−finite measure space and let us take* $f \in L_1(X,\mathscr{A},\mu)$. *Then* *(a)* $\int\limits_X \left(|f|-\lambda\right)^+ \mathrm{d}\mu = \int\limits_0^\infty \left(f^*(t)-\lambda\right)^+ \mathrm{d}t$ *for all* $\lambda > 0$*, where* $(x)^+ = \max\{x,0\}$*;*

(b) $D_f(\lambda) = -\frac{\mathrm{d}}{\mathrm{d}\lambda}\int\limits_X \left(|f|-\lambda\right)^+ \mathrm{d}\mu$*;*

(c) $f^{**}(t) = \inf\limits_{\lambda\geq 0}\left\{\lambda + \frac{1}{t}\int\limits_X (|f|-\lambda)^+ \mathrm{d}\mu\right\}, t > 0.$

Proof. To show item (a), we note that

$$\begin{aligned}\int\limits_X \left(|f|-\lambda\right)^+ \mathrm{d}\mu &= \int\limits_{\{|f|>\lambda\}} \left(|f|-\lambda\right)\mathrm{d}\mu \\ &= \int\limits_{\{|f|>\lambda\}} |f|\,\mathrm{d}\mu - \lambda\mu\Big(\{|f|>\lambda\}\Big) \\ &= \int\limits_0^\infty \chi_{(0,|f|)}(\lambda)|f|\,\mathrm{d}\mu - \lambda\mu\Big(\{|f|>\lambda\}\Big).\end{aligned}$$

Then

$$\frac{\mathrm{d}}{\mathrm{d}\lambda}\int\limits_X \left(|f|-\lambda\right)^+ \mathrm{d}\mu = -\mu\Big(\{|f|>\lambda\}\Big). \tag{4.18}$$

Moreover

$$\int\limits_0^\infty \left(f^*(t)-\lambda\right)^+ \mathrm{d}t = \int\limits_{\{f^*(t)>\lambda\}} f^*(t)\,\mathrm{d}t - \lambda m\Big(\{f^*(t)>\lambda\}\Big),$$

and thus

$$\frac{\mathrm{d}}{\mathrm{d}\lambda}\int\limits_0^\infty \left(f^*(t)-\lambda\right)^+ \mathrm{d}t = -m\Big(\{f^*(t)>\lambda\}\Big), \tag{4.19}$$

since

$$m\Big(\{f^*(t)>\lambda\}\Big) = \mu\Big(\{|f|>\lambda\}\Big),$$

then

$$\int\limits_X \left(|f|-\lambda\right)^+ \mathrm{d}\mu = \int\limits_0^\infty \left(f^*(t)-\lambda\right)^+ \mathrm{d}t. \tag{4.20}$$

(b) By (4.18) we have

$$D_f(\lambda) = -\frac{\mathrm{d}}{\mathrm{d}\lambda} \int_X \left(|f| - \lambda\right)^+ \mathrm{d}\mu.$$

(c) Observe that

$$\int_0^t \left(f^*(s) - \lambda\right)^+ \mathrm{d}s \le \int_0^\infty \left(f^*(s) - \lambda\right)^+ \mathrm{d}s,$$

thus by part (a) we have

$$\int_0^t f^*(s)\,\mathrm{d}s \le \lambda t + \int_0^\infty \left(f^*(s) - \lambda\right)^+ \mathrm{d}s = \lambda t + \int_X \left(|f| - \lambda\right)^+ \mathrm{d}\mu.$$

Then

$$f^{**}(t) = \frac{1}{t}\int_0^t f^*(s)\,\mathrm{d}s \le \lambda + \frac{1}{t}\int_X \left(|f| - \lambda\right)^+ \mathrm{d}\mu,$$

from this we get

$$f^{**}(t) \le \inf_{\lambda \ge 0} \left\{ \lambda + \frac{1}{t}\int_X \left(|f| - \lambda\right)^+ \mathrm{d}\mu \right\}. \tag{4.21}$$

On the other hand, assume that $\mu(E) = t$ for some $E \in \mathscr{A}$. In particular, let us take $E = \{x : |f(x)| > \lambda\}$. Since

$$\int_X |fg|\,\mathrm{d}\mu \le \int_0^\infty f^*(s) g^*(s)\,\mathrm{d}s,$$

if $g = \chi_E$, then

$$\int_E |f|\,\mathrm{d}\mu \le \int_0^t f^*(s)\,\mathrm{d}s,$$

thus

$$\int_E |f|\,\mathrm{d}\mu - \lambda\mu(E) \le \int_0^t f^*(s)\,\mathrm{d}s - \lambda\mu(E)$$

$$\int_E \left(|f| - \lambda\right)\mathrm{d}\mu \le \int_0^t f^*(s)\,\mathrm{d}s - \lambda t$$

$$\lambda + \frac{1}{t}\int\limits_X \left(|f| - \lambda\right)^+ \mathrm{d}\mu \le \frac{1}{t}\int\limits_0^t f^*(s)\,\mathrm{d}s$$

$$\inf_{\lambda>0}\left\{\lambda + \frac{1}{t}\int\limits_X \left(|f| - \lambda\right)^+ \mathrm{d}\mu\right\} \le f^{**}(t). \tag{4.22}$$

By (4.21) and (4.22) the result holds. □

As a corollary we obtain.

Corollary 4.27 *Let $f, g \in L_1(X, \mathscr{A}, \mu)$. Then for all $t > 0$*

$$t|f^{**} - g^{**}| \le \int\limits_X \big||f| - |g|\big|\,\mathrm{d}\mu \le \int\limits_X |f - g|\,\mathrm{d}\mu.$$

Proof. Let us define

$$\alpha_f(\lambda) = \lambda + \frac{1}{t}\int\limits_X \left(|f| - \lambda\right)^+ \mathrm{d}\mu.$$

Then

$$t\Big(\alpha_f(\lambda) - \alpha_g(\lambda)\Big) = \int\limits_X \left(|f| - \lambda\right)^+ \mathrm{d}\mu - \int\limits_X \left(|g| - \lambda\right)^+ \mathrm{d}\mu,$$

thus

$$\begin{aligned} t|\alpha_f(\lambda) - \alpha_g(\lambda)| &\le \int\limits_X \left(|f| - \lambda\right)^+ \mathrm{d}\mu - \int\limits_X \left(|g| - \lambda\right)^+ \mathrm{d}\mu \\ &\le \int\limits_X \big||f| - |g|\big|\,\mathrm{d}\mu. \end{aligned}$$

From this, we have

$$t|\inf \alpha_f(\lambda) - \inf \alpha_g(\lambda)| \le \sup |\alpha_f(\lambda) - \alpha_g(\lambda)| \le \int\limits_X \big||f| - |g|\big|\,\mathrm{d}\mu,$$

yielding

$$t|f^{**}(t) - g^{**}(t)| \le \int\limits_X \big||f| - |g|\big|\,\mathrm{d}\mu \le \int\limits_X |f - g|\,\mathrm{d}\mu.$$

□

Corollary 4.28 *If $\{f_n\}_{n\in\mathbb{N}}$ and f are integrable on X and*

$$\lim_{n\to\infty}\int_X |f_n(x)-f(x)|\,\mathrm{d}\mu = 0.$$

Then

$$\lim_{n\to\infty} f_n^*(t) = f^*(t).$$

at every point of continuity of $f^*(t)$.

Proof. By Corollary 4.27 we have

$$|f_n^{**}(t)-f^{**}(t)| \le \int_X |f_n(x)-f(x)|\,\mathrm{d}\mu \to 0,$$

as $n\to\infty$. From this we also have

$$\lim_{n\to\infty}\int_0^t f_n^*(s)\,\mathrm{d}s = \int_0^t f^*(s)\,\mathrm{d}s.$$

On the other hand, using the monotonicity of f_n^*, for any $x_0\in X$ and any $r>0$, we set

$$\frac{1}{r}\int_{x_0}^{x_0+r} f_n^*(t)\,\mathrm{d}t \le f_n^*(x_0) \le \frac{1}{r}\int_{x_0-r}^{x_0} f_n^*(t)\,\mathrm{d}t.$$

Passing to the limit as $n\to\infty$, we derive

$$\frac{1}{r}\int_{x_0}^{x_0+r} f^*(t)\,\mathrm{d}t \le \liminf_{n\to\infty} f_n^*(x_0) \le \limsup_{n\to\infty} f_n^*(x_0) \le \frac{1}{r}\int_{x_0-r}^{x_0} f_n^*(t)\,\mathrm{d}t.$$

Next

$$\lim_{r\to\infty}\frac{1}{r}\int_{x_0}^{x_0+r} f^*(t)\,\mathrm{d}t \le \liminf_{n\to\infty} f_n^*(x_0) \le \limsup_{n\to\infty} f_n^*(x_0) \le \lim_{r\to\infty}\frac{1}{r}\int_{x_0-r}^{x_0} f^*(t)\,\mathrm{d}t,$$

which entails

$$f(x_0) \le \liminf_{n\to\infty} f_n^*(x_0) \le \limsup_{n\to\infty} f_n^*(x_0) \le f^*(x_0).$$

Finally

$$\lim_{n\to\infty} f_n^*(x_0) = f^*(x_0).$$

□

We end this section with a result that will be used in Theorem 6.23 in discussing duality in Lorentz spaces. The result shows under what conditions a function in $[0,\infty)$ coincides with the decreasing rearrangement of a measurable function.

Theorem 4.29. *Let $(X,\mathscr{A},\mu)$ be nonatomic measure space. Let φ be a right continuous and decreasing function on $[0,\infty)$. Then there exists a measurable function f on X with $f^*(t)=\varphi(t)$ for all $t>0$.*

Proof. Let us consider the following function

$$\sum_{k=1}^{n} c_k \chi_{E_k}$$

where $E_k \in \mathscr{A}$, $\mu(E_k)>0$ and $E_j \cap E_k = \emptyset$ if $j \neq k$. We can choose $c_1 > c_2 > c_3 > \ldots > c_n$, $c_{n+1}=0$. Let $d_k = \mu(E_1)+\ldots+\mu(E_k)$, $1 \le k \le n$ and define $d_0 = 0$. Then, the distribution function $D_f(\lambda)$ has the form

$$D_f(\lambda) = \begin{cases} d_k \text{ if } c_{k+1} \le \lambda \le c_k,\ 1 \le k \le n \\ 0 \ \text{ if } c_n \le \lambda. \end{cases}$$

It follows that

$$f^*(t) = \begin{cases} c_k \text{ if } \quad d_{k-1} \le t \le d_k,\ 1 \le k \le n \\ 0 \ \text{ if } \quad d_n \le t. \end{cases}$$

Next, let us write

$$\varphi(t) = \begin{cases} c_k \text{ if } \quad d_{k-1} \le t \le d_k \\ 0 \ \text{ if } \quad d_n \le t. \end{cases}$$

Note that φ is a right-continuous function and if $t_0 \le t$, then $\varphi(t) \le \varphi(t_0)$ which means that φ is a decreasing function, then

$$f^*(t) = \varphi(t).$$

for all $t>0$. For general continuous functions we use approximation. □

4.3 Rearrangement of the Fourier Transform

In this section we will take the definition of the n-dimensional Fourier transform of f as

$$\mathscr{F}f(x) = \widehat{f}(x) = \int_{\mathbb{R}^n} \mathrm{e}^{-\mathrm{i}x\cdot t} f(t)\,\mathrm{d}t.$$

We use the common notation $\widehat{f}$ whenever possible.

We begin with an estimate of the rearrangement of a function that arise as the Fourier transform of a characteristic function

Lemma 4.30. *Let* $f(x) = \operatorname{sinc}(x)$ *(where* $\operatorname{sinc}(x) = \sin(x)/x$ *is the so-called* cardinal sine function*), then*

$$f^*(t) \geq (3\pi + t)^{-1}.$$

Proof. We estimate the distribution function $D_f(\lambda)$ of f as follows. For $\lambda > 0$,

$$\begin{aligned}
D_f(\lambda) &= m\left(\{x : \left|\frac{\sin x}{x}\right| > \lambda\}\right) \\
&= m\left(\{x : |\sin x| > \lambda x\}\right) \\
&= m\left(\{x > 0 : |\sin x| > \lambda x\}\right) \\
&= 2\sum_{n=1}^{\infty} m\left(\{x \in \left((n-1)\pi, n\pi\right) : |\sin x| > \lambda x\}\right).
\end{aligned}$$

For $x \in \left((n-1)\pi, n\pi\right)$ the condition $|\sin x| > \lambda x$ is weaker than the condition $|\sin x| > n\pi\lambda$, thus we have

$$\begin{aligned}
D_f(\lambda) &\geq 2\sum_{n=1}^{\infty} m\left(\{x \in \left((n-1)\pi, n\pi\right) : |\sin x| > n\pi\lambda\}\right) \\
&= 2\sum_{n=1}^{\infty} m\left(\{x \in \left(0, \pi\right) : \sin x > n\pi\lambda\}\right) \\
&= 4\sum_{n=1}^{\infty} m\left(\{x \in \left(0, \pi/2\right) : \sin x > n\pi\lambda\}\right),
\end{aligned}$$

using the symmetry of $\sin x$. Since the condition $\sin x > n\pi\lambda$ is never satisfied for $n\pi\lambda > 1$, we may restrict the sum to those n for which $n\pi\lambda \leq 1$. To this end, we let N be the integer satisfying $\frac{1}{\pi\lambda} - 1 < N < \frac{1}{\pi\lambda}$.

Also, $\sin x \geq \frac{2x}{\pi}$ for $0 \leq x \leq \pi/2$ that is,

$$\left\{x \in \left(0, \pi/2\right) : \frac{2x}{\pi} > \pi\lambda\right\} \subseteq \left\{x \in \left(0, \pi/2\right) : \sin x > \pi n\lambda\right\},$$

we have

$$D_f(\lambda) \geq 4\sum_{n=1}^{N} m\left(\{x \in \left(0, \pi/2\right) : \sin x > n\pi\lambda\}\right)$$

$$\begin{aligned}
&\geq 4\sum_{n=1}^{N} m\left(\{x\in\left(0,\pi/2\right):\frac{2x}{\pi}>n\pi\lambda\}\right)\\
&=4\sum_{n=1}^{N} m\left(\frac{n\lambda\pi^2}{2},\frac{\pi}{2}\right)\\
&=4\sum_{n=1}^{N}\left(\frac{\pi}{2}-\frac{n\lambda\pi^2}{2}\right)\\
&=2\pi N-\lambda\pi^2(N+1)N\\
&=2\pi\left(N-\frac{\lambda\pi N(N+1)}{2}\right).
\end{aligned}$$

The definition of N completes the estimate of $D_f(\lambda)$.

$$D_f(\lambda)\geq 2\pi\left(\frac{1}{\pi\lambda}-1-\frac{\pi\lambda}{2}\frac{1}{\pi\lambda}\left(\frac{1}{\pi\lambda}+1\right)\right)=\frac{1}{\lambda}-3\pi.$$

Now, for $t>0$, we get

$$f^*(t)=\inf\{\lambda:D_f(\lambda)\leq t\}\geq\inf\{\lambda:\frac{1}{\lambda}-3\pi\leq t\}=\left(3\pi+t\right)^{-1}.$$

□

Corollary 4.31 *If $z>0$ and $f=\chi_{(0,1/z)}$, then $(\widehat{f}\,)^*(t)\geq\left(3\pi z+t/2\right)^{-1}$.*

Proof. The (one-dimensional) Fourier transform of f is

$$\widehat{f}(x)=\int_0^{1/z}e^{-ixt}\,\mathrm{d}t=\frac{e^{-ix/z}-1}{-ix},$$

if $F(x)=\frac{\sin x}{x}$ as in Lemma 4.30 we have

$$\begin{aligned}
|\widehat{f}(x)|&=\left|\frac{1}{z}\frac{e^{-ix/2z}}{x/2z}\frac{e^{ix/2z}-e^{-ix/2z}}{2i}\right|\\
&=\frac{1}{2}\left|\frac{\sin(x/2z)}{x/2z}\right|=\frac{1}{z}\left|F(x/2z)\right|
\end{aligned}$$

since we are taking the rearrangement with respect to Lebesgue measure, it respect dilation. That is, if $g_a(x)=g(ax)$, then $g_a^*(t)=g^*(|a|t)$. These properties together with Theorem 4.5 (e) and Lemma 4.30 show that

$$(\widehat{f})^*(t)=\frac{1}{z}F^*(t/2z)\geq\frac{1}{z}\frac{1}{3\pi+t/2z}=\frac{1}{3\pi z+t/2}.$$

□

Lemma 4.32. *Suppose that $f : \mathbb{R} \to \mathbb{C}$ is a compactly supported L_1-function and k is a positive integer. For any $\varepsilon > 0$ there exists a compactly supported L_1-function g such that $g^*(t) = f^*(t/k)$ for $t > 0$ and*

$$(\widehat{f})^*(t/k) - \varepsilon \leq (\widehat{g})^*(t) \leq (\widehat{f})^*(t/k) + \varepsilon \quad \textit{for} \quad t > 0. \tag{4.23}$$

Proof. We show that for T and N sufficiently large,

$$g(t) = \sum_{j=1}^{k} e^{ijNt} f(t + jT)$$

will do. It is clear that such a g is compactly supported and in L_1. Choose T so large that the supports of $f(t + jT)$, $j = a, \ldots, k$ are disjoint. Then no matter what N is for all $\lambda > 0$ we have

$$m\Big(\{t : |g(t)| > \lambda\}\Big) = \sum_{j=1}^{k} m\Big(\{t : |f(t + jT)| > \lambda\}\Big) = k\, m\Big(\{t : |f(t)| > \lambda\}\Big)$$

using the translation invariance of Lebesgue measure. We use this to express the rearrangement of g in terms of the rearrangement of f.

$$\begin{aligned} g^*(t) &= \inf\Big\{\lambda : m\Big(\{s : |g(s)| > \lambda\}\Big) \leq t\Big\} \\ &= \inf\Big\{\lambda : m\Big(\{s : |f(s)| > \lambda\}\Big) \leq t/k\Big\} \\ &= f^*(t/k). \end{aligned}$$

Now, we turn to the Fourier transform of g and the choice of N. By the Riemann Lebesgue lemma (cf. Appendix C, p. 445) we have

$$\lim_{|x| \to \infty} \big|\widehat{f}(x)\big| = 0,$$

so, we may choose N so large that

$$\big|\widehat{f}(x)\big| < \frac{\varepsilon}{k}, \quad \text{whenever} \quad |x| \geq N/2.$$

Since

$$\widehat{g}(x) = \sum_{j=1}^{k} e^{i(x - jN)jT} \widehat{f}(x - jT),$$

we see that if $x \in \big(jN - N/2, jN + N/2\big)$ for some j then only the jth term of the sum can contribute more than ε/k, thus

$$\big|\widehat{f}(x - jN)\big| - \varepsilon \leq \big|\widehat{g}(x)\big| \leq \big|\widehat{f}(x - jN)\big| + \varepsilon \tag{4.24}$$

and if $x \notin \left(jN - N/2, jN + N/2\right)$ for any j then none of the terms in the sum can contribute more than ε/k so $\left|\widehat{g}(x)\right| < \varepsilon$.

Thus, for $\lambda > 0$ we have

$$\begin{aligned}
m\Big(\{x : |\widehat{g}(x)| > \lambda\}\Big) \\
&= \sum_{j=1}^{k} m\Big(\{x \in \left(jN - N/2, jN + N/2\right) : |\widehat{g}(x)| > \lambda\}\Big) \\
&\leq \sum_{j=1}^{k} m\Big(\{x \in \left(jN - N/2, jN + N/2\right) : |\widehat{f}(x - jN)| > \lambda - \varepsilon\}\Big) \\
&= \sum_{j=1}^{k} m\Big(\{x \in \left(-N/2, N/2\right) : |\widehat{f}(x)| > \lambda - \varepsilon\}\Big) \\
&\leq km\Big(\{x : |\widehat{f}(x)| > \lambda - \varepsilon\}\Big).
\end{aligned}$$

This implies that if $(\widehat{g})^*(t) > \varepsilon$ then

$$\begin{aligned}
(\widehat{g})^*(t) &= \inf\left\{\lambda : m\Big(\{x : |\widehat{g}(x)| > \lambda\}\Big) \leq t\right\} \\
&\leq \inf\left\{\lambda : m\Big(\{x : |\widehat{f}(x)| > \lambda - \varepsilon\}\Big) \leq t/k\right\} \\
&= \inf\left\{\lambda - \varepsilon : m\Big(\{x : |\widehat{f}(x)| > \lambda - \varepsilon\}\Big) \leq t/k\right\} + \varepsilon \\
&= \left(\widehat{f}\right)^*(t/k) + \varepsilon.
\end{aligned}$$

Of course, if $(\widehat{g})^*(t/k) \leq \varepsilon$, then we also have

$$\left(\widehat{g}\right)^*(t) \leq \left(\widehat{f}\right)^*(t/k) + \varepsilon,$$

thus, we have established the second inequality in (4.24).

To prove the first inequality in (4.24) we observe that for all $\lambda > 0$ (4.32) implies that

$$\begin{aligned}
m\Big(\{x : |\widehat{g}(x)| > \lambda\}\Big) &\geq \sum_{j=1}^{k} m\Big(\{x \in \left(jN - N/2, jN + N/2\right) : |\widehat{g}(x)| > \lambda\}\Big) \\
&\geq \sum_{j=1}^{k} m\Big(\{x \in \left(jN - N/2, jN + N/2\right) : |\widehat{f}(x - jN)| > \lambda + \varepsilon\}\Big)
\end{aligned}$$

$$= \sum_{j=1}^{k} m\Big(\{x \in \big(-N/2, N/2\big) : |\widehat{f}(x)| > \lambda + \varepsilon\}\Big)$$
$$\leq km\Big(\{x : |\widehat{f}(x)| > \lambda + \varepsilon\}\Big),$$

where the last equality uses the fact that

$$\big|\widehat{f}(x)\big| < \varepsilon \quad \text{for} \quad |x| \geq N/2.$$

Now

$$\begin{aligned}
(\widehat{g})^*(t) &= \inf\Big\{\lambda : m\Big(\{x : |\widehat{g}(x)| > \lambda\}\Big) \leq t\Big\} \\
&\geq \inf\Big\{\lambda : m\Big(\{x : |\widehat{f}(x)| > \lambda + \varepsilon\}\Big) \leq t/k\Big\} \\
&= \inf\Big\{\lambda + \varepsilon : m\Big(\{x : |\widehat{f}(x)| > \lambda + \varepsilon\}\Big) \leq t/k\Big\} - \varepsilon \\
&\geq \inf\Big\{\lambda : m\Big(\{x : |\widehat{f}(x)| > \lambda\}\Big) \leq t/k\Big\} - \varepsilon \\
&= \Big(\widehat{f}\Big)^*(t/k) - \varepsilon,
\end{aligned}$$

as required. This completes the proof. □

Corollary 4.33 *Given $z > 0$, $r > 0$, and $\varepsilon > 0$ there exists a compactly supported L_1-function $g : \mathbb{R} \to \mathbb{C}$ such that*

$$g^* = \chi_{\left[0, \frac{1}{z}\right]} \quad \textit{and} \quad (\widehat{g})^*(t) + \varepsilon \geq \left(3\pi(r+1)z + \frac{y}{2r}\right)^{-1}.$$

Proof. Let k be the positive integer that satisfies $k - 1 < r \leq k$ and set $f = \chi_{(0, \frac{1}{kz})}$.

Choose g by Lemma 4.32 so that

$$g^*(t) = f^*(t/k) = \chi_{[0, \frac{1}{kz})}(t/k) = \chi_{[0, \frac{1}{z})}(t)$$

and

$$\Big(\widehat{g}\Big)^*(t) \geq \Big(\widehat{f}\Big)^*(t/k) - \varepsilon \geq \left(3\pi kz + \frac{t}{2k}\right)^{-1} - \varepsilon \geq \left(3\pi(r+1)z + \frac{t}{2r}\right)^{-1} - \varepsilon.$$

Here we have used Corollary 4.31 to estimate $\Big(\widehat{f}\Big)^*$. □

We recall that the *convolution of f and g*, denoted by $f * g$, is given by

$$(f*g)(x) = \int_{\mathbb{R}} f(x-y)g(y)\,dy.$$

For a thorough study of the convolution, see Section 11.1.

Theorem 4.34. *Let f and g be two functions in $\mathfrak{F}(\mathbb{R}^n, \mathscr{L})$ where $\sup_{x\in\mathbb{R}^n} f(x) \le M$ and f vanishes outside a measurable set E with $m(E) = s$. Let $h = f*g$. Then, for $t > 0$*

$$h^{**}(t) \le Msg^{**}(s)$$

and

$$h^{**}(t) \le Msg^{**}(t).$$

Proof. For $\alpha > 0$, define

$$g_\alpha(x) = \begin{cases} g(x) & \text{if } |g(x)| \le \alpha \\ \alpha \operatorname{sgn}\Big(g(x)\Big) & \text{if } |g(x)| > \alpha. \end{cases}$$

Define g^α by the equation

$$g^\alpha(x) = g(x) - g_\alpha(x).$$

Then, define functions h_1 and h_2 by

$$h = f*g = f*g_\alpha + f*g^\alpha =: h_1 + h_2.$$

From elementary estimates involving the convolution we obtain

$$\sup_{x\in\mathbb{R}^n}\{h_2(x)\} = \sup_{x\in E}\{f*g^\alpha(x)\} \le \left(\sup_{x\in E}\{f(x)\}\right)\|g^\alpha\|_1 \le M\int_\alpha^\infty D_g(\lambda)\,\mathrm{d}\lambda.$$

Since $g^\alpha(x) = 0$ whenever $|g(x)| \le \alpha$. Also

$$\sup_{x\in\mathbb{R}^n}\{h_1(x)\} = \sup_{x\in E}\{f*g_\alpha(x)\} \le \|f\|_1\left(\sup_{x\in E}\{g_\alpha(x)\}\right) \le Ms\alpha$$

and

$$\|h_2\|_1 \le \|f\|_1\|g^\alpha\|_1 \le Ms\int_\alpha^\infty D_g(\lambda)\,\mathrm{d}\lambda.$$

Now set $\alpha = g^*(s)$ in (4.25) and (4.25) and obtain

$$h^{**}(t) = \frac{1}{t}\int_0^t h^*(y)\,\mathrm{d}y \le \|h\|_\infty$$

$$\begin{aligned}
&\leq \|h_1\|_\infty + \|h_2\|_\infty \\
&\leq Msg^*(s) + M \int_{g^*(s)}^{\infty} D_g(\lambda)\,\mathrm{d}\lambda \\
&= M\left[sg^*(s) + \int_{g^*(s)}^{\infty} D_g(\lambda)\,\mathrm{d}\lambda\right] \\
&= Msg^{**}(s).
\end{aligned}$$

The last equality follows from Theorem 4.34 and thus, the first inequality of the lemma is established.

To prove the second inequality, set $\alpha = g^*(t)$ and use (4.25) and (4.25) to obtain

$$\begin{aligned}
th^{**}(t) = \int_0^t h^*(y)\,\mathrm{d}y &\leq \int_0^t h_1^*(y)\,\mathrm{d}y + \int_0^t h_2^*(y)\,\mathrm{d}y \\
&\leq t\|h_1\|_\infty + \int_0^\infty h_2^*(y)\,\mathrm{d}y \\
&= t\|h_1\|_\infty + \|h_2\|_1 \\
&\leq tMsg^*(t) + Ms \int_{g^*(t)}^{\infty} D_g(\lambda)\,\mathrm{d}\lambda \\
&= Ms\left[tg^*(t) + \int_{g^*(t)}^{\infty} D_g(\lambda)\,\mathrm{d}\lambda\right] \\
&= Mstg^*(t),
\end{aligned}$$

and our conclusion follows by dividing by t. □

Theorem 4.35. *If h, f, and g are in $\mathfrak{F}\left(\mathbb{R}^n, \mathscr{L}\right)$ such that $h = f * g$, then for any $t > 0$*

$$h^{**}(t) \leq tf^{**}(t)g^{**}(t) + \int_t^\infty f^*(u)g^*(u)\,\mathrm{d}u.$$

Proof. Fix $t > 0$. Select a doubly infinite sequence $\{y_n\}_{n\in\mathbb{N}}$ whose indices ranges from $-\infty$ to $+\infty$ such that

$$y_0 = f^*(t), \quad y_n = y_{n+1}, \quad \lim_{n\to\infty} y_n = \infty, \quad \lim_{n\to-\infty} y_n = 0.$$

Let

$$f(z) = \sum_{n=-\infty}^{\infty} f_n(z).$$

Where

$$f_n(z) = \begin{cases} 0 & \text{if } |f(z)| \le y_{n-1} \\ f(z) - y_{n-1}\operatorname{sgn} f(z) & \text{if } y_{n-1} < |f(z)| \le y_n \\ y_n - y_{n-1}\operatorname{sgn} f(z) & \text{if } y_n < |f(z)|. \end{cases}$$

Clearly, the series converges absolutely and therefore,

$$\begin{aligned} h = f * g &= \left(\sum_{n=-\infty}^{\infty} f_n \right) * g \\ &= \left(\sum_{n=-\infty}^{0} f_n \right) * g + \left(\sum_{n=1}^{\infty} f_n \right) * g \\ &= h_1 + h_2. \end{aligned}$$

with

$$h^{**}(t) \le h_1^{**}(t) + h_2^{**}(t).$$

To value $h_2^{**}(t)$ we use the second inequality of Theorem 4.34 with $E_n = \{z : |f(z)| > y_{n-1}\}$ and $\alpha = y_n - y_{n-1}$ to obtain

$$\begin{aligned} h_2^{**}(t) &\le \sum_{n=1}^{\infty} (y_n - y_{n-1}) D_f(y_{n-1}) g^{**}(t) \\ &= g^{**}(t) \sum_{n=1}^{\infty} D_f(y_{n-1})(y_n - y_{n-1}). \end{aligned}$$

The series on the right is an infinite Riemann sum for the integral

$$\int_{f^*(t)}^{\infty} D_f(y)\, \mathrm{d}y,$$

and provides an arbitrarily close approximation with an appropriate choice of the sequence $\{y_n\}_{n \in \mathbb{N}}$. Therefore,

$$h_2^{**}(t) \le g^{**}(t) \int_{f^*(t)}^{\infty} D_f(y)\, \mathrm{d}y. \tag{4.25}$$

By the first inequality of Theorem 4.34.

$$h_1^{**}(t) \le \sum_{n=1}^{\infty} (y_n - y_{n-1}) D_f(y_{n-1}) g^{**}\Big(D_f(y_{n-1})\Big).$$

The sum on the right is an infinite Riemann sum tending (with proper choice of y_n) to the integral

$$\int_0^{f^*(t)} D_f(y) g^{**}\Big(D_f(y)\Big)\, dy.$$

We shall evaluate the integral by making the substitution $y = f^*(u)$ and then integrating by parts. In order to justify the change of variable in the integral, consider a Riemann sum

$$\sum_{n=1}^{\infty} D_f(y_{n-1}) g^{**}\Big(D_f(y_{n-1})\Big)(y_n - y_{n-1})$$

that provides a close approximation to

$$\int_0^{f^*(t)} D_f(y) g^{**}\Big(D_f(y)\Big)\, dy.$$

By adding more points to the Riemann sum if necessary, we may assume that the left-hand end point of each interval on which D_f is constant is included among the y_n. Then, the Riemann sum is not changed if each y_n that is contained in the interior of an interval on which D_f is constant is deleted. It is now an easy matter to verify that for each of the remaining y_n there is precisely one element, u_n such that $y_n = f^*(u_n)$ and that $D_f\Big(f^*(u_n)\Big) = u_n$. Thus, we have

$$\sum_{n=1}^{\infty} D_f(y_{n-1}) g^{**}\Big(D_f(y_{n-1})\Big)(y_n - y_{n-1}),$$

which, by adding more points if necessary, provides a close approximate to

$$-\int_t^{\infty} u g^{**}(u)\, d f^*(u).$$

Therefore, we have

$$\begin{aligned} h_1^{**}(t) &\le \int_0^{f^*(t)} D_f(y) g^{**}\Big(D_f(y)\Big)\, dy \\ &= -\int_t^{\infty} u g^{**}(u)\, d f^*(u) \end{aligned}$$

$$= -ug^{**}(u)f^*(u)\Big|_t^\infty + \int_t^\infty f^*(u)g^*(u)\,\mathrm{d}u$$

$$\le tg^{**}(t)f^*(t) + \int_t^\infty f^*(u)g^*(u)\,\mathrm{d}u. \tag{4.26}$$

To justify the integration by parts, let λ be an arbitrary large number and choose u_m such that $t = u_1 \le u_2 \le \ldots \le u_{m+1} = \lambda$.

Observe that

$$\lambda g^{**}(\lambda)f^*(\lambda) - tg^{**}(t)f^*(t)$$
$$= \sum_{n=1}^m u_{n+1}g^{**}(u_{n+1})\Big[f^*(u_{n+1}) - f^*(u_n)\Big] + \sum_{n=1}^m f^*(u_n)\Big[g^{**}(u_{n+1})u_{n+1} - g^{**}(u_n)u_n\Big]$$
$$= \sum_{n=1}^m u_{n+1}g^{**}(u_{n+1})\Big[f^*(u_{n+1}) - f^*(u_n)\Big] + \sum_{n=1}^m f^*(u_n)\left[\int_{u_n}^{u_{n+1}} g^*(s)\,\mathrm{d}s\right]$$
$$\le \sum_{n=1}^m u_{n+1}g^{**}(u_{n+1})\Big[f^*(u_{n+1}) - f^*(u_n)\Big] + \sum_{n=1}^m f^*(u_n)g^*(u_n)[u_{n+1} - u_n].$$

This shows that

$$\lambda g^{**}(\lambda)f^*(\lambda) - tg^{**}(t)f^*(t) \le \int_t^\lambda ug^{**}(u)\,\mathrm{d}f^*(u) + \int_t^\lambda f^*(u)g^*(u)\,\mathrm{d}u.$$

To establish the opposite inequality, write

$$\lambda g^{**}(\lambda)f^*(\lambda) - tg^{**}(t)f^*(t)$$
$$= \sum_{n=1}^m u_n g^{**}(u_n)\Big[f^*(u_{n+1}) - f^*(u_n)\Big] + \sum_{n=1}^m f^*(u_{n+1})\Big[g^{**}(u_{n+1})u_{n+1} - g^{**}(u_n)u_n\Big]$$

$$= \sum_{n=1}^m u_n g^{**}(u_n)\Big[f^*(u_{n+1}) - f^*(u_n)\Big] + \sum_{n=1}^m f^*(u_{n+1})\left[\int_{u_n}^{u_{n+1}} g^*(s)\,\mathrm{d}s\right]$$
$$\ge \sum_{n=1}^m u_n g^{**}(u_n)\Big[f^*(u_{n+1}) - f^*(u_n)\Big] + \sum_{n=1}^m f^*(u_{n+1})g^*(u_{n+1})[u_{n+1} - u_n].$$

Now let $\lambda \to \infty$ to obtain the desired equality. Thus, from (4.25), (4.26), and Theorem 4.34

$$h_1^{**}(t)+h_2^{**}(t) \le g^{**}(t)\left[tf^*(t)+\int_{f^*(t)}^{\infty} D_f(y)\,\mathrm{d}y\right]+\int_t^{\infty} f^*(u)g^*(u)\,\mathrm{d}u$$

$$\le tf^{**}(t)g^{**}(t)+\int_t^{\infty} f^*(u)g^*(u)\,\mathrm{d}u.$$

□

Theorem 4.36. *Under the hypotheses of Theorem 4.35, we have*

$$h^{**}(t) \le \int_t^{\infty} f^{**}(u)g^{**}(u)\,\mathrm{d}u.$$

Proof. We may as well assume the integral on the right is finite and then conclude

$$\lim_{u\to\infty} u f^{**}(u)g^{**}(u)=0. \tag{4.27}$$

By Theorem 4.35 and the fact that $f^* \le f^{**}$, we have

$$h^{**}(t) \le tf^{**}(t)g^{**}(t)+\int_t^{\infty} f^*(u)g^*(u)\,\mathrm{d}u$$

$$\le tf^{**}(t)g^{**}(t)+\int_t^{\infty} f^{**}(u)g^*(u)\,\mathrm{d}u \tag{4.28}$$

Note that since f^* and g^* are nonincreasing

$$\frac{d}{du}f^{**}(u)=\frac{d}{du}\left[\frac{1}{u}\int_0^u f^*(s)\,\mathrm{d}s\right]$$

$$=-\frac{1}{u^2}\int_0^u f^*(s)\,\mathrm{d}s+\frac{1}{u}f^*(u)$$

$$=\frac{1}{u}\left[f^*(u)-\frac{1}{u}\int_0^u f^*(s)\,\mathrm{d}s\right]$$

$$=\frac{1}{u}\left[f^*(u)-f^{**}(u)\right]$$

and

$$\frac{d}{du} u g^{**}(u) = \frac{d}{du}\left[\int_0^u g^*(s)\,ds\right] = g^*(u),$$

for almost all (in fact, all but countably many) u. Since f^{**} and g^{**} are absolutely continuous, we perform integration by parts and employ (4.27) and (4.28) to obtain

$$\begin{aligned} h^{**}(t) &\le t f^{**}(t) g^{**}(t) + \Big[u f^{**}(u) g^{**}(u)\Big]_t^\infty + \int_t^\infty \Big[f^{**}(u) - f^*(u)\Big] g^{**}(u)\,du \\ &= \int_t^\infty \Big[f^{**}(u) - f^*(u)\Big] g^{**}(u)\,du \\ &\le \int_t^\infty f^{**}(u) g^{**}(u)\,du. \end{aligned}$$

□

4.4 Problems

4.37. Try to understand geometrically the meaning of the maximal function f^{**} given by (4.17).

4.38. Let $\big([0,\infty], \mathscr{L}, m\big)$ be the usual Lebesgue measure space and let f be defined as

$$f(x) = \begin{cases} 0, & x = 0, \\ \log\left(\frac{1}{1-x}\right), & 0 < x < 1, \\ \infty, & 1 \le x \le 2, \\ \log\left(\frac{1}{x-2}\right), & 2 < x < 3, \\ 0, & x \ge 3. \end{cases}$$

Show that

(a) $D_f(\lambda) = 1 + 2e^{-\lambda}$.

(b)

$$f^*(t) = \begin{cases} \infty, & 0 \le t \le 1, \\ \log\left(\frac{2}{t-1}\right), & 1 < t < 3, \\ 0, & t \ge 3. \end{cases}$$

Remark: From the above example, we get that even if f is infinite over some nondegenerate interval the distribution and the decreasing rearrangement functions are still well defined.

4.39. Let us take $f : [0,\infty) \to \mathbb{R}$ as

$$f(x) = \begin{cases} x(1-x), & x \in [0,1), \\ 0, & x > 1. \end{cases}$$

Find $f^*(t)$.

4.40. On $\mathbb{R}^n$, let $\delta^\varepsilon(f)(x) = f(\varepsilon x)$, $\varepsilon > 0$, be the dilation operator. Show that
(a) $D_{\delta^\varepsilon(f)}(\lambda) = \varepsilon^{-n} D_f(\lambda)$.
(b) $(\delta^\varepsilon(f))^*(t) = f^*(\varepsilon^n t)$.

4.41. Suppose that μ denotes the Lebesgue measure on the σ-algebra $\mathscr{B}$ of Borel subset of $\mathbb{R}$ and for each $a > 0$ put $f_a(x) = \mathrm{e}^{-a|x|}$ and $g_a(x) = \mathrm{e}^{-ax^2}$ for $x \in \mathbb{R}$.

(a) Calculate D_{f_a} and f_a^* for $a > 0$.
(b) Calculate D_{g_a} and g_a^* for $a > 0$.

4.42. If $f \in L_p(\mathbb{R}^n, \mathscr{A}, m)$, $1 < p < \infty$, prove that

$$\left(\int_0^\infty (f^{**}(t))^p \, \mathrm{d}t \right)^{1/p} \leq \frac{p}{p-1} \left(\int_{\mathbb{R}^n} |f(x)|^p \, \mathrm{d}x \right)^{1/p}.$$

Remark: Try to use a different approach from the techniques used in the proof of Theorems 3.64 and 10.5.

4.43. Prove or disprove the following statements. Let $(X, \mathscr{A}, \mu)$ be a measure space:

1. $f\chi_E$ and $f\chi_{(0,\mu(E))}$ are equimeasurable for any $\mathscr{A}$-measurable set E.
2. Let φ be an absolutely continuous function on $[0,+\infty)$ with $\varphi(0) = 0$. If E is an $\mathscr{A}$-measurable function, then

$$\int_E \varphi(|f|) \, \mathrm{d}\mu = \int_0^{\mu(E)} \varphi(f^*(t)) \, \mathrm{d}t.$$

4.44. Let $(X, \mathscr{A}, \mu)$ be a finite nonatomic measure space, suppose that f belongs to $\mathscr{F}(X, \mathscr{A})$ and let t be any number satisfying $0 \leq t \leq \mu(X)$. Prove that there is a measurable set E_t with $\mu(E_t) = t$ such that

$$\int_{E_t} |f| \, \mathrm{d}\mu = \int_0^t f^*(s) \, \mathrm{d}s.$$

Moreover, the sets E_t can be constructed so as to increase with t:

$$0 \leq s \leq t \leq \mu(X), \text{ then } E_s \subset E_t.$$

Hint: See [1] Lemma 2.5 on p. 46.

4.5 Notes and Bibliographic References

Section 4.3 is largely based on the paper Sinnamon [70].

The decreasing rearrangement of functions was introduced in Hardy and Littlewood [28]. It appeared in book form in Hardy, Littlewood, and Pólya [30]. Exposition on the topic has also been given by Grothendieck [24], Lorentz [43], Day [15], Chong and Rice [7], Ryff [63], among others. For the history of the rearrangement, we refer to the book by Bennett and Sharpley [1] and the references cited there.

The introduction of the function f^{**} is due to Calderón [4].

Chapter 5
Weak Lebesgue Spaces

Mathematics is the most beautiful and most powerful creation of the human spirit.
STEFAN BANACH

Abstract In this chapter we study the so-called *weak Lebesgue spaces* which are one of the first generalizations of the Lebesgue spaces and a prototype of the so-called Lorentz spaces which will be studied in a subsequent chapter. In the framework of weak Lebesgue spaces we will study, among other topics, embedding results, convergence in measure, interpolation results, and the question of normability of the space. We also show a Fatou type lemma for weak Lebesgue spaces as well as the completeness of the quasi-norm. The Lyapunov inequality and the Hölder inequality are shown to hold.

5.1 Weak Lebesgue Spaces

We start with a simple observation that will be used when defining the weak Lebesgue space.

Lemma 5.1. *Let $(X, \mathscr{A}, \mu)$ be a measure space and f be an $\mathscr{A}$-measurable function that satisfies*

$$\mu(\{x \in X : |f(x)| > \lambda\}) \leq \left(\frac{C}{\lambda}\right)^p \tag{5.1}$$

for some $C > 0$. Then

$$\inf\left\{C > 0 : D_f(\lambda) \leq \left(\frac{C}{\lambda}\right)^p\right\} = \left(\sup_{\lambda>0} \lambda^p D_f(\lambda)\right)^{1/p} = \sup_{\lambda>0} \lambda\,(D_f(\lambda))^{1/p},$$

where D_f is the distribution function given by (4.1).

R.E. Castillo, H. Rafeiro, *An Introductory Course in Lebesgue Spaces*, CMS Books in Mathematics, DOI 10.1007/978-3-319-30034-4_5

Proof. Let us define

$$\lambda = \inf\left\{C > 0 : D_f(\alpha) \le \left(\frac{C}{\alpha}\right)^p\right\},$$

and

$$B = \left(\sup_{\alpha>0} \alpha^p D_f(\alpha)\right)^{1/p}.$$

Since f satisfies (5.1) then

$$D_f(\alpha) \le \left(\frac{C}{\alpha}\right)^p,$$

for some $C > 0$, then

$$\left\{C > 0 : D_f(\alpha) \le \left(\frac{C}{\alpha}\right)^p \quad \forall \alpha > 0\right\} \neq \emptyset.$$

On the other hand $\alpha^p D_f(\alpha) \le B^p$, thus $\{\alpha^p D_f(\alpha) : \alpha > 0\}$ is bounded above by B^p and so $B \in \mathbb{R}$.

Therefore

$$\lambda = \inf\left\{C > 0 : D_f(\alpha) \le \left(\frac{C}{\alpha}\right)^p \quad \alpha > 0\right\} \le B. \tag{5.2}$$

Now, let $\varepsilon > 0$, then there exists C such that

$$\lambda \le C < \lambda + \varepsilon,$$

and thus

$$D_f(\lambda) \le \frac{C^p}{\lambda^p} < \frac{(\lambda+\varepsilon)^p}{\lambda^p},$$

from which we get

$$\sup_{\lambda>0} \lambda^p D_f(\lambda) < (\lambda+\varepsilon)^p.$$

By the arbitrariness of $\varepsilon > 0$, we obtain $B < \lambda$ which, together with (5.2), we obtain $B = \lambda$. □

We now introduce the weak Lebesgue space.

Definition 5.2. Let $0 < p < \infty$. The weak Lebesgue space, denoted by weak-L_p or by $L_{(p,\infty)}(X, \mathscr{A}, \mu)$, is defined as the set of all μ-measurable functions f such that

$$\|f\|_{L_{(p,\infty)}} = \|f\|_{(p,\infty)} = \|f \mid L_{(p,\infty)}\| := \sup_{\lambda>0} \lambda \left(D_f(\lambda)\right)^{1/p} < \infty.$$

Two functions will be considered equal if they are equal μ-a.e., as in the case of L_p spaces.

$\oslash$

The weak Lebesgue spaces are larger than the Lebesgue spaces, which is just a restatement of the Chebyshev inequality, namely:

Theorem 5.3 *For any $0 < p < \infty$ and any $f \in L_p$ we have*

$$L_p \subset L_{(p,\infty)}. \tag{5.3}$$

and hence

$$\left\| f \mid L_{(p,\infty)} \right\| \leq \| f \mid L_p \|. \tag{5.4}$$

Proof. If $f \in L_p$, then

$$\lambda^p \mu\left(\{x \in X : |f(x)| > \lambda\}\right) \leq \int_{\{|f|>\lambda\}} |f|^p \mathrm{d}u \leq \int_X |f|^p \mathrm{d}u = \| f \mid L_p \|^p, \; .$$

therefore

$$\mu\left(\{x \in X : |f(x)| > \lambda\}\right) \leq \left(\frac{\| f \mid L_p \|}{\lambda} \right)^p. \tag{5.5}$$

Hence $f \in L_{(p,\infty)}$, which means that (5.3).
Next, from (5.5) we have

$$\left(\sup_{\lambda>0} \{\lambda^p D_f(\lambda)\} \right)^{1/p} \leq \| f \mid L_p \|$$

from which (5.4) follows. $\square$

The inclusion (5.3) is strict. We give a counter-example for a particular measure space. Let $f(x) = x^{-1/p}$ on $(0,\infty)$ with the Lebesgue measure. Note

$$m\left(\left\{ x \in (0,\infty) : \frac{1}{|x|^{1/p}} > \lambda \right\} \right) = m\left(\left\{ x \in (0,\infty) : |x| < \frac{1}{\lambda^p} \right\} \right) = 2\lambda^{-p}.$$

Thus $f \in L_{(p,\infty)}(0,\infty)$, but

$$\int_0^\infty \left(\frac{1}{x^{1/p}} \right)^p \mathrm{d}x = \int_0^\infty \frac{\mathrm{d}x}{x},$$

which is a divergent integral, thus $f \notin L_p(0,\infty)$.

We now show that the functional $f \mapsto \sup_{\lambda>0} \lambda D_f(\lambda)^{1/p}$ satisfies a quasi-triangle inequality and is also a homogeneous functional.

Theorem 5.4 *Let* $f, g \in L_{(p,\infty)}$. *Then*

(a) $\left\| cf \mid L_{(p,\infty)} \right\| = |c| \left\| f \mid L_{(p,\infty)} \right\|$ *for any constant c,*

(b) $\left\| f+g \mid L_{(p,\infty)} \right\| \leq 2 \left(\left\| f \mid L_{(p,\infty)} \right\|^p + \left\| g \mid L_{(p,\infty)} \right\|^p \right)^{1/p}$.

Proof. (a) For $c > 0$ we have

$$\mu\left(\{x \in X : |cf(x)| > \lambda\}\right) = \mu\left(\left\{x \in X : |f(x)| > \frac{\lambda}{c}\right\}\right),$$

in other words, $D_{cf}(\lambda) = D_f\left(\frac{\lambda}{c}\right)$.

We then have

$$\begin{aligned}
\|cf \mid L_{(p,\infty)}\| &= \left(\sup_{\lambda>0} \lambda^p D_{cf}(\lambda)\right)^{1/p} \\
&= \left(\sup_{\lambda>0} \lambda^p D_f\left(\frac{\lambda}{c}\right)\right)^{1/p} \\
&= \left(\sup_{cw>0} c^p w^p D_f(w)\right)^{1/p} \\
&= c\left(\sup_{cw>0} w^p D_f(w)\right)^{1/p},
\end{aligned}$$

which means

$$\|cf \mid L_{(p,\infty)}\| = c\|f \mid L_{(p,\infty)}\|.$$

(b) Note that

$$\left\{x \in X : |f(x)+g(x)| > \lambda\right\} \subseteq \left\{x \in X : |f(x)| > \frac{\lambda}{2}\right\} \cup \left\{x \in X : |g(x)| > \frac{\lambda}{2}\right\}.$$

Hence

$$\begin{aligned}
&\mu\left(\{x \in X : |f(x)+g(x)| > \lambda\}\right) \\
&\qquad \leq \mu\left(\left\{x \in X : |f(x)| > \frac{\lambda}{2}\right\}\right) + \mu\left(\left\{x \in X : |g(x)| > \frac{\lambda}{2}\right\}\right),
\end{aligned}$$

then

$$\lambda^p D_{f+g}(\lambda) \leq \lambda^p D_f\left(\frac{\lambda}{2}\right) + \lambda^p D_g\left(\frac{\lambda}{2}\right)$$

and

$$\lambda^p D_{f+g}(\lambda) \le 2^p \left[\sup_{\lambda>0} \lambda^p D_f(\lambda) + \sup_{\lambda>0} \lambda^p D_g(\lambda) \right],$$

therefore

$$\|f+g \mid L_{(p,\infty)}\| \le 2 \left(\|f \mid L_{(p,\infty)}\|^p + \|g \mid L_{(p,\infty)}\|^p \right)^{1/p},$$

which ends the proof. □

5.2 Convergence in Measure

Next, we discuss some convergence notions.

We now discuss some convergence issues. We will see that the convergence in measure is a more general property than convergence in either L_p or $L_{(p,\infty)}$, namely:

Theorem 5.5 *Let $0 < p \le \infty$ and f_n, f be in $L_{(p,\infty)}$.*

(a) If f_n, f are in L_p and $f_n \to f$ in L_p, then $f_n \to f$ in $L_{(p,\infty)}$.
(b) If $f_n \to f$ in $L_{(p,\infty)}$ then $f_n \xrightarrow{\mu} f$.

Proof. (*a*) Fix $0 < p < \infty$. Theorem 5.3 gives that for all $\varepsilon > 0$ we have:

$$\mu\left(\{x \in X : |f_n(x) - f(x)| > \varepsilon\}\right) \le \frac{1}{\varepsilon^p} \int_X |f_n - f|^p \mathrm{d}\mu$$

$$\varepsilon^p \mu\left(\{x \in X : |f_n(x) - f(x)| > \varepsilon\}\right) \le \|f_n - f \mid L_p\|^p$$

$$\sup_{\lambda>0} \lambda^p D_{f_n - f}(\lambda) \le \|f_n - f \mid L_p\|^p,$$

and thus

$$\|f_n - f \mid L_{(p,\infty)}\| \le \|f_n - f \mid L_p\|.$$

This shows that convergence in L_p implies convergence in weak Lebesgue spaces. The case $p = \infty$ is tautological.

(*b*) Give $\varepsilon > 0$ find an $n_0 \in \mathbb{N}$ such that for $n > n_0$, we have

$$\|f_n - f \mid L_{(p,\infty)}\| = \left(\sup_{\lambda>0} \lambda^p D_{f_n - f}(\lambda) \right)^{1/p} < \varepsilon^{\frac{1}{p}+1},$$

then taking $\lambda = \varepsilon$, we conclude that

$$\varepsilon^p \mu\left(\{x \in X : |f_n(x) - f(x)| > \varepsilon\}\right) < \varepsilon^{p+1},$$

for $n > n_0$, hence $\mu\left(\{x \in X : |f_n(x) - f(x)| > \varepsilon\}\right) < \varepsilon$ for $n > n_0$. □

We now show a sequence which convergences in measure to the zero function, but do not convergence in the $L_{(p,\infty)}$ sense.

Example 5.6. Fix $0 < p < \infty$. On $[0,1]$ define the functions

$$f_{k,j} = k^{1/p}\chi_{\left(\frac{j-1}{k},\frac{j}{k}\right)} \quad k \geq 1, \quad 1 \leq j \leq k.$$

Consider the sequence $\{f_{1,1}, f_{2,1}, f_{2,2}, f_{3,1}, f_{3,2}, f_{3,3}, \ldots\}$.
Observe that

$$m\left(\{x \in [0,1] : f_{k,j}(x) > 0\}\right) = \frac{1}{k},$$

thus

$$\lim_{k\to\infty} m\left(\{x \in [0,1] : f_{k,j}(x) > 0\}\right) = 0,$$

that is $f_{k,j} \xrightarrow{m} 0$.

Likewise, observe that

$$\begin{aligned}\|f_{k,j} \mid L_{(p,\infty)}\| &= \left(\sup_{\lambda>0} \lambda^p m\left(\{x \in [0,1] : f_{k,j}(x) > \lambda\}\right)\right)^{1/p} \\ &\geq \left(\sup_{k\geq 1} \frac{k-1}{k}\right)^{1/p} = 1,\end{aligned}$$

which implies that $f_{k,j}$ does not converge to 0 in $L_{(p,\infty)}$. ⊘

From the previous result and the so-called Riesz Theorem it turns out that every convergent sequence in $L_{(p,\infty)}$ has a subsequence that converges μ-a.e. to the same limit. Due to the importance of Riesz Theorem we will prove it below.

Theorem 5.7 (Riesz). *Let f_n and f be complex-valued measurable functions on a measure space $(X, \mathscr{A}, \mu)$ and suppose $f_n \xrightarrow{\mu} f$. Then some subsequence of f_n converges to f μ-a.e.*

Proof. For all $k = 1, 2, \ldots$ choose inductively n_k such that

$$\mu\left(\{x \in X : |f_n(x) - f(x)| > 2^{-k}\}\right) < 2^{-k}, \tag{5.6}$$

and such that $n_1 < n_2 < \ldots < n_k < \ldots$. Define the sets

$$A_k = \left\{x \in X : |f_{n_k}(x) - f(x)| > 2^{-k}\right\}.$$

Equation (5.6) implies that

$$\mu\left(\bigcup_{k=m}^{\infty} A_k\right) \leq \sum_{k=m}^{\infty} \mu(A_k) \leq \sum_{k=m}^{\infty} 2^{-k} = 2^{1-m}, \tag{5.7}$$

for all $m = 1,2,3,...$ It follows from (5.7) that

$$\mu\left(\bigcup_{k=1}^{\infty} A_k\right) \le 1 < \infty. \tag{5.8}$$

Using (5.7) and (5.8), we conclude that the sequence of the measure of the sets $\{\bigcup_{k=m}^{\infty} A_k\}_{m\in\mathbb{N}}$ converges as $m \to \infty$ to

$$\mu\left(\bigcap_{m=1}^{\infty}\bigcup_{k=m}^{\infty} A_k\right) = 0. \tag{5.9}$$

To finish the proof, observe that the null set in (5.9) contains the set of all $x \in X$ for which $f_{n_k}(x)$ does not converge to $f(x)$. □

In many situations we are given a sequence of functions and we would like to extract a convergent subsequence. One way to achieve this is via the next theorem which is a useful variant of theorem 5.7, relying on the notion of Cauchy sequence in measure.

Theorem 5.8. *Let $(X,\mathscr{A},\mu)$ be a measure space and let $\{f_n\}_{n\in\mathbb{N}}$ be a complex valued sequence on X, that is Cauchy in measure. Then some subsequence of f_n converges μ-a.e.*

Proof. The proof is very similar to that of Theorem 5.7. For all $k = 1,2,\ldots$ choose an increasing sequence n_k inductively such that

$$\mu\left(\{x\in X : |f_{n_k}(x) - f_{n_{k+1}}(x)| > 2^{-k}\}\right) < 2^{-k}. \tag{5.10}$$

Define

$$A_k = \left\{x \in X : |f_{n_k}(x) - f_{n_{k+1}}(x)| > 2^{-k}\right\}.$$

As shown in the proof of Theorem 5.7, from the condition (5.10) we get

$$\mu\left(\bigcap_{m=1}^{\infty}\bigcup_{k=m}^{\infty} A_k\right) = 0, \tag{5.11}$$

for $x \notin \bigcup\limits_{k=m}^{\infty} A_k$ and $i \ge j \ge j_0 \ge m$ (and j_0 large enough) we have

$$|f_{n_i}(x) - f_{n_j}(x)| \le \sum_{l=j}^{i-1} |f_{n_l}(x) - f_{n_{l+1}}(x)| \le \sum_{l=j}^{i} 2^{-l} \le 2^{1-j} \le 2^{1-j_0}.$$

This implies that the sequence $\{f_{n_i}(x)\}_{i\in\mathbb{N}}$ is Cauchy for every x in the set $\left(\bigcup_{k=m}^{\infty} A_k\right)^{\complement}$ and therefore converges for all such x. We define a function

$$f(x) = \begin{cases} \lim_{j \to \infty} f_{n_j}(x) & \text{when } x \notin \bigcap_{m=1}^{\infty} \bigcup_{k=m}^{\infty} A_k, \\ 0 & \text{when } x \in \bigcap_{m=1}^{\infty} \bigcup_{k=m}^{\infty} A_k, \end{cases}$$

which implies that $f_{n_j} \to f$ μ-almost everywhere. □

The next result shows that, under some measure finiteness condition on the function f, if the function f belongs to the weak Lebesgue space $L_{(p,\infty)}$ it will also belong to the Lebesgue space L_q for $q < p$.

Theorem 5.9 *If $f \in L_{(p,\infty)}$ and $\mu(\{x \in X : f(x) \neq 0\}) < \infty$, then $f \in L_q$ for all $q < p$. On the other hand, if $f \in L_{(p,\infty)} \cap L_\infty$ then $f \in L_q$ for all $q > p$.*

Proof. If $p < \infty$, we write

$$\begin{aligned} \int_X |f(x)|^q \mathrm{d}\mu &= q \int_0^\infty \lambda^{q-1} D_f(\lambda) \mathrm{d}\lambda \\ &= q \int_0^1 \lambda^{q-1} D_f(\lambda) \mathrm{d}\lambda + q \int_1^\infty \lambda^{q-1} D_f(\lambda) \mathrm{d}\lambda. \end{aligned}$$

We have that $\mu(\{x \in X : |f(x)| > \lambda\}) \leq C$, since $\mu(\{x \in X : |f(x)| > \lambda\}) \leq \mu(\{x \in X : f(x) \neq 0\})$, from which we get

$$\int_X |f(x)|^q \mathrm{d}\mu \leq qC \int_0^1 \lambda^{q-1} \mathrm{d}\lambda + qC \int_1^\infty \lambda^{q-p-1} \mathrm{d}\lambda = C + \left.\frac{qC\lambda^{q-p}}{q-p}\right|_1^\infty < \infty,$$

in other words, $f \in L_q$.

Let $f \in L_{(p,\infty)} \cap L_\infty$. Then

$$\begin{aligned} \int_X |f(x)|^q \mathrm{d}\mu &= q \int_0^\infty \lambda^{q-1} D_f(\lambda) \mathrm{d}\lambda \\ &= q \int_0^M \lambda^{q-1} D_f(\lambda) \mathrm{d}\lambda + q \int_M^\infty \lambda^{q-1} D_f(\lambda) \mathrm{d}\lambda, \end{aligned}$$

where $M = \operatorname{ess\,sup} |f(x)|$. Note that

$$\mu(\{x \in X : |f(x)| > \lambda\}) = 0 \quad \text{for} \quad \lambda > M,$$

since $f \in L_{(p,\infty)} \cap L_\infty$, therefore

$$q \int_M^\infty \lambda^{q-1} D_f(\lambda) \mathrm{d}\lambda = 0 \quad \text{and} \quad D_f(\lambda) \leq \frac{\|f \mid L_{(p,\infty)}\|^p}{\lambda^p}.$$

From the above considerations, we have

$$\begin{aligned}
\int_X |f(x)|^q \mathrm{d}\mu &= q \int_0^M \lambda^{q-1} D_f(\lambda) \mathrm{d}\lambda \\
&\leq q \|f \mid L_{(p,\infty)}\|^p \int_0^M \lambda^{q-p-1} \mathrm{d}\lambda \\
&= \frac{q \|f \mid L_{(p,\infty)}\|^p M^{q-p}}{q-p},
\end{aligned}$$

which is finite, therefore $f \in L_q$. □

5.3 Interpolation

It is a useful fact that if a function is in $L_p(X,\mu) \cap L_q(X,\mu)$, then it also lies in $L_r(X,\mu)$ for all $p < r < q$. We now show that a similar result holds for the case of weak Lebesgue spaces, namely:

Theorem 5.10 *Let $f \in L_{(p_0,\infty)} \cap L_{(p_1,\infty)}$ with $p_0 < p < p_1$. Then $f \in L_p$.*

Proof. Let us write

$$f = f\chi_{\{|f| \leq 1\}} + f\chi_{\{|f|>1\}} = f_1 + f_2.$$

Observe that $f_1 \leq f$ and $f_2 \leq f$. In particular $f_1 \in L_{(p_0,\infty)}$ and $f_2 \in L_{(p_1,\infty)}$. Also, write that f_1 is bounded and

$$\mu\left(\{x \in X : f_2(x) \neq 0\}\right) = \mu\left(\{x \in X : |f(x)| > 1\}\right) < C < \infty.$$

Therefore by Theorem 5.9, we have $f_1 \in L_p$ and $f_2 \in L_p$. Since L_p is a linear vector space, we conclude that $f \in L_p$. □

The previous result can be improved in the sense that we can explicitly obtain a bound for the L_r norm of a function using the weak Lebesgue norms.

Theorem 5.11 *Let $0 < p < q \leq \infty$ and let f in $L_{(p,\infty)} \cap L_{(q,\infty)}$. Then f is in L_r for all $p < r < q$ and*

$$\|f \mid L_r\| \leq \left(\frac{r}{r-p} + \frac{r}{q-r}\right)^{1/r} \|f \mid L_{(p,\infty)}\|^{\frac{\frac{1}{r}-\frac{1}{q}}{\frac{1}{p}-\frac{1}{q}}} \|f \mid L_{(q,\infty)}\|^{\frac{\frac{1}{p}-\frac{1}{r}}{\frac{1}{p}-\frac{1}{q}}} \tag{5.12}$$

with the suitable interpolation when $q = \infty$.

Proof. Let us take first $q < \infty$. We know that

$$D_f(\lambda) \le \min\left\{ \frac{\|f \mid L_{(p,\infty)}\|^p}{\lambda^p}, \frac{\|f \mid L_{(q,\infty)}\|^q}{\lambda^q} \right\}, \tag{5.13}$$

and set

$$B = \left(\frac{\|f \mid L_{(q,\infty)}\|^q}{\|f \mid L_{(p,\infty)}\|^p} \right)^{\frac{1}{q-p}}. \tag{5.14}$$

By (5.13), (5.14), we have

$$\begin{aligned}
\|f \mid L_r\|^r &= r \int_0^\infty \lambda^{r-1} D_f(\lambda) \mathrm{d}\lambda \\
&\le r \int_0^\infty \lambda^{r-1} \min\left(\frac{\|f \mid L_{(p,\infty)}\|^p}{\lambda^p}, \frac{\|f \mid L_{(q,\infty)}\|^q}{\lambda^q} \right) \mathrm{d}\lambda \\
&= r \int_0^B \lambda^{r-1-p} \|f \mid L_{(p,\infty)}\|^p \mathrm{d}\lambda + r \int_B^\infty \lambda^{r-1-q} \|f \mid L_{(q,\infty)}\|^q \mathrm{d}\lambda \qquad (5.15) \\
&= \frac{r}{r-p} \|f \mid L_{(p,\infty)}\|^p B^{r-p} + \frac{r}{q-r} \|f \mid L_{(q,\infty)}\|^q B^{r-q} \\
&= \left(\frac{r}{r-p} + \frac{r}{q-r} \right) \left(\|f \mid L_{(p,\infty)}\|^p \right)^{\frac{q-r}{q-p}} \left(\|f \mid L_{(q,\infty)}\|^q \right)^{\frac{r-p}{q-p}}.
\end{aligned}$$

Observe that the integrals converge, since $r - p > 0$ and $r - q < 0$.
The case $q = \infty$ is easier. Since $D_f(\lambda) = 0$ for $\lambda > \|f\|_{L_\infty}$ we need to use only the inequality

$$D_f(\lambda) \le \lambda^{-p} \|f \mid L_{(p,\infty)}\|^p,$$

for $\lambda \le \|f\|_{L_\infty}$ in estimating the first integral in (5.15). We obtain

$$\|f \mid L_r\|^r \le \frac{r}{r-p} \|f \mid L_{(p,\infty)}\|^p \|f \mid L_\infty\|^{r-p},$$

Which is nothing other than (5.12) when $q = \infty$. This completes the proof. □

Note that (5.12) holds with constant 1 if $L_{(p,\infty)}$ and $L_{(q,\infty)}$ are replaced by L_p and L_q, respectively. It is often convenient to work with functions that are only locally in some L_p space. This leads to the following definition.

Definition 5.12. For $0 < p < \infty$, the space $L^p_{\text{loc}}(\mathbb{R}^n, \mathscr{L}, m)$ or simply $L^p_{\text{loc}}(\mathbb{R}^n)$ is the set of all Lebesgue-measurable functions f on $\mathbb{R}^n$ that satisfy

$$\int_K |f(x)|^p \mathrm{d}x < \infty, \tag{5.16}$$

for any compact subset K of $\mathbb{R}^n$. Functions that satisfy (5.16) with $p = 1$ are called locally integrable functions on $\mathbb{R}^n$.

$\oslash$

The union of all $L_p(\mathbb{R}^n)$ spaces for $1 \leq p \leq \infty$ is contained in $L^1_{\text{loc}}(\mathbb{R}^n)$.

More generally, for $0 < p < q < \infty$ we have the following:

$$L_q(\mathbb{R}^n) \subseteq L^q_{\text{loc}}(\mathbb{R}^n) \subseteq L^p_{\text{loc}}(\mathbb{R}^n).$$

Functions in $L_p(\mathbb{R}^n)$ for $0 < p < 1$ may not be locally integrable. For example, take $f(x) = |x|^{-\alpha-n}\chi_{\{x:|x|\leq 1\}}$ which is in $L_p(\mathbb{R}^n)$ when $p < n/(n+\alpha)$, and observe that f is not integrable over any open set in $\mathbb{R}^n$ containing the origin.

In what follows we will need the following useful result.

Theorem 5.13 *Let $\{a_j\}_{j\in\mathbb{N}}$ be a sequence of positives reals.*

(a) $\left(\sum_{j=1}^{\infty} a_j\right)^{\theta} \leq \sum_{j=1}^{\infty} a_j^{\theta}$ *for any* $0 \leq \theta \leq 1$. *If* $\sum_{j=1}^{\infty} a_j^{\theta} < \infty$.

(b) $\sum_{j=1}^{\infty} a_j^{\theta} \leq \left(\sum_{j=1}^{\infty} a_j\right)^{\theta}$ *for any* $1 \leq \theta < \infty$. *If* $\sum_{j=1}^{\infty} a_j < \infty$.

(c) $\left(\sum_{j=1}^{N} a_j\right)^{\theta} \leq N^{\theta-1} \sum_{j=1}^{N} a_j^{\theta}$ *when* $1 \leq \theta < \infty$.

(d) $\left(\sum_{j=1}^{N} a_j^{\theta}\right) \leq N^{1-\theta} \left(\sum_{j=1}^{N} a_j\right)^{\theta}$ *when* $0 \leq \theta \leq 1$.

Proof. (a) We proceed by induction. Note that if $0 \leq \theta \leq 1$, then $\theta - 1 \leq 0$, also $a_1 + a_2 \geq a_1$ and $a_1 + a_2 \geq a_2$ from this we have $(a_1 + a_2)^{\theta-1} \leq a_1^{\theta-1}$ and $(a_1 + a_2)^{\theta-1} \leq a_2^{\theta-1}$ therefore

$$a_1(a_1 + a_2)^{\theta-1} \leq a_1^{\theta} \quad \text{and} \quad a_2(a_1 + a_2)^{\theta-1} \leq a_2^{\theta}.$$

Hence

$$a_1(a_1 + a_2)^{\theta-1} + a_2(a_1 + a_2)^{\theta-1} \leq a_1^{\theta} + a_2^{\theta},$$

next, pulling out the common factor on the left-hand side of the above inequality, we have

$$(a_1 + a_2)^{\theta-1}(a_1 + a_2) \leq a_1^{\theta} + a_2^{\theta},$$
$$(a_1 + a_2)^{\theta} \leq a_1^{\theta} + a_2^{\theta}.$$

Now, suppose that

$$\left(\sum_{j=1}^{n} a_j\right)^{\theta} \leq \sum_{j=1}^{n} a_j^{\theta},$$

holds. Since

$$\sum_{j=1}^{n} a_j + a_{n+1} \geq a_{n+1},$$

and

$$\sum_{j=1}^{n} a_j + a_{n+1} \geq \sum_{j=1}^{n} a_j,$$

we have

$$\left(\sum_{j=1}^{n} a_j + a_{n+1}\right)^{\theta-1} \leq a_{n+1}^{\theta-1},$$

and

$$\left(\sum_{j=1}^{n} a_j + a_{n+1}\right)^{\theta-1} \leq \left(\sum_{j=1}^{n} a_j\right)^{\theta-1}.$$

Hence

$$\left(\sum_{j=1}^{n} a_j + a_{n+1}\right)^{\theta-1}\left(\sum_{j=1}^{n} a_j + a_{n+1}\right) \leq a_{n+1}^{\theta} + \left(\sum_{j=1}^{n} a_j\right)^{\theta}$$

$$\left(\sum_{j=1}^{n} a_j + a_{n+1}\right)^{\theta} \leq a_{n+1}^{\theta} + \left(\sum_{j=1}^{n} a_j\right)^{\theta}$$

$$\leq a_{n+1}^{\theta} + \sum_{j=1}^{n} a_j^{\theta} = \sum_{j=1}^{n+1} a_j^{\theta}.$$

Since $\sum_{j=1}^{\infty} a_j^{\theta} < \infty$, we have

$$\left(\sum_{j=1}^{\infty} a_j\right)^{\theta} \leq \sum_{j=1}^{\infty} a_j^{\theta}.$$

(b) Since $\sum_{j=1}^{\infty} a_j < \infty$, then $\lim_{j\to\infty} a_j = 0$, which implies that there exists $n_0 \in \mathbb{N}$ such that

$$0 < a_j < 1 \quad \text{if} \quad j \geq n_0, \quad \text{since} \quad 1 \leq \theta < \infty,$$

we obtain

$$a_j^{\theta} < a_j \quad \text{for all} \quad j \geq n_0,$$

and from this we have

$$\sum_{j=1}^{\infty} a_j^{\theta} < \infty.$$

Consider the sequence $\{a_j^{\theta}\}_{j\in\mathbb{N}}$, since $1 \le \theta$, then $0 < \dfrac{1}{\theta} \le 1$ and by item (a)

$$\left(\sum_{j=1}^{\infty} a_j^{\theta}\right)^{\frac{1}{\theta}} \le \sum_{j=1}^{\infty} \left(a_j^{\theta}\right)^{\frac{1}{\theta}} = \sum_{j=1}^{\infty} a_j,$$

and thus

$$\sum_{j=1}^{\infty} a_j^{\theta} \le \left(\sum_{j=1}^{\infty} a_j\right)^{\theta}.$$

(c) By Hölder's inequality we have

$$\sum_{j=1}^{N} a_j \le \left(\sum_{j=1}^{N} 1\right)^{1-\frac{1}{\theta}} \left(\sum_{j=1}^{N} a_j^{\theta}\right)^{\frac{1}{\theta}} = N^{\frac{\theta-1}{\theta}} \left(\sum_{j=1}^{N} a_j^{\theta}\right)^{\frac{1}{\theta}},$$

then

$$\left(\sum_{j=1}^{N} a_j\right)^{\theta} \le N^{\theta-1} \sum_{j=1}^{N} a_j^{\theta}.$$

(d) On more time, by Hölder's inequality

$$\sum_{j=1}^{N} a_j^{\theta} \le \left(\sum_{j=1}^{N} 1\right)^{1-\theta} \left(\sum_{j=1}^{N} \left(a_j^{\theta}\right)^{\frac{1}{\theta}}\right)^{\theta} = N^{1-\theta} \left(\sum_{j=1}^{N} a_j\right)^{\theta}.$$

This ends the proof. □

We now obtain some bound for the norm of the sum of functions in weak Lebesgue spaces.

Theorem 5.14 *Let $f_1, \ldots, f_N$ belong to $L_{(p,\infty)}$. Then*

(a) $$\left\| \sum_{j=1}^{N} f_j \mid L_{(p,\infty)} \right\| \le N \sum_{j=1}^{N} \|f_j \mid L_{(p,\infty)}\| \quad \textit{for} \quad 1 \le p < \infty.$$

(b) $$\left\| \sum_{j=1}^{N} f_j \mid L_{(p,\infty)} \right\| \le N^{\frac{1}{p}} \sum_{j=1}^{N} \|f_j \mid L_{(p,\infty)}\| \quad \textit{for} \quad 0 < p < 1.$$

Proof. First of all, note that for $\alpha > 0$ and $N \geq 1$

$$|f_1| + \ldots + |f_N| \geq |f_1 + f_2 + \ldots + f_N| > \alpha \geq \frac{\alpha}{N}.$$

Thus

$$\left\{x \in X : |f_1 + f_2 + \ldots + f_N| > \alpha\right\}$$
$$\subset \left\{x \in X : |f_1| > \frac{\alpha}{N}\right\} \cup \left\{x \in X : |f_2| > \frac{\alpha}{N}\right\} \cup \ldots \cup \left\{x \in X : |f_N| > \frac{\alpha}{N}\right\}.$$

Then

$$\mu\left(\{x \in X : |f_1 + f_2 + \ldots + f_N| > \alpha\}\right) \leq \sum_{j=1}^{N} \mu\left(\left\{x \in X : |f_j| > \frac{\alpha}{N}\right\}\right),$$

that is

$$D_{\Sigma f_j}(\alpha) \leq \sum_{j=1}^{N} D_{f_j}\left(\frac{\alpha}{N}\right).$$

Hence

$$\begin{aligned}
\left\| \sum_{j=1}^{N} f_j \mid L_{(p,\infty)} \right\|^p &= \sup_{\alpha>0} \alpha^p D_{\Sigma f_j}(\alpha) \\
&\leq \sum_{j=1}^{N} \sup_{\alpha>0} \alpha^p D_{f_j}\left(\frac{\alpha}{N}\right) \\
&= \sum_{j=1}^{N} \sup_{\alpha>0} \alpha^p D_{N f_j}(\alpha) \\
&= \sum_{j=1}^{N} \|N f_j \mid L_{(p,\infty)}\|^p \\
&= N^p \sum_{j=1}^{N} \|f_j \mid L_{(p,\infty)}\|^p,
\end{aligned}$$

thus

$$\left\| \sum_{j=1}^{N} f_j \mid L_{(p,\infty)} \right\| \leq N \left(\sum_{j=1}^{N} \|f_j \mid L_{(p,\infty)}\|^p \right)^{\frac{1}{p}}.$$

By Theorem 5.13 (a) since $0 < \frac{1}{p} < 1$ we have

$$\left\| \sum_{j=1}^{N} f_j \mid L_{(p,\infty)} \right\| \leq N \left(\sum_{j=1}^{N} \|f_j \mid L_{(p,\infty)}\| \right).$$

(b) As in item (a) we have

$$\left\|\sum_{j=1}^{N} f_j \mid L_{(p,\infty)}\right\|^p \leq N^p \left(\sum_{j=1}^{N} \|f_j \mid L_{(p,\infty)}\|^p\right).$$

Since $0 < p < 1$, then $1 < \frac{1}{p}$, next by Theorem 5.13 (c) we have

$$\begin{aligned}\left\|\sum_{j=1}^{N} f_j \mid L_{(p,\infty)}\right\| &\leq N \left(\sum_{j=1}^{N} \|f_j \mid L_{(p,\infty)}\|^p\right)^{\frac{1}{p}} \\ &\leq N(N^{\frac{1}{p}-1}) \sum_{j=1}^{N} \left(\|f_j \mid L_{(p,\infty)}\|^p\right)^{\frac{1}{p}} \\ &= N^{\frac{1}{p}} \sum_{j=1}^{N} \|f_j \mid L_{(p,\infty)}\|,\end{aligned}$$

which ends the proof. □

Theorem 5.15 *Given a measurable function f on (X,μ) and $\lambda > 0$, define $f_\lambda := f\chi_{\{|f|>\lambda\}}$ and $f^\lambda := f - f_\lambda = f\chi_{\{|f|\leq\lambda\}}$.*

(a) Then

$$D_{f_\lambda}(\alpha) = \begin{cases} D_f(\alpha) & \text{when } \alpha > \lambda, \\ D_f(\lambda) & \text{when } \alpha \leq \lambda. \end{cases}$$

and

$$D_{f^\lambda}(\alpha) = \begin{cases} 0 & \text{when } \alpha \geq \lambda, \\ D_f(\alpha) - D_f(\lambda) & \text{when } \alpha < \lambda. \end{cases}$$

(b) If $f \in L_p(X,\mu)$, then

$$\|f_\lambda \mid L_p\|^p = p\int_\lambda^\infty \alpha^{p-1} D_f(\alpha)\,\mathrm{d}\alpha + \lambda^p D_f(\lambda),$$

$$\|f^\lambda \mid L_p\|^p = p\int_0^\lambda \alpha^{p-1} D_f(\alpha)\,\mathrm{d}\alpha - \lambda^p D_f(\lambda),$$

$$\int_{\lambda<|f|\leq\delta} |f|^p\,\mathrm{d}\mu = p\int_\lambda^\delta \alpha^{p-1} D_f(\alpha)\,\mathrm{d}\alpha - \delta^p D_f(\alpha) + \lambda^p D_f(\lambda).$$

(c) If f is in $L_{(p,\infty)}$ then f^λ is in $L_q(X,\mu)$ for any $q > p$ and f_λ is in $L_q(X,\mu)$ for any $q < p$. Thus $L_{(p,\infty)} \subseteq L_{p_0} + L_{p_1}$ when $0 < p_0 < p < p_1 \leq \infty$.

Proof. (a) Note

$$D_{f_\lambda}(\alpha) = \mu\left(\{x : |f(x)|\chi_{\{|f|>\lambda\}}(x) > \alpha\}\right) = \mu\left(\{x : |f(x)| > \alpha\} \cap \{x : |f| > \lambda\}\right),$$

if $\alpha > \lambda$, then $\{x : |f(x)| > \alpha\} \subseteq \{x : |f| > \lambda\}$, thus

$$\begin{aligned} D_{f_\lambda}(\alpha) &= \mu\left(\{x : |f(x)| > \alpha\} \cap \{x : |f| > \lambda\}\right) \\ &= \mu\left(\{x : |f(x)| > \alpha\}\right) \\ &= D_f(\alpha). \end{aligned}$$

If $\alpha \leq \lambda$, then $\{x : |f(x)| > \lambda\} \subseteq \{x : |f| > \alpha\}$, thus

$$D_{f_\lambda}(\alpha) = \mu\left(\{x : |f(x)| > \alpha\} \cap \{x : |f| > \lambda\}\right) = \mu\left(\{x : |f(x)| > \lambda\}\right) = D_f(\lambda),$$

which entails

$$D_{f_\lambda}(\alpha) = \begin{cases} D_f(\alpha) \text{ when } \alpha > \lambda, \\ D_f(\lambda) \text{ when } \alpha \leq \lambda. \end{cases} \tag{5.17}$$

Next, consider

$$\begin{aligned} D_{f^\lambda}(\alpha) &= \mu\left(\{x : |f(x)|\chi_{\{|f|\leq\lambda\}}(x) > \alpha\}\right) \\ &= \mu\left(\{x : |f(x)| > \alpha\} \cap \{x : |f| \leq \lambda\}\right), \end{aligned}$$

if $\alpha \geq \lambda$ then $\{x : |f| > \alpha\} \cap \{x : |f(x)| \leq \lambda\} = \emptyset$, thus $D_{f^\lambda}(\alpha) = 0$.
If $\alpha < \lambda$, then

$$\begin{aligned} D_{f^\lambda}(\alpha) &= \mu\left(\{x : |f(x)| > \alpha\} \cap \{x : |f(x)| \leq \lambda\}\right) \\ &= \mu\left(\{x : |f(x)| > \alpha\} \cap \{x : |f(x)| > \lambda\}^{\complement}\right) \\ &= \mu\left(\{x : |f(x)| > \alpha\} \setminus \{x : |f(x)| > \lambda\}\right) \\ &= \mu\left(\{x : |f(x)| > \alpha\}\right) - \mu\left(\{x : |f(x)| > \lambda\}\right) \\ &= D_f(\alpha) - D_f(\lambda), \end{aligned}$$

and hence

$$D_{f^\lambda}(\alpha) = \begin{cases} 0, & \text{when } \alpha \geq \lambda, \\ D_f(\alpha) - D_f(\lambda), & \text{when } \alpha < \lambda. \end{cases} \tag{5.18}$$

(b) If $f \in L_p(X, \mu)$, then

$$\begin{aligned} \|f_\lambda \mid L_p\|^p &= p \int_0^\infty \alpha^{p-1} D_{f_\lambda}(\alpha)\,\mathrm{d}\alpha \\ &= p \int_0^\lambda \alpha^{p-1} D_{f_\lambda}(\alpha)\,\mathrm{d}\alpha + p \int_\lambda^\infty \alpha^{p-1} D_{f_\lambda}(\alpha)\,\mathrm{d}\alpha, \end{aligned}$$

By (5.17) we have

$$\|f_\lambda \mid L_p\|^p = p\int_0^\lambda \alpha^{p-1} D_f(\lambda)\,\mathrm{d}\alpha + p\int_\lambda^\infty \alpha^{p-1} D_f(\alpha)\,\mathrm{d}\alpha$$
$$= \lambda^p D_f(\lambda) + p\int_\lambda^\infty \alpha^{p-1} D_f(\alpha)\,\mathrm{d}\alpha.$$

We also have

$$\|f^\lambda \mid L_p\|^p = p\int_0^\infty \alpha^{p-1} D_{f^\lambda}(\alpha)\,\mathrm{d}\alpha$$
$$= p\int_0^\lambda \alpha^{p-1} D_{f^\lambda}(\alpha)\,\mathrm{d}\alpha + p\int_\lambda^\infty \alpha^{p-1} D_{f^\lambda}(\alpha)\,\mathrm{d}\alpha,$$

by (5.18) we obtain

$$\|f^\lambda \mid L_p\|^p = p\int_0^\lambda \alpha^{p-1}\left(D_f(\alpha) - D_f(\lambda)\right)\mathrm{d}\alpha$$
$$= p\int_0^\lambda \alpha^{p-1} D_f(\alpha)\,\mathrm{d}\alpha - \lambda^p D_f(\lambda).$$

Next,

$$\int_{\lambda<|f|\le\delta} |f|^p\,\mathrm{d}\mu$$
$$= \int_{|f|>\lambda} |f|^p\,\mathrm{d}\mu - \int_{|f|>\delta} |f|^p\,\mathrm{d}\mu$$
$$= \int_X |f|^p \chi_{\{|f|>\lambda\}}\,\mathrm{d}\mu - \int_X |f|^p \chi_{\{|f|>\delta\}}\,\mathrm{d}\mu$$
$$= \int_X |f_\lambda|^p\,\mathrm{d}\mu - \int_X |f_\delta|^p\,\mathrm{d}\mu$$
$$= p\int_\lambda^\infty \alpha^{p-1} D_f(\alpha)\,\mathrm{d}\alpha + \lambda^p D_f(\lambda) - p\int_\delta^\infty \alpha^{p-1} D_f(\alpha)\,\mathrm{d}\alpha - \delta^p D_f(\delta)$$

$$= p\left(\int_{\lambda}^{\infty} \alpha^{p-1} D_f(\alpha)\,\mathrm{d}\alpha - \int_{\delta}^{\infty} \alpha^{p-1} D_f(\alpha)\,\mathrm{d}\alpha\right) + \lambda^p D_f(\lambda) - \delta^p D_f(\delta)$$

$$= p\int_{\lambda}^{\delta} \alpha^{p-1} D_f(\alpha)\,\mathrm{d}\alpha - \delta^p D_f(\alpha) + \lambda^p D_f(\lambda).$$

(c) We know that

$$D_f(\alpha) \leq \frac{\|f \mid L_{(p,\infty)}\|^p}{\alpha^p},$$

then if $q > p$

$$\begin{aligned}
\|f^\lambda \mid L_q\|^q &= q\int_0^{\lambda} \alpha^{q-1} D_f(\alpha)\,\mathrm{d}\alpha - \lambda^q D_f(\lambda) \\
&\leq q\int_0^{\lambda} \alpha^{q-1} \frac{\|f \mid L_{(p,\infty)}\|^p}{\alpha^p}\,\mathrm{d}\alpha - \lambda^q D_f(\lambda) \\
&= q\|f \mid L_{(p,\infty)}\|^p \frac{\lambda^{q-p}}{q-p} - \lambda^q D_f(\lambda) \\
&\leq q\|f \mid L_{(p,\infty)}\|^p \frac{\lambda^{q-p}}{q-p}
\end{aligned}$$

which is finite, therefore $f^\lambda \in L_q$ if $q > p$.

Now, if $q < p$, then

$$\begin{aligned}
\|f_\lambda \mid L_q\|^q &= q\int_{\lambda}^{\infty} \alpha^{q-1} D_f(\alpha)\,\mathrm{d}\alpha + \lambda^q D_f(\lambda) \\
&\leq q\|f \mid L_{(p,\infty)}\|^p \int_{\lambda}^{\infty} \alpha^{q-p-1}\,\mathrm{d}\alpha + \lambda^q D_f(\lambda) \\
&= q\frac{\lambda^{q-p}}{p-q}\|f \mid L_{(p,\infty)}\|^p + \lambda^q D_f(\lambda),
\end{aligned}$$

which is finite, thus $f_\lambda \in L_q$ if $q < p$.

Finally, since $f \in L_{(p,\infty)}$ and

$$f = f^\lambda + f_\lambda,$$

where $f^\lambda \in L_{p_1}$ if $p < p_1$ and $f_\lambda \in L_{p_0}$ if $p_0 < p$. Then $L_{(p,\infty)} \subseteq L_{p_0} + L_{p_1}$ when $0 < p_0 < p < p_1 \leq \infty$. $\square$

Theorem 5.16 *Let (X,μ) be a measure space and let E be a subset of X with $\mu(E)<\infty$. Then*

(a) for $0<q<p$ we have

$$\int_E |f(x)|^q \,\mathrm{d}\mu \le \frac{p}{p-q}\left[\mu(E)\right]^{1-\frac{q}{p}} \|f \mid L_{(p,\infty)}\|^q \quad \text{for} \quad f\in L_{(p,\infty)}.$$

(b) if $\mu(X)<\infty$ and $0<q<p$, we have

$$L_p(X,\mu)\subseteq L_{(p,\infty)}(X,\mu)\subseteq L_q(X,\mu).$$

Proof. Let $f\in L_{(p,\infty)}$, then

$$\begin{aligned}
&\int_E |f|^q \,\mathrm{d}\mu \\
&= q\int_0^\infty \lambda^{q-1}\mu\left(\{x\in E: |f(x)|>\lambda\}\right)\mathrm{d}\lambda \\
&\le q\int_0^{\mu(E)^{-\frac{1}{p}}\|f|L_{(p,\infty)}\|} \lambda^{q-1}\mu(E)\,\mathrm{d}\lambda + q\int_{\mu(E)^{-\frac{1}{p}}\|f|L_{(p,\infty)}\|}^\infty \lambda^{q-1}D_f(\lambda)\,\mathrm{d}\lambda \\
&\le q\int_0^{\mu(E)^{-\frac{1}{p}}\|f|L_{(p,\infty)}\|} \lambda^{q-1}\mu(E)\,\mathrm{d}\lambda + q\int_{\mu(E)^{-\frac{1}{p}}\|f|L_{(p,\infty)}\|}^\infty \lambda^{q-1}\frac{\|f \mid L_{(p,\infty)}\|^p}{\lambda^p}\,\mathrm{d}\lambda \\
&= \left(\mu(E)^{-\frac{1}{p}}\|f \mid L_{(p,\infty)}\|\right)^q \mu(E) + \frac{q}{p-q}\left(\mu(E)^{-\frac{1}{p}}\|f \mid L_{(p,\infty)}\|\right)^{q-p} \|f \mid L_{(p,\infty)}\|^p \\
&= \mu(E)^{1-\frac{q}{p}}\|f \mid L_{(p,\infty)}\|^q + \frac{q}{p-q}\mu(E)^{1-\frac{q}{p}}\|f \mid L_{(p,\infty)}\|^q \\
&= \frac{p}{p-q}\mu(E)^{1-\frac{q}{p}}\|f \mid L_{(p,\infty)}\|^q,
\end{aligned}$$

i.e.

$$\int_E |f|^q \,\mathrm{d}\mu \le \frac{p}{p-q}\mu(E)^{1-\frac{q}{p}}\|f \mid L_{(p,\infty)}\|^q.$$

(b) If $\mu(X) < \infty$, then

$$\int_X |f|^q \, \mathrm{d}\mu \leq \frac{p}{p-q} \mu(X)^{1-\frac{q}{p}} \left\| f \mid L_{(p,\infty)} \right\|^q.$$

Hence $L_p \subseteq L_{(p,\infty)} \subseteq L_q$. □

Corollary 5.17 *Let (X,μ) be a measurable space and let E be a subset of X with $\mu(E) < \infty$. Then*

$$\left\| f \mid L_{\frac{p}{2}} \right\| \leq (4\mu(E))^{1/p} \left\| f \mid L_{(p,\infty)} \right\|,$$

and thus $L_{(p,\infty)} \subseteq L_{p/2}$.

Proof. Since $0 < \frac{p}{2} < p$ we can apply Theorem 5.16 to obtain

$$\begin{aligned} \int_E |f|^{p/2} \, \mathrm{d}\mu &\leq \frac{p}{p-\frac{p}{2}} \mu(E)^{1-\frac{p/2}{p}} \left\| f \mid L_{(p,\infty)} \right\|^{p/2} \\ &= 2\mu(E)^{1/2} \left\| f \mid L_{(p,\infty)} \right\|^{p/2} \end{aligned}$$

from which we get

$$\left\| f \mid L_{\frac{p}{2}} \right\| \leq (4\mu(E))^{1/p} \left\| f \mid L_{(p,\infty)} \right\|,$$

which entails that $L_{(p,\infty)} \subseteq L_{p/2}$. □

5.4 Normability

Let $(X, \mathscr{A}, \mu)$ be a measure space and let $0 < p < \infty$. Pick $0 < r < p$ and define

$$|||f \mid L_{(p,\infty)}||| = \sup_{0<\mu(E)<\infty} \mu(E)^{-\frac{1}{r}+\frac{1}{p}} \left(\int_E |f|^r \, \mathrm{d}\mu \right)^{\frac{1}{r}},$$

where the supremum is taken over all measurable subsets E of X of finite measure.

Theorem 5.18 *Let f be in $L_{(p,\infty)}$. Then*

$$\left\| f \mid L_{(p,\infty)} \right\| \leq |||f \mid L_{(p,\infty)}||| \leq \left(\frac{p}{p-r} \right)^{\frac{1}{r}} \left\| f \mid L_{(p,\infty)} \right\|.$$

Proof. By Theorem 5.16 with $q = r$ we have

$$
\begin{aligned}
|||f \mid L_{(p,\infty)}||| &= \sup_{0<\mu(E)<\infty} \mu(E)^{-\frac{1}{r}+\frac{1}{p}} \left(\int_E |f|^r \mathrm{d}\mu \right)^{\frac{1}{r}} \\
&\leq \sup_{0<\mu(E)<\infty} \mu(E)^{-\frac{1}{r}+\frac{1}{p}} \left(\frac{p}{p-r} \mu(E)^{1-\frac{r}{p}} \| f \mid L_{(p,\infty)} \|^r \right)^{\frac{1}{r}} \\
&= \sup_{0<\mu(E)<\infty} \mu(E)^{-\frac{1}{r}+\frac{1}{p}} \left(\frac{p}{p-r} \right)^{\frac{1}{r}} \mu(E)^{\frac{1}{r}-\frac{1}{p}} \| f \mid L_{(p,\infty)} \| \\
&= \left(\frac{p}{p-r} \right)^{\frac{1}{r}} \| f \mid L_{(p,\infty)} \|.
\end{aligned}
$$

On the other hand, by definition

$$
\mu(E)^{-\frac{1}{r}+\frac{1}{p}} \left(\int_E |f|^r \mathrm{d}\mu \right)^{\frac{1}{r}} \leq |||f \mid L_{(p,\infty)}|||,
$$

for all $E \in \mathscr{A}$ such that $\mu(E) < \infty$. Now, let us consider $A = \{x : |f(x)| > \alpha\}$ for $f \in L_{(p,\infty)}$. Observe that $\mu(A) < \infty$. Then

$$
\begin{aligned}
|||f \mid L_{(p,\infty)}|||^p &\geq \left(\mu(A)^{-\frac{1}{r}+\frac{1}{p}} \left(\int_A |f|^r \mathrm{d}\mu \right)^{\frac{1}{r}} \right)^p \\
&\geq D_f(\alpha)^{-\frac{p}{r}+1} \left(\int_A \alpha^r \mathrm{d}\mu \right)^{\frac{p}{r}} \\
&= D_f(\alpha)^{-\frac{p}{r}+1} \cdot \alpha^p \cdot D_f(\alpha)^{\frac{p}{r}} \\
&= \alpha^p D_f(\alpha).
\end{aligned}
$$

That is

$$
\alpha^p D_f(\alpha) \leq |||f \mid L_{(p,\infty)}|||,
$$

and thus

$$
\sup_{\alpha>0} \alpha^p D_f(\alpha) \leq |||f \mid L_{(p,\infty)}|||,
$$

This ends the proof. □

The next result is a Fatou type lemma adapted to the framework of weak Lebesgue spaces.

Lemma 5.19 (Fatou's type lemma). *For all measurable function g_n on X we have*

$$\left\| \liminf_{n\to\infty} |g_n| \mid L_{(p,\infty)} \right\| \leq C_p \liminf_{n\to\infty} \left\| g_n \mid L_{(p,\infty)} \right\|,$$

for some constant C_p that depends only on $p \in (0,\infty)$.

Proof. By Fatou's lemma, we have

$$\begin{aligned}
\left\| \liminf_{n\to\infty} |g_n| \mid L_{(p,\infty)} \right\| &\leq ||| \liminf_{n\to\infty} |g_n| \mid L_{(p,\infty)} ||| \\
&= \sup_{0<\mu(E)<\infty} \mu(E)^{-\frac{1}{r}+\frac{1}{p}} \left(\int_E \left(\liminf_{n\to\infty} |g_n| \right)^r \mathrm{d}\mu \right)^{\frac{1}{r}} \\
&\leq \sup_{0<\mu(E)<\infty} \mu(E)^{-\frac{1}{r}+\frac{1}{p}} \left(\int_E \liminf_{n\to\infty} |g_n|^r \mathrm{d}\mu \right)^{\frac{1}{r}} \\
&\leq \sup_{0<\mu(E)<\infty} \mu(E)^{-\frac{1}{r}+\frac{1}{p}} \left(\liminf_{n\to\infty} \int_E |g_n|^r \mathrm{d}\mu \right)^{\frac{1}{r}} \\
&\leq \liminf_{n\to\infty} \sup_{0<\mu(E)<\infty} \mu(E)^{-\frac{1}{r}+\frac{1}{p}} \left(\int_E |g_n|^r \mathrm{d}\mu \right)^{\frac{1}{r}} \\
&\leq \liminf_{n\to\infty} \left(\frac{p}{p-r} \right)^{\frac{1}{r}} \left\| g_n \mid L_{(p,\infty)} \right\| \\
&= \left(\frac{p}{p-r} \right)^{\frac{1}{r}} \liminf_{n\to\infty} \left\| g_n \mid L_{(p,\infty)} \right\|.
\end{aligned}$$

This ends the proof. □

The following result is an improvement of Lemma 5.19.

Lemma 5.20. *For all measurable functions g_n on X we have*

$$\left\| \liminf_{n\to\infty} |g_n| \mid L_{(p,\infty)} \right\| \leq \liminf_{n\to\infty} \left\| g_n \mid L_{(p,\infty)} \right\|.$$

Proof. Since

$$D_{\liminf_{n\to\infty} |g_n|}(\lambda) \leq \liminf_{n\to\infty} D_{g_n}(\lambda),$$

then

$$\left\{ C > 0 : \liminf_{n\to\infty} D_{g_n}(\lambda) \leq \frac{C^p}{\lambda^p} \right\} \subseteq \left\{ C > 0 : D_{\liminf_{n\to\infty} |g_n|}(\lambda) \leq \frac{C^p}{\lambda^p} \right\}.$$

From the above considerations, we have

$$
\begin{aligned}
\left\| \liminf_{n\to\infty} |g_n| \mid L_{(p,\infty)} \right\| &= \inf\left\{ C > 0 : D_{\liminf_{n\to\infty}|g_n|}(\lambda) \le \frac{C^p}{\lambda^p} \right\} \\
&\le \inf\left\{ C > 0 : \liminf_{n\to\infty} D_{g_n}(\lambda) \le \frac{C^p}{\lambda^p} \right\} \\
&= \liminf_{n\to\infty} \left(\inf\left\{ C > 0 : D_{g_n}(\lambda) \le \frac{C^p}{\lambda^p} \right\} \right) \\
&= \liminf_{n\to\infty} \left\| g_n \mid L_{(p,\infty)} \right\|,
\end{aligned}
$$

which finishes the proof. □

Theorem 5.21 *Let $0 < p < 1$, $0 < s < \infty$, and $(X, \mathscr{A}, \mu)$ be a measurable space*

(a) Let f be a measurable function on X. Then

$$
\int_{\{|f| \le s\}} |f| \,\mathrm{d}\mu \le \frac{s^{1-p}}{1-p} \left\| f \mid L_{(p,\infty)} \right\|^p .
$$

(b) Let f_j, $1 \le j \le m$, be measurable functions on X. Then

$$
\left\| \max_{1 \le j \le m} |f_j| \mid L_{(p,\infty)} \right\|^p \le \sum_{j=1}^{m} \left\| f_j \mid L_{(p,\infty)} \right\|^p .
$$

(c)

$$
\left\| f_1 + \ldots + f_m \mid L_{(p,\infty)} \right\|^p \le m \frac{2-p}{1-p} \sum_{j=1}^{m} \left\| f_j \mid L_{(p,\infty)} \right\|^p .
$$

The latter estimate is refereed to as the p–normability of $L_{(p,\infty)}$ for $p < 1$.

Proof. By Theorem 5.15 (b) with $p = 1$, we have

$$
\begin{aligned}
\int_{\{|f| \le s\}} |f| \,\mathrm{d}\mu &= \int_X |f| \chi_{\{|f| \le s\}} \,\mathrm{d}\mu \\
&= \int_X |f^s| \,\mathrm{d}\mu \\
&= \int_0^s D_f(\alpha) \,\mathrm{d}\alpha - s D_f(s) \\
&\le \int_0^s \frac{\alpha^p D_f(\alpha)}{\alpha^p} \,\mathrm{d}\alpha
\end{aligned}
$$

$$\leq \|f \mid L_{(p,\infty)}\|^p \int_0^s \frac{d\alpha}{\alpha^p}$$

$$= \frac{s^{1-p}}{1-p} \|f \mid L_{(p,\infty)}\|^p.$$

(b) Let $\max\limits_{1\leq j\leq k} |f_j(x)| = f_k(x)$ for some $1 \leq k \leq m$. Then

$$\begin{aligned} D_{\max|f_j|}(\alpha) &= \mu\left(\{x : \max_{1\leq j\leq m} |f_j(x)| > \alpha\}\right) \\ &= \mu\left(\{x : f_k(x) > \lambda\}\right) = D_{f_k}(\alpha) \quad \text{for some} \quad 1 \leq k \leq m \\ &\leq \sum_{j=1}^{m} D_{f_j}(\alpha). \end{aligned}$$

We now obtain

$$\alpha^p D_{\max|f_j|}(\alpha) \leq \sum_{j=1}^{m} \sup \alpha^p D_{f_j}(\alpha),$$

and thus

$$\|\max_{1\leq j\leq m} |f_j| \mid L_{(p,\infty)}\|^p \leq \sum_{j=1}^{m} \|f_j \mid L_{(p,\infty)}\|^p.$$

(c) Observe that

$$\max_{1\leq j\leq m} |f_j| \leq |f_1| + |f_2| + \ldots + |f_m|,$$

from which

$$\left\{x : \max_{1\leq j\leq m} |f_j(x)| > \alpha\right\} \subset \left\{x : |f_1| + \ldots + |f_m| > \alpha\right\},$$

and then

$$\begin{aligned} &\left\{x : |f_1| + \ldots + |f_m| > \alpha\right\} \\ &= \left(\left\{x : |f_1| + \ldots + |f_m| > \alpha\right\} \cap \left\{x : \max_{1\leq j\leq m} |f_j(x)| \leq \alpha\right\}\right) \cup \left\{x : \max_{1\leq j\leq m} |f_j(x)| > \alpha\right\}. \end{aligned}$$

We now have

$$\begin{aligned} D_{f_1+\ldots+f_m}(\alpha) &= \mu\left(\{x : |f_1 + \ldots + f_m| > \alpha\}\right) \\ &\leq \mu\left(\{x : |f_1| + \ldots + |f_m| > \alpha\}\right) \\ &\leq \mu\left(\{x : |f_1| + \ldots + |f_m| > \alpha\} \cap \{x : \max_{1\leq j\leq m} |f_j(x)| \leq \alpha\}\right) \end{aligned}$$

$$+\mu\left(\{x: \max_{1\le j\le m}|f_j(x)|>\alpha\}\right)$$

$$=\mu\left(\left\{x\in\{x: \max_{1\le j\le m}|f_j(x)|\le\alpha\}: |f_1|+\ldots+|f_m|>\alpha\right\}\right)$$

$$+\mu\left(\{x: \max_{1\le j\le m}|f_j(x)|>\alpha\}\right)$$

$$=\mu\left(\left\{x: \left(|f_1|+\ldots+|f_m|\right)\chi_{\{x:\max|f_j|\le\alpha\}}>\alpha\right\}\right)$$

$$+\mu\left(\{x: \max_{1\le j\le m}|f_j(x)|>\alpha\}\right).$$

By Chebyshev's inequality

$$D_{f_1+\ldots+f_m}(\alpha)\le\frac{1}{\alpha}\int\limits_{\{x:\max|f_j|\le\alpha\}}\left(|f_1|+\ldots+|f_m|\right)\mathrm{d}\mu+D_{\max|f_j|}(\alpha)$$

$$=\sum_{j=1}^{m}\frac{1}{\alpha}\int\limits_{\{x:\max|f_j|\le\alpha\}}|f_j|\,\mathrm{d}\mu+D_{\max|f_j|}(\alpha)$$

$$\le\sum_{j=1}^{m}\frac{1}{\alpha}\int\limits_{\{x:\max|f_j|\le\alpha\}}\max_{1\le j\le m}|f_j|\,\mathrm{d}\mu+D_{\max|f_j|}(\alpha).$$

By item (a) we have

$$\sum_{j=1}^{m}\frac{1}{\alpha}\int\limits_{\{x:\max|f_j|\le\alpha\}}\max_{1\le j\le m}|f_j|\,\mathrm{d}\mu+D_{\max|f_j|}(\alpha)$$

$$\le\sum_{j=1}^{m}\frac{1}{\alpha}\frac{\alpha^{1-p}}{1-p}\left\|\max_{1\le j\le m}|f_j|\mid L_{(p,\infty)}\right\|^p+\frac{\left\|\max_{1\le j\le m}|f_j|\mid L_{(p,\infty)}\right\|^p}{\alpha^p}$$

$$\le\sum_{j=1}^{m}\frac{\alpha^{-p}}{1-p}\left\|\max_{1\le j\le m}|f_j|\mid L_{(p,\infty)}\right\|^p+\sum_{j=1}^{m}\frac{\left\|\max_{1\le j\le m}|f_j|\mid L_{(p,\infty)}\right\|^p}{\alpha^p}.$$

Finally by item (b) we obtain

$$\alpha^pD_{f_1+\ldots+f_m}(\alpha)\le\sum_{j=1}^{m}\left(\frac{1}{1-p}+1\right)\left\|\max_{1\le j\le m}|f_j|\mid L_{(p,\infty)}\right\|^p$$

$$=\sum_{j=1}^{m}\left(\frac{2-p}{1-p}\right)\left\|\max_{1\le j\le m}|f_j|\mid L_{(p,\infty)}\right\|^p$$

$$\leq \frac{2-p}{1-p} \sum_{j=1}^{m} \sum_{j=1}^{m} \|f_j \mid L_{(p,\infty)}\|^p$$
$$= m\frac{2-p}{1-p} \sum_{j=1}^{m} \|f_j \mid L_{(p,\infty)}\|^p.$$

This ends the proof. □

We now show a log-convex type inequality for the weak Lebesgue spaces, the so-called Lyapunov's inequality.

Theorem 5.22 (Lyapunov's inequality) *Let (X,μ) be a measurable space. Suppose that $0 < p_0 < p < p_1 < \infty$ and $\frac{1}{p} = \frac{1-\theta}{p_0} + \frac{\theta}{p_1}$ for some $\theta \in [0,1]$. If $f \in L_{(p_0,\infty)} \cap L_{(p_1,\infty)}$ then $f \in L_{(p,\infty)}$ and*

$$\left\| f \mid L_{(p,\infty)} \right\| \leq \left\| f \mid L_{(p_0,\infty)} \right\|^{1-\theta} \left\| f \mid L_{(p_1,\infty)} \right\|^{\theta}.$$

Proof. Observe that

$$\alpha^p D_f(\alpha) = \alpha^{p(1-\theta+\theta)} \Big[D_f(\alpha)\Big]^{p\left(\frac{1}{p}\right)}$$

$$= \alpha^{p(1-\theta)} \alpha^{p\theta} \Big[D_f(\alpha)\Big]^{p\left(\frac{1-\theta}{p_0}+\frac{\theta}{p_1}\right)}$$

$$= \alpha^{p(1-\theta)} \Big[D_f(\alpha)\Big]^{p\left(\frac{1-\theta}{p_0}\right)} \alpha^{p\theta} \Big[D_f(\alpha)\Big]^{\frac{p\theta}{p_1}}$$

$$= \Big[\alpha^{p_0} D_f(\alpha)\Big]^{p\left(\frac{1-\theta}{p_0}\right)} \Big[\alpha^{p_1} D_f(\alpha)\Big]^{\frac{p\theta}{p_1}}.$$

Thus

$$\alpha^p D_f(\alpha) \leq \Big[\|f \mid L_{(p_0,\infty)}\|^{p_0}\Big]^{p\left(\frac{1-\theta}{p_0}\right)} \Big[\|f \mid L_{(p_1,\infty)}\|^{p_1}\Big]^{\frac{p\theta}{p_1}},$$

from which

$$\sup_{\alpha>0} \alpha^p D_f(\alpha) \leq \Big[\|f \mid L_{(p_0,\infty)}\|^{p_0}\Big]^{p\left(\frac{1-\theta}{p_0}\right)} \Big[\|f \mid L_{(p_1,\infty)}\|^{p_1}\Big]^{\frac{p\theta}{p_1}}$$

and this entails

$$\|f \mid L_{(p,\infty)}\| \leq \|f \mid L_{(p_0,\infty)}\|^{1-\theta} \|f \mid L_{(p_1,\infty)}\|^{\theta},$$

which ends the proof. □

We now arrive at the Hölder inequality in weak Lebesgue spaces.

Theorem 5.23 (Hölder's inequality). *Let f_j be in $L_{(p_j,\infty)}$ where $0 < p_j < \infty$ and $1 \le j \le k$. Let*

$$\frac{1}{p} = \frac{1}{p_1} + \ldots + \frac{1}{p_k}.$$

Then

$$\left\| f_1 \times \cdots \times f_k \mid L_{(p,\infty)} \right\| \le p^{-\frac{1}{p}} \prod_{j=1}^{k} p_j^{\frac{1}{p_j}} \prod_{j=1}^{k} \left\| f_j \mid L_{(p_j,\infty)} \right\|.$$

Proof. Let us consider $\left\| f_j \mid L_{(p_j,\infty)} \right\| = 1$, $1 \le j \le k$, and let $x_1, \ldots, x_n$ be a positive real numbers such that

$$\frac{1}{x_1} \cdots \frac{1}{x_k} = \alpha,$$

then

$$D_{f_1 \ldots f_k}(\alpha) = D_{f_1 \ldots f_k}\left(\frac{1}{x_1} \cdots \frac{1}{x_k} \right)$$

$$\le D_{f_1}\left(\frac{1}{x_1} \right) + D_{f_2}\left(\frac{1}{x_2} \right) + \ldots + D_{f_k}\left(\frac{1}{x_k} \right). \tag{5.19}$$

Since

$$1 = \left\| f_j \mid L_{(p_j,\infty)} \right\|^{p_j} \ge \sup_j \left(\frac{1}{x_j} \right)^{p_j} D_{f_j}\left(\frac{1}{x_j} \right),$$

then

$$\left(\frac{1}{x_j} \right)^{p_j} D_{f_j}\left(\frac{1}{x_j} \right) \le 1,$$

thus

$$D_{f_j}\left(\frac{1}{x_j} \right) \le x_j^{p_j} \quad \text{for} \quad 1 \le j \le k.$$

Hence, we can write (5.19) as follows

$$D_{f_1 \ldots f_k}\left(\frac{1}{x_1} \cdots \frac{1}{x_k} \right) \le x_1^{p_1} + x_2^{p_2} + \ldots + x_k^{p_k}.$$

Next, let us define

$$F(x_1, \ldots, x_k) = x_1^{p_1} + x_2^{p_2} + \ldots + x_k^{p_k}.$$

In what follows, we will use the Lagrange multipliers in order to obtain the minimum value of F subject to the constrain

$$\frac{1}{x_1} \cdots \frac{1}{x_k} = \alpha.$$

That is

$$f(x_1,x_2,\ldots,x_k) = x_1^{p_1} + x_2^{p_2} + \ldots + x_k^{p_k}$$

$$g(x_1,x_2,\ldots,x_k) = x_1x_2\ldots x_k - \frac{1}{\alpha}.$$

Then, next

$$\nabla F = \lambda \nabla g.$$

And thus

$$\begin{aligned} p_1x_1^{p_1-1} &= \lambda(x_2x_3\ldots x_k) \\ p_2x_2^{p_2-1} &= \lambda(x_1x_3\ldots x_k) \\ &\vdots \\ p_jx_j^{p_j-1} &= \lambda(x_1x_3\ldots x_k), \end{aligned}$$

thus

$$\begin{aligned} p_1x_1^{p_1} &= \lambda(x_1x_2\ldots x_k) \\ p_2x_2^{p_2} &= \lambda(x_1x_2\ldots x_k) \\ &\vdots \\ p_jx_j^{p_j} &= \lambda(x_1x_2\ldots x_k). \end{aligned}$$

Observe that

$$x_1x_2\ldots x_k = \frac{1}{\alpha}. \tag{5.20}$$

On the other hand, note that

$$p_1x_1^{p_1} = p_jx_j^{p_j} \quad \text{for} \quad 2 \le j \le k. \tag{5.21}$$

Now replacing (5.21) into (5.20) we have

$$\begin{aligned} &x_1\left(\frac{p_1}{p_2}\right)^{\frac{1}{p_2}} x_1^{\frac{p_1}{p_2}} \left(\frac{p_1}{p_3}\right)^{\frac{1}{p_3}} x^{\frac{p_1}{p_3}} \ldots \left(\frac{p_1}{p_k}\right)^{\frac{1}{p_k}} x_1^{\frac{p_1}{p_k}} \\ &= x_1\left(\frac{p_1}{p_2}\right)^{\frac{1}{p_2}} \left(\frac{p_1}{p_3}\right)^{\frac{1}{p_3}} \ldots \left(\frac{p_1}{p_k}\right)^{\frac{1}{p_k}} x_1^{\frac{p_1}{p_2}+\frac{p_1}{p_3}+\ldots+\frac{p_1}{p_k}} = \frac{1}{\alpha}, \end{aligned} \tag{5.22}$$

but

$$\left(\frac{p_1}{p_1}\right)^{\frac{1}{p_1}} = 1,$$

then we can write (5.22) as follows

$$\left(\frac{p_1}{p_1}\right)^{\frac{1}{p_1}} \left(\frac{p_1}{p_2}\right)^{\frac{1}{p_2}} \left(\frac{p_1}{p_3}\right)^{\frac{1}{p_3}} \ldots \left(\frac{p_1}{p_k}\right)^{\frac{1}{p_k}} x_1^{\frac{p_1}{p_1}+\frac{p_1}{p_2}+\frac{p_1}{p_3}+\ldots+\frac{p_1}{p_k}} = \frac{1}{\alpha}.$$

And, thus

$$\frac{p_1^{\frac{1}{p_1}+\frac{1}{p_2}+\ldots+\frac{1}{p_k}}}{\prod_{j=1}^{k} p_j^{\frac{1}{p_j}}} x_1^{p_1\left(\frac{1}{p_1}+\frac{1}{p_2}+\ldots+\frac{1}{p_k}\right)} = \frac{1}{\alpha}.$$

Then

$$p_1^{\frac{1}{p}} x^{\frac{p_1}{p}} = \frac{\prod_{j=1}^{k} p_j^{\frac{1}{p_j}}}{\alpha},$$

hence

$$x_1^{p_1} = \frac{1}{p_1 \alpha^p}\left[\prod_{j=1}^{k} p_j^{\frac{1}{p_j}}\right]^p.$$

Therefore the $x_1 \ldots x_k$ such that

$$\begin{cases} x_1^{p_1} = \dfrac{1}{p_1 \alpha^p}\left[\prod_{j=1}^{k} p_j^{\frac{1}{p_j}}\right]^p \\ x_j^{p_j} = \dfrac{p_1}{p_j} x_1^{p_1} \end{cases} \tag{5.23}$$

are the unique critical real point.

For this critical real point, using (5.23) we have

$$\begin{aligned} x_1^{p_1} + x_2^{p_2} + \ldots + x_k^{p_k} &= x_1^{p_1} + \frac{p_1}{p_2} x_1^{p_1} + \ldots + \frac{p_1}{p_k} x_1^{p_1} \\ &= p_1 x_1^{p_1}\left[\frac{1}{p_1} + \ldots + \frac{1}{p_k}\right] \\ &= \frac{1}{\alpha^p}\left[\prod_{j=1}^{k} p_j^{\frac{1}{p_j}}\right]^p \frac{1}{p}. \end{aligned}$$

On the other hand, observe that one can make the function

$$F(x_1,\ldots,x_k) = x_1^{p_1} + \ldots + x_k^{p_k},$$

subject to the constrain

$$x_1 x_2 \ldots x_k = \frac{1}{\alpha},$$

as big as one wish. Indeed if $x_1 = \frac{M}{\alpha}$, $x_2 = \frac{1}{M}$ and $x_j = 1$ for $3 \leq j \leq k$. Then

$$\begin{aligned} F(x_1,\ldots,x_k) &= x_1^{p_1} + x_2^{p_2} + \ldots + x_k^{p_k} \\ &= \left(\frac{M}{\alpha}\right)^{p_1} + \left(\frac{1}{M}\right)^{p_1} + 1 + \ldots + 1 \\ &= \left(\frac{M}{\alpha}\right)^{p_1} + \left(\frac{1}{M}\right)^{p_1} + k - 2 \to \infty, \end{aligned}$$

as $M \to \infty$, therefore the critical part (5.23) is a minimum. Then

$$D_{f_1 \ldots f_k}(\alpha) \leq \frac{1}{p\alpha^p} \left[\prod_{j=1}^{k} p_j^{\frac{1}{p_j}}\right]^p$$

$$\alpha^p D_{f_1 \ldots f_k}(\alpha) \leq \frac{1}{p} \left[\prod_{j=1}^{k} p_j^{\frac{1}{p_j}}\right]^p,$$

thus, we have

$$\alpha^p D_{f_1 \ldots f_k}(\alpha) \leq \frac{1}{p} \left[\prod_{j=1}^{k} p_j^{\frac{1}{p_j}}\right]^p$$

$$\left\|f_1 \ldots f_k \mid L_{(p,\infty)}\right\| \leq \left(\frac{1}{p}\right)^{\frac{1}{p}} \left(\prod_{j=1}^{k} p_j^{\frac{1}{p_j}}\right) \prod_{j=1}^{k} \left\|f_j \mid L_{(p_j,\infty)}\right\|, \tag{5.24}$$

since $\left\|f_j \mid L_{(p_j,\infty)}\right\| = 1$.

In general, if $\left\|f_j \mid L_{(p_j,\infty)}\right\| \neq 1$, $1 \leq j \leq k$ choose $g_j = \dfrac{f_j}{\|f_j\|_{L_{(p_j,\infty)}}}$ and use (5.24). □

We now show that the weak Lebesgue space is complete using the quasi-norm.

Theorem 5.24. *The weak Lebesgue space with the quasi-norm $\|\cdot \mid L_{(p,\infty)}\|$ is complete for all $0 < p < \infty$.*

Proof. Let $\{f_n\}_{n\in\mathbb{N}}$ be a Cauchy sequence in $(L_{(p,\infty)}, \|\cdot \mid L_{(p,\infty)}\|)$. Then for every $\varepsilon > 0$ there exists an $n_0 \in \mathbb{N}$ such that

$$\|f_n - f_m \mid L_{(p,\infty)}\| < \varepsilon^{\frac{1}{p}+1}$$

if $m,n \geq n_0$, that is,

$$\left(\sup_{\lambda>0} \lambda^p D_{f_n - f_m}(\lambda)\right)^{1/p} = \|f_n - f_m \mid L_{(p,\infty)}\| < \varepsilon^{\frac{1}{p}+1}.$$

Taking $\lambda = \varepsilon$ we have

$$\varepsilon^p \mu\left(\{x \in X : |f_n(x) - f_m(x)| > \varepsilon\}\right) < \varepsilon^{p+1},$$

for $m,n \geq n_0$. Hence

$$\mu\left(\{x \in X : |f_n(x) - f_m(x)| > \varepsilon\}\right) < \varepsilon,$$

for $m,n \geq n_0$. This means that $\{f_n\}_{n\in\mathbb{N}}$ is a Cauchy sequence in the measure μ. We therefore apply Theorem 5.8 and conclude that there exists an $\mathscr{A}$-measurable function f such that some subsequence of $\{f_n\}_{n\in\mathbb{N}}$ converges to f μ-a.e. Let $\{f_{n_k}\}_{k\in\mathbb{N}}$ be such subsequence of $\{f_n\}_{n\in\mathbb{N}}$ of $\{f_n\}_{n\in\mathbb{N}}$ then $f_{n_k} \to f$ μ-a.e. as $k \to \infty$. If we apply twice the Lemma 5.20 we obtain

$$\begin{aligned}\|f \mid L_{(p,\infty)}\| &= \left\|\liminf |f_{n_k}| \mid L_{(p,\infty)}\right\| \\ &\leq \liminf \left\|f_{n_k} \mid L_{(p,\infty)}\right\|,\end{aligned}$$

which is finite, thus $f \in L_{(p,\infty)}$.

We also have

$$\begin{aligned}\|f - f_n \mid L_{(p,\infty)}\| &= \left\|\liminf |f_{n_k} - f_n| \mid L_{(p,\infty)}\right\| \\ &\leq \liminf \left\|f_{n_k} - f_n \mid L_{(p,\infty)}\right\| \\ &< \varepsilon^{\frac{1}{p}+1},\end{aligned}$$

if $n_k, n \geq n_0$, which proves that $L_{(p,\infty)}$ is complete for $0 < p < \infty$. □

5.5 Problems

5.25. Let $(\mathbb{R}, \mathscr{L}, m)$ be the Lebesgue measure space. If $A \in \mathscr{L}$ with $m(A) < \infty$ and $f = \chi_A$, show that

$$\lim_{p\to\infty} \frac{\|f\|_p^p}{\|f\|_{(p,\infty)}^p} = 1.$$

5.26. Let $([0,a], \mathscr{L}, m)$ be the Lebesgue measure space. Define $g(x) = x$ and show that

$$\lim_{n\to\infty} \frac{\|g\|_n^n}{\|g\|_{(n,\infty)}^n} = \mathrm{e}.$$

5.27. Let $([0,a],\mathscr{L},m)$ be the Lebesgue measure space. Define $g(x)=x^n$ and show that

$$\lim_{n\to\infty}\frac{\|g\|_{n^2}^{n^2}}{\|g\|_{(n^2,\infty)}^{n^2}}=\mathrm{e}.$$

5.28. Let $([0,1],\mathscr{L},m)$ be the Lebesgue measure space. Define $g(x)=\mathrm{e}^x$ and show that

$$\lim_{p\to\infty}\frac{\|g\|_p^p}{\|g\|_{(p,\infty)}^p}=\mathrm{e}.$$

5.29. Let $([1,\infty],\mathscr{L},m)$ be the Lebesgue measure space. Define $g(x)=\frac{1}{x^2}$ and show that

$$\lim_{p\to\infty}\frac{\|g\|_p^p}{\|g\|_{(p,\infty)}^p}=\mathrm{e}.$$

5.30. Let $([-a,a],\mathscr{L},m)$ be the Lebesgue measure space. Define $g(x)=a^2-x^2$ and show that

$$\lim_{p\to\infty}\frac{\|g\|_p^p}{\|g\|_{(p,\infty)}^p}=\sqrt{\frac{\pi\mathrm{e}}{2}}.$$

5.6 Notes and Bibliographic References

This chapter is based upon Castillo, Vallejo Narvaez, and Ramos Fernández [5].

Chapter 6
Lorentz Spaces

Abstract The spaces considered in the previous chapters are one-parameter dependent. We now study the so-called Lorentz spaces which are a scale of function spaces which depend now on two parameters. Our first task therefore will be to define the Lorentz spaces and derive some of their properties, like completeness, separability, normability, duality among other topics, e.g., Hölder's type inequality, Lorentz sequence spaces, and the spaces $L\exp$ and $L\log L$, which were introduced by Zygmund and Titchmarsh.

6.1 Lorentz Spaces

We start to recall that, by Remark 4.14, we can calculate the Lebesgue norm of a function using the notion of nonincreasing rearrangement in the following way

$$\int_{\mathbb{R}^n} |f(x)|^p \,\mathrm{d}\mu(x) = \int_0^\infty f^*(t)^p \,\mathrm{d}t = \int_0^\infty \left(t^{\frac{1}{p}} f^*(t)\right)^p \frac{\mathrm{d}t}{t}.$$

Using the right-hand side representation we can try to replace the power p by q to obtain some kind of generalization. It turns out that this indeed will produce a new scale of function spaces which have the Lebesgue spaces as a special case.

Definition 6.1. Let $(X, \mathscr{A}, \mu)$ be a measure space. For any $f \in \mathfrak{F}(X, \mathscr{A})$ and any two extended real numbers p and q in the set $[1, \infty]$ put

R.E. Castillo, H. Rafeiro, *An Introductory Course in Lebesgue Spaces*, CMS Books in Mathematics, DOI 10.1007/978-3-319-30034-4_6

$$\left\| f \mid L_{(p,q)} \right\| = \|f\|_{(p,q)} = \begin{cases} \left(\int\limits_0^\infty \left(t^{1/p} f^*(t) \right)^q \frac{\mathrm{d}t}{t} \right)^{1/q}, & q < \infty \\ \sup\limits_{t>0} t^{1/p} f^*(t), & q = \infty. \end{cases} \tag{6.1}$$

The functionals $\| \cdot \|_{(p,q)}$ are thus extended nonnegative valued functions on $\mathfrak{F}(X, \mathscr{A})$ and the Lorentz space will be defined in terms of these functions just as the Lebesgue spaces were defined in terms of the functional $\| \cdot \|_p$. ⊘

In view of Remark 4.14 (ii), it is obvious that

$$\|f\|_p = \|f\|_{(p,p)}, \tag{6.2}$$

for any function $f \in \mathfrak{F}(X, \mathscr{A})$ and any $p \in [1, \infty)$ and this equality also holds for $p = \infty$. An easy calculation, using the fact that

$$\chi_A^*(t) = \chi_{[0,\mu(A)]}(t),$$

will show that

$$\|\chi_A\|_{(p,q)} = \begin{cases} \left(\frac{p}{q}\right)^{1/q} \left(\mu(A)\right)^{1/p}, & 1 \le p, q < \infty \\ \infty, & p = \infty, q < \infty \\ \left(\mu(A)\right)^{1/p}, & 1 \le p < \infty, q = \infty \\ 1, & p = q = \infty, \end{cases}$$

for any set $A \in \mathscr{A}$ for which $0 < \mu(A) < \infty$.

We are now in conditions to introduce the Lorentz spaces

Definition 6.2. For any measure space $(X, \mathscr{A}, \mu)$ and any two extended real numbers p and q in the interval $[0, \infty]$ the set

$$\mathbf{L}_{(p,q)}(X, \mathscr{A}, \mu) = \left\{ f \in \mathfrak{F}(X, \mathscr{A}) : \|f\|_{(p,q)} < \infty \right\},$$

is called the *pre-Lorentz spaces* associated with $(X, \mathscr{A}, \mu)$. ⊘

The space $\mathbf{L}_{(p,q)}(X, \mathscr{A}, \mu)$ with $1 \le q < \infty$ and $p = \infty$ will not be of any interest. In fact, if f is a function in $\mathfrak{F}(X, \mathscr{A})$ with the property that $\|f\|_{(\infty,q)} < \infty$ for some $q \in [1, \infty)$, then

$$\int\limits_0^\infty [f^*(t)]^q \frac{\mathrm{d}t}{t} \ge \int\limits_0^s [f^*(t)]^q \frac{\mathrm{d}t}{t} \ge [f^*(s)]^q \int\limits_0^s \frac{\mathrm{d}t}{t},$$

since $f^*(t) \geq f^*(s)$ whenever $0 \leq t \leq s$ and therefore $f^*(s) = 0$ for all $s > 0$. Thus $f = 0$ μ-a.e. by (6.2). For this reason we have $\mathbf{L}_{(p,q)}(X, \mathscr{A}, \mu) = \{0\}$ for every $0 < q < \infty$.

Theorem 6.3 *Let $(X, \mathscr{A}, \mu)$ be a measure space and let p, q and r be three extended real numbers satisfying $0 < p \leq \infty$ and $0 < q < r \leq \infty$. Then*

(a) $\|f\|_{(p,r)} \leq \left(\frac{q}{p}\right)^{\frac{1}{q}-\frac{1}{r}} \|f\|_{(p,q)}$ *for any $f \in \mathfrak{F}(X, \mathscr{A})$.*

(b) $\mathbf{L}_{(p,q)}(X, \mathscr{A}, \mu) \subseteq \mathbf{L}_{(p,r)}(X, \mathscr{A}, \mu)$.

Proof. It is clearly sufficient to prove (i). Indeed for $r = \infty$, we have

$$\begin{aligned} \|f\|_{(p,q)}^q &= \int_0^\infty \left(t^{1/p} f^*(t)\right)^q \frac{dt}{t} \\ &\geq \int_0^s \left(t^{1/p} f^*(s)\right)^q \frac{dt}{t} \\ &= \left(\frac{p}{q}\right) s^{\frac{q}{p}} \left(f^*(s)\right)^q, \end{aligned}$$

for any $s > 0$ thus

$$\|f\|_{(p,\infty)} = \sup_{t>0} t^{1/p} f^*(t) \leq \left(\frac{q}{p}\right)^{\frac{1}{q}} \|f\|_{(p,q)}. \tag{6.3}$$

On the other hand, if $1 \leq q < r < \infty$, then

$$\begin{aligned} \|f\|_{(p,r)}^r &= \int_0^\infty \left(t^{1/p} f^*(t)\right)^{r-q} \left(t^{1/p} f^*(t)\right)^q \frac{dt}{t} \\ &\leq \|f\|_{(p,\infty)}^{r-q} \int_0^\infty \left(t^{1/p} f^*(t)\right)^q \frac{dt}{t} \\ &\leq \left(\frac{q}{p}\right)^{\frac{r-q}{q}} \|f\|_{(p,q)}^{r-q} \|f\|_{(p,q)}^q \\ &= \left(\frac{q}{p}\right)^{\frac{r-q}{q}} \|f\|_{(p,q)}^r, \end{aligned}$$

by definition of $\|\cdot\|_{(p,\infty)}$ and (6.3), and this completes the proof of Theorem 6.3. □

A natural interrogation is to ask if the functional $\|\cdot\|_{(p,q)}$ defined in (6.1) is a norm on $\mathbf{L}_{(p,q)}(X, \mathscr{A}, \mu)$. The following result gives us some light on this regard.

Theorem 6.4. *If $(X, \mathscr{A}, \mu)$ is a measure space and if $p \in [1, \infty]$. Then*

(a) $\|f+g\|_{(p,q)} \leq 2^{1/p}\left(\|f\|_{(p,q)} + \|g\|_{(p,q)}\right)$ *for any two functions $f, g \in \mathfrak{F}(X, \mathscr{A})$.*

(b) $\mathbf{L}_{(p,q)}(X,\mathscr{A},\mu)$ *is a vector space.*

(c) If f *is a function in* $\mathfrak{F}(X,\mathscr{A})$*, then* $\|f\|_{(p,q)}=0$ *if and only if* $f=0$ μ*-a.e.*

Proof. (a) Assume that f and g are two functions in $\mathfrak{F}(X,\mathscr{A})$. The particular case of theorem1.2.3 implies that

$$\begin{aligned}\|f+g\|_{(p,\infty)} &= \sup_{t>0} t^{1/p}(f+g)^*(t)\\ &\leq \sup_{t>0} t^{1/p}\Big[f^*(t/2)+g^*(t/2)\Big]\\ &\leq 2^{1/p}\Big(\|f\|_{(p,\infty)}+\|g\|_{(p,\infty)}\Big)\end{aligned}$$

and, if $q<\infty$, that

$$\begin{aligned}\|f+g\|_{(p,q)} &= \left(\int_0^\infty \Big(t^{1/p}(f+g)^*(t)\Big)^q \frac{\mathrm{d}t}{t}\right)^{1/q}\\ &\leq \left(\int_0^\infty \Big[t^{1/p}\Big(f^*(t/2)+g^*(t/2)\Big)\Big]^q \frac{\mathrm{d}t}{t}\right)^{1/q}.\end{aligned}$$

By Minkowski's inequality we have

$$\begin{aligned}\|f+g\|_{(p,q)} &\leq \left(\int_0^\infty \Big(t^{1/p}f^*(t/2)\Big)^q \frac{\mathrm{d}t}{t}\right)^{1/q} + \left(\int_0^\infty \Big(t^{1/p}g^*(t/2)\Big)^q \frac{\mathrm{d}t}{t}\right)^{1/q}\\ &= 2^{1/p}\left[\left(\int_0^\infty \Big(t^{1/p}f^*(t)\Big)^q \frac{\mathrm{d}t}{t}\right)^{1/q} + \left(\int_0^\infty \Big(t^{1/p}g^*(t)\Big)^q \frac{\mathrm{d}t}{t}\right)^{1/q}\right]\\ &= 2^{1/p}\Big(\|f\|_{(p,q)}+\|g\|_{(p,q)}\Big).\end{aligned}$$

(b) Item (a) implies that if f and g belong to $\mathbf{L}_{(p,q)}(X,\mathscr{A},\mu)$, then so does $f+g$. Since it is easy to see that $\|cf\|_{(p,q)}=|c|\|f\|_{(p,q)}$ for any $c\in\mathbb{R}$ and any $f\in\mathfrak{F}(X,\mathscr{A})$ this shows that $\mathbf{L}_{(p,q)}(X,\mathscr{A},\mu)$ is a vector space.

(c) Suppose that $\|f\|_{(p,q)}=0$, then

$$0 = \int_0^\infty \left(t^{1/p} f^*(t)\right)^q \frac{dt}{t} \geq \int_0^s \left(t^{1/p} f^*(s)\right)^q \frac{dt}{t}$$

$$= \left(\frac{p}{q}\right) s^{q/p} \left(f^*(s)\right)^q \geq 0$$

hence $f^*(s) = 0$ for all $s > 0$, from this and Remark 4.14 (ii) we have $\int_X |f| \, d\mu = \int_0^\infty f^*(s) \, ds = 0$ which implies $f = 0$ μ-a.e.

□

Part (a) of this theorem leaves unanswered the question of whether $\|\cdot\|_{(p,q)}$ is a norm on $\mathbf{L}_{(p,q)}(X, \mathscr{A}, \mu)$. It turns out that $\|\cdot\|_{(p,q)}$ is a norm on $\mathbf{L}_{(p,q)}(X, \mathscr{A}, \mu)$ if $1 \leq q \leq p < \infty$ (see corollary 6.8). Next we will see that, in general $\|\cdot\|_{(p,q)}$ is not a norm but equivalent to one if $1 < p < q \leq \infty$.

Theorem 6.5. *Let $(X, \mathscr{A}, \mu)$ be a nonatomic measure space. Then*

$$\|\cdot\|_{(p,q)} : \mathbf{L}_{(p,q)}(X, \mathscr{A}, \mu) \to \mathbb{R}^+$$

is not a norm for:

(a) $1 \leq p < q \leq \infty$.
(b) $0 < p < 1$, $0 < q \leq \infty$.
(c) $0 < p < 1 \leq q < \infty$.

Proof. (a) We start with the case when $1 \leq p < q < \infty$. Take

$$f(x) = (1+\varepsilon)\chi_{[0,a+h]}(x) + \chi_{[a+h,a+2h]}(x),$$

and

$$g(x) = \chi_{[0,h]}(x) + (1+\varepsilon)\chi_{[a+h,a+2h]}(x),$$

where a, h, $\varepsilon > 0$. It is easy to see that $f^*(t) = g^*(t) = f(x)$ and since

$$(f+g)^*(t) = (2+2\varepsilon)\chi_{[0,a]}(t) + (2+\varepsilon)\chi_{[a,a+2h]}(t),$$

it follows that we can evaluate the norm of f, g and $f+g$ by

$$\|f\|_{(p,q)}^q = \|g\|_{(p,q)}^q = \frac{p}{q}\left[(1+\varepsilon)^q (a+h)^{q/p} + (2a+h)^{q/p} - (a+h)^{q/p}\right],$$

and

$$\|f+g\|_{(p,q)}^q = \frac{p}{q}\left[(2+2\varepsilon)^q a^{q/p} + (2+\varepsilon)^q \left((a+2h)^{q/p} - a^{q/p}\right)\right],$$

respectively. Now, let us assume that the triangle inequality holds, that is

$$\|f+g\|_{(p,q)} \le \|f\|_{(p,q)} + \|g\|_{(p,q)}.$$

Then the inequality

$$(2+2\varepsilon)^q a^{q/p} + (2+\varepsilon)^q\left((a+2h)^{q/p} - a^{q/p}\right)$$
$$\le 2^q\left((1+\varepsilon)^q(a+h)^{q/p} + (a+2h)^{q/p} - (a+h)^{q/p}\right),$$

holds and it can be written as

$$(a+2h)^{q/p} - (a+h)^{q/p} \le \frac{(1+\varepsilon)^q - (1+\varepsilon/2)^q}{(1+\varepsilon/2)^q - 1}\left((a+h)^{q/p} - a^{q/p}\right).$$

Taking $\varepsilon \to 0$, we obtain that

$$(a+2h)^{q/p} + a^{q/p} \le 2(a+h)^{q/p}. \tag{6.4}$$

If we define a function f as

$$f(x) = \int_0^x t^{\frac{q}{p}-1}\,\mathrm{d}t,$$

then we can rewrite inequality (6.4) as

$$f(a+2h) + f(a) \le 2f(a+h),$$

which implies, together with the fact that f is continuous, that it is a concave function. By the concavity of f the derivative $f'(x) = x^{\frac{q}{p}-1}$ must be a decreasing function, that is, it must be that $q \le p$. This contradicts $q > p$, and we concluded that the triangle inequality does not hold. For $q = \infty$, take measurable sets $A \subset B \subset X$ such that

$$a = \left(\frac{\mu(B)}{\mu(A)}\right)^{1/p} > 1,$$

and $\mu(B\backslash A) \le \mu(A)$. If we let

$$f(x) = a\chi_A(x) + \chi_{B\backslash A}(x),$$

and

$$g(x) = \chi_A(x) + a\chi_{B\backslash A}(x),$$

then

$$f^*(t) = a\chi_{[0,\mu(A)]}(t) + \chi_{[\mu(A),\mu(B)]}(t),$$

and

$$g^*(t) = a\chi_{[0,\mu(B\backslash A)]}(t) + \chi_{[\mu(B\backslash A),\mu(B)]}(t).$$

Thus

$$\|f\|_{(p,\infty)} = \max\left(a\Big(\mu(A)\Big)^{1/p}, \Big(\mu(B)\Big)^{1/p}\right) = \Big(\mu(B)\Big)^{1/p}$$

$$\|g\|_{(p,\infty)} = \max\left(a\Big(\mu(B\backslash A)\Big)^{1/p}, \Big(\mu(B)\Big)^{1/p}\right) = \Big(\mu(B)\Big)^{1/p},$$

and since $f(x)+g(x) = (a+1)\chi_B$, we have $\|f+g\|_{(p,\infty)} = (a+1)\chi_B$. Then

$$\begin{aligned}\|f+g\|_{(p,\infty)} &= (a+1)\Big(\mu(B)\Big)^{1/p} \\ &> 2\Big(\mu(B)\Big)^{1/p} \\ &= \|f\|_{(p,\infty)} + \|g\|_{(p,\infty)},\end{aligned}$$

which shows that the triangle inequality does not hold. Hence the proof of the first case is complete.

(b) Let $0 < p, q < \infty$ and A, B be two measurable sets such that $A \cap B \neq \emptyset$. Then

$$\|\chi_A + \chi_B\|_{(p,q)} = \left(\frac{p}{q}\right)^{1/q}\Big(\mu(A)+\mu(B)\Big)^{1/p},$$

and

$$\|\chi_A\|_{(p,q)} + \|\chi_B\|_{(p,q)} = \left(\frac{p}{q}\right)^{1/q}\left(\Big[\mu(A)\Big]^{1/p} + \Big[\mu(B)\Big]^{1/p}\right).$$

The triangle inequality gives

$$\Big(\mu(A)+\mu(B)\Big)^{1/p} \leq \Big[\mu(A)\Big]^{1/p} + \Big[\mu(B)\Big]^{1/p},$$

and it fails for any $0 < q < \infty$ if $0 < p < 1$.
If $q = \infty$, we get the same norms.

(c) Let $0 < p < \infty$ and $0 < q < 1$. Define

$$f(x) = \begin{cases} 2 \text{ if } 0 < x < 2^{-p} \\ 0 \text{ otherwise,} \end{cases}$$

and

$$g(x) = \begin{cases} 4 \text{ if } 0 < x < 2^{-2p} \\ 0 \text{ otherwise.} \end{cases}$$

Then

$$\|f\|_{(p,q)} = \|g\|_{(p,q)} = \left(\frac{p}{q}\right)^{1/q},$$

and since

$$f(x)+g(x) = \begin{cases} 6 \text{ if } 0 < x < 2^{-2p} \\ 2 \text{ if } 2^{-2p} < x < 2^{-p}, \end{cases}$$

the decreasing rearrangement of $f+g$ is equal to $f+g$, i.e., $(f+g)^* = f+g$. Thus

$$\|f+g\|_{(p,q)} = \left(\frac{p}{q}\right)^{1/q} \left(2^{2-2q} + 2^{1-q}\right)^{1/q}.$$

Assume that the triangle inequality holds. Then

$$2^{2-2q} + 2^{1-q} \leq 2^q,$$

which can be written as

$$4 \leq 2^q(2^{2q}-2) < 2(2^2-2) = 4.$$

Hence, we have a contradiction and the assumption that the triangle inequality holds is wrong. This proves the third case. □

The following result gives us a characterization of $\mathbf{L}_{(p,q)}(X,\mathscr{A},\mu)$ in terms of the distribution function.

Theorem 6.6. *Let $(X,\mathscr{A},\mu)$ be a σ-finite measure space. For $0 < p < \infty$ and $0 < q \leq \infty$ we have the identity*

$$\|f\|_{(p,q)} = p^{1/q} \left(\int_0^\infty \left[\lambda \left(D_f(\lambda) \right)^{1/p} \right]^q \frac{d\lambda}{\lambda} \right)^{1/q}.$$

Proof. Case $q = \infty$. For this case let us define

$$C = \sup_{\lambda > 0} \left\{ \lambda^p D_f(\lambda) \right\}^{1/p},$$

then

$$D_f(\lambda) \leq \frac{C^p}{\lambda^p}.$$

Choosing $t = \frac{C^p}{\lambda^p}$ we have $\lambda = \frac{C}{t^{1/p}}$, and thus it is clear that

$$f^*(t) = \inf\left\{ \lambda > 0 : D_f(\lambda) \leq t \right\} \leq \frac{C}{t^{1/p}}.$$

Hence $t^{1/p} f^*(t) \leq C$, for all $t > 0$, then

$$\sup_{t>0} t^{1/p} f^*(t) \leq C. \tag{6.5}$$

On the other hand, given $\lambda > 0$ choose ε satisfying $0 < \varepsilon < \lambda$, Theorem 4.5(b) yields $f^*(D_f(\lambda) - \varepsilon) > \lambda$ which implies that

$$\begin{aligned}\sup_{t>0} t^{\frac{1}{p}} f^*(t) &\geq (D_f(\lambda) - \varepsilon)^{\frac{1}{p}} f^*\left(D_f(\lambda) - \varepsilon\right) \\ &> (D_f(\lambda) - \varepsilon)^{\frac{1}{p}} \lambda.\end{aligned}$$

We first let $\varepsilon \to 0$ and take the supremum over all $\lambda > 0$ to obtain

$$\begin{aligned}\sup_{t>0} t^{\frac{1}{p}} f^*(t) &\geq \sup_{\lambda>0} \lambda \left(D_f(\lambda)\right)^{\frac{1}{p}} \\ &= \sup_{\lambda>0} \left\{\lambda^p D_f(\lambda)\right\}^{\frac{1}{p}}.\end{aligned} \tag{6.6}$$

Combining (6.5) and (6.6) we obtain

$$\|f\|_{(p,\infty)} = \sup_{t>0} t^{1/p} f^*(t) = \sup_{\lambda>0} \left\{\lambda^p D_f(\lambda)\right\}^{1/p}.$$

Case $0 < q < \infty$. In this case we use Theorem 4.5(b) and the Fubini theorem, indeed

$$\begin{aligned}\|f\|_{(p,q)}^q &= \int_0^\infty \left(t^{1/p} f^*(t)\right)^q \frac{dt}{t} \\ &= \int_0^\infty \left(f^*(t)\right)^q t^{\frac{q}{p}-1}\, dt \\ &= \int_0^\infty \left(\int_0^{f^*(t)} q\lambda^{q-1}\, d\lambda\right) t^{\frac{q}{p}-1}\, dt \\ &= \int_0^\infty \left(\int_0^\infty q\lambda^{q-1} \chi_{\{\lambda>0: f^*(t)>\lambda\}}(\lambda)\, d\lambda\right) t^{\frac{q}{p}-1}\, dt \\ &= \int_0^\infty \left(\int_0^\infty q\lambda^{q-1} t^{\frac{q}{p}-1} \chi_{\{t\geq 0: D_f(\lambda)>t\}}(t)\, dt\right) d\lambda \\ &= \int_0^\infty q\lambda^{q-1} \left(\int_0^\infty t^{\frac{q}{p}-1} \chi_{\left(0, D_f(\lambda)\right)}(t)\, dt\right) d\lambda\end{aligned}$$

$$= \int_0^\infty q\lambda^{q-1} \left(\int_0^{D_f(\lambda)} t^{\frac{q}{p}-1} \, dt \right) d\lambda$$

$$= p \int_0^\infty \lambda^{q-1} \left(D_f(\lambda)\right)^{\frac{q}{p}} d\lambda$$

$$= p \int_0^\infty \left[\lambda \left(D_f(\lambda) \right)^{\frac{1}{p}} \right]^q \frac{d\lambda}{\lambda}.$$

Finally

$$\|f\|_{(p,q)} = p^{1/q} \left(\int_0^\infty \left[\lambda \left(D_f(\lambda) \right)^{\frac{1}{p}} \right]^q \frac{d\lambda}{\lambda} \right)^{1/q}$$

□

The following result together with Theorem 6.12 (Hardy) will help us to prove the triangle inequality for $\| \cdot \|_{(p,q)}$.

Theorem 6.7. *Suppose that $(X, \mathscr{A}, \mu)$ is a nonatomic σ–finite measure space, that p and q are two numbers satisfying $1 \le q \le p < \infty$. In addition, let q' be the conjugate exponent to q and let $\mathfrak{F}_0$ be the set of nonnegative-value nonincreasing function on $[0, \infty)$. If h is any function in $\mathbf{L}_{(p,q)}(X, \mathscr{A}, \mu)$, then*

$$\|h\|_{(p,q)} = \sup \left\{ \int_0^\infty h^*(t) t^{\frac{1}{p}-\frac{1}{q}} k(t) \, dt : k \in \mathfrak{F}_0 \quad \text{and} \quad \|k\|_{L_{q'(0,\infty)}} = 1 \right\}.$$

Proof. By Hölder inequality we have

$$\int_0^\infty h^*(t) t^{\frac{1}{p}-\frac{1}{q}} k(t) \, dt \le \left(\int_0^\infty \left(t^{1/p} h^*(t) \right)^q \frac{dt}{t} \right)^{1/q} \left(\int_0^\infty \left(k(t) \right)^{q'} dt \right)^{1/q'},$$

then

$$\sup \left\{ \int_0^\infty h^*(t) t^{\frac{1}{p}-\frac{1}{q}} k(t) \, dt : k \in \mathfrak{F}_0 \quad \text{and} \quad \|k\|_{L_{q'(0,\infty)}} = 1 \right\} \le \|h\|_{(p,q)}. \tag{6.7}$$

Now, let $k(t) = c\left[t^{\frac{1}{p}-\frac{1}{q}}h^*(t)\right]^{q-1}$, clearly k is a nonnegative valued non-decreasing function on $[0,\infty)$ and

$$\left(\int_0^\infty \left(k(t)\right)^{q'}\,\mathrm{d}t\right)^{1/q'} = c\|h\|_{(p,q)}^{q/q'},$$

taking $c = \|h\|_{(p,q)}^{1-q}$, we obtain $\|k\|_{L_{q'}(0,\infty)} = 1$.
On the other hand, since $k(t) = c\left[t^{\frac{1}{p}-\frac{1}{q}}h^*(t)\right]^{q-1}$ we have

$$t^{\frac{1}{p}-\frac{1}{q}}h^*(t)k(t) = c\left[t^{\frac{1}{p}-\frac{1}{q}}h^*(t)\right]^{q-1},$$

then

$$\int_0^\infty h^*(t)t^{\frac{1}{p}-\frac{1}{q}}k(t)\,\mathrm{d}t = \|h\|_{(p,q)},$$

therefore

$$\|h\|_{(p,q)} = \sup\left\{\int_0^\infty h^*(t)t^{\frac{1}{p}-\frac{1}{q}}k(t)\,\mathrm{d}t : k \in \mathfrak{F}_0 \quad \text{and} \quad \|k\|_{L_{q'}(0,\infty)} = 1\right\} \tag{6.8}$$

Finally by (6.6) and (6.7) the result follows. □

Corollary 6.8 *Let $(X,\mathscr{A},\mu)$ be a σ–finite measure space. Suppose $1 \le q \le p < \infty$ with $q' = \frac{q}{q-1}$ and that f and g are two functions in $\mathbf{L}_{(p,q)}(X,\mathscr{A},\mu)$. Then*

$$\|f+g\|_{(p,q)} \le \|f\|_{(p,q)} + \|g\|_{(p,q)}. \tag{6.9}$$

Proof. The hypothesis $q \le p$ implies that $t^{\frac{1}{p}-\frac{1}{q}}$ is decreasing, hence $t^{\frac{1}{p}-\frac{1}{q}}k(t)$ is decreasing. We may apply Theorem 4.19 and Hölder's inequality to obtain

$$\begin{aligned}
\int_0^\infty t^{\frac{1}{p}-\frac{1}{q}}(f+g)^*(t)k(t)\,\mathrm{d}t &\le \int_0^\infty t^{\frac{1}{p}-\frac{1}{q}}(f)^*(t)k(t)\,\mathrm{d}t + \int_0^\infty t^{\frac{1}{p}-\frac{1}{q}}(g)^*(t)k(t)\,\mathrm{d}t \\
&\le \left(\int_0^\infty \left(t^{1/p}f^*(t)\right)^q \frac{\mathrm{d}t}{t}\right)^{1/q} \|k\|_{L_{q'}(0,\infty)} \\
&\quad + \left(\int_0^\infty \left(t^{1/p}g^*(t)\right)^q \frac{\mathrm{d}t}{t}\right)^{1/q} \|k\|_{L_{q'}(0,\infty)}
\end{aligned}$$

$$= \|f\|_{(p,q)} + \|g\|_{(p,q)}.$$

Since $\|k\|_{L_{q'}(0,\infty)} = 1$, this together with Theorem 6.6 establishes (6.9). □

The following result is a Hölder's type inequality.

Theorem 6.9. *Let $(X, \mathscr{A}, \mu)$ be a measure space, let p and q be two extended real numbers in $[1,\infty]$ and let p' and q' be their conjugate exponents. If $f \in \mathbf{L}_{(p,q)}(X, \mathscr{A}, \mu)$ and $g \in \mathbf{L}_{(p',q')}(X, \mathscr{A}, \mu)$, then*

$$\|fg\|_1 \leq \|f\|_{(p,q)} \|g\|_{(p',q')}.$$

Proof. By Theorem 4.15 and Hölder's inequality we have

$$\begin{aligned}
\int_X |fg| \, \mathrm{d}\mu &\leq \int_0^\infty f^*(t) g^*(t) \, \mathrm{d}t \\
&= \int_0^\infty \left(t^{\frac{1}{p} - \frac{1}{q}} f^*(t) \right) \left(t^{\frac{1}{p'} - \frac{1}{q'}} g^*(t) \right) \mathrm{d}t \\
&\leq \left(\int_0^\infty \left(t^{1/p} f^*(t) \right)^q \frac{\mathrm{d}t}{t} \right)^{1/q} \left(\int_0^\infty \left(t^{1/p'} g^*(t) \right)^{q'} \frac{\mathrm{d}t}{t} \right)^{1/q'} \\
&= \|f\|_{(p,q)} \|g\|_{(p',q')}.
\end{aligned}$$

□

6.2 Normability

One can associate the so-called Lorentz spaces with the pre-Lorentz spaces. In order to do that let us define a relation $\sim$ on $\mathbf{L}_{(p,q)}(X, \mathscr{A}, \mu)$ as follows:

$$f \sim g \quad \text{if and only if} \quad f = g \quad \mu - a.e.$$

It is not hard to prove that $\sim$ is an equivalence relation. Let us write

$$[f] = \left\{ g \in \mathfrak{F}(X, \mathscr{A}) : f = g \quad \mu - a.e \right\}$$

for each function $f \in \mathfrak{F}(X, \mathscr{A})$ and

$$L_{(p,q)}(X, \mathscr{A}, \mu) = \left\{ [f] : f \in \mathbf{L}_{(p,q)}(X, \mathscr{A}, \mu) \right\}.$$

It was stated in Theorem 6.4 (c) and Corollary 6.8 that $\|\cdot\|_{(p,q)}$ is norm on $\mathbf{L}_{(p,q)}(X,\mathscr{A},\mu)$ provided that $1 \le q \le p \le \infty$ but, in general it is not a norm for the remaining case. The reason for this is that the nonincreasing rearrangement operator is not sub-additive in the sense that, in general, the inequality $(f+g)^* \le f^* + g^*$ does not hold for any two measurable functions f and g. This means that one should not expect to be able to define a norm in terms of the nonincreasing rearrangement operator since the triangle inequality is not likely to be satisfied. The aim of the present section is to define an operator that is related to the nonincreasing rearrangement operator and that is sub-additive and which for $p > 1$, defines a norm on $L_{(p,q)}(X,\mathscr{A},\mu)$ equivalent to $\|\cdot\|_{(p,q)}$.

One can use the maximal function (see Definition 4.24) of f in place of f^* to define another two parameter family of functions on $\mathfrak{F}(X,\mathscr{A})$.

Definition 6.10. For $1 \le p < \infty$ and $1 \le q \le \infty$, the Lorentz spaces $L_{(p,q)}(X,\mathscr{A},\mu)$ is defined as

$$L_{(p,q)}(X,\mathscr{A},\mu) = \left\{ f \in \mathfrak{F}(X,\mathscr{A}) : \|f\|_{pq} < \infty \right\}$$

where $\|\cdot\|_{pq}$ is defined by

$$\|f \mid L_{pq}\| = \|f\|_{pq} = \begin{cases} \left(\int\limits_0^\infty \left(t^{1/p} f^{**}(t) \right)^q \frac{\mathrm{d}t}{t} \right)^{1/q}, & 1 \le p < \infty, 1 \le q < \infty \\ \sup\limits_{t>0} t^{1/p} f^{**}(t), & 1 \le p < \infty, q = \infty. \end{cases} \tag{6.10}$$

⊘

Remark 6.11. If $(X,\mathscr{A},\mu)$ is a nonatomic measure space, it follows from Remark 4.18 that

$$\int\limits_0^t (f+g)^*(s)\,\mathrm{d}s \le \int\limits_0^t f^*(s)\,\mathrm{d}s + \int\limits_0^t g^*(s)\,\mathrm{d}s \tag{6.11}$$

where $t = \mu(E)$ for $E \in \mathscr{A}$. Now, using Definition 4.24 and (6.11) we have

$$\begin{aligned} (f+g)^{**}(t) &= \frac{1}{t} \int\limits_0^t (f+g)^*(s)\,\mathrm{d}s \\ &\le \frac{1}{t} \int\limits_0^t f^*(s)\,\mathrm{d}s + \frac{1}{t} \int\limits_0^t g^*(s)\,\mathrm{d}s \\ &= f^{**}(t) + g^{**}(t), \end{aligned}$$

that is

$$(f+g)^{**}(t) \le f^{**}(t) + g^{**}(t) \tag{6.12}$$

for any two function $f, g \in \mathfrak{F}(X, \mathscr{A})$.

The sub-additivity of the maximal operator (6.12) means that if the set

$$\left\{ f \in \mathfrak{F}(X, \mathscr{A}) : \|f\|_{pq} < \infty \right\}$$

is a vector space, then $\|\cdot\|_{pq}$ is a norm on this spaces.

Now this spaces turns out to be identical to $\mathbf{L}_{(p,q)}(X, \mathscr{A}, \mu)$ provided that $p > 1$ (see Theorem 6.6) and these spaces are therefore normed space. On the other hand, just as f^{**} tends to be a more complicated function than f^*, the quantity $\|f\|_{pq}$ tends to be more difficult to work with than $\|f\|_{(p,q)}$. For example, if A is a set in $\mathscr{A}$ for which $0 < \mu(A) < \infty$, then

$$\|\chi_A\|_{pq} = \begin{cases} \left[\mu(A)\right]^{1/p} \left(\frac{p^2}{q(p-1)}\right)^{1/q}, & 1 \le q < \infty, 1 \le p < \infty \\ \infty, & p = 1, q < \infty. \\ \infty, & p = \infty, q < \infty \\ \left(\mu(A)\right)^{1/p}, & 1 \le p < \infty, q = \infty \\ 1, & p = q = \infty. \end{cases} \tag{6.13}$$

For fixed p and q in $[1, \infty]$ there are several inequalities relating the functions $\|\cdot\|_{(p,q)}$ and $\|.\|_{pq}$ and the key to deducing them is the following integral inequality due to Hardy.

The following integral inequality comes in different shapes; we will use the one given below since it is the key to deduce inequalities related to the functionals $\|\cdot\|_{(p,q)}$ and $\|.\|_{pq}$ for fixed p and q in $[1, \infty]$.

Theorem 6.12 (G. H. Hardy). *If f is a nonnegative "valued-measurable function on $[0, \infty)$ and if q and r are two numbers satisfying $1 \le q < \infty$ and $0 < r < \infty$, then*

(a)

$$\int_0^\infty \left(\int_0^t f(s)\, \mathrm{d}s \right)^q t^{-r-1}\, \mathrm{d}t \le \left(\frac{q}{r}\right)^q \int_0^\infty \left(sf(s)\right)^q s^{-r-1}\, \mathrm{d}s.$$

(b)

$$\int_0^\infty \left(\int_t^\infty f(s)\, \mathrm{d}s \right)^q t^{r-1}\, \mathrm{d}t \le \left(\frac{q}{r}\right)^q \int_0^\infty \left(sf(s)\right)^q s^{r-1}\, \mathrm{d}s.$$

Proof. (a) If $q = 1$, then by Fubini's theorem we have

$$\int_0^\infty \left(\int_0^t f(s)\, \mathrm{d}s \right) t^{-r-1}\, \mathrm{d}t = \int_0^\infty \int_s^\infty t^{-r-1} f(s)\, \mathrm{d}t\, \mathrm{d}s$$

$$= \frac{1}{r}\int_0^\infty \left[sf(s)\right]s^{-r-1}\,\mathrm{d}s$$

$$= \left(\frac{q}{r}\right)^q \int_0^\infty (sf(s))^q s^{-r-1}\,\mathrm{d}s.$$

Now, suppose that $q > 1$ and let p be the conjugate exponent of q. Then by Hölder's inequality with respect to the measure $s^{\frac{r}{q}-1}\,\mathrm{d}s$ we have

$$\begin{aligned}
\left(\int_0^t f(s)\,\mathrm{d}s\right)^q &= \left(\int_0^t f(s)s^{1-\frac{r}{q}}s^{\frac{r}{q}-1}\,\mathrm{d}s\right)^q \\
&\le \left(\int_0^t \left[f(s)\right]^q s^{q-r}s^{\frac{r}{q}-1}\,\mathrm{d}s\right)\left(\int_0^t s^{\frac{r}{q}-1}\,\mathrm{d}s\right)^{q/p} \\
&= \left(\frac{q}{r}t^{\frac{r}{q}}\right)^{q/p}\int_0^t \left[f(s)\right]^q s^{q-r}s^{\frac{r}{q}-1}\,\mathrm{d}s \\
&= \left(\frac{q}{r}\right)^{q/p} t^{\frac{r}{p}}\int_0^t \left[f(s)\right]^q s^{q-r}s^{\frac{r}{q}-1}\,\mathrm{d}s.
\end{aligned}$$

By integrating both sides from zero to infinity and using Fubini's theorem we have

$$\begin{aligned}
\int_0^\infty\left(\int_0^t f(s)\,\mathrm{d}s\right)^q t^{-r-1}\,\mathrm{d}t &\le \left(\frac{q}{r}\right)^{q/p}\int_0^\infty t^{-r-1+r(1-\frac{1}{q})}\int_0^t \left[f(s)\right]^q s^{q-r}s^{\frac{r}{q}-1}\,\mathrm{d}s\,\mathrm{d}t \\
&= \left(\frac{q}{r}\right)^{q/p}\int_0^\infty \left[f(s)\right]^q s^{q-r}s^{\frac{r}{q}-1}\left(\int_s^\infty t^{-1-\frac{r}{q}}\,\mathrm{d}t\right)\mathrm{d}s \\
&= \left(\frac{q}{r}\right)^{\frac{q}{p}+1}\int_0^\infty \left[sf(s)\right]^q s^{-r+\frac{r}{q}-1}s^{-r/q}\,\mathrm{d}s \\
&= \left(\frac{q}{r}\right)^q\int_0^\infty \left[sf(s)\right]^q s^{-r-1}\,\mathrm{d}s.
\end{aligned}$$

Hence

$$\int_0^\infty \left(\int_0^t f(s)\,\mathrm{d}s \right)^q t^{-r-1}\,\mathrm{d}t \le \left(\frac{q}{r}\right)^q \int_0^\infty \Big(sf(s)\Big)^q s^{-r-1}\,\mathrm{d}s.$$

(b) If $q = 1$ one more time by Fubini's theorem we have

$$\begin{aligned}
\int_0^\infty \left(\int_t^\infty f(s)\,\mathrm{d}s \right)^q t^{r-1}\,\mathrm{d}t &= \int_0^\infty \int_0^s t^{r-1} f(s)\,\mathrm{d}t\,\mathrm{d}s \\
&= \frac{1}{r} \int_0^\infty \Big(sf(s)\Big) s^{r-1}\,\mathrm{d}s \\
&= \left(\frac{q}{r}\right)^q \int_0^\infty \Big(sf(s)\Big)^q s^{r-1}\,\mathrm{d}s.
\end{aligned}$$

Next, let $q > 1$ and p be its conjugate exponent. Then from Hölder's inequality with respect to the measure $s^{-\frac{r}{q}-1}\,\mathrm{d}s$ we have

$$\begin{aligned}
\left(\int_t^\infty f(s)\,\mathrm{d}s \right)^q &= \left(\int_t^\infty f(s) s^{\frac{r}{q}+1} s^{-\frac{r}{q}-1}\,\mathrm{d}s \right) \\
&\le \left(\int_t^\infty \Big[f(s)\Big]^q s^{r+q} s^{-\frac{r}{q}-1}\,\mathrm{d}s \right) \left(\int_t^\infty s^{-\frac{r}{q}-1}\,\mathrm{d}s \right)^{q/p} \\
&= \left(\frac{q}{r}\right)^{\frac{q}{p}} t^{-\frac{r}{p}} \left(\int_t^\infty \Big[sf(s)\Big]^q s^{r-\frac{r}{q}-1}\,\mathrm{d}s \right).
\end{aligned}$$

By integrating both sides from zero to infinity and using Fubini's theorem we obtain

$$\begin{aligned}
\int_0^\infty \left(\int_t^\infty f(s)\,\mathrm{d}s \right)^q t^{r-1}\,\mathrm{d}t &\le \left(\frac{q}{r}\right)^{q/p} \int_0^\infty t^{-\frac{r}{p}+r-1} \left(\int_t^\infty \Big[sf(s)\Big]^q s^{r-\frac{r}{q}-1}\,\mathrm{d}s \right) \mathrm{d}t \\
&= \left(\frac{q}{r}\right)^{q/p} \int_0^\infty t^{\frac{r}{q}-1} \left(\int_t^\infty \Big[sf(s)\Big]^q s^{r-\frac{r}{q}-1}\,\mathrm{d}s \right) \mathrm{d}t \\
&= \left(\frac{q}{r}\right)^{q/p} \int_0^\infty \Big[sf(s)\Big]^q s^{r-\frac{r}{q}-1} \left(\int_0^s t^{\frac{r}{q}-1}\,\mathrm{d}t \right) \mathrm{d}s
\end{aligned}$$

$$= \left(\frac{q}{r}\right)^{\frac{q}{p}+1} \int_0^\infty \left[sf(s)\right]^q s^{r-\frac{r}{q}-1} s^{r/q}\,\mathrm{d}s$$

$$= \left(\frac{q}{r}\right)^q \int_0^\infty \left[sf(s)\right]^q s^{r-1}\,\mathrm{d}s.$$

Hence

$$\int_0^\infty \left(\int_t^\infty f(s)\,\mathrm{d}s\right)^q t^{r-1}\,\mathrm{d}t \le \left(\frac{q}{r}\right)^q \int_0^\infty \left(sf(s)\right)^q s^{r-1}\,\mathrm{d}s.$$

□

We now prove that the norm $\|\cdot\|_{pp}$ is equivalent with the Lebesgue norm $\|\cdot\|_p$, i.e., the diagonal case of Lorentz spaces coincides with the Lebesgue space.

Theorem 6.13 *If* $1 < p < \infty$ *and* $\frac{1}{p} + \frac{1}{q} = 1$, *then*

$$\|f\|_p \le \|f\|_{pp} \le \frac{p}{p-1}\|f\|_p.$$

Proof. Since $f^* \le f^{**}$ by Remark 4.14 (ii) we have

$$\|f\|_p^p = \int_0^\infty \left[f^*(t)\right]^p \mathrm{d}t \tag{6.14}$$

$$= \int_0^\infty \left[t^{1/p} f^*(t)\right] \frac{\mathrm{d}t}{t}$$

$$\le \int_0^\infty \left[t^{1/p} f^{**}(t)\right]^p \mathrm{d}t$$

$$= \|f\|_{pp}. \tag{6.15}$$

On the other hand, the second inequality follows immediately from the definition of f^{**} and the Hardy's inequality with $r = p-1$, that is,

$$\left(\int_0^\infty \left(t^{\frac{1}{p}-1} \int_0^t f(s)\,\mathrm{d}s\right)^p \frac{\mathrm{d}t}{t}\right)^{1/p} \le \frac{p}{p-1}\left(\int_0^\infty \left(f(s)\right)^p \mathrm{d}s\right)^{1/p}$$

thus

$$\|f\|_{pq} \le \frac{p}{p-1}\|f\|_p. \tag{6.16}$$

By (6.16) and (6.15) we obtain

$$\|f\|_p \le \|f\|_{pp} \le \frac{p}{p-1}\|f\|_p,$$

therefore the two norms are equivalent. □

We also have that the norms $\|\cdot\|_{(p,q)}$ and $\|\cdot\|_{pq}$ are equivalent.

Theorem 6.14 *If $(X,\mathscr{A},\mu)$ is a measure space and if p and q are two extended real numbers satisfying $1 < p \le \infty$ and $1 \le q \le \infty$, then*

$$\|f\|_{(p,q)} \le \|f\|_{pq} \le \frac{p}{p-1}\|f\|_{(p,q)}$$

for any $f \in \mathfrak{F}(X,\mathscr{A})$, where $\frac{p}{p-1}$ is to be interpreted as 1 if $p = \infty$.

Proof. Since $f^* \le f^{**}$, then

$$\begin{aligned}\|f\|_{(p,q)}^q &= \int_0^\infty \left(t^{1/p} f^*(t)\right)^q \frac{\mathrm{d}t}{t} \\ &\le \int_0^\infty \left(t^{1/p} f^{**}(t)\right)^q \frac{\mathrm{d}t}{t} \\ &= \|f\|_{pq}^q.\end{aligned}$$

Next, if $q < \infty$ and $p = \infty$, then, as was pointed out following Definition 6.2, either $f = 0$ a.e. or else $\|f\|_{(p,q)} = \infty$, and the second inequality is obvious in either case. If both q and p are finite, by Hardy's inequality Theorem (6.12) we have

$$\begin{aligned}\|f\|_{pq}^q &= \int_0^\infty \left(t^{1/p} f^{**}(t)\right)^q \frac{\mathrm{d}t}{t} \\ &= \int_0^\infty \left(t^{\frac{1}{p}-1} \int_0^t f^*(s)\,\mathrm{d}s\right)^q \frac{\mathrm{d}t}{t} \\ &= \int_0^\infty \left(\int_0^t f^*(s)\,\mathrm{d}s\right)^q t^{\frac{q}{p}-q-1}\,\mathrm{d}t \\ &\le \left(\frac{p}{p-1}\right)^q \|f\|_{(p,q)}^q.\end{aligned}$$

And, finally, if $q = \infty$, then

$$t^{1/p} f^{**}(t) = t^{\frac{1}{p}-1} \int_0^t f^*(s)\,\mathrm{d}s$$

$$= t^{\frac{1}{p}-1} \int_0^t s^{1/p} f^*(s) s^{-1/p} \, ds$$

$$\leq t^{\frac{1}{p}-1} \|f\|_{(p,q)} \int_0^t s^{-1/p} \, ds$$

$$= \frac{p}{p-1} \|f\|_{(p,q)},$$

for all $t > 0$ regardless of whether p is finite or infinite. □

The significance of this theorem is, of course, that for $1 < p \leq \infty$ and $1 \leq q \leq \infty$ the space $\mathbf{L}_{(p,q)}(X, \mathscr{A}, \mu)$ could equally well be defined to consist of those functions $f \in \mathfrak{F}(X, \mathscr{A})$ for which $\|f\|_{(p,q)} < \infty$ or for which $\|f\|_{pq} < \infty$. Note that for such p and q, the Theorem 6.4(c) and (6.12) imply that the function $\|\cdot\|_{pq}$ determines a norm on $L_{(p,q)}(X, \mathscr{A}, \mu)$.

The following two lemmas give embedding information regarding Lorentz spaces, namely they provide some comparison between the spaces $L_{(p,q)}(X, \mathscr{A}, \mu)$ and $L_{(p,r)}(X, \mathscr{A}, \mu)$.

Lemma 6.15. *Let $f \in L_{(p,q)}$. Then*

$$f^{**}(t) \leq \left(\frac{q}{p}\right)^{1/q} \frac{\|f\|_{pq}}{t^{1/p}}. \tag{6.17}$$

Proof. Note that

$$\|f\|_{pq}^q = \int_0^\infty \left(t^{1/p} f^{**}(t)\right)^q \frac{dt}{t}$$

$$\geq \int_0^t \left(f^{**}(s)\right)^q s^{\frac{q}{p}-1} \, ds$$

$$\geq \left[f^{**}(t)\right]^q \int_0^t s^{\frac{q}{p}-1} \, ds$$

$$= \frac{p}{q} \left[f^{**}(t)\right]^q t^{q/p},$$

from which (6.17) follows. □

Corollary 6.16 *Let $f \in L_{(p,q)}$. Then $\|f\|_{(p,\infty)} \leq \|f\|_{p\infty} \leq \left(\frac{q}{p}\right)^{1/q} \|f\|_{pq}$.*

We now show an embedding result in the framework of Lorentz spaces, namely $L_{(p,q)} \hookrightarrow L_{(p,r)}$ whenever $q < r$.

Lemma 6.17 (Calderón). *If* $1 < p < \infty$ *and* $1 \leq q < r \leq \infty$, *then*

$$\|f\|_{pr} \leq \left(\frac{q}{p}\right)^{\frac{1}{q}-\frac{1}{r}} \|f\|_{pq}.$$

Proof. Using Lemma 6.15 we have

$$\begin{aligned}
\|f\|_{pr}^r &= \int_0^\infty \left(f^{**}(t)\right)^r t^{\frac{r}{p}-1}\,\mathrm{d}t \\
&= \int_0^\infty \left(f^{**}(t)\right)^q \left(f^{**}(t)\right)^{r-q} t^{\frac{r}{p}-1}\,\mathrm{d}t \\
&\leq \int_0^\infty \left(f^{**}(t)\right)^q \left[\left(\frac{q}{p}\right)^{1/q} \frac{\|f\|_{pq}}{t^{1/p}}\right]^{r-q} t^{\frac{r}{p}-1}\,\mathrm{d}t \\
&= \left(\frac{q}{p}\right)^{\frac{r}{q}-1} \left(\|f\|_{pq}\right)^{r-q} (\|f\|_{pq})^q \\
&= \left(\frac{q}{p}\right)^{\frac{1}{q}-1} \left(\|f\|_{pq}\right)^r,
\end{aligned}$$

which ends the proof. □

6.3 Completeness

We are now ready to prove completeness which follows, as in the ordinary L_p case, from the Riesz theorem.

Theorem 6.18 (Completeness). *The normed space* $(L_{(p,q)}(X,\mathscr{A},\mu), \|\cdot\|_{pq})$ *is complete (Banach space) for all* $0 < p < \infty$ *and* $0 < q \leq \infty$.

Proof. Let $\{f_n\}_{n\in\mathbb{N}}$ be an arbitrary Cauchy sequence in $L_{(p,q)}(X,\mathscr{A},\mu)$. Then

$$\|f_m - f_n\|_{pq} \to 0 \quad \text{as} \quad n,m \to \infty,$$

and by Corollary 6.16 we have

$$\|f_m - f_n\|_{(p,\infty)} \leq \left(\frac{q}{p}\right)^{1/q} \|f_m - f_n\|_{pq} \to 0$$

as $n,m \to \infty$.

Thus

$$\sup_{t>0} t^{1/p} (f_m - f_n)^*(t) = \|f_m - f_n\|_{(p,\infty)} \to 0$$

as $n,m \to \infty$.

By Theorem 6.6 (case $q=\infty$) we have

$$\sup_{\lambda>0}\left\{\lambda^p D_{f_m-f_n}(\lambda)\right\}^{1/p} = \sup_{t>0} t^{1/p}\left(f_m-f_n\right)^*(t) \to 0$$

as $m,n\to\infty$, then

$$\sup_{\lambda>0}\left\{\lambda^p \mu\left(\{x\in X : |f_m(x)-f_n(x)|>\lambda\}\right)\right\}^{1/p} = \sup_{\lambda>0}\left\{\lambda^p D_{f_m-f_n}(\lambda)\right\}^{1/p} \to 0$$

as $m,n\to\infty$, this implies that

$$\mu\left(\{x\in X : |f_m(x)-f_n(x)|>\lambda\}\right) \to 0$$

as $m,n\to\infty$ for any $\lambda>0$.

We showed that $\{f_n\}_{n\in\mathbb{N}}$ is a Cauchy sequence in the measure μ. We can therefore apply F. Riesz's theorem and conclude that there exists an $\mathscr{A}$-measurable function f such that f_n converges to f in the measure μ. This implies again by a theorem of F. Riesz that there is a subsequence $\{f_{n_k}\}_{k\in\mathbb{N}}$ of $\{f_n\}_{n\in\mathbb{N}}$ which converges to f μ-a.e. on X.

Let $\varepsilon>0$ be arbitrary. Since $\{f_n\}_{n\in\mathbb{N}}$ is Cauchy there exists an $n_0\in\mathbb{N}$ such that

$$\|f_n - f_{n_0}\|_{pq} < \varepsilon \qquad (n>n_0)$$

and $f_{n_k}-f_{n_0}$ converge to $f-f_{n_0}$ μ-a.e. on X.

It follows now by Theorem 4.5 (g) that

$$(f-f_{n_0})^*(t) \le \liminf_{k\to\infty}(f_{n_k}-f_{n_0})^*(t)$$

for all $t>0$. Using the Fatou Lemma we have

$$(f-f_{n_0})^{**}(t) \le \liminf(f_{n_k}-f_{n_0})^{**}(t)$$

for all $t>0$. One more time by Fatou's Lemma we have

$$\|f-f_{n_0}\|_{pq} = \left(\int_0^\infty \left(t^{1/p}(f-f_{n_0})^{**}(t)\right)^q \frac{dt}{t}\right)^{1/q}$$

$$
\begin{aligned}
&\leq \left(\int_0^\infty \left(t^{1/p} \liminf_{k\to\infty} (f_{n_k} - f_{n_0})^{**}(t) \right)^q \frac{dt}{t} \right)^{1/q} \\
&\leq \liminf_{k\to\infty} \left(\int_0^\infty \left(t^{1/p} (f_{n_k} - f_{n_0})^{**}(t) \right)^q \frac{dt}{t} \right)^{1/q} \\
&= \liminf_{k\to\infty} \|f_{n_k} - f_{n_0}\|_{pq} < \varepsilon
\end{aligned}
$$

whenever $n_k > n_0$. Since $f = (f - f_{n_0}) + f_{n_0} \in L_{(p,q)}(X, \mathscr{A}, \mu)$ and this proves that $L_{(p,q)}(X, \mathscr{A}, \mu)$ is complete for all $0 < p < \infty$ and $0 < q \leq \infty$. □

6.4 Separability

To show that $L_{(p,q)}(X, \mathscr{A}, \mu)$ is separable we need to show that the set of all simple functions is dense in $L_{(p,q)}(X, \mathscr{A}, \mu)$. This desirable property means that any function in $L_{(p,q)}(X, \mathscr{A}, \mu)$ can be approximated by a simple function in the norm of $L_{(p,q)}(X, \mathscr{A}, \mu)$.

Theorem 6.19. *The set of all simple functions S is dense in $L_{(p,q)}(X, \mathscr{A}, \mu)$ for $0 < p < \infty$ and $0 < q < \infty$.*

Proof. Let $q < \infty$ and $f \in L_{(p,q)}(X, \mathscr{A}, \mu)$ be arbitrary. We can without loss of generality assume that f is positive, and then there exists a sequence of simple integrable functions such that $0 \leq s_n \leq f$ for all $n \in \mathbb{N}$ and $s_n \to f$ as $n \to \infty$. Hence, by Theorem 4.11 we have

$$
\begin{aligned}
(f - s_n)^*(t) &\leq f^*(t/2) + s_n^*(t/2) \\
&\leq 2f^*(t/2),
\end{aligned}
$$

since $s_n^*(t/2) \leq f^*(t/2)$ for all $n \in \mathbb{N}$, and thus, if we apply Lebesgue's dominated convergence theorem, and Theorem 6.14 we have

$$
\lim_{n\to\infty} \|f - s_n\|_{pq} \leq \frac{p}{p-1} \lim_{n\to\infty} \|f - s_n\|_{(p,q)} = 0.
$$

Since f was arbitrary this shows that $\overline{S} = L_{(p,q)}(X, \mathscr{A}, \mu)$ that is, S is dense in $L_{(p,q)}(X, \mathscr{A}, \mu)$. □

The separability of $L_{(p,q)}(X, \mathscr{A}, \mu)$ will now follow by showing that the set S from the previous theorem is countable and this is the case if and only if the measure is separable. Before we give the proof of this fact, we state the following definition.

Definition 6.20. A measure μ is separable if there exists a countable family $\mathfrak{H}$ of sets from $\mathscr{A}$ of finite measure such that for any $\varepsilon > 0$ and any set $A \in \mathscr{A}$ of finite measure we can find a set $B \in \mathfrak{H}$ with $\mu(A\triangle B) < \varepsilon$. ⊘

Theorem 6.21 (Separability). *The Lorentz $L_{(p,q)}(X, \mathscr{A}, \mu)$ space is separable for $0 < p < \infty$ and $0 < q < \infty$ if and only if the measure μ is separable.*

Proof. Assume first that μ is a separable measure and let A be any measurable set with finite measure and $\varepsilon > 0$ be arbitrary. Then there exists a countable family $\mathfrak{H}$ of subsets of X of finite measure and a set $B \in \mathfrak{H}$ such that

$$\mu\Big(A\Delta B\Big) = \mu\Big[(A\backslash B) \cup (B\backslash A)\Big] < \varepsilon.$$

It follows that

$$\begin{aligned}
\|\chi_A - \chi_B\|_{pq} &= \left(\int_0^\infty \Big(t^{1/p}(\chi_A - \chi_B)^{**}(t)\Big)^q \frac{\mathrm{d}t}{t}\right)^{1/q} \\
&= \left(\int_0^\infty t^{\frac{q}{p}-1}\chi^{**}_{A\Delta B}(t)\,\mathrm{d}t\right)^{1/q} \\
&= \left(\frac{p^2}{q(p-1)}\right)^{1/q}\Big(\mu\Big(A\Delta B\Big)\Big)^{1/p} \\
&< \left(\frac{p^2}{q(p-1)}\right)^{1/q}\varepsilon^{1/p}.
\end{aligned}$$

Hence, for any characteristic function of any $\mathscr{A}$-measurable set A of finite measure we can always find another characteristic function of a set $B \in \mathfrak{H}$ such that the norm of the difference between the two functions is as small as we wish.

Let s be a simple function of the following form

$$s = \sum_{j=1}^{m} \alpha_j \chi_{A_j}$$

where $\alpha_1 > \alpha_2 > \ldots > \alpha_m > 0$ and

$$A_j = \{x \in X : s(x) = \alpha_j\}$$

for all $j = 1, 2, \ldots, m$.
We can then define a new simple function

$$\widetilde{s} = \sum_{j=1}^{m} \alpha_j \chi_{B_j}$$

where $B_j \in \mathfrak{H}$ is chosen such that

$$\mu(A_j \Delta B_j) < \left(\frac{p^2}{q(p-1)}\right)^{1/q} \varepsilon^{1/p}$$

for all $j = 1, 2, \ldots, m$. It follows that

$$\|s - \widetilde{s}\|_{pq} = \left\| \sum_{j=1}^{m} \alpha_j \chi_{A_j \Delta B_j} \right\|_{pq}$$
$$< m \left(\frac{p^2}{q(p-1)}\right)^{1/q} \varepsilon^{1/p}.$$

That is, for any simple function s defined on $\mathscr{A}$-measurable sets $A_1, A_2, \ldots, A_m$, we can always find another simple functions $\widetilde{s}$, defined on sets $B_1, B_2, \ldots, B_m$ where $B_j \in \mathfrak{H}$; $j = 1, 2, \ldots, m$. Since the set of all simple functions is dense in $L_{(p,q)}(X, \mathscr{A}, \mu)$ it follows that the set

$$\widetilde{S} = \left\{ s = \sum_{j=1}^{m} \alpha_j \chi_{B_j} : B_j \in \mathfrak{H},\ \alpha_j \in \mathbb{R},\ m = 1, 2, 3, \ldots \right\}$$

is dense in $L_{(p,q)}(X, \mathscr{A}, \mu)$.

Moreover, since the countable set

$$\widetilde{S}_{\mathbb{Q}} = \left\{ \widetilde{s} = \sum_{j=1}^{m} \alpha_j \chi_{B_j} : B_j \in \mathfrak{H},\ \alpha_j \in \mathbb{Q},\ m = 1, 2, 3, \ldots \right\}$$

is dense in $\widetilde{S}$, it follows that $\widetilde{S}_{\mathbb{Q}}$ is dense in $L_{(p,q)}(X, \mathscr{A}, \mu)$. Hence $L_{(p,q)}(X, \mathscr{A}, \mu)$ is separable.

Now, assume that μ is not separable. Then there exists an $\varepsilon > 0$ and uncountable family of sets $\mathfrak{H}$ such that for A, B in $\mathfrak{H}$

$$\mu\left(A \Delta B\right) \geq \varepsilon.$$

Thus the set

$$H = \{f = \chi_A : A \in \mathfrak{H}\}$$

is uncountable and for $f, g \in H$ we have

$$\begin{aligned}\|f-g\|_{pq} &= \|\chi_A - \chi_B\|_{pq} \\ &= \|\chi_{A\Delta B}\|_{pq} \\ &= \left(\frac{p^2}{q(p-1)}\right)^{1/q} \left(\mu\left(A\Delta B\right)\right)^{1/p} \\ &\geq \left(\frac{p^2}{q(p-1)}\right)^{1/q} \varepsilon^{1/p}.\end{aligned}$$

where $A \neq B$. Hence, we have an uncountable set $H \subset L_{(p,q)}(X,\mathscr{A},\mu)$ such that for two functions $f,g \in H$, $\|f-g\|_{pq}$ is not as small as we wish, that is $L_{(p,q)}(X,\mathscr{A},\mu)$ is not separable. □

6.5 Duality

We will now take a closer look on the space of all bounded linear functionals on $L_{(p,q)}(X,\mathscr{A},\mu)$ which we will denote by

$$\left[L_{(p,q)}(X,\mathscr{A},\mu)\right]^*.$$

In the next theorem we collect the duality of Lorentz spaces depending on the different parameters.

Theorem 6.22. *Suppose that $(X,\mathscr{A},\mu)$ is a nonatomic σ-finite measure space. Then:*

(a) $\left(L_{(p,q)}(X,\mathscr{A},\mu)\right)^* = \{0\}$ *when* $0 < p < 1$, $0 < q \leq \infty$,

(b) $\left(L_{(p,q)}(X,\mathscr{A},\mu)\right)^* = L_\infty(X,\mathscr{A},\mu)$ *when* $p = 1$, $0 < q < 1$,

(c) $\left(L_{(p,q)}(X,\mathscr{A},\mu)\right)^* = \{0\}$ *when* $p = 1$, $1 < q < \infty$,

(d) $\left(L_{(p,q)}(X,\mathscr{A},\mu)\right)^* \neq \{0\}$ *when* $p = 1$, $q = \infty$,

(e) $\left(L_{(p,q)}(X,\mathscr{A},\mu)\right)^* = L_{(p',\infty)}(X,\mathscr{A},\mu)$ *when* $1 < p < \infty$, $0 < q \leq 1$,

(f) $\left(L_{(p,q)}(X,\mathscr{A},\mu)\right)^* = L_{(p',q')}(X,\mathscr{A},\mu)$ *when* $1 < p < \infty$, $1 < q < \infty$,

(g) $\left(L_{(p,q)}(X,\mathscr{A},\mu)\right)^* \neq \{0\}$ *when* $1 < p < \infty$, $q = \infty$,

(h) $\left(L_{(p,q)}(X,\mathscr{A},\mu)\right)^* \neq \{0\}$ *when* $p = q = \infty$.

Proof. Since X is σ-finite, we have that $X = \bigcup_{n=1}^{\infty} X_n$, where X_n is an increasing sequence of sets with $\mu\left(X_n\right) < \infty$ for all $n \in \mathbb{N}$. Given $T \in \left(L_{(p,q)}(X,\mathscr{A},\mu)\right)^*$ where $0 < p < \infty$ and $0 < q < \infty$. Let us define $\sigma(E) = T\left(\chi_E\right)$ for all $E \in \mathscr{A}$.

Next, we like to show that:

(i) σ defines a signed measure on X, and

(ii)

$$|\sigma(E)| \leq \left(\tfrac{p}{q}\right)^{1/q} \left(\mu(E)\right)^{1/p} \|T\| \quad \text{when} \quad q < \infty,$$

and

$$|\sigma(E)| \leq \|T\| \left(\mu(E)\right)^{1/p} \quad \text{for} \quad q = \infty.$$

(i) Note that

$$\sigma(\emptyset) = T\left(\chi_\emptyset\right) = T(0) = 0$$

Since T is a linear functional. On the other hand, let $\{E_n\}_{n\in\mathbb{N}} \in \mathscr{A}$ such that $E_n \cap E_m = \emptyset$ if $n \neq m$. Then

$$\sigma\left(\bigcup_{n=1}^{\infty} E_n\right) = T\left(\chi_{\bigcup_{n=1}^{\infty} E_n}\right) = T\left(\sum_{n=1}^{\infty} \chi_{E_n}\right)$$
$$\sum_{n=1}^{\infty} T\left(\chi_{E_n}\right) = \sum_{n=1}^{\infty} \sigma\left(E_n\right).$$

(ii) Observe that

$$\left|\sigma(E)\right| = \left|T\left(\chi_E\right)\right| \leq \|\chi_E\|_{(p,q)} \|T\| \tag{6.18}$$

for $q < \infty$, and

$$\|\chi_E\|_{(p,q)} = \left(\int_0^\infty \left(t^{1/p}\chi_E^*(t)\right)^q \frac{dt}{t}\right)^{1/q}$$
$$= \left(\int_0^\infty \left(t^{1/p}\chi_{\left(0,\mu(E)\right)}(t)\right)^q \frac{dt}{t}\right)^{1/q}$$
$$= \left(\int_0^{\mu(E)} t^{\frac{q}{p}-1}\, dt\right)^{1/q}$$

$$= \left(\frac{p}{q}\right)^{1/q} \left(\mu(E)\right)^{1/p}. \tag{6.19}$$

Putting (6.19) into (6.18) we have

$$|\sigma(E)| \leq \left(\frac{p}{q}\right)^{1/q} \left(\mu(E)\right)^{1/p} \|T\|,$$

for $q < \infty$.
If $q = \infty$, then

$$\|\chi_E\|_{(p,\infty)} = \sup_{t>0} t^{1/p} \chi_E^*(t) = \sup_{t>0} t^{1/p} \chi_{\left(0,\mu(E)\right)}(t) = \left[\mu\left(E\right)\right]^{1/p},$$

thus

$$|\sigma(E)| \leq \|T\| \left(\mu(E)\right)^{1/p},$$

for $q = \infty$.

Once we proved (i) and (ii) we easily see that σ is absolutely continuous with respect to the measure μ. By Radon-Nikodym theorem, there exists a complex-valued measurable function g (which satisfies $\int_{X_n} |g| \, \mathrm{d}\mu < \infty$ for all n) such that

$$\sigma(E) = T(\chi_E) = \int_X |g| \chi_E \, \mathrm{d}\mu, \tag{6.20}$$

Linearity implies that (6.20) holds for any simple function on X. The continuity of T and the density of the simple functions on $L_{(p,q)}(X, \mathscr{A}, \mu)$ (when $q < \infty$) give

$$T(f) = \int_X |gf| \, \mathrm{d}\mu, \tag{6.21}$$

for every $f \in L_{(p,q)}(X, \mathscr{A}, \mu)$. We now examine each case (a), (e), (f) separately, for the remaining cases see Grafakos [22] and the reference therein.

(a) We first consider the case $0 < p < 1$. Let $f = \sum_n a_n \chi_{E_n}$ be a simple function on X (take f to be countably simple when $q = \infty$). If X is nonatomic, we can split each E_n as $E_n = \bigcup_{j=1}^N E_{jn}$, where E_{jn} are disjoint sets and $\mu(E_{jn}) = \frac{\mu(E_n)}{N}$.

Let $f_j = \sum_n a_n \chi_{E_{jn}}$, then

$$\|f\|_{(p,q)} = \left(\int_0^\infty \left(t^{1/p} \left(\sum_n a_n \chi_{E_n} \right)^* (t) \right)^q \frac{\mathrm{d}t}{t} \right)^{1/q}$$

$$= \left(\sum_n a_n \int_0^\infty \left(t^{1/p} (\chi_{E_n})^*(t) \right)^q \frac{\mathrm{d}t}{t} \right)^{1/q}$$

$$= \left(\sum_n a_n \int_0^\infty \left(t^{1/p} \chi_{\left(0,\mu(E_n)\right)}(t) \right)^q \frac{\mathrm{d}t}{t} \right)^{1/q}$$

$$= \left(\sum_n a_n \int_0^{\mu(E_n)} t^{\frac{q}{p}-1}\,\mathrm{d}t \right)^{1/q}$$

$$= \sum_n a_n \left(\frac{p}{q} \right)^{1/q} \left(\mu(E_n) \right)^{1/p}$$

$$= \sum_n a_n \left(\frac{p}{q} \right)^{1/q} \left(N\mu(E_{jn}) \right)^{1/p}$$

$$= N^{1/p} \sum_n a_n \left(\frac{p}{q} \right)^{1/q} \left(\mu(E_{jn}) \right)^{1/p}$$

$$= N^{1/p} \left(\int_0^\infty \left(t^{1/p} \left(\sum_n a_n \chi_{E_{jn}} \right)^*(t) \right)^q \frac{\mathrm{d}t}{t} \right)^{1/q}$$

$$= N^{1/p} \|f_j\|_{(p,q)}.$$

Thus $\|f_j\|_{(p,q)} = N^{-1/p}\|f\|_{(p,q)}$. Now, if $T \in \left(L_{(p,q)}(X, \mathscr{A}, \mu) \right)^*$, it follows that

$$\begin{aligned} |T(t)| &= \left| T\left(\sum_n a_n \chi_{E_n} \right) \right| \\ &= \left| T\left(\sum_n a_n \chi_{\bigcup_{j=1}^N E_{jn}} \right) \right| \\ &= \left| T\left(\sum_n a_n \sum_{j=1}^N \chi_{E_{jn}} \right) \right| \\ &= \left| T\left(\sum_n \sum_{j=1}^N a_n \chi_{E_{jn}} \right) \right| \\ &= \left| T\left(\sum_{j=1}^N \sum_n a_n \chi_{E_{jn}} \right) \right| \\ &= \left| T\left(\sum_{j=1}^N f_j \right) \right| \end{aligned}$$

$$\begin{aligned}
&\leq \sum_{j=1}^{N} |T(f_j)| \\
&\leq \sum_{j=1}^{N} \|T\| \|f_j\|_{(p,q)} \\
&= \|T\| \sum_{j=1}^{N} \|f_j\|_{(p,q)} \\
&= N^{1-\frac{1}{p}} \|T\| \|f_j\|_{(p,q)}.
\end{aligned}$$

Let $N \to \infty$ and use that $p < 1$ to obtain that $T = 0$.

(e) We now take up case $p > 1$ and $0 < q \leq 1$. By Theorem 4.15 and Theorem 6.3 we see that if $g \in L_{(p',\infty)}(X, \mathscr{A}, \mu)$, then

$$\begin{aligned}
\left| \int_X fg \, \mathrm{d}\mu \right| &\leq \int_0^\infty t^{1/p} f^*(t) t^{1/p'} g^*(t) \frac{\mathrm{d}t}{t} \\
&\leq \|f\|_{(p,1)} \|g\|_{(p',\infty)} \\
&\leq \left(\frac{q}{p}\right)^{\frac{1}{q}-1} \|f\|_{(p,q)} \|g\|_{(p',\infty)},
\end{aligned}$$

from which we have

$$\|T\| \leq \left(\frac{q}{p}\right)^{\frac{1}{q}-1} \|g\|_{(p',q)}. \tag{6.22}$$

Conversely, suppose that $T \in \Big(L_{(p,q)}(X, \mathscr{A}, \mu)\Big)^*$ when $1 < p < \infty$ and $0 < q \leq 1$. Let g satisfy (6.21). Taking $f = \overline{g}|g|^{-1}\chi_{\{|g|>\lambda\}}$ then

$$\begin{aligned}
\int_X fg \, \mathrm{d}\mu &= \int_{\{|g|>\lambda\}} g\overline{g}|g|^{-1} \, \mathrm{d}\mu \\
&= \int_{\{|g|>\lambda\}} |g|^2 |g|^{-1} \, \mathrm{d}\mu \\
&= \int_{\{|g|>\lambda\}} |g| \, \mathrm{d}\mu,
\end{aligned}$$

and

$$\lambda\mu\Big(\{|g| > \lambda\}\Big) \leq \left| \int_X fg \, \mathrm{d}\mu \right| \leq \|T\| \|f\|_{(p,q)}. \tag{6.23}$$

Since

$$|f| = |\overline{g}||g|^{-1}\chi_{\{|g|>\lambda\}} = \chi_{\{|g|>\lambda\}},$$

hence

$$f^*(t) = |f|^*(t) = \chi_{\left(0,\mu(\{|g|>\lambda\})\right)}$$

thus

$$\begin{aligned}
\|f\|_{(p,q)} &= \left(\int_0^\infty \left(t^{1/p} f^*(t)\right)^q \frac{\mathrm{d}t}{t}\right)^{1/q} \\
&= \left(\int_0^\infty \left(t^{1/p} \chi_{\left(0,\mu(\{|g|>\lambda\})\right)}\right)^q \frac{\mathrm{d}t}{t}\right)^{1/q} \\
&= \int_0^{\mu\left(\{|g|>\lambda\}\right)} t^{\frac{q}{p}-1}\,\mathrm{d}t \\
&= \left(\frac{p}{q}\right)^{1/q} \left(\mu\left(\{|g|>\lambda\}\right)\right)^{1/p}.
\end{aligned}$$

Now, back to (6.22) we have

$$\begin{aligned}
\lambda\mu\left(\{|g|>\lambda\}\right) &\leq \left(\frac{q}{p}\right)^{\frac{1}{q}-1} \|T\| \left(\mu\left(\{|g|>\lambda\}\right)\right)^{1/p} \\
\lambda\left[\mu\left(\{|g|>\lambda\}\right)\right]^{1/p'} &\leq \left(\frac{q}{p}\right)^{\frac{1}{q}-1} \|T\| \\
\lambda\left(D_g(\lambda)\right)^{1/p'} &\leq \left(\frac{q}{p}\right)^{\frac{1}{q}-1} \|T\| \\
\sup_{t>0} t^{1/p'} g^*(t) &\leq \left(\frac{q}{p}\right)^{\frac{1}{q}-1} \|T\| \\
\|g\|_{(p',\infty)} &\leq \left(\frac{q}{p}\right)^{\frac{1}{q}-1} \|T\|.
\end{aligned} \tag{6.24}$$

Finally by (6.22) and (6.24) we have

$$\|g\|_{(p',\infty)} \approx \|T\|.$$

(6) Using Theorem 4.15 and Hölder's inequality, we obtain

$$|T(f)| = \left| \int_X fg \, d\mu \right|$$
$$\le \int_0^\infty t^{1/p} f^*(t) t^{1/p'} g^*(t) \frac{dt}{t}$$
$$\le \|f\|_{(p,q)} \|g\|_{(p',q')}.$$

Hence T is bounded and if we take the supremum on both sides over all functions f with norm 1 we have

$$\|T\| \le \|g\|_{(p',q')}. \tag{6.25}$$

Thus, for every g in $L_{(p',q')}$ we can find a linear bounded functional on the space $L_{(p,q)}(X, \mathscr{A}, \mu)$.

Conversely, let T be in $\left[L_{(p,q)}(X, \mathscr{A}, \mu)\right]^*$. Note that T is given by integration against a locally integrable function g. It remains to prove that $g \in L_{(p',q')}(X, \mathscr{A}, \mu)$. Using Theorem 4.26 for all f in $L_{(p,q)}(X, \mathscr{A}, \mu)$ we have

$$\int_0^\infty f^*(t) g^*(t) \, dt = \sup \left| \int_X f \widetilde{g} \, d\mu \right| \le \|T\| \|f\|_{(p,q)},$$

where the supremum is taken over all $\mathscr{A}$-measurable functions $\widetilde{g}$ equimeasurables with g. Next, by Theorem 4.29 there exists a measurable function on X such that

$$f^*(t) = \int_{t/2}^\infty h(s) \frac{ds}{s}.$$

where $h(s) = s^{\frac{q'}{p'}-1} \left(g^*(s)\right)^{q'-1}$. Then by Theorem 6.12 (b) with $r = p$

$$\|f\|_{(p,q)}^q = \int_0^\infty \left(t^{1/p} f^*(t)\right)^q \frac{dt}{t}$$
$$= \int_0^\infty \left(\int_{t/2}^\infty h(s) \frac{ds}{s} \right)^q t^{\frac{q}{p}-1} \, dt$$
$$= 2^{q/p} \int_0^\infty \left(\int_u^\infty h(s) \frac{ds}{s} \right)^q u^{\frac{q}{p}-1} \, du$$

$$\leq \left(\frac{2^{1/p}q}{p}\right)^q \int_0^\infty u^{\frac{q}{p}-1}\left(h(u)\right)^q \mathrm{d}u$$

$$= \left(\frac{2^{1/p}q}{p}\right)^q \int_0^\infty u^{\frac{q}{p}-1+\frac{qq'}{p'}-q}\left(g^*(u)\right)^{q(q'-1)} \mathrm{d}u$$

$$= \left(\frac{2^{1/p}q}{p}\right)^q \int_0^\infty u^{\frac{q'}{p'}-1}\left(g^*(u)\right)^{q'} \mathrm{d}u$$

$$= \left(\frac{2^{1/p}q}{p}\right)^q \int_0^\infty \left(u^{\frac{1}{p'}-1}g^*(u)\right)^{q'} \frac{\mathrm{d}u}{u}$$

$$= \left(\frac{2^{1/p}q}{p}\right)^q \|g\|_{(p',q')}^{q'},$$

thus

$$\|f\|_{(p,q)} \leq \left(\frac{2^{1/p}q}{p}\right)\|g\|_{(p',q')}^{q'/q}. \tag{6.26}$$

On the other hand, we have

$$\int_0^\infty f^*(t)g^*(t)\,\mathrm{d}t \geq \int_0^\infty \int_{t/2}^t s^{\frac{q'}{p'}-1}\left(g^*(s)\right)^{q'-1}\frac{\mathrm{d}s}{s}g^*(t)\,\mathrm{d}t$$

$$\geq \int_0^\infty \left(g^*(t)\right)^{q'} \int_{t/2}^t s^{\frac{q'}{p'}-1}\frac{\mathrm{d}s}{s}\,\mathrm{d}t = \int_0^\infty \left(g^*(t)\right)^{q'} \int_{t/2}^t s^{\frac{q'}{p'}-2}\,\mathrm{d}s\,\mathrm{d}t$$

$$= \frac{p'}{q'-p'}\left(1-2^{1-\frac{q'}{p'}}\right)\int_0^\infty t^{\frac{q'}{p'}-1}\left(g^*(t)\right)^{q'}\mathrm{d}t$$

$$= \frac{p'}{q'-p'}\left(1-2^{1-\frac{q'}{p'}}\right)\|g\|_{(p',q')}^{q'}. \tag{6.27}$$

Combining (6.26) and (6.27) we obtain

$$\|g\|_{(p',q')} \leq \left[\frac{q'-p'}{p'\left(1-2^{1-\frac{q'}{p'}}\right)}\right]^{1/q'} \|T\|. \tag{6.28}$$

Finally by (6.25) and (6.28) we have the required conclusion. □

Theorem 6.23 (Duality). *Let $1 < p < \infty$ and $1 \leq q < \infty$ or $p = q = 1$. Then the space of all bounded linear functionals on $L_{(p,q)}(X,\mathscr{A},\mu)$, denoted by $\left[L_{(p,q)}(X,\mathscr{A},\mu)\right]^*$ is isomorphic to $L_{(p',q')}(X,\mathscr{A},\mu)$ where $\frac{1}{p}+\frac{1}{p'}=1$ and $\frac{1}{q}+\frac{1}{q'}=1$.*

Proof. If $p=q$ then we have that $L_{(p,p)}(X,\mathscr{A},\mu)=L_p(X,\mathscr{A},\mu)$ which is isomorphic to $L_{p'}(X,\mathscr{A},\mu)$ for all $1\leq p<\infty$.
Thus, we need to only consider the case when $1<p<\infty$, $1<q<\infty$ and $p\neq q$. To prove that $\left[L_{(p,q)}(X,\mathscr{A},\mu)\right]^*$ is isomorphic to $L_{(p',q')}(X,\mathscr{A},\mu)$ we must show that for each element in $L_{(p',q')}(X,\mathscr{A},\mu)$ there exists a unique corresponding element in $\left[L_{(p,q)}(X,\mathscr{A},\mu)\right]^*$ and vice versa.

We start with the case when $1<p,q<\infty$, $p\neq q$. Let $g\in L_{p',q'}(X,\mathscr{A},\mu)$ be arbitrary and define the functional T as

$$T(f)=\int_X fg\,\mathrm{d}\mu,$$

for all $f\in L_{(p,q)}(X,\mathscr{A},\mu)$. By Theorem 4.15 and Hölder's inequality we get that

$$\begin{aligned}
|T(f)| &= \left|\int_X fg\,\mathrm{d}\mu\right| \\
&\leq \int_X |fg|\,\mathrm{d}\mu \\
&\leq \int_0^\infty f^*(t)g^*(t)\,\mathrm{d}t \\
&\leq \int_0^\infty f^{**}(t)g^{**}(t)\,\mathrm{d}t \\
&\leq \int_0^\infty t^{1/p}f^{**}(t)t^{1/p'}g^{**}(t)\frac{\mathrm{d}t}{t} \\
&\leq \left(\int_0^\infty \left(t^{1/p}f^{**}(t)\right)^q\frac{\mathrm{d}t}{t}\right)^{1/q}\left(\int_0^\infty \left(t^{1/p'}g^{**}(t)\right)^{q'}\frac{\mathrm{d}t}{t}\right)^{1/q'} \\
&= \|f\|_{pq}\|g\|_{p'q'}.
\end{aligned}$$

Hence, T is bounded and if we take the supremum on both sides over all functions f with norm 1 we have that

$$\|T\|\leq\|g\|_{p'q'}.$$

If $1<p<\infty$ and $q=1$ we can use that

$$\int_0^\infty t^{1/p}f^{**}(t)t^{1/p'}g^{**}(t)\frac{\mathrm{d}t}{t}\leq\int_0^\infty t^{1/p}f^{**}(t)\left(\sup_{s>0}s^{1/p}g^{**}(s)\right)\frac{\mathrm{d}t}{t}$$

to obtain that

$$\|T\| \leq \|g\|_{p'\infty}.$$

Hence, for all functions in $L_{(p',q')}(X,\mathscr{A},\mu)$ we can find a linear bounded functional on $L_{(p,q)}(X,\mathscr{A},\mu)$, that is an element in $\left[L_{(p,q)}(X,\mathscr{A},\mu)\right]^*$. In Theorem 6.22 we showed that

$$\sigma(E) = T\left(\chi_E\right)$$

is absolutely continuous with respect to μ and by Radon-Nikodym theorem there exists a unique function $g \in L_1(X,\mathscr{A},\mu) = L_{(1,1)}(X,\mathscr{A},\mu)$ such that

$$\sigma(E) = T\left(\chi_E\right) = \int_X g\chi_E \,\mathrm{d}\mu.$$

By the linearity of the integral and density of simple functions it follows that

$$T(f) = \int_X gf \,\mathrm{d}\mu$$

for all $f \in L_{(p,q)}(X,\mathscr{A},\mu)$.

Once again, by Theorem 4.29 there exists a measurable function f on X such that

$$f^*(t) = \int_{t/2}^{\infty} h(s)\frac{\mathrm{d}s}{s},$$

where $h(s) = s^{\frac{q'}{p'}-1}\left(g^*(s)\right)^{q'-1}$.

Then by Theorem and Theorem 6.12 with $r = p$ we have

$$\begin{aligned}
\|f\|_{pq}^q &\leq \left(\frac{p}{p-1}\|f\|_{(p,q)}\right)^q \\
&= \left(\frac{p}{p-1}\right)^q \int_0^{\infty} \left(t^{1/p} f^*(t)\right)^q \frac{\mathrm{d}t}{t} \\
&= \left(\frac{p}{p-1}\right)^q \int_0^{\infty} \left(\int_{t/2}^{\infty} h(s)\frac{\mathrm{d}s}{s}\right)^q t^{\frac{q}{p}-1}\,\mathrm{d}t \\
&= 2^{q/p}\left(\frac{p}{p-1}\right)^q \int_0^{\infty} \left(\int_u^{\infty} h(s)\frac{\mathrm{d}s}{s}\right)^q u^{\frac{q}{p}-1}\,\mathrm{d}u
\end{aligned}$$

$$\leq \left(\frac{2^{1/p}q}{p-1}\right)^q \int_0^\infty u^{\frac{q}{p}-1}\left(h(u)\right)^q \mathrm{d}u$$

$$= \left(\frac{2^{1/p}q}{p-1}\right)^q \int_0^\infty u^{\frac{q}{p}-1+\frac{qq'}{p'}-q}\left(g^*(u)\right)^{q(q'-1)} \mathrm{d}u$$

$$= \left(\frac{2^{1/p}q}{p-1}\right)^q \int_0^\infty u^{\frac{q'}{p'}-1}\left(g^*(u)\right)^{q'} \mathrm{d}u$$

$$= \left(\frac{2^{1/p}q}{p-1}\right)^q \int_0^\infty \left(u^{\frac{1}{p'}} g^*(u)\right)^{q'} \frac{\mathrm{d}u}{u}$$

$$= \left(\frac{2^{1/p}q}{p-1}\right)^q \|g\|^{q'}_{(p',q')},$$

thus

$$\|f\|_{pq} \leq \left(\frac{2^{1/p}q}{p-1}\right)^q \|g\|^{q'/q}_{(p'q')}.$$

On the other hand, using the definition of the norm of T and Theorem 4.23 we have

$$\|T\| = \sup_{f\in L_{(p,q)}} \frac{|T(f)|}{\|f\|_{pq}} = \sup_{f\in L_{(p,q)}} \frac{\int_0^\infty f^*(t)g^*(t)\,\mathrm{d}t}{\|f\|_{pq}},$$

thus

$$\int_0^\infty f^*(t)g^*(t)\,\mathrm{d}t \leq \|T\|\|f\|_{pq}.$$

Now, observe that

$$\int_0^\infty f^*(t)g^*(t)\,\mathrm{d}t \geq \int_0^\infty \left(\int_{t/2}^\infty s^{\frac{q'}{p'}-1}\left(g^*(s)\right)^{q'-1}\frac{\mathrm{d}s}{s}\right) g^*(t)\,\mathrm{d}t$$

$$\geq \int_0^\infty \left(g^*(t)\right)^{q'} \int_{t/2}^t s^{\frac{q'}{p'}-1}\frac{\mathrm{d}s}{s}\,\mathrm{d}t$$

$$= \int_0^\infty \left(g^*(t)\right)^{q'} \int_{t/2}^t s^{\frac{q'}{p'}-2}\,\mathrm{d}s\,\mathrm{d}t$$

$$
\begin{aligned}
&= \frac{p'}{q'-p'}\left(1-2^{1-\frac{q'}{p'}}\right)\int_0^\infty t^{\frac{q'}{p'}-1}\left(g^*(t)\right)^{q'}\,dt \\
&= \frac{p'}{q'-p'}\left(1-2^{1-\frac{q'}{p'}}\right)\|g\|_{(p',q')}^{q'} \\
&\geq \frac{p'}{q'-p'}\left(1-2^{1-\frac{q'}{p'}}\right)\left(\frac{1}{p}\|g\|_{p'q'}\right)^{q'}.
\end{aligned}
$$

Finally we have

$$
\begin{aligned}
\frac{\left(\frac{1}{p}\|g\|_{p'q'}\right)^{q'}}{\|f\|_{pq}}\frac{p'}{q'-p'}\left(1-2^{1-\frac{q'}{p'}}\right) &\leq \|T\| \\
C\|g\|_{p'q'}^{q'-\frac{q'}{p}} &\leq \|T\| \\
C\|g\|_{p'q'} &\leq \|T\|,
\end{aligned}
$$

where

$$
C = \left(\frac{1}{p}\right)^q \frac{p-1}{2^{1/p}q}\left(\frac{p'}{q'-p'}\right)\left(1-2^{1-\frac{q'}{p'}}\right).
$$

This shows that $\|g\|_{p'q'} < \infty$ and thus $g \in L_{(p',q')}(X,\mathscr{A},\mu)$. Also we have that

$$
C\|g\|_{p'q'} \leq \|T\| \leq \|g\|_{p'q'},
$$

hence $L^*_{(p,q)}$ and $L_{(p',q')}$ are isomorphic for $1 < q < p < \infty$. □

6.6 $L_1 + L_\infty$ Space

We now introduce a space based upon the concept of sum space.

Definition 6.24. Let $(X,\mathscr{A},\mu)$ be a σ-finite measure space. The space $L_1 + L_\infty$ consists of all functions $f \in \mathfrak{F}(X,\mathscr{A})$ that are representable as a sum $f = g + h$ of functions $g \in L_1$ and $h \in L_\infty$. For each $f \in L_1 + L_\infty$, let

$$
\|f\|_{L_1+L_\infty} := \inf_{f=g+h}\left\{\|g\|_{L_1} + \|h\|_{L_\infty}\right\}. \tag{6.29}
$$

where the infimum is taken over all representations $f = g + h$, where $g \in L_1$ and $h \in L_\infty$. ⊘

The next result provides an analogous description of the norm in $L_1 + L_\infty$.

Theorem 6.25. *Let $(X,\mathscr{A},\mu)$ be a σ-finite measure space and suppose f belongs to $\mathfrak{F}(X,\mathscr{A})$. Then*

$$\inf_{f=g+h}\left\{\|g\|_{L_1} + \|h\|_{L_\infty}\right\} = \int_0^t f^*(s)\,\mathrm{d}s = t f^{**}(t), \tag{6.30}$$

for all $t > 0$.

Proof. For $t > 0$ denote

$$\alpha_t = \inf_{f=g+h}\left\{\|g\|_{L_1} + \|h\|_{L_\infty}\right\}.$$

Next, we like to show that

$$\int_0^t f^*(s)\,\mathrm{d}s \le \alpha_t. \tag{6.31}$$

We may assume that f belongs to $L_1 + L_\infty$. Since, otherwise the infimum α_t is infinite and there is nothing to prove. In this case f may be expressed as a sum $f = g + h$ with $g \in L_1$ and $h \in L_\infty$. The sub-additivity of f^{**} (see (6.12)) gives

$$\int_0^t f^*(s)ds = \int_0^t (g+h)^*(s)ds \le \int_0^t g^*(s)ds + \int_0^t h^*(s)\,\mathrm{d}s.$$

Since $h^*(s) \le h^*(0)$ for $s > 0$, we have

$$\int_0^t f^*(s)\,\mathrm{d}s \le \int_0^\infty g^*(s)\,\mathrm{d}s + \int_0^t h^*(0)\,\mathrm{d}s = \int_X |g|\,\mathrm{d}\mu + t\|h\|_\infty = \|g\|_{L_1} + t\|h\|_\infty.$$

Taking the infimum over all possible representations $f = g + h$, we obtain

$$\int_0^t f^*(s)\,\mathrm{d}s \le \alpha_t. \tag{6.32}$$

For the reverse inequality of (6.32) it suffices to construct functions $g \in L_1$ and $h \in L_\infty$ such that $f = g + h$ and assume that $\int_0^t f^*(s)\,\mathrm{d}s.\infty$.

Let $E = \{x \in X : |f(x)| > f^*(t)\}$ and $\mu(E) = t_0$. Since $D_f(f^*(t)) \le t$, then $t_0 = \mu(E) = D_f(f^*(t)) \le t$, thus $t_0 \le t$, then by the Hardy-Littlewood inequality we have

$$\begin{aligned}\int_E |f|\,\mathrm{d}\mu &= \int_X |f|\chi_E\,\mathrm{d}\mu \\ &\le \int_0^\infty f^*(s)\chi_{(0,\mu(E))}(s)\,\mathrm{d}s\end{aligned}$$

$$= \int_0^{t_0} f^*(s)\,\mathrm{d}s$$
$$\leq \int_0^{t} f^*(s)\,\mathrm{d}s < \infty,$$

thus f is integrable over E. Now, let us define

$$g(x) = \max\left\{|f(x)| - f^*(t), 0\right\} \operatorname{sgn} f(x)$$

and

$$h(x) = \max\left\{|f(x)|, f^*(t)\right\} \operatorname{sgn} f(x).$$

The L_1 norm of g can be calculated as

$$\|g\|_{L_1} = \int_X |g|\,\mathrm{d}\mu = \int_E |g|\,\mathrm{d}\mu + \int_{E^{\complement}} |g|\,\mathrm{d}\mu.$$

Note that the second integral is null on $E^{\complement}$ and so

$$\begin{aligned}
\|g\|_{L_1} &= \int_E |g|\,\mathrm{d}\mu \\
&= \int_E |f(x)| d\mu - f^*(t) \int_E \mathrm{d}\mu \\
&= \int_E |f(x)| d|m - t_0 f^*(t) \\
&\leq \int_E |f(x)| d\mu < \infty,
\end{aligned}$$

thus g belongs to L_1. Next, observe that

$$\begin{aligned}
&\mu\left(\{x \in X : |h(x)| > f^*(t)\}\right) \\
&= \mu\left(\{x \in E : |h(x)| > f^*(t)\}\right) + \mu\left(\left\{x \in E^{\complement} : |h(x)| > f^*(t)\right\}\right) \\
&= \mu\left(\{x \in E : f^*(t) > f^*(t)\}\right) + \mu\left(\left\{x \in E^{\complement} : |f(x)| > f^*(t)\right\}\right) \\
&= 0
\end{aligned}$$

since the sets in the penultimate equality are the empty set, therefore

$$\|h\|_\infty = \inf\left\{M > 0 : \mu\left(\{x \in X : |h(x)| > M\}\right) = 0\right\} = f^*(t),$$

hence h belongs to L_∞.

On the other hand, observe that

$$\begin{aligned} g(x) + h(x) &\\ &= \max\{|f(x)| - f^*(t), 0\} \operatorname{sgn} f(x) + \min\{|f(x)|, f^*(t)\} \operatorname{sgn} f(x) \\ &= \begin{cases} |f(x)| - f^*(t) + f^*(t) & if\ |f(x)| > f^*(t) \\ 0 + |f(x)| & if\ |f(x)| \leq f^*(t) \end{cases} \\ &= |f(x)|. \end{aligned}$$

Since

$$\|g\|_{L_1} = \int_E |g| d\mu = \int_E |f(x)| d\mu - t_0 f^*(t) \leq \int_0^{t_0} f^*(s)\,\mathrm{d}s - t_0 f^*(t)$$

and thus

$$\begin{aligned} \|g\|_{L_1} + t\|h\|_\infty &\leq \int_0^{t_0} f^*(s)\,\mathrm{d}s + t\|h\|_\infty - t_0 f^*(t) \\ &\leq \int_0^{t_0} f^*(s)\,\mathrm{d}s + (t - t_0) f^*(t) \\ &= \int_0^{t_0} f^*(s)\,\mathrm{d}s + \int_{t_0}^{t} f^*(t)\,\mathrm{d}s \\ &\leq \int_0^{t_0} f^*(s)\,\mathrm{d}s + \int_{t_0}^{t} f^*(s)\,\mathrm{d}s \\ &= \int_0^{t} f^*(s)\,\mathrm{d}s \end{aligned}$$

which entails

$$\alpha_t \leq \int_0^t f^*(s)\,\mathrm{d}s. \tag{6.33}$$

Combining (6.31) and (6.33) we have

$$\int_0^t f^*(s)\,\mathrm{d}s = \inf_{f=g+h} \left\{ \|g\|_{L_1} + t\|h\|_\infty \right\}$$

and

$$tf^{**}(t) = \int_0^t f^*(s)ds = \inf_{f=g+h} \left\{ \|g\|_{L_1} + t\|h\|_\infty \right\}.$$

□

Note that we can calculate the $L_1 + L_\infty$ norm of a function via:

$$f^{**}(1) = \inf_{f=g+h} \left\{ \|g\|_{L_1} + \|h\|_\infty \right\} = \|f\|_{L_1+L_\infty}$$

and

$$\int_0^1 f^*(s)\,\mathrm{d}s = \|f\|_{L_1+L_\infty}.$$

We now introduce the notion of maximal function $f \mapsto Mf$, which will be studied in more detail in Chapter 9.

Definition 6.26. Let $f \in L_{1,loc}(\mathbb{R}^n)$. The *Hardy-Littlewood maximal function* is defined as

$$Mf(x) = \sup_{0<r<\infty} \frac{1}{m\Big(B(x,r)\Big)} \int_{B(x,r)} |f(y)|\,\mathrm{d}y. \tag{6.34}$$

where $B(x,r) = \Big\{ y \in \mathbb{R}^n : |y-x| < r \Big\}$ is an open ball in $\mathbb{R}^n$. ⊘

With the notion of maximal function at hand, we now prove that the maximal function given in (4.17) and the new one given in (6.34) are related in the following sense.

Theorem 6.27. *There exists a constant c depending only on n, such that*

$$(Mf)^*(f) \le cf^{**}(t)$$

for every locally integrable function f on $\mathbb{R}^n$.

Proof. Fix $t > 0$. For the left-hand side inequality we may suppose $f^{**}(t) < \infty$, otherwise there is nothing to prove. In this case by Theorem 6.25, given $\varepsilon > 0$ there are functions $g_t \in L_1$ and $h_t \in L_\infty$ such that $f = g_t + h_t$ and

$$\|g_T\|_{L_1} + t\|h_T\|_\infty \le tf^{**}(t) + \varepsilon, \tag{6.35}$$

then by Theorem 7.29 and Theorem 3.38, for any $s > 0$

$$(Mf)^*(s) \le (Mg_t)^*\left(\frac{s}{2}\right) + (Mh_t)^*\left(\frac{s}{2}\right)$$
$$\le \frac{C}{s}\|g_t\|_{L_1} + \|h_t\|_\infty = \frac{C}{s}\left(\|g_t\|_{L_1} + s\|h_t\|_\infty\right).$$

□

Theorem 6.28 (Hardy-Littlewood inequality). *Let $1 < p \le \infty$ and suppose that $f \in L_p(\mathbb{R}^n)$. Then $Mf \in L_p(\mathbb{R}^n)$ and*

$$\|Mf\|_p \le c\|f\|_p$$

where c is a constant depending only on p and n

We wish to point out that using the rearrangement (6.30) a proof of Theorem 6.28 can be obtained directly and without using any covering technique.

Proof (Proof of Theorem 6.28). If $f \in L_\infty(\mathbb{R}^n)$, by Theorem

$$\|Mf\|_{L_\infty} = \sup_{t>0}(Mf)^*(t) \le c\sup_{t>0} f^{**}(t) \le c\|f\|_{L_\infty}.$$

Now, if $f \in L_p(\mathbb{R}^n)$ with $1 < p < \infty$ then one more time by Theorem 6.27 we have

$$\|Mf\|_p = \left(\int_0^\infty ((Mf)^*(t))^p \,\mathrm{d}t\right)^{\frac{1}{p}} \le c\left(\int_0^\infty \left(\frac{1}{t}\int_0^t f^*(s)\,\mathrm{d}s\right)^p \mathrm{d}t\right)^{\frac{1}{p}}$$
$$\le \frac{cp}{p-1}\left(\int_0^\infty (f^*(t))^p \,\mathrm{d}t\right)^{\frac{1}{p}} = \frac{cp}{p-1}\|f\|_p.$$

□

6.7 $L\exp$ and $L\log L$ Spaces

We now introduce another function spaces.

Definition 6.29. The Zygmund space $L\exp$ consists of all $f \in \mathfrak{F}(X,\mathscr{A})$ for which there is a constant $\alpha = \alpha(f)$ such that

$$\int_E \exp(\alpha|f(x)|)\,\mathrm{d}\mu(x) < \infty \tag{6.36}$$

for all $E \in \mathscr{A}$. The Zygmund space $L\log L$ consists of all $f \in \mathfrak{F}(X,\mathscr{A})$ for which

$$\int_X |f(x)| \log^+ |f(x)| \, d\mu(x) < \infty \tag{6.37}$$

where $\log^+ x = \max\{\log x, 0\}$. ⊘

The quantities introduced in (6.36) and (6.37) are evidently far from satisfying the properties of norm.

The expression introduced in the next theorem, defined in terms of the decreasing rearrangement, will prove more manageable.

Theorem 6.30. *Let $f \in \mathfrak{F}(X, \mathscr{A})$ and $A \in \mathscr{A}$ such that $0 < \mu(E) < \infty$. Then*

(a) $\int_E |f(x)| \log^+ |f(x)| \, d\mu(x) < \infty$ *if and only if* $\int_0^{\mu(E)} f^*(t) \log\left(\frac{\mu(E)}{t}\right) dt < \infty$.

(b) $\int_E \exp\left(\alpha |f(x)|\right) d\mu(x) < \infty$ *for some constant $\alpha = \alpha(f)$ if and only if there is a constant $c = c(f)$ such that*

$$f^*(t) \leq c \left(1 + \log\left(\frac{\mu(E)}{t}\right)\right)$$

for $0 < t < \mu(E)$.

Proof. To this end, we first apply Theorem 4.13 to obtain

$$\int_E |f(x)| \log^+ |f(x)| \, d\mu = \int_0^\infty f^*(t) \log^+ f^*(t) \, dt,$$

thus $\int_0^{\mu(E)} f^*(t) \log^+ f^*(t) dt < \infty$. On the other hand, note that

$$f^*(t) \leq f^{**}(t) = \frac{1}{t} \int_0^t f^*(s) \, ds \leq \frac{1}{t} \int_0^{\mu(E)} f^*(s) \, ds = \frac{1}{t} \int_E |f| d\mu = \frac{\|f\|_{L_1}}{t},$$

since $\log^+ x$ is an increasing function and $\log \frac{1}{x}$ is a decreasing function. On the one hand, if

$$\int_0^{\mu(E)} f^*(t) \log\left(\frac{\mu(E)}{t}\right) dt < \infty,$$

then

$$\int_0^{\mu(E)} f^*(t)\log^+ f^*(t)\,\mathrm{d}t \le \int_0^{\mu(E)} f^*(t)\log^+ \frac{\|f\|_{L_1}}{t}\,\mathrm{d}t = \int_0^{\min\{\|f\|_{L_1},\mu(E)\}} f^*(t)\log\left(\frac{\|f\|_{L_1}}{t}\right)\mathrm{d}t,$$

since $\log^+ \frac{\|f\|_{L_1}}{t} = \log\left(\frac{\|f\|_{L_1}}{t}\right)$ if $0 < t < \|f\|_{L_1}$ and 0 otherwise.

Next,

$$\begin{aligned}
\int_0^{\min\{\|f\|_{L_1},\mu(E)\}} f^*(t)\log\left(\frac{\|f\|_{L_1}}{t}\right)\mathrm{d}t &\le \int_0^{\|f\|_{L_1}} f^*(t)\log\left(\frac{\|f\|_{L_1}}{t}\right)\mathrm{d}t \\
&= \left(\int_0^{\mu(E)} + \int_{\mu(E)}^{\|f\|_{L_1}}\right) f^*(t)\log\left(\frac{\|f\|_{L_1}}{t}\right)\mathrm{d}t \\
&\le \int_0^{\mu(E)} f^*(t)\log\left(\frac{\|f\|_{L_1}}{t}\right)\mathrm{d}t + \|f\|_{L_1}\int_{\mu(E)}^{\|f\|_{L_1}} \frac{1}{t}\log\left(\frac{\|f\|_{L_1}}{t}\right)\mathrm{d}t \\
&= \int_0^{\mu(E)} f^*(t)\log\left(\frac{\|f\|_{L_1}}{t}\right)\mathrm{d}t + \frac{\|f\|_{L_1}}{2}\left(\log\frac{\|f\|_{L_1}}{\mu(E)}\right)^2.
\end{aligned}$$

Thus $\int_E |f(x)|\log^+|f(x)|\,\mathrm{d}\mu < \infty$. On the other hand, if $\int_E |f(x)|\log^+|f(x)|\,\mathrm{d}\mu < \infty$ we consider the following set

$$A = \left\{t \in [0,\mu(E)] : f^*(t) > \left(\frac{\mu(E)}{t}\right)^{\frac{1}{2}}\right\} \quad \text{and} \quad B = [0,\mu(E)]\setminus A,$$

either of which may be empty. Next, we can write

$$\int_0^{\mu(E)} f^*(t)\log\left(\frac{\mu(E)}{t}\right)\mathrm{d}t = \left(\int_A + \int_B\right) f^*(t)\log\left(\frac{\mu(E)}{t}\right)\mathrm{d}t$$

$$\leq \int_0^{\mu(E)} f^*(t)\log(f^*(t))^2\,\mathrm{d}t + (\mu(E))^{\frac{1}{2}} \int_0^{\mu(E)} t^{-\frac{1}{2}}\log\left(\frac{\mu(E)}{t}\right)\mathrm{d}t$$

$$= 2\int_0^{\mu(E)} f^*(t)\log(f^*(t))\,\mathrm{d}t + \frac{(\mu(E))^{\frac{3}{2}}}{2},$$

hence $\int_0^{\mu(E)} f^*(t)\log\left(\frac{\mu(E)}{t}\right)\mathrm{d}t < \infty$. Turning now to the equivalence in (b) we suppose first that $f^*(t) \leq C\left(1+\log\left(\frac{\mu(E)}{t}\right)\right)$ for some constant $C > 0$ with $0 < t < \mu(E)$. Then

$$\int_0^{\mu(E)} \exp\left(\alpha f^*(t)\right)\mathrm{d}t \leq \int_0^{\mu(E)} \exp\left[\alpha C\left(1+\log\left(\frac{\mu(E)}{t}\right)\right)\right]\mathrm{d}t$$

$$= e^{C\alpha}\left[\mu(E)\right]^{C\alpha} \int_0^{\mu(E)} t^{-C\alpha}\,\mathrm{d}t.$$

from which $\int_0^{\mu(E)} \exp\left(\alpha f^*(t)\right)\mathrm{d}t < \infty$ for any constant $\alpha < 1/C$.

Conversely, suppose that

$$M = \int_0^{\mu(E)} \exp\left(\alpha f^*(t)\right)\mathrm{d}t < \infty.$$

Clearly $M \geq \mu(E)$. Since f^* is decreasing we have

$$f^*(t) = f^*(t)\frac{1}{t}\int_0^t \mathrm{d}s \leq \frac{1}{t}\int_0^t f^*(s)\,\mathrm{d}s.$$

Then by Jensen's inequality we have

$$\exp\left(\alpha f^*(t)\right) \leq \exp\left(\frac{1}{t}\int_0^t \alpha f^*(s)\,\mathrm{d}s\right) \leq \frac{1}{t}\int_0^t \exp\left(\alpha f^*(s)\right)\mathrm{d}s,$$

which entails

$$f^*(t) \leq \frac{1}{\alpha}\log\left(\frac{M}{t}\right) \leq \frac{1}{\alpha}\left(1+\log\left(\frac{M}{\mu(E)}\right)\right)\left(1+\log\left(\frac{\mu(E)}{t}\right)\right).$$

The proof is now complete. □

An integration by parts shows that

$$\int_0^{\mu(E)} f^*(t)\log\left(\frac{\mu(E)}{t}\right)\mathrm{d}t = \int_0^{\mu(E)} f^{**}(t)\,\mathrm{d}t. \tag{6.38}$$

The latter quantity involving the sub-additive function $f \to f^{**}$ satisfies the triangle inequality and so may be used to directly define a norm in $L\log L$.
On the other hand the expression

$$f^*(t) \leq C\left(1+\log\left(\frac{\mu(E)}{t}\right)\right) \qquad 0<t<\mu(E) \tag{6.39}$$

for the space $L\exp$ involve f^* rather than the sub-additive f^{**}. This present no problem, however

$$\begin{aligned} f^{**}(t) &= \frac{1}{t}\int_0^t f^*(s)\,\mathrm{d}s \\ &\leq \frac{C}{t}\int_0^t \left(1+\log\left(\frac{\mu(E)}{s}\right)\right)\mathrm{d}s \\ &\leq 2C\left(1+\log\left(\frac{\mu(E)}{t}\right)\right). \end{aligned} \tag{6.40}$$

Hence (6.39) (with constant C) implies (6.40) (with constant $2C$) and so (6.39) and (6.40) are equivalent. We use the relation (6.40) to define a norm on $L\exp$ as follows.

Definition 6.31. Let $f \in \mathfrak{F}(X,\mathscr{A})$. Set

$$\|f\|_{L\log L} = \int_0^{\mu(E)} f^*(t)\log\left(\frac{\mu(E)}{t}\right)dt = \int_0^{\mu(E)} f^{**}(t)\,\mathrm{d}t. \tag{6.41}$$

and

$$\|f\|_{L\exp}\|f\|_{L\exp} = \sup_{0<t<\mu(E)} \frac{f^{**}(t)}{\left(1+\log\left(\frac{\mu(E)}{t}\right)\right)}. \tag{6.42}$$

⊘

It follows from Theorem 6.30 and the observation made above that $L\log L$ and $L\exp$ consist of all function $f \in \mathfrak{F}(X, \mathscr{A})$ for which the representative quantities (6.41) and (6.42) are finite. Since $f \to f^{**}$ is sub-additive, it is easy to prove directly that this qualities define norms under which $L\log L$ and $L\exp$ are rearrangement-invariant Banach spaces. The following result gives a Hölder type inequality.

Theorem 6.32. *Let $(X, \mathscr{A}, \mu)$ be a σ-finite measure space and $f \in L\log L$ and $g \in L\exp$. Then*

$$\int_E |fg| \, \mathrm{d}\mu \le 2\|f\|_{L\log L}\|g\|_{L\exp}$$

for any $E \in \mathscr{A}$.

Proof. Let $E \in \mathscr{A}$ and $f \in L\log L$, $g \in L\exp$. Then from Theorem 4.13 and the fact that $f^* \le f^{**}$ we have

$$\begin{aligned}
\int_E |fg| d\mu &= \int_X |fg\chi_E| \, \mathrm{d}\mu \le \int_0^\infty f^*(t) g^*(t) \chi_{(0,\mu(E))}(t) \, \mathrm{d}t = \int_0^{\mu(E)} f^*(t) g^*(t) \, \mathrm{d}t \\
&\le \int_0^{\mu(E)} f^*(t) \left(1 + \log \frac{\mu(E)}{t}\right) \frac{g^{**}(t)}{1 + \log \frac{\mu(E)}{t}} \, \mathrm{d}t \\
&\le \left[\int_0^{\mu(E)} f^*(t) \left(1 + \log \frac{\mu(E)}{t}\right) \mathrm{d}t \right] \|g\|_{L\exp} \\
&\le 2 \left(\int_0^{\mu(E)} f^{**}(t) \, \mathrm{d}t \right) \|g\|_{L\exp} \\
&= 2\|f\|_{L\log L}\|g\|_{L\exp},
\end{aligned}$$

and the assertion of the theorem holds. □

6.8 Lorentz Sequence Spaces

In this section we will investigate the Lorentz sequence spaces.

For $X = \mathbb{N}$ with $A = 2^{\mathbb{N}}$, the power set of X and μ = counting measure, the distribution function of any complex-valued function $a = \{a(n)\}_{n \ge 1}$ can be written as

$$D_a(\lambda) = \mu(\{n \in \mathbb{N} : |a(n)| > \lambda\}) \quad (\lambda \ge 0).$$

The decreasing rearrangement a^* of a is given as

$$a^*(t) = \inf\{\lambda > 0 : D_a(\lambda) \le t\} \quad (t \ge 0).$$

We can interpret the decreasing rearrangement of a with $D_a(\lambda) < \infty$, $\lambda > 0$ as a sequence $\{a^*(n)\}$ if we define for $n-1 \le t \le n$

$$a^*(n) = a^*(t) = \inf\{\lambda > 0 : D_a(\lambda) \le n-1\} \tag{6.43}$$

Then the sequence $a^* = \{a^*(n)\}$ is obtained by permuting $\{|a(n)|\}_{n \in S}$ where $S = \{n : a(n) \neq 0\}$, in the decreasing order with $a^*(n) = 0$ for $n > \mu(S)$ if $\mu(S) < \infty$.

Definition 6.33. The Lorentz sequence space $\ell_{(p,q)}$, $1 < p \le \infty$, $1 \le q \le \infty$, is the set of all complex sequences $a = \{a(n)\}$ such that $\|a\|^s_{(p,q)} < \infty$ where

$$\|a\|^s_{(p,q)} = \begin{cases} \left(\sum_{n=1}^{\infty} \left(n^{1/p} a^*(n) \right)^q \frac{1}{n} \right)^{1/q}, & 1 < p < \infty, 1 \le q < \infty \\ \sup_{n \ge 1} n^{1/p} a^*(n), & 1 < p \le \infty, q = \infty. \end{cases}$$

where a^* is given in (6.43).

⊘

The Lorentz sequence space $\ell_{(p,q)}$, $1 < p \le \infty$, $1 < p \le \infty$, $q = \infty$, is a linear space and $\|\cdot\|^s_{(p,q)}$ is a quasi-norm. Moreover, $\ell_{(p,q)}$ $1 < p \le \infty$, $1 \le q \le \infty$, is complete with respect to the quasi-norm $\|\cdot\|^s_{(p,q)}$.

The Lorentz sequence space $\ell_{(p,q)}$ and $L_{(p,q)}$ when $X = \mathbb{N}$, $\mathscr{A} = 2^{\mathbb{N}}$ and $\mu(\{n\}) = 1$ are equivalent for $0 < p < \infty$, $0 < q < \infty$. In fact, if we let $a^*(n) = f^*(t)$ for $n-1 \le t < n$, we have

$$\begin{aligned} \|f\|_{(p,q)} &= \left(\int_0^{\infty} \left(t^{1/p} f^*(t) \right)^q \frac{dt}{t} \right)^{1/q} \\ &= \left(\sum_{n=1}^{\infty} [a^*(n)]^q \int_{n-1}^{n} t^{q/p-1} \, dt \right)^{1/q} \end{aligned}$$

and since

$$\left(\frac{1}{2}\right)^{q/p} n^{q/p-1} \le \int_{n-1}^{n} t^{q/p-1} \, dt \le 2n^{q/p-1}$$

we obtain

$$\left(\frac{1}{2}\right)^{q/p} \sum_{n=1}^{\infty} (a^*(n))^q n^{q/p-1} \le \sum_{n=1}^{\infty} (a^*(n))^q \int_{n-1}^{n} t^{q/p-1} \, dt \le 2 \sum_{n=1}^{\infty} (a^*(n))^q n^{q/p-1},$$

from which we get

$$\left(\frac{1}{2}\right)^{1/p}\left(\sum_{n=1}^{\infty}(n^{1/p}a^*(n))^q\frac{1}{n}\right)^{1/q} \leq \|f\|_{(p,q)} \leq 2^{1/q}\left(\sum_{n=1}^{\infty}(n^{1/p}a^*(n))^q\frac{1}{n}\right)^{1/q}$$

thus

$$\left(\frac{1}{2}\right)^{1/p}\|a\|^s_{(p,q)} \leq \|f\|_{(p,q)} \leq 2^{1/q}\|a\|^s_{(p,q)}.$$

Observe that the space $\ell_{(p,q)}$ is not empty when $p=\infty$. For example, all sequences which only have a finite number of nonzero elements are in $\ell_{(p,q)}$ for all $0<q\leq\infty$. This show that there is a fundamental difference between $L_{(\infty,q)}$ and $\ell_{(\infty,q)}$.

The following result will be of great utility in our study, and we include a short proof for the benefit of the reader.

Theorem 6.34. *If $a=\{a(n)\}_{n\in\mathbb{N}}$ and $b=\{b(n)\}_{n\in\mathbb{N}}$ are complex sequences, $b\in\ell_{(p,q)}$ with $1<p\leq\infty$, $1\leq q\leq\infty$ and $|a(n)|\leq|b(n)|$ for all $n\in\mathbb{N}$, then $a\in\ell_{(p,q)}$ and $\|a\|^s_{(p,q)}\leq\|b\|^s_{(p,q)}$.*

Proof. If $|a(n)|\leq|b(n)|$ for all $n\in\mathbb{N}$, then

$$\left\{n\in\mathbb{N}:|a(n)|>\lambda\right\}\subset\left\{n\in\mathbb{N}:|b(n)|>\lambda\right\},$$

by the monotonicity of the measure, we have $D_a(\lambda)\leq D_b(\lambda)$ for all $\lambda>0$. Thus, for any $m\in\mathbb{N}$, we obtain

$$\{\lambda>0:D_b(\lambda)\leq m-1\}\subset\{\lambda>0:D_a(\lambda)\leq m-1\}$$

and hence $a^*(m)\leq b^*(m)$. From this last fact, the result follows easily. □

The aim of this section is to present basic results about Lorentz sequence spaces. The Lorentz sequence space $\ell_{(p,q)}$ is a normed linear space if and only if $1\leq q\leq p<\infty$. Moreover, $\ell_{(p,q)}$ is normable when $1<p<q\leq\infty$, that is there exists a norm equivalent to $\|\cdot\|^s_{(p,q)}$. For the remaining cases $\ell_{(p,q)}$ cannot be equipped with an equivalent norm.

The normable case for $p<q$ comes up in the following way

$$\|a\|^*_{(p,q)}=\begin{cases}\left(\sum_{n=1}^{\infty}\left(a^{**}(n)\right)^q n^{q/p-1}\right)^{1/q}, & q<\infty\\ \sup_{n\geq1}\left\{n^{1/p}a^{**}(n)\right\}, & q=\infty\end{cases}$$

where $a^{**} = \{a^{**}(n)\}_n$ is called the maximal sequence of $a^* = \{a^*(n)\}_n$ and it is defined as

$$a^{**}(n) = \frac{1}{n}\sum_{k=1}^{n} a^*(k).$$

We now prove that the functionals $\|\cdot\|^*$ and $\|\cdot\|^s$ are equivalent.

Theorem 6.35. *Let $a = \{a(n)\}_{n\geq 1}$ be a complex sequence, $1 < p \leq q < \infty$ then*

$$\|a\|^s_{(p,q)} \leq \|a\|^*_{(p,q)} \leq \left(\frac{p}{p-1}\right)^q \|a\|^s_{(p,q)}.$$

Proof. The inequality

$$\|a\|^s_{(p,q)} \leq \|a\|^*_{(p,q)}$$

is an easy consequence of the fact that

$$a^*(n) \leq a^{**}(n)$$

for all $n \in \mathbb{N}$. Hence, we just need to show that

$$\|a\|^*_{(p,q)} \leq \left(\frac{p}{p-1}\right)^q \|a\|^s_{(p,q)}.$$

In fact, let $r = q - \frac{q}{p}$. Using the fact that the function $g(t) = t^{r/q-1}$ is decreasing we apply the Hölder inequality to obtain

$$\begin{aligned}
\left(\sum_{k=1}^{n} a^*(k)\right)^q &= \left(\sum_{k=1}^{n} a^*(k)k^{1-\frac{r}{q}}k^{\frac{r}{q}-1}\right)^q \\
&\leq \left(\sum_{k=1}^{n} (a^*(k))^q k^{q-r} k^{\frac{r}{q}-1}\right)\left(\sum_{k=1}^{n} k^{\frac{r}{q}-1}\right)^{q/q'} \\
&\leq \left(\sum_{k=1}^{n} (a^*(k))^q k^{q-r} k^{\frac{r}{q}-1}\right)\left(\int_0^n t^{\frac{r}{q}-1}\,\mathrm{d}t\right)^{q/q'} \\
&= \left(\frac{q}{r}\right)^{q/q'} n^{r/q'} \sum_{k=1}^{n} (a^*(k))^q k^{q-r} k^{\frac{r}{q}-1}.
\end{aligned}$$

Moreover since $f(t) = t^{-1-\frac{r}{q}}$ is decreasing and $\int_k^\infty f(t)\,\mathrm{d}t < \infty$ we have

$$\sum_{n=k-1}^{k} f(n) \leq \int_{k-2}^{k} f(t)\,\mathrm{d}t,$$

from which we get

$$\sum_{k=m+1}^{\infty}\left(\sum_{n=k-1}^{k} f(n)\right) \leq \sum_{k=m+1}^{\infty}\left(\int_{k-2}^{k} f(t)\,\mathrm{d}t\right),$$

and

$$\sum_{n=m}^{\infty} f(n) \leq \int_{m-1}^{\infty} f(t)\,\mathrm{d}t.$$

Next using the above inequality and Fubini's Theorem we have

$$\begin{aligned}
&\sum_{n=1}^{\infty}\left(\sum_{k=1}^{n} a^*(k)\right)^q n^{-r-1}\\
&\leq \left(\frac{q}{r}\right)^{q/q'} \sum_{n=1}^{\infty} n^{-r-1+r\left(1-\frac{1}{q}\right)}\left(\sum_{k=1}^{n}(a^*(k))^q k^{q-r} k^{\frac{r}{q}-1}\right)\\
&= \left(\frac{q}{r}\right)^{q/q'} \sum_{k=1}^{\infty}(a^*(k))^q k^{q-r} k^{\frac{r}{q}-1}\left(\sum_{n=k}^{\infty} n^{-1-\frac{r}{q}}\right)\\
&= \left(\frac{q}{r}\right)^{\frac{q}{q'}+1} \sum_{k=1}^{\infty}(a^*(k))^q k^{q-r-1}\\
&= \left(\frac{p}{p-1}\right)^q \sum_{k=1}^{\infty}(a^*(k))^q k^{\frac{q}{p}-1}.
\end{aligned}$$

That is

$$\|a\|^*_{(p,q)} \leq \left(\frac{p}{p-1}\right)^q \|a\|^s_{(p,q)}.$$

□

Note that if $\{a_k\}_{k=1,2,\ldots,N} \in \ell_{(p,q)}$, then

$$\left\|\sum_{k=1}^{N} a_k\right\|^s_{(p,q)} \leq \left\|\sum_{k=1}^{N} a_k\right\|^*_{(p,q)} \tag{6.44}$$

$$\leq \left(\frac{p}{p-1}\right)^q \sum_{k=1}^{N} \|a_k\|^s_{(p,q)} \tag{6.45}$$

That is, $\|\cdot\|^s_{(p,q)}$ is a quasi-norm. On the other hand, if (6.44) holds, then $\|\cdot\|^s_{(p,q)}$ is equivalent to a norm, this norm is called decomposition norm and it is defined as

$$\|a\|_{pq} = \inf\left\{\sum_{k=1}^{N} \|a_k\|^s_{(p,q)} : a = \sum_{k=1}^{N} a_k\right\}.$$

The functional $\|\cdot\|_{pq}$ is an equivalent norm to $\|\cdot\|^{s}_{(p,q)}$ when $1 \leq p,q \leq \infty$. Moreover

$$\|\cdot\|_{pq} = \|\cdot\|^{s}_{(p,q)} \text{ if } 1 \leq q \leq p.$$

The following result is due to Hardy and Littlewood.

Theorem 6.36. *If* $a = \{a(k)\}_{k\in\mathbb{N}}$ *and* $b = \{b(k)\}_{k\in\mathbb{N}}$ *are complex sequences, then*

$$\sum_{k=1}^{\infty} |a(k)b(k)| \leq \sum_{k=1}^{\infty} a^*(k)b^*(k). \tag{6.46}$$

Proof. It will be enough to show (6.46) for nonnegative sequences. For $n \in \mathbb{N}$ fixed, we set $E = \{1,2,\cdots,n\}$. Let us consider $c(1),c(2),\cdots,c(m)$ the different elements of the set $\{a(k) : k \in E\}$. Then it is clear that $m \leq n = \mu(E)$. Thus, for $j \in \{1,2,\cdots,m\}$ we can define the sets

$$F_j = \{k \in E : a(k) = c(j)\}$$

Note that the sets F_j are pairwise disjoint and $\cup_{j=1}^{m} F_j = E$. Observe that

$$\sum_{k\in E} a(k) = \sum_{j=1}^{m} c(j)\mu(F_j). \tag{6.47}$$

Furthermore, for any $k \in E$ there exists an unique $j_k \in \{1,2,\cdots,m\}$ such that $k \in F_{j_k}$ and therefore

$$a(k) = \sum_{j=1}^{m} c(j)\chi_{F_j}(k),$$

then

$$a^*(k) = \sum_{j=1}^{m} c(j)\chi_{[1,\mu(F_j)]}(k).$$

Therefore from (6.47), we have

$$\begin{aligned}
\sum_{k\in E} a(k) &= \sum_{j=1}^{m} c(j)\mu(F_j) \\
&\leq \sum_{j=1}^{m} c(j) \sum_{k=1}^{\mu(E)} \chi_{[1,\mu(F_j)]}(k) \\
&= \sum_{k=1}^{\mu(E)} \sum_{j=1}^{m} c(j)\chi_{[1,\mu(F_j)]}(k) \\
&= \sum_{k=1}^{\mu(E)} a^*(k).
\end{aligned}$$

That is

$$\sum_{k\in E} a(k) \leq \sum_{k=1}^{\infty} a^*(k), \tag{6.48}$$

and since $n \in \mathbb{N}$ was arbitrary, we conclude

$$\sum_{k=1}^{\infty} a(k) \leq \sum_{k=1}^{\infty} a^*(k).$$

On the other hand, employing inequality (6.48) we obtain

$$\begin{aligned}
\sum_{k\in\mathbb{N}} a(k)b(k) &= \sum_{k\in\mathbb{N}} \left(\sum_{j=1}^{m} c(j)\chi_{F_j}(k) \right) b(k) \\
&= \sum_{j=1}^{m} c(j) \sum_{k\in\mathbb{N}} b(k)\chi_{F_j}(k) \\
&= \sum_{j=1}^{m} c(j) \sum_{k\in F_j} b(k) \\
&\leq \sum_{j=1}^{m} c(j) \sum_{k=1}^{\mu(F_j)} b^*(k) \\
&\leq \sum_{k=1}^{\infty} \sum_{j=1}^{m} c(j)\chi_{[1,\mu(F_j)]}(k)b^*(k) \\
&= \sum_{k=1}^{\infty} a^*(k)b^*(k).
\end{aligned}$$

Hence

$$\sum_{k\in\mathbb{N}} a(k)b(k) \leq \sum_{k=1}^{\infty} a^*(k)b^*(k),$$

which ends the proof. □

We now show a Hölder type inequality for Lorentz sequence spaces.

Theorem 6.37. *Let* $a = \{a(k)\} \in \ell_{(p,t)}$ *and* $b = \{b(k)\} \in \ell_{(q,r)}$ *where* $\frac{1}{t} + \frac{1}{r} = 1$, *then*

$$\sum_{k=1}^{\infty} \left|a(k)b(k)\right| \leq \|a\|_{(p,t)}^{s} \|b\|_{(q,r)}^{s}. \tag{6.49}$$

Proof. By virtue of Theorem 6.36 and by Hölder's inequality we have

$$\begin{aligned}
\sum_{k\in\mathbb{N}} \left|a(k)b(k)\right| &\leq \sum_{k=1}^{\infty} a^*(k)b^*(k) \\
&= \sum_{k=1}^{\infty} k^{\frac{1}{p}-\frac{1}{t}} a^*(k) k^{\frac{1}{q}-\frac{1}{r}} b^*(k)
\end{aligned}$$

$$\leq \left(\sum_{k=1}^{\infty}\left(k^{\frac{1}{p}-\frac{1}{t}}a^*(k)\right)^t\right)^{1/t}\left(\sum_{k=1}^{\infty}\left(k^{\frac{1}{q}-\frac{1}{r}}b^*(k)\right)^r\right)^{1/r}$$

$$=\left(\sum_{k=1}^{\infty}(a^*(k))^t k^{t/p-1}\right)^{1/t}\left(\sum_{k=1}^{\infty}(b^*(k))^r k^{r/q-1}\right)^{1/r},$$

from which (6.49) follows. □

6.9 Problems

6.38. Let μ denote the Lebesgue measure on the σ-algebra $\mathscr{B}$ of Borel set of $[0,1]$ and put $f(x)=x$ and $g(x)=1-x$ for $x\in[0,1]$. Express $\|f\|_{(p,q)}$ and $\|g\|_{(p,q)}$ for $p\in[0,1]$, $q\in[1,\infty)$ in terms of the gamma function.

Hint. First express these numbers in terms of the Beta function.

6.39. Suppose that μ denotes the Lebesgue measure on the σ-algebra $\mathscr{B}$ of Borel subset of $\mathbb{R}$ and for each $a>0$ put $f_a(x)=\mathrm{e}^{-a|x|}$ and $g_a(x)=\mathrm{e}^{-ax^2}$ for $x\in\mathbb{R}$.

(a) Calculate $\|f_a\|_{(p,q)}$ for $a>0$ and $p,q\in[1,\infty]$.
(b) Calculate $\|g_a\|_{(p,q)}$ for $a>0$ and $p,q\in[1,\infty]$.

6.40. On $\mathbb{R}^n$, let $\delta^\varepsilon(f)(x)=f(\varepsilon x)$, $\varepsilon>0$, be the dilation operator. Show that $\|\delta^\varepsilon(f)\|_{(p,q)}=\varepsilon^{-n/p}\|f\|_{(p,q)}$.

6.41. Show that

$$\sup_{t>0} t^{1/p}f^*(t)\leq\|f^*\|_p.$$

6.42. Let $(X,\mathscr{A},\mu)$ be a measure space, let p and q be two extended real numbers in $[1,\infty]$, and let p' and q' their conjugate exponents. If $f\in L_{(p,q)}(X,\mathscr{A},\mu)$ and $g\in L_{(p',q')}(X,\mathscr{A},\mu)$, prove that

$$\int_0^\infty f^{**}(t)g^{**}(t)\,\mathrm{d}t\leq\|f\|_{pq}\|g\|_{p'q'}.$$

6.43. Let f and g be nonnegative μ-measurable functions on $\mathbb{R}^+$. Prove that

$$\int_{\mathbb{R}} fg\,\mathrm{d}\mu\leq\frac{1}{2}\int_0^\infty f^{**}(t)g^{**}(t)\,\mathrm{d}t$$

the constant $1/2$ is optimal.

6.44. Let $1 < p_0 < \infty$ and assume that $f \in L(p,\infty)(\mathbb{R}^+)$ for every $1 < p \leq p_0 < \infty$. Prove that

$$\lim_{p\to 1} \|f\|_{p\infty} = \|f\|_1.$$

6.45. Let $1 < p_0 < \infty$ and assume that $\varphi \in L_{(p,1)}(\mathbb{R}^+)$ for every $1 < p_0 \leq p < \infty$. Prove that

$$\lim_{p\to\infty} \frac{1}{p}\|\varphi\|_{(p,1)} = \|\varphi\|_\infty.$$

6.46. Let $f \in L_p(X, \mathscr{A}, \mu)$. Prove that

$$\|Mf\|_{L_{(p,\infty)}} \leq C\|f\|_p$$

where C is a positive constant and M stands for the Hardy-Littlewood maximal operator (9.2).

6.47. Let $f \in L_1(X, \mathscr{A}, \mu)$. Prove that

$$\|I_\alpha f\|_{L_{(\frac{n}{n-\alpha},\infty)}} \leq C\|f\|_1$$

where C is a positive constant and I^α stands for the Riesz potential operator (11.9).

6.48. We say that $h \in \frac{L^q}{\log^\alpha L}(\Omega)$, $\alpha > 0$, if

$$\int_\Omega \frac{|h(x)|^q}{\log^\alpha(\mathrm{e}+|h(x)|)}\,\mathrm{d}x < \infty.$$

Show that $h \in \frac{L}{\log^\alpha L}[(0,1/\mathrm{e})]$ if and only if $\alpha + \beta > 1$, where

$$h(x) = \frac{1}{x|\log x|^\beta}, \quad \beta \in \mathbb{R}.$$

6.10 Notes and Bibliographic References

The Lorentz spaces were introduced in Lorentz [44, 45]. It seems that the first expository paper on the topic is Hunt [34].

The duality problem regarding Lorentz spaces was investigated in Cwikel [10], Cwikel and Fefferman [11, 12]

The space $L_1 + L_\infty$ was studied in Gould [21] and Luxemburg and Zaanen [47].

The spaces $L\exp$ and $L\log L$ were introduced independently by Zygmund [85] and Titchmarsh [78, 79].

Chapter 7
Nonstandard Lebesgue Spaces

Give more spaces to functions.
ALOIS KUFNER

Abstract In recent years, it had become apparent that the plethora of existing function spaces were not sufficient to model a wide variety of applications, e.g., in the modeling of electrorheological fluids, thermorheological fluids, in the study of image processing, in differential equations with nonstandard growth, among others. Thus, naturally, new fine scales of function spaces have been introduced, namely variable exponent spaces and grand spaces. In this chapter we study variable exponent Lebesgue spaces and grand Lebesgue spaces. In variable exponent Lebesgue spaces we study the problem of normability, denseness, completeness, embedding, among others. We give a brisk introduction to grand Lebesgue spaces via Banach function space theory, dealing with the problem of normability, embeddings, denseness, reflexivity, and the validity of a Hardy inequality in the aforementioned spaces.

⚠ In previous chapters we tried to give detailed proofs of the results, but in this chapter we will be much more concise, approaching the reader to a style more close to a research paper than to a textbook exposition.

7.1 Variable Exponent Lebesgue Spaces

Our goal in this section is to define the so-called *variable exponent Lebesgue spaces* $L_{p(\cdot)}(\Omega)$, introduce an appropriate norm and study some fundamental properties of the space, for simplicity we will work only on a measurable subset Ω of $\mathbb{R}^n$ with the Lebesgue measure.

By $\mathscr{P}(\Omega)$ we denote the family of all measurable functions $p : \Omega \longrightarrow [1,\infty]$. For $p \in \mathscr{P}(\Omega)$ we define the following sets

$$\begin{aligned}
\Omega_1(p) &:= \Omega_1 = \{x \in \Omega : p(x) = 1\},\\
\Omega_\infty(p) &:= \Omega_\infty = \{x \in \Omega : p(x) = \infty\},\\
\Omega_+(p) &:= \Omega_* = \{x \in \Omega : 1 < p(x) < \infty\}.
\end{aligned}$$

R.E. Castillo, H. Rafeiro, *An Introductory Course in Lebesgue Spaces*, CMS Books in Mathematics, DOI 10.1007/978-3-319-30034-4_7

Definition 7.1. By $L_{p(\cdot)}(\Omega)$ we denote the *variable exponent Lebesgue space* as the set of all measurable functions $f : \Omega \longrightarrow \mathbb{R}$ such that

$$\rho_{p(\cdot)}(f) := \int_{\Omega \setminus \Omega_\infty} |f(x)|^{p(x)}\, dx < \infty \tag{7.1}$$

and

$$\operatorname*{ess\,sup}_{x \in \Omega_\infty} |f(x)| < \infty,$$

where the measurable function $p : \Omega \longrightarrow (0, \infty]$ is called *variable exponent*. The functional $\rho_{p(\cdot)}$ is known as a *modular*. ⊘

For the variable exponent p we define the following numbers

$$p_-(\Omega) = p_- := \operatorname*{ess\,inf}_{\Omega_*} p(x), \quad p_+(\Omega) = p_+ := \operatorname*{ess\,sup}_{\Omega_*} p(x) \tag{7.2}$$

if $m(\Omega_*) > 0$, and $p_- = p_+ = 1$ if $m(\Omega_*) = 0$. For $p \in \mathscr{P}(\Omega)$ we define the *dual exponent* or the *conjugate exponent* has

$$p'(x) = \begin{cases} \infty, & x \in \Omega_1, \\ \frac{p(x)}{p(x)-1}, & x \in \Omega_*, \\ 1, & x \in \Omega_\infty, \end{cases}$$

which implies the pointwise inequality

$$\frac{1}{p(x)} + \frac{1}{p'(x)} = 1.$$

If a measurable function $p : \mathbb{R}^n \longrightarrow [1, \infty)$ satisfies

$$1 < p_-, \quad p_+ < \infty, \tag{7.3}$$

then the conjugate function

$$p'(x) := \frac{p(x)}{p(x) - 1}$$

is well defined and moreover it satisfies (7.3).

Working with the definition of p_-, p_+ and the conjugate exponent, we have the following relations

1. $(p'(\cdot))_+ = (p_-)'$;
2. $(p'(\cdot))_- = (p_+)'$.

A natural question is whether the space $L_{p(\cdot)}(\Omega)$ is, in general, linear. The answer is affirmative whenever $p_+ < \infty$.

Lemma 7.2. *The space $L_{p(\cdot)}(\Omega)$ is linear if and only if $p_+ < \infty$.*

Proof. NECESSITY. Suppose that $p_+ = \infty$. We will show that there exists a function $f_0 \in L_{p(\cdot)}(\Omega)$ such that $2f_0 \notin L_{p(\cdot)}(\Omega)$. Let $A_m = \{x \in \Omega \backslash \Omega_\infty : m-1 \leq p(x) \leq m\}$. Since $p_+ = \infty$, there exists a sequence $m_k \to \infty, k \in \mathbb{N}$ such that $m(A_{m_k}) > 0$. We now construct a step function f_0; i.e., $f_0(x) = c_m$ for $x \in A_m$, where c_m is given by the relation

$$\int_{A_m} c_m^{p(x)}\, dx = m^{-2},$$

this defines c_m univocally if $m(A_m) \neq 0$. We then have

$$\rho_{p(\cdot)}(f_0) = \sum_{m=1}^{\infty} \int_{A_m} c_m^{p(x)}\, dx = \sum_{m=1}^{\infty} m^{-2} < \infty$$

which entails that $f_0 \in L_{p(\cdot)}(\Omega)$. On the other hand,

$$\begin{aligned}\rho_{p(\cdot)}(2f_0) &\geq \sum_{k=1}^{\infty} \int_{A_{m_k}} (2c_{m_k})^{p(x)}\, dx \\ &\geq \sum_{k=1}^{\infty} 2^{m_k-1} \int_{A_{m_k}} c_{m_k}^{p(x)}\, dx \\ &= \sum_{k=1}^{\infty} 2^{m_k-1} m_k^{-2} = \infty,\end{aligned}$$

which means that $2f_0 \notin L_{p(\cdot)}(\Omega)$.

SUFFICIENCY. Let $p_+ < \infty$. We have

$$\rho_{p(\cdot)}(cf) \leq \max\{|c|^{p_+}, 1\}\rho_{p(\cdot)}(f)$$

and

$$\rho_{p(\cdot)}(f+g) \leq 2^{p_+}[\rho_{p(\cdot)}(f) + \rho_{p(\cdot)}(g)]$$

for all function f and g in $L_{p(\cdot)}(\Omega)$. □

The next result tells us that the definition of the variable Lebesgue space is not void, in the sense that it always contains the set of step functions, whenever $p_+ < \infty$.

Lemma 7.3. *Let $p_+ < \infty$. Then the set of step functions belongs to the space $L_{p(\cdot)}(\Omega)$.*

Proof. Let $f(x) = \sum_{k=1}^{N} c_k \chi_{\Omega_k}(x)$ be a step function where Ω_k are pairwise disjoint. We then have

$$\rho_{p(\cdot)}(f) = \sum_{k=1}^{N} \int_{\Omega_k} |c_k|^{p(x)}\, dx \leq \sum_{k=1}^{N} \max\{1, |c_k|^{p_+}\} m(\Omega_k) < \infty,$$

which shows the validity of the lemma. □

7.1.1 Luxemburg-Nakano Type Norm

From Lemma 7.2 we already know that the space is linear if and only if $p_+ < \infty$. We now want to introduce a norm in the variable exponent Lebesgue space, but after a moment's reflection it is clear that the norm cannot be introduced in a similar manner as in the case of constant p. We will use the so-called *Luxemburg norm*, also known as *Luxemburg-Nakano norm*. Before proceeding in doing that, we will prove some auxiliary lemmas that will be used in the problem of introducing a norm in the aforementioned space.

Lemma 7.4. *Let $f \in L_{p(\cdot)}(\Omega)$, $0 \leq p(x) \leq \infty$. The function*

$$F(\lambda) := \rho_{p(\cdot)}\left(\frac{f}{\lambda}\right), \quad \lambda > 0, \tag{7.4}$$

take finite values for all $\lambda \geq 1$. Moreover, this function is continuous, decreasing, and $\lim_{\lambda\to\infty} F(\lambda) = 0$. If $p_+ < \infty$, the same is true for all $\lambda > 0$.

Proof. By definition we have that $F(1) < \infty$. It is clear that the function (7.4) is decreasing, which immediately entails that $F(\lambda) < \infty$ for all $\lambda \geq 1$. The continuity follows from

$$\begin{aligned}\lim_{\lambda\to\lambda_0} |F(\lambda) - F(\lambda_0)| &\leq \lim_{\lambda\to\lambda_0} \int_{\Omega\setminus\Omega_\infty} |f(x)|^{p(x)} |\lambda^{-p(x)} - \lambda_0^{-p(x)}| \,\mathrm{d}x \\ &\leq \int_{\Omega\setminus\Omega_\infty} \lim_{\lambda\to\lambda_0} |f(x)|^{p(x)} |\lambda^{-p(x)} - \lambda_0^{-p(x)}| \,\mathrm{d}x\end{aligned} \tag{7.5}$$

where we used the Lebesgue dominated convergence theorem since $\lambda^{-p(x)} \leq 1$ for $\lambda \geq 1$. Using again the Lebesgue dominated convergence theorem we obtain $\lim_{\lambda\to\infty} F(\lambda) = 0$.

When $p_+ < \infty$, for $\lambda < 1$ we have that $F(\lambda) \leq F(1)\lambda^{-p_+} < \infty$. The continuity follow, once again, from (7.5) since $\lambda^{-p(x)} \leq c\lambda_0^{-p_+}$ for λ near λ_0. □

We now introduce a norm in the space $L_{p(\cdot)}(\Omega)$.

Theorem 7.5. *Let $0 \leq p(x) \leq \infty$. For any $f \in L_{p(\cdot)}(\Omega)$ the functional*

$$\|f\|_{(p)} := \inf\left\{\lambda > 0 : \int_{\Omega\setminus\Omega_\infty} \left|\frac{f(x)}{\lambda}\right|^{p(x)} \mathrm{d}x \leq 1\right\} \tag{7.6}$$

takes finite values and

$$\rho_{p(\cdot)}\left(\frac{f}{\|f\|_{(p)}}\right) \leq 1, \quad \|f\|_{(p)} \neq 0. \tag{7.7}$$

If the exponent satisfies $p_+ < \infty$ *or* $\|f\|_{(p)} \geq 1$, *then*

$$\rho_{p(\cdot)}\left(\frac{f}{\|f\|_{(p)}}\right) = 1, \ \|f\|_{(p)} \neq 0. \tag{7.8}$$

Moreover, if $1 \leq p(x) \leq p_+ < \infty$, $x \in \Omega \backslash \Omega_\infty$, *we have that*

$$\|f\|_{L_{p(\cdot)}(\Omega)} = \|f\|_{(p)} + \operatorname*{ess\,sup}_{x \in \Omega_\infty} |f(x)| \tag{7.9}$$

is a norm in the space $L_{p(\cdot)}(\Omega)$.

Proof. By Lemma 7.4 we have that $\|f\|_{(p)}$ is finite whenever $f \in L_{p(\cdot)}(\Omega)$ and (7.7)–(7.8) are consequences of the definition given in (7.6) and Lemma 7.4. To show that (7.9) is a norm, it suffices to show the triangle inequality for $\|f\|_{(p)}$, which follows from the inequality

$$|\lambda y_1 + (1-\lambda) y_2|^p \leq \lambda |y_1|^p + (1-\lambda)|y_2|^p, \tag{7.10}$$

for $0 \leq \lambda \leq 1$ and $p \geq 1$, since $t \mapsto t^p$ is a convex function. □

We now obtain upper and lower bounds for the modular $\rho_{p(\cdot)}$ via the functional $\|\cdot\|_{(p)}$.

Corollary 7.6. *The functional* (7.6) *and the modular* $\rho_{p(\cdot)}$ *are related by the following estimates*

$$\left(\frac{\|f\|_{(p)}}{\lambda}\right)^{p_+} \leq \rho_{p(\cdot)}\left(\frac{f}{\lambda}\right) \leq \left(\frac{\|f\|_{(p)}}{\lambda}\right)^{p_-}, \ \lambda \geq \|f\|_{(p)}, \tag{7.11}$$

$$\left(\frac{\|f\|_{(p)}}{\lambda}\right)^{p_-} \leq \rho_{p(\cdot)}\left(\frac{f}{\lambda}\right) \leq \left(\frac{\|f\|_{(p)}}{\lambda}\right)^{p_+}, \ 0 < \lambda \leq \|f\|_{(p)}, \tag{7.12}$$

where the extreme cases $p_- = 0$ *or* $p_+ = \infty$ *are admitted.*

Proof. Let us rewrite (7.11) and (7.12) as

$$\lambda^{p_+} \leq \rho_{p(\cdot)}\left(\frac{\lambda}{\|f\|_{(p)}} f\right) \leq \lambda^{p_-}, \ 0 < \lambda \leq 1, \tag{7.13}$$

and

$$\lambda^{p_-} \leq \rho_{p(\cdot)}\left(\frac{\lambda}{\|f\|_{(p)}} f\right) \leq \lambda^{p_+}, \ \lambda \geq 1. \tag{7.14}$$

We now have that(7.13) and (7.14) are a consequence of (7.8) if $p_+ < \infty$ or $p_+ = \infty$ with $\|f\|_{(p)} \geq 1$. If $p_+ = \infty$ and $\|f\|_{(p)} \leq 1$, the right-hand side of the inequality in (7.13) is a consequence of (7.7), and the left-hand side of (7.14) holds since $\|g\|_{(p)} = \lambda \geq 1$ for $g(x) = \lambda f(x)/\|f\|_{(p)}$. □

Corollary 7.7. *Let p be a measurable function, $0 \le p_- \le p(x) \le p_+ < \infty$, $x \in \Omega \backslash \Omega_\infty$, we have the following estimates*

$$\|f\|_{(p)}^{p_+} \le \rho_{p(\cdot)}(f) \le \|f\|_{(p)}^{p_-}, \ \|f\|_{(p)} \le 1, \tag{7.15}$$

$$\|f\|_{(p)}^{p_-} \le \rho_{p(\cdot)}(f) \le \|f\|_{(p)}^{p_+}, \ \|f\|_{(p)} \ge 1. \tag{7.16}$$

Corollary 7.7 states that in questions related to convergence, $\rho_{p(\cdot)}(\cdot)$ and $\|\cdot\|_{(p)}$ are equivalent. This observation is quite useful due to the fact that the norm is given by a supremum and calculating explicitly the norm can be impossible, except in trivial cases.

With these estimates at hand, we can get an upper and lower bound for the norm of an indicator function of a set.

Corollary 7.8. *Let E be a measurable set in $\Omega \backslash \Omega_\infty$. If $0 < p_- \le p_+ < \infty$ we have the estimate*

$$m(E)^{1/p_-} \le \|\chi_E\|_{(p)} \le m(E)^{1/p_+},$$

when $m(E) \le 1$. In the case $m(E) \ge 1$, the signs of the inequality are reversed. As a particular case, we have that $\|\chi_E\|_{(p)} = 1$ is equivalent to $m(E) = 1$.

Example 7.9. An example that illustrates (7.7) instead of (7.8) is the following. Let

$$\Omega = [0,1], \ p(x) = \frac{1}{x}, \ \Omega_\infty = \{0\}, \ f(x) = 4^{-x}x^{-x/2}.$$

We have that $\|f\|_{(p)} = \|f\|_{L_{p(\cdot)}(\Omega)} = 1$, since $F(1) = \int\limits_0^1 |f(x)|^{p(x)}\,dx < 1$, but $F(\lambda) \equiv \infty$ for all $\lambda < 1$. ⊘

Remark 7.10. The space $L_{p(\cdot)}(\Omega)$ is *ideal*; i.e., it is a complete space and the inequality $|f(x)| \le |g(x)|$, $g \in L_{p(\cdot)}(\Omega)$ implies that $\|f\|_{L_{p(\cdot)}(\Omega)} \le \|g\|_{L_{p(\cdot)}(\Omega)}$ (the completeness will be showed in § 7.1.4).

Let $1 \le p(x) \le \infty$ be such that $p_+ < \infty$. The semi-norm $\|f\|_{(p)}$ can be represented in the form

$$\|f\|_{(p)} = \int\limits_{\Omega \backslash \Omega_\infty} \Phi(x) f(x)\,dx \ , \ \Phi(x) \in L_{p'(\cdot)}(\Omega) \tag{7.17}$$

where $\Phi(x) = \left|\frac{f(x)}{\|f\|_{(p)}}\right|^{p(x)-1} \frac{f(x)}{|f(x)|}$, $x \notin \Omega_\infty$ and $\|\Phi\|_{(p')} \le 1$. In reality (7.17) is simply (7.8), the inequality $\|\Phi\|_{(p')} \le 1$ is immediate.

The next lemma, albeit simple, is also a useful tool dealing with the estimation of norms in variable exponent Lebesgue spaces. It states that if the modular of a dilated function is bounded then the function is bounded in norm, with certain upper bound.

Lemma 7.11. *Let $0 < p_- \le p_+ \le \infty$. If*

$$\rho_{p(\cdot)}\left(\frac{f}{a}\right) \le b, \quad a > 0,\ b > 0, \tag{7.18}$$

then $\|f\|_{(p)} \le ab^{\nu}$ with $\nu = 1/p_-$ if $b \ge 1$ and $\nu = 1/p_+$ if $b \le 1$.

Proof. By (7.18) we have the inequality $\rho_{p(\cdot)}(f/(ab^{\nu})) \le 1$, and now by the definition (7.6) we get that $\|f\|_{(p)} \le ab^{\nu}$. □

The next result generalizes the property

$$\left\|f^{\gamma}\right\|_{p} = \|f\|_{\gamma p}^{\gamma}$$

for the variable setting.

Lemma 7.12. *Let $0 < \gamma(x) \le p(x) \le p_+ < \infty$, $x \in \Omega\backslash\Omega_\infty$. Then*

$$\|f\|_{(p)}^{\gamma_-} \le \|f^{\gamma}\|_{(\frac{p}{\gamma})} \le \|f\|_{(p)}^{\gamma_+},\ \|f\|_{(p)} \ge 1, \tag{7.19}$$

$$\|f\|_{(p)}^{\gamma_+} \le \|f^{\gamma}\|_{(\frac{p}{\gamma})} \le \|f\|_{(p)}^{\gamma_-},\ \|f\|_{(p)} \le 1, \tag{7.20}$$

where $f^{\gamma} = |f(x)|^{\gamma(x)}$. If p and γ are continuous functions, there exists a point $x_0 \in \Omega\backslash\Omega_\infty$ such that

$$\|f^{\gamma}\|_{(\frac{p}{\gamma})} = \|f\|_{(p)}^{\gamma(x_0)}. \tag{7.21}$$

Proof. Let $\lambda = \|f\|_{(p)}$, $\mu = \|f^{\gamma}\|_{(\frac{p}{\gamma})}$. Since $\Omega_\infty(\frac{p}{\gamma}) = \Omega_\infty(p)$, by (7.8) we have

$$\int\limits_{\Omega\backslash\Omega_\infty} \left|\frac{|f(x)|^{\gamma(x)}}{\mu}\right|^{\frac{p(x)}{\gamma(x)}} dx = \int\limits_{\Omega\backslash\Omega_\infty} \left|\frac{f(x)}{\lambda}\right|^{p(x)} dx = 1.$$

Therefore

$$\int\limits_{\Omega\backslash\Omega_\infty} |f(x)|^{p(x)} \frac{\lambda^{p(x)} - \mu^{\frac{p(x)}{\gamma(x)}}}{\lambda^{p(x)}\mu^{p(x)/\gamma(x)}} dx = 0. \tag{7.22}$$

Suppose that $\lambda \ge 1$. We now show the right-hand side inequality in (7.19), i.e., $\mu \le \lambda^{\gamma_+}$. Suppose that $\mu > \lambda^{\gamma_+}$, then $\mu^{\frac{p(x)}{\gamma(x)}} > \lambda^{\frac{p(x)}{\gamma(x)}\gamma_+}$. This means that the numerator in (7.22) is non-positive in almost every point, which is impossible. A similar supposition: $\mu < \lambda^{\gamma_-}$ gives a nonnegative numerator, which is also impossible. The case $\lambda \le 1$ is similar.

Now, if p and γ are continuous functions, then from (7.22) we get that the numerator of the fraction must be zero in some point, which implies (7.21). □

We now obtain that under some circumstances it is possible to realize the value $\|f\|_{(p)}$ in the following sense.

Corollary 7.13. *Let $0 \leq p_- \leq p(x) \leq p_+ < \infty$, $x \in \Omega \backslash \Omega_\infty$. If p is a continuous function in $\Omega \backslash \Omega_\infty$, there exists a point $x_0 \in \Omega \backslash \Omega_\infty$ (which depends on f) such that*

$$\|f\|_{(p)} = \left\{ \int_{\Omega \backslash \Omega_\infty} |f(x)|^{p(x)} \, dx \right\}^{\frac{1}{p(x_0)}}. \tag{7.23}$$

Proof. Taking $\gamma(x) = p(x)$ in the equality (7.21) we get (7.23). □

Definition 7.14. We define the *sum space* $L_p(\Omega) + L_q(\Omega)$ as

$$L_p(\Omega) + L_q(\Omega) := \{ f = g + h : g \in L_p(\Omega), h \in L_q(\Omega) \},$$

which is a Banach space with the norm

$$\|f\|_{L_p(\Omega)+L_q(\Omega)} = \inf_{f=g+h} \{ \|g\|_{L_p(\Omega)} + \|h\|_{L_q(\Omega)} \}.$$

The *intersection space* $L_p(\Omega) \cap L_q(\Omega)$ is defined as

$$\|f\|_{L_p(\Omega) \cap L_q(\Omega)} = \max\{ \|f\|_{L_p(\Omega)}, \|f\|_{L_q(\Omega)} \}$$

which is a Banach space. ⊘

We now show that the variable exponent Lebesgue space is embedded between the sum and intersection spaces of the spaces L_{p_-} and L_{p_+}.

Lemma 7.15. *Let $1 \leq p_- \leq p(x) \leq p_+ \leq \infty$, $x \in \Omega$, $m(\Omega_\infty) = 0$. Then*

$$L_{p(\cdot)}(\Omega) \subseteq L_{p_-}(\Omega) + L_{p_+}(\Omega). \tag{7.24}$$

Moreover,

$$\|f\|_{L_{p(\cdot)}(\Omega)} \leq \max\{ \|f\|_{p_-}, \|f\|_{p_+} \}.$$

The result follows from the splitting $f(x) = f_1(x) + f_2(x)$ where $f_1(x) = f(x)$ if $|f(x)| \leq 1$ and $f_1(x) = 0$ otherwise.

The Lemma 7.15 admits the following natural generalization.

Lemma 7.16. *Let $1 \leq p_1(x) \leq p(x) \leq p_2(x) \leq \infty$ and $m(\Omega_\infty(p_2)) = 0$. Then*

$$L_{p(\cdot)}(\Omega) \subseteq L_{p_1(\cdot)}(\Omega) + L_{p_2(\cdot)}(\Omega).$$

In the previous lemmas, splitting the function in an appropriate way we were able to obtain embedding results. We now want to obtain embedding results where the splitting is applied to the underlying set Ω.

Lemma 7.17. *Let $\Omega = \Omega_1 \cup \Omega_2$ and let p be a function in Ω, $p(x) \geq 1$ with $p_+ < \infty$. Then*

$$\max\{ \|f\|_{L_{p(\cdot)}(\Omega_1)}, \|f\|_{L_{p(\cdot)}(\Omega_2)} \} \leq \|f\|_{L_{p(\cdot)}(\Omega)} \leq \|f\|_{L_{p(\cdot)}(\Omega_1)} + \|f\|_{L_{p(\cdot)}(\Omega_2)} \tag{7.25}$$

for all functions $f \in L_{p(\cdot)}(\Omega)$.

Proof. Let us take $m(\Omega_\infty) = 0$ for simplicity. Without loss of generality, let $a = \|f\|_{L_{p(\cdot)}(\Omega_1)}$, $b = \|f\|_{L_{p(\cdot)}(\Omega_2)}$ with $a \geq b$. We have

$$\int_\Omega \left|\frac{f(x)}{\max\{a,b\}}\right|^{p(x)} \mathrm{d}x \geq \int_{\Omega_1} \left|\frac{f(x)}{a}\right|^{p(x)} \mathrm{d}x = 1.$$

Therefore $\|f\|_{L_{p(\cdot)}(\Omega)} \geq \max\{a,b\}$.

To show the right-hand side inequality, we write

$$\frac{f(x)}{a+b} = \frac{a}{a+b}\frac{\chi_1(x)f(x)}{a} + \frac{b}{a+b}\frac{\chi_2(x)f(x)}{b}$$

where $\chi_i(x)$ are the characteristic functions of the sets $\Omega_i, i = 1,2$. Using (7.10) we get

$$\int_\Omega \left|\frac{f(x)}{a+b}\right|^{p(x)} \mathrm{d}x \leq 1,$$

which shows the right-hand side inequality in (7.25).

For the case $m(\Omega_\infty) > 0$, the arguments are similar if we take into account the fact that the lemma was already proved for the case $\Omega \backslash \Omega_\infty = \Omega_1^* \cup \Omega_2^*$ where $\Omega_i^* = \Omega_i \backslash \Omega_\infty$, $i = 1,2$. □

7.1.2 Another Version of the Luxemburg-Nakano Norm

The Luxemburg-Nakano type norm can be introduced directly with respect to all the set Ω in the following form

$$\|f\|_p^1 = \inf\left\{\lambda > 0 : \rho_{p(\cdot)}\left(\frac{f}{\lambda}\right) + \operatorname*{ess\,sup}_{x\in\Omega_\infty}\left|\frac{f(x)}{\lambda}\right| \leq 1\right\}, \tag{7.26}$$

which is well defined for $f \in L_{p(\cdot)}(\Omega)$ and any variable exponent p with $0 \leq p(x) \leq \infty$. It is a norm if $1 \leq p(x) \leq \infty$, which can be shown in the same way as Theorem 7.5. In an analogous way to (7.8) it is possible to show that

$$\int_{\Omega\backslash\Omega_\infty} \left|\frac{f(x)}{\|f\|_p^1}\right|^{p(x)} \mathrm{d}x + \frac{\|f\|_{L_\infty(\Omega_\infty)}}{\|f\|_p^1} = 1 \tag{7.27}$$

if $p_+ < \infty$ or $p_+ = \infty$, but $\|f\|_p^1 \geq 1$.

Theorem 7.18. *The norms* (7.9) *and* (7.27) *are equivalent, i.e.*

$$\frac{1}{2}\|f\|_{L_{p(\cdot)}(\Omega)} \leq \|f\|_p^1 \leq \|f\|_{L_{p(\cdot)}(\Omega)} \tag{7.28}$$

where $f \in L_{p(\cdot)}(\Omega)$, $1 \leq p(x) \leq \infty$, $p_+ < \infty$.

Proof. The right-hand side inequality in (7.28) is equivalent to

$$\inf\{\lambda > 0 : F(\lambda) + c/\lambda \leq 1\} \leq \lambda_0 + c,$$

where $F(\lambda)$ is defined by (7.4) and

$$c = \|f\|_{L_\infty(\Omega_\infty)}, \quad \lambda_0 = \|f\|_{(p)}.$$

From the above, it is sufficient to show that $F(\lambda_0 + c) + \frac{c}{\lambda_0+c} \leq 1$, or in other words: $F(\lambda_0 + c) \leq \frac{\lambda_0}{\lambda_0+c}$. Since $F(\lambda_0 + c) = \rho_{p(\cdot)}\left(\frac{f}{\|f\|_{(p)}+c}\right)$, by (7.11) we obtain that $F(\lambda_0 + c) \leq \frac{\|f\|_{(p)}}{\|f\|_{(p)}+c} = \frac{\lambda_0}{\lambda_0+c}$.

The left-hand side in (7.28) is a consequence of the inequalities

$$\inf\left\{\lambda > 0 : F(\lambda) + \frac{c}{\lambda} \leq 1\right\} \geq \inf\{\lambda > 0 : F(\lambda) \leq 1\} = \lambda_0,$$

and

$$\inf\left\{\lambda > 0 : F(\lambda) + \frac{c}{\lambda} \leq 1\right\} \geq \inf\left\{\lambda > 0 : \frac{c}{\lambda} \leq 1\right\} = c,$$

since the left-hand side inequality is not less that $\frac{\lambda_0+c}{2}$. □

7.1.3 Hölder Inequality

We now proceed to get Hölder's inequality and after that we will get the Minkowski inequality using F. Riesz construction via Hölder's inequality.

Theorem 7.19 (Hölder's inequality). *Let* $f \in L_{p(\cdot)}(\Omega)$, $\varphi \in L_{p'(\cdot)}(\Omega)$ *and* $1 \leq p(x) \leq \infty$. *Then*

$$\int_\Omega |f(x)\varphi(x)|\,\mathrm{d}x \leq k\|f\|_{L_{p(\cdot)}(\Omega)}\|\varphi\|_{L_{p'(\cdot)}(\Omega)} \tag{7.29}$$

with $k = \frac{1}{p_-} + \frac{1}{(p')_-} = \sup\frac{1}{p(x)} + \sup\frac{1}{p'(x)}$.

Proof. Let us note that, under the conditions of the theorem, the functionals $\|f\|_{L_{p(\cdot)}(\Omega)}$ and $\|\varphi\|_{p'(\cdot)}$ are not necessarily norms and the classes $L_{p(\cdot)}$ and $L_{p'(\cdot)}$ are not necessarily linear, but they always exist by Theorem 7.5.

To show (7.29), we use the Young inequality

$$ab \leq \frac{a^p}{p} + \frac{b^{p'}}{p'} \tag{7.30}$$

with $a > 0, b > 0, \frac{1}{p} + \frac{1}{p'} = 1$ and $1 < p < \infty$. The inequality (7.30) is valid for $p = 1$ in the form $ab \le \frac{a^p}{p}$ if $b \le 1$ and for $p = \infty$ in the form $ab \le \frac{b^{p'}}{p'}$ if $a \le 1$. Therefore,

$$\left| \frac{f(x)\varphi(x)}{\|f\|_{p(\cdot)}\|\varphi\|_{p'(\cdot)}} \right| \le \frac{1}{p(x)} \left| \frac{f(x)}{\|f\|_{p(\cdot)}} \right|^{p(x)} + \frac{1}{p'(x)} \left| \frac{\varphi(x)}{\|\varphi\|_{p'(\cdot)}} \right|^{p'(x)},$$

where $x \in \Omega \backslash \Omega_\infty(p) \cup \Omega_\infty(p')$, meanwhile for $x \in \Omega_\infty(p)$ and $x \in \Omega_\infty(p')$ we have to omit the first and second terms respectively in the right-hand side, since $\left| \frac{f(x)}{\|f\|_{p(\cdot)}} \right| \le 1$ for $x \in \Omega_\infty(p)$ and $\left| \frac{\varphi(x)}{\|\varphi\|_{p'(\cdot)}} \right| \le 1$ for $x \in \Omega_\infty(p')$. Integrating over Ω and estimating p and p', we arrive at (7.29). □

In the constant exponent case $p(x) \equiv p$, the Hölder inequality has a generalization of the form

$$\|uv\|_r \le \|u\|_p \|v\|_q, \quad \frac{1}{p} + \frac{1}{q} = \frac{1}{r},$$

which is an immediate consequence of the Hölder inequality and the relation

$$\||u|^r\|_p = \|u\|_{pr}^r. \tag{7.31}$$

In the variable exponent Lebesgue space the relation (7.31) is no more valid in general, cf. Lemma 7.12 and (7.81). Nonetheless, the inequality is valid.

Lemma 7.20. *Let* $\frac{1}{p(x)} + \frac{1}{q(x)} \equiv \frac{1}{r(x)}$, $p(x) \ge 1$, $q(x) \ge 1$, $r(x) \ge 1$ *and let* $R = \sup_{x \in \Omega \backslash \Omega_\infty(r)} r(x) < \infty$. *Then*

$$\|uv\|_{L_{r(\cdot)}(\Omega)} \le c \|u\|_{L_{p(\cdot)}(\Omega)} \|v\|_{L_{q(\cdot)}(\Omega)} \tag{7.32}$$

for all functions $u \in L_{p(\cdot)}$ *and* $v \in L_{q(\cdot)}$ *with* $c = c_1 + c_2$, $c_1 = \sup_{x \in \Omega \backslash \Omega_\infty(r)} \frac{r(x)}{p(x)}$ *and* $c_2 = \sup_{x \in \Omega \backslash \Omega_\infty(r)} \frac{r(x)}{q(x)}$.

Proof. To show (7.32) we use the inequality

$$(AB)^r \le \frac{r}{p} A^p + \frac{r}{q} B^q$$

with $A > 0, B > 0, p > 0, q > 0$ and $\frac{1}{p} + \frac{1}{q} = \frac{1}{r}$, see Problem 1.48.

Integrating the inequality

$$|u(x)v(x)|^{r(x)} \le \frac{r(x)}{p(x)} |u(x)|^{p(x)} + \frac{r(x)}{q(x)} |v(x)|^{q(x)}$$

we get

$$\int\limits_{\Omega \backslash \Omega_\infty(r)} |u(x)v(x)|^{r(x)} \, dx \le c_1 \int\limits_{\Omega \backslash \Omega_\infty(p)} |u(x)|^{p(x)} \, dx + c_2 \int\limits_{\Omega \backslash \Omega_\infty(q)} |v(x)|^{q(x)} \, dx \tag{7.33}$$

since $\Omega_\infty(r) = \Omega_\infty(p) \cap \Omega_\infty(q)$. From (7.33) and (7.7) it follows

$$\int\limits_{\Omega\setminus\Omega_\infty(r)} \left|\frac{u(x)v(x)}{\|u\|_{(p)}\|v\|_{(q)}}\right|^{r(x)} \mathrm{d}x$$

$$\leq c_1 \int\limits_{\Omega\setminus\Omega_\infty(p)} \left|\frac{u(x)}{\|u\|_{(p)}}\right|^{p(x)} \mathrm{d}x + c_2 \int\limits_{\Omega\setminus\Omega_\infty(q)} \left|\frac{v(x)}{\|v\|_{(q)}}\right|^{q(x)} \mathrm{d}x \leq c_1 + c_2.$$

From Lemma 7.11 we now get $\|uv\|_{(r)} \leq (c_1+c_2)\|u\|_{(p)}\|v\|_{(q)}$, since $c_1 + c_2 \geq 1$. $\square$

The inequality (7.32) is also valid in the form

$$\rho_{r(\cdot)}(uv) \leq c\|u\|_{L_{p(\cdot)}(\Omega)}\|v\|_{L_{q(\cdot)}(\Omega)}$$

if $\|u\|_{L_{p(\cdot)}(\Omega)} \leq 1$ and $\|v\|_{L_{q(\cdot)}(\Omega)} \leq 1$, which follows from the Hölder inequality (7.29) and the estimate (7.20).

7.1.4 Convergence and Completeness

Theorem 7.21. *Let $1 \leq p(x) \leq p_+ < \infty$. The space $L_{p(\cdot)}(\Omega)$ is complete.*

Proof. The space $L_{p(\cdot)}(\Omega)$ is the sum of $L_{p(\cdot)}(\Omega_*) + L_\infty(\Omega_\infty)$ where each space is understood as the space of functions which are 0 outside the sets Ω_* and Ω_∞, respectively. Therefore, we only need to show the completeness of the space $L_{p(\cdot)}(\Omega_*)$.

Let $\{f_k\}$ be a Cauchy sequence in $L_{p(\cdot)}(\Omega_*)$ such that for any positive number s exists N_s $(N_1 < N_2 < \ldots)$ such that

$$\|f_{N_{s+1}} - f_{N_s}\|_{L_{p(\cdot)}(\Omega_*)} < 2^{-s}, \quad s = 1,2,3,\ldots.$$

Then

$$\sum_{s=1}^{\infty} \|f_{N_{s+1}} - f_{N_s}\|_{L_{p(\cdot)}(\Omega_*)} < \infty.$$

Let $\Omega_r = \{x \in \Omega_* : |x| < r\}, r > 0$. By Hölder's inequality (7.29) we obtain

$$\sum_{s=1}^{\infty} \int\limits_{\Omega_r} \left|f_{N_{s+1}}(x) - f_{N_s}(x)\right| \mathrm{d}x \leq c_r \sum_{s=1}^{\infty} \|f_{N_{s+1}} - f_{N_s}\|_{L_{p(\cdot)}(\Omega_*)} < \infty \tag{7.34}$$

where $c_r = \left(\frac{1}{p_-} + \frac{1}{(p')_-}\right)\|\chi_{\Omega_r}\|_{L_{p'(\cdot)}(\Omega_*)} < \infty$. By (7.34), $\{f_{N_s}(x)\}$ is a Cauchy sequence in $L_1(\Omega_r)$. Therefore, there exists the limit $f(x) = \lim_{s\to\infty} f_{N_s}(x)$ for almost all $x \in \Omega_r$, which entails that the same happens for almost all $x \in \Omega_*$ since $r > 0$ is arbitrary. Now we only need to show that

$$\lim_{k\to\infty} \|f_k - f\|_{L_{p(\cdot)}(\Omega_*)} = 0.$$

Since $\{f_k\}$ is a Cauchy sequence, we have that $\|f_k - f_{N_s}\|_{L_{p(\cdot)}(\Omega_*)} < \varepsilon$ whenever k and s are sufficiently large. Now by (7.15) we get

$$\int_{\Omega_*} |f_k(x) - f_{N_s}(x)|^{p(x)}\,\mathrm{d}x \le \varepsilon^{p_-} \le \varepsilon.$$

Invoking Fatou's Lemma we obtain

$$\begin{aligned}\int_{\Omega_*} |f_k(x) - f(x)|^{p(x)}\,\mathrm{d}x &\le \liminf_{s\to\infty} \int_{\Omega_*} |f_k(x) - f_{N_s}(x)|^{p(x)}\,\mathrm{d}x \\ &\le \sup_s \int_{\Omega_*} |f_k(x) - f_{N_s}(x)|^{p(x)}\,\mathrm{d}x \\ &< \varepsilon\end{aligned}$$

which ends the proof. □

Lemma 7.22. *Let $0 < p_- \le p(x) \le p_+ < \infty$, $x \in \Omega\backslash\Omega_\infty$. The convergence*

$$\int_{\Omega\backslash\Omega_\infty} |f_m(x) - f(x)|^{p(x)}\,\mathrm{d}x + \operatorname*{ess\,sup}_{x\in\Omega_\infty} |f(x) - f_m(x)| < \varepsilon$$

is equivalent to the norm convergence

$$\|f - f_m\|_{(p)} + \operatorname*{ess\,sup}_{x\in\Omega_\infty} |f(x) - f_m(x)| < \varepsilon.$$

Proof. Follows from Corollary 7.7. □

7.1.5 Embeddings and Dense Sets

Theorem 7.23. *Let $0 \le r(x) \le p(x) \le \infty$ and let $m(\Omega\backslash\Omega_\infty(r)) < \infty$. If $\Omega_\infty(r) \subseteq \Omega_\infty(p)$ and*

$$R := \sup_{x\in\Omega_\infty(p)\backslash\Omega_\infty(r)} r(x),$$

then $L_{p(\cdot)}(\Omega) \subseteq L_{r(\cdot)}(\Omega)$ and

$$\rho_{r(\cdot)}(f) \le \rho_{p(\cdot)}(f) + m(\Omega_\infty(p)\backslash\Omega_\infty(r))\|f\|^R_{L_\infty(\Omega_\infty(p)\backslash\Omega_\infty(r))} + m(\Omega\backslash\Omega_\infty(r)) \quad (7.35)$$

for any $f \in L_{p(\cdot)}(\Omega)$. (In the case $\Omega_\infty(p) = \Omega_\infty(r)$, the second term in the right-hand side should be omitted and R can be infinite). If, moreover, $1 \le r(x) \le p(x)$ and $\Omega_\infty(p) = \Omega_\infty(r)$, the inequality for norms is also valid:

$$\|f\|_{(r)} \le c_0^\nu \|f\|_{(p)} \tag{7.36}$$

where

$$c_0 = c_2 + (1 - c_1)m(\Omega \backslash \Omega_\infty(p)),\; c_1 = \inf_{x\in\Omega\backslash\Omega_\infty(p)} \frac{r(x)}{p(x)},\; c_2 = \sup_{x\in\Omega\backslash\Omega_\infty(p)} \frac{r(x)}{p(x)},$$

$\nu = \frac{1}{r_0}$ *if* $c_0 \ge 1$ *and* $\nu = \frac{1}{R}$ *if* $c_0 \le 1$.

Proof. The estimate (7.35) is derived from the equality $\rho_{r(\cdot)}(f) = \int\limits_{\Omega_1} + \int\limits_{\Omega_2} + \int\limits_{\Omega_3}$ with $\Omega_1 = \{x \in \Omega\backslash\Omega_\infty(p) : |f(x)| \ge 1\}$, $\Omega_2 = \{x \in \Omega_\infty(p)\backslash\Omega_\infty(r) : |f(x)| \ge 1\}$, $\Omega_3 = \{x \in \Omega\backslash\Omega_\infty(r) : |f(x)| \le 1\}$.

The classical technique to show the inequality (7.36) for norms is based on the Hölder inequality with the exponents $p_1(x) = \frac{p(x)}{r(x)}$ and $p_2(x) = \frac{r(x)}{p(x)-r(x)}$ which is no more appropriate for the variable setting since we can have $p(x) = r(x)$ in some arbitrary set. Using the inequality $(AB)^r \le \frac{r}{p}A^p + \frac{r}{q}B^q$ and taking $A = |f(x)|/\|f\|_{(p)}$ and $B = 1$, we get, via (7.7), that

$$\int\limits_{\Omega\backslash\Omega_\infty} \left|\frac{f(x)}{\|f\|_{(p)}}\right|^{r(x)} \mathrm{d}x \le c_0.$$

Therefore, by Lemma 7.11 we get (7.36). □

We now show the denseness of the bounded functions with compact support.

Lemma 7.24. *Let $m(\Omega_\infty(p)) = 0$, $1 \le p(x) \le p_+ < \infty$. The set of bounded functions with compact support is dense in $L_{p(\cdot)}(\Omega)$.*

Proof. For $f \in L_{p(\cdot)}(\Omega)$ we define $f_{N,m}$ as

$$f_{N,m}(x) = \begin{cases} f(x), & \text{when } |f(x)| \le N \text{ and } |x| \le m; \\ 0, & \text{otherwise.} \end{cases}$$

By Lemma 7.22, we have

$$\int\limits_{\Omega} |f(x) - f_{N,m}(x)|^{p(x)} \mathrm{d}x \le \int\limits_{\omega_m} |g(x)|\,\mathrm{d}x + \int\limits_{\Omega_N} |g(x)|\,\mathrm{d}x \to 0$$

when $m \to \infty, N \to \infty$, with $\omega_m = \{x \in \Omega : |x| \ge m\}$, $\Omega_N = \{x \in \Omega : f(x) \ge N\}$ and $g(x) = |f(x)|^{p(x)} \in L_1(\Omega)$. □

Theorem 7.25. *Let $p \in \mathscr{P}(\Omega) \cap L_\infty(\Omega)$. Then the set $C(\Omega) \cap L_{p(\cdot)}(\Omega)$ is dense in $L_{p(\cdot)}(\Omega)$. Moreover, if Ω is open, then the set of all functions infinitely differentiable with compact support $C_c^\infty(\Omega)$ is dense in $L_{p(\cdot)}(\Omega)$.*

Proof. Let $f \in L_{p(\cdot)}(\Omega)$ and $\varepsilon > 0$. From Lemma 7.24 there exists a bounded function $g \in L_{p(\cdot)}(\Omega)$ such that

$$\|f-g\|_{L_{p(\cdot)}(\Omega)} < \varepsilon. \tag{7.37}$$

By Luzin's Theorem, there exists a function $h \in C(\Omega)$ and an open set U such that

$$m(U) < \min\left\{1, \left(\frac{\varepsilon}{2\|g\|_\infty}\right)^{p_+}\right\},$$

$g(x) = h(x)$ for all $x \in \Omega \setminus U$ and $\sup|h(x)| = \sup_{\Omega\setminus U}|g(x)| \leq \|g\|_\infty$. Then,

$$\rho_{p(\cdot)}\left(\frac{g-h}{\varepsilon}\right) \leq \max\left\{1, \left(\frac{2\|g\|_\infty}{\varepsilon}\right)^{p_+}\right\} m(U) \leq 1$$

i.e., $\|g-h\|_{L_{p(\cdot)}(\Omega)} \leq \varepsilon$, which together with (7.37) implies that

$$\|f-h\|_{L_{p(\cdot)}(\Omega)} \leq 2\varepsilon. \tag{7.38}$$

On the other hand, let us assume that Ω is open. Since $p \in L_\infty(\Omega)$, we have that $C_c^\infty(\Omega) \subset L_{p(\cdot)}(\Omega)$ and $\rho_{p(\cdot)}\left(\frac{h}{\varepsilon}\right) < \infty$, in this way there is an open and bounded set $G \subset \Omega$ such that $\rho_{p(\cdot)}\left(\frac{h\chi_{\Omega\setminus G}}{\varepsilon}\right) \leq 1$. In other words,

$$\|h - h\chi_G\|_{L_{p(\cdot)}(\Omega)} \leq \varepsilon. \tag{7.39}$$

By the Weierstrass approximation theorem, let m be a polynomial which satisfies the condition $\sup|h(x) - m(x)| \leq \varepsilon \min\{1, |G|^{-1}\}$. Therefore $\rho_{p(\cdot)}\left(\frac{h\chi_G - m\chi_G}{\varepsilon}\right) \leq \min\{1, |G|^{-1}\}|G| \leq 1$, from which

$$\|h\chi_G - m\chi_G\|_{L_{p(\cdot)}(\Omega)} \leq \varepsilon. \tag{7.40}$$

Finally, similar considerations to the ones that were used to get (7.39) permit to conclude that for a sufficient small number a, the compact set $K_a = \{x \in G : \operatorname{dist}(x, \partial G) \geq a\}$ satisfies that $\|m\chi_G - m\chi_{K_a}\|_{L_{p(\cdot)}(\Omega)} \leq \varepsilon$. Taking $\varphi \in C_c^\infty(G)$ such that $0 \leq \varphi(x) \leq 1$ for $x \in G$ and $\varphi(x) = 1$ for $x \in K_a$ we obtain

$$\|m\chi_G - m\varphi\|_{L_{p(\cdot)}(\Omega)} \leq \|m\chi_G - m\chi_{K_a}\|_{L_{p(\cdot)}(\Omega)} \leq \varepsilon,$$

from which, together with (7.38) and (7.40), we conclude that

$$\|f - m\varphi\|_{L_{p(\cdot)}(\Omega)} \leq 4\varepsilon.$$

Clearly $m\varphi \in C_c^\infty(\Omega)$, which concludes the proof. □

By $L_c^\infty(\mathbb{R}^n)$ we denote the class of all bounded functions in $\mathbb{R}^n$ with compact support. From Theorem 7.25 we get the result.

Lemma 7.26. *Let $p : \mathbb{R}^n \longrightarrow [0,\infty)$ be a measurable function such that $1 < p_- \leq p_+ < \infty$. Then $L_c^\infty(\mathbb{R}^n)$ is dense in $L_{p(\cdot)}(\mathbb{R}^n)$ and in $L_{p'(\cdot)}(\mathbb{R}^n)$.*

We now show that the set of step functions is dense in the framework of variable exponent spaces with finite exponent.

Theorem 7.27. *Let $p : \mathbb{R}^n \longrightarrow [0,\infty)$ be a measurable function such that $1 < p_- \leq p_+ < \infty$. The set S of step functions is dense in $L_{p(\cdot)}(\Omega)$.*

Proof. It follows from Lemma 7.3 and from Theorem 7.25 together with the fact that continuous functions in compact sets are uniformly approximated by step functions. □

Theorem 7.28. *Under the conditions of Lemma 7.24 the space $L_{p(\cdot)}(\Omega)$ is separable.*

Proof. By Theorem 7.25 it is sufficient to show that any continuous function f with compact support $F \subset \overline{\Omega}$ can be approximated by functions in some enumerable set. We know that such functions can be approximated uniformly by polynomials $r_m(x)$ with rational coefficients. Taking $f_m(x) = r_m(x)$ for $x \in F$ and $f_m(x) = 0$ for $x \notin F$, we see that the functions $f_m(x)$ approximate uniformly the function $f(x)$, which ends the proof. □

7.1.6 Duality

We now characterize the dual space of variable exponent Lebesgue spaces, which is similar to the classical Lebesgue space, viz. the dual space of L_p is $L_{p'}$, where p' is the conjugate exponent. For simplicity, we will work with $m(\Omega) < \infty$. For $m(\Omega) = \infty$ see Cruz-Uribe and Fiorenza [9].

Theorem 7.29. *Let $1 < p_- \leq p(x) \leq p_+ < \infty$ and $m(\Omega) < \infty$. Then*

$$\left[L_{p(\cdot)}(\Omega)\right]^* = L_{p'(\cdot)}(\Omega).$$

Proof. The inclusion $L_{p'(\cdot)}(\Omega) \subseteq \left[L_{p(\cdot)}(\Omega)\right]^*$ is an immediate consequence of the Hölder inequality (7.29). We now show the opposite inclusion $\left[L_{p(\cdot)}(\Omega)\right]^* \subseteq L_{p'(\cdot)}(\Omega)$. Let $\Phi \in \left[L_{p(\cdot)}(\Omega)\right]^*$, then we define the set function μ as $\mu(E) = \Phi(\chi_E)$ for all measurable sets E such that $E \subset \Omega$. Since $\chi_{E\cup F} = \chi_E + \chi_F - \chi_{E\cap F}$ we have that μ is an additive function. In fact it is σ-additive. To show that, let

$$E = \bigcup_{j=1}^{\infty} E_j$$

where $E_j \subset \Omega$ are pairwise disjoint set, and let

$$F_k = \bigcup_{j=1}^{k} E_j.$$

Then

$$\left\|\chi_E - \chi_{F_k}\right\|_{L_{p(\cdot)}(\Omega)} \leq C\left\|\chi_E - \chi_{F_k}\right\|_{p_+} = C \cdot m(E \backslash F_k)^{1/p_+}.$$

Since $m(E) < \infty$, $m(E \backslash F_k)$ tends to 0 when $k \to \infty$, therefore $\chi_{F_k} \to \chi_E$ in norm. From the continuity of Φ we have that $\Phi(\chi_{F_k}) \to \Phi(\chi_E)$, which is equivalent to

$$\sum_{j=1}^{\infty} \mu(E_j) = \mu(E)$$

and from this we get that μ is σ-additive. The function μ is a measure in Ω and, moreover, is absolutely continuous: if $E \subset \Omega$ and $m(E) = 0$, therefore $\mu(E) = \Phi(\chi_E) = 0$, since $|\Phi(f)| \leq \|\Phi\| \|f\|_{L_{p(\cdot)}(\Omega)}$.

By the Radon-Nikodym Theorem, there exists $g \in L_1(\Omega)$ such that

$$\Phi(\chi_E) = \mu(E) = \int_{\Omega} \chi_E(x) g(x)\, \mathrm{d}x.$$

By the linearity of Φ, for a step function $f = \sum_{i=1}^{n} a_i \chi_{E_i}$, $E_i \subset \Omega$, we get

$$\Phi(f) = \int_{\Omega} f(x) g(x)\, \mathrm{d}x.$$

Using a density argument, similar to the constant case, we get the result. □

Corollary 7.30. *Let $1 < p_- \leq p(x) \leq p_+ < \infty$ and $m(\Omega) < \infty$. Then the space $L_{p(\cdot)}(\Omega)$ is reflexive.*

7.1.7 Associate Norm

We now introduce a norm inspired by the Riesz representation theorem for linear functionals in L_p. Let

$$\mathfrak{L}_{p(\cdot)}(\Omega) := \left\{ f \in \mathfrak{F}(\Omega, \mathscr{L}) : \left| \int_{\Omega} f(x)\varphi(x)\, \mathrm{d}x \right| < \infty,\ \forall \varphi \in L_{p'(\cdot)}(\Omega) \right\} \tag{7.41}$$

with $1 \leq p(x) \leq \infty$. This space coincides with the space $L_{p(\cdot)}(\Omega)$ under certain natural conditions in the variable exponent p and it is in fact the associate space of

$L_{p'(\cdot)}(\Omega)$ (see Definition (7.43) for the notion of *associate space* in the context of Banach Function Spaces).

The inclusion

$$L_{p(\cdot)}(\Omega) \subseteq \mathfrak{L}_{p(\cdot)}(\Omega), \ 1 \leq p(x) \leq \infty \tag{7.42}$$

is an immediate consequence of the Hölder inequality (7.29). Observe that the space defined in (7.41) is always linear. From Lemma 7.2, we have that this space cannot coincide with the space $L_{p(\cdot)}(\Omega)$ if $p_+ = \infty$.

Let us introduce the following notation

$$p_-^1 = \operatorname*{ess\,inf}_{x\in\Omega\setminus\Omega_1(p)} p(x), \quad (p')_-^1 = \operatorname*{ess\,inf}_{x\in\Omega\setminus\Omega_1(p')} p'(x).$$

We have

$$\Omega_1(p) = \Omega_\infty(p'), \ \Omega_1(p') = \Omega_\infty(p), \ (p')_+ = \frac{p_-^1}{p_-^1 - 1}, \ (p')_-^1 = \frac{p_+}{p_+ - 1}.$$

The space introduced in (7.41) can be equipped with the next natural norms

$$\|f\|_p^* = \sup_{\delta_{p'(\cdot)}(\varphi)\leq 1} \left| \int_\Omega f(x)\varphi(x)\,\mathrm{d}x \right|, \tag{7.43}$$

and

$$\|f\|_p^{**} = \sup_{\|\varphi\|_{p'(\cdot)}\leq 1} \left| \int_\Omega f(x)\varphi(x)\,\mathrm{d}x \right|, \tag{7.44}$$

where we take $\delta_{p(\cdot)}(\varphi)$ as

$$\delta_{p(\cdot)}(\varphi) = \left(\int_{\Omega\setminus\Omega_\infty} |\varphi(x)|^{p(x)}\,\mathrm{d}x \right)^{\frac{1}{p_+}} + \operatorname*{ess\,sup}_{x\in\Omega_\infty} |\varphi(x)|$$

and we assume that $(p')_+ < \infty$ (i.e., $p_-^1 > 1$) in (7.43), while $p(x)$ can be taken arbitrary ($1 \leq p(x) \leq \infty$) in the case (7.44). Sometimes the norm (7.44) is called *Orlicz type norm.*

Note that by (7.11) we have

$$\|f\|_{L_{p(\cdot)}(\Omega)} \leq \|f\|_p^{**}$$

in the case $1 \leq p(x) \leq p_+ < \infty$ and $m(\Omega_\infty) = 0$.

Lemma 7.31. *Let* $f \in \mathfrak{L}_{p(\cdot)}(\Omega)$, $(p')_-^1 > 1$. *Then* $\|f\|_p^* < \infty$ *and*

$$\int_\Omega |f(x)\varphi(x)|\,\mathrm{d}x \leq \|f\|_p^* \|\varphi\|_{p'}^1 \leq \|f\|_p^* \|\varphi\|_{L_{p'(\cdot)}(\Omega)} \tag{7.45}$$

for all $\varphi \in L_{p'(\cdot)}(\Omega)$, *where* $\|\varphi\|_{p'}^1$ *is the norm* (7.26). *Moreover, the functional* (7.43) *is a norm in* $\mathfrak{L}_{p(\cdot)}(\Omega)$.

Proof. Suppose that $\|f\|_p^* = \infty$. Then there exists a function $f_0(x) \in \mathfrak{L}_{p(\cdot)}(\Omega)$ and a sequence $\varphi_k \in L_{p'(\cdot)}(\Omega)$ such that $\delta_{p'(\cdot)}(\varphi_k) \leq 1$ and

$$\int_\Omega f_0(x)\varphi_k(x)\,\mathrm{d}x \geq 2^{Qk}, k = 1,2,\ldots$$

($f_0 \geq 0, \varphi_k \geq 0$). Therefore, $j_m = \sum_{k=1}^m 2^{-Qk}\varphi_k(x)$ is an increasing sequence. Direct calculations show that $\delta_{p'(\cdot)}(j_m) \leq 1$ and

$$\int_\Omega f_0(x) j_m(x)\,\mathrm{d}x = \sum_{k=1}^m 2^{-Qk} \int_\Omega f_0(x)\varphi_k(x)\,\mathrm{d}x \geq m. \tag{7.46}$$

The sequence $j_m(x)$ converges monotonically to the function

$$j(x) = \sum_{k=1}^\infty 2^{-Qk}\varphi_k(x).$$

Moreover,

$$\int_{\Omega\setminus\Omega_\infty(p')} |j(x)|^{p'(x)}\,\mathrm{d}x = \lim_{m\to\infty} \int_{\Omega\setminus\Omega_\infty(p')} |j_m(x)|^{p'(x)}\,\mathrm{d}x \leq 1,$$

by the Lebesgue monotone convergence theorem and, since

$$\sup_{x\in\Omega_\infty(p')} j(x) = \sum_{k=1}^\infty 2^{-Qk} < \infty$$

we get that $j \in L_{p'(\cdot)}(\Omega)$. By the Lebesgue monotone convergence theorem and by (7.46) we obtain that $\int_\Omega f_0(x)\,j(x)\,\mathrm{d}x = \infty$ which is a contradiction due to the fact that $f_0(x) \in \mathfrak{L}_{p(\cdot)}(\Omega)$.

Therefore, $\|f\|_p^* < \infty$ and by the definition (7.43) we get

$$\left|\int_\Omega f(x)\varphi(x)\,\mathrm{d}x\right| \leq A\|f\|_p^*$$

where $A > 0$ and $\delta_{p'(\cdot)}(\varphi/A) \leq 1$. Taking infimum with respect to A, we get the left-hand side of (7.45) due to the definition (7.26). The right-hand side of the inequality follows from (7.28). We only need to verify the norm axioms. The homogeneity and the triangle inequality are evident. Taking $\|f\|_p^* = 0$, then $\int_\Omega f(x)\varphi(x)\,dx = 0$ for all $\varphi \in L_{p'(\cdot)}(\Omega)$ which entails that all function $\varphi(x) \in \mathscr{S}$ by the Lemma 7.3. Therefore $f(x) \equiv 0$. □

We now show that the norms (7.43) and (7.44) are equivalent.

Lemma 7.32. *Let $1 \le p(x) \le \infty$, $p_-^1 > 1$, and $p_+ < \infty$. The norms (7.43) and (7.44) are equivalent in functions $f \in \mathfrak{L}_{p(\cdot)}(\Omega)$:*

$$2^{1-(p')_+/(p')_-^1}\|f\|_p^{**} \le \|f\|_p^* \le \|f\|_p^{**}. \tag{7.47}$$

The norms coincide in the cases:
(1) $m(\Omega_1(p)) = 0$,
(2) $p(x) = const$ for $x \in \Omega\backslash(\Omega_\infty \cup \Omega_1)$.

Proof. To obtain the right-hand side inequality, we show that

$$\left\{\varphi : \delta_{p'(\cdot)}(\varphi) \le 1\right\} \subseteq \left\{\varphi : \|\varphi\|_{L_{p'(\cdot)}(\Omega)} \le 1\right\} \tag{7.48}$$

for $\varphi \in L_{p'(\cdot)}(\Omega)$. Let $\delta_{p'(\cdot)}(\varphi) \le 1$. We have that $\rho_{p'(\cdot)}(\varphi) \le 1$ whenever $\|\varphi\|_{(p')} \le 1$ by (7.15)–(7.16). Then, by (7.15) we have that $\|\varphi\|_{(p')} \le \left(\rho_{p'(\cdot)}(\varphi)\right)^{1/(p')_+} \le 1$ which implies the inequality

$$\|\varphi\|_{L_{p'(\cdot)}(\Omega)} \le \left[\rho_{p'(\cdot)}(\varphi)\right]^{1/(p')_+} + \operatorname*{ess\,sup}_{x\in\Omega_\infty(p')} |\varphi(x)| = \delta_{p'(\cdot)}(\varphi) \le 1$$

whence (7.47) is proved.

Furthermore, let $c = 2^{1-(p')_+/(p')_-} \le 1$. We will show that

$$\left\{\varphi : \|\varphi\|_{L_{p'(\cdot)}(\Omega)} \le 1\right\} \subseteq \left\{\varphi : \delta_{p'(\cdot)}(c\varphi) \le 1\right\},$$

which shows the left-hand side inequality in (7.47). We have $\|\varphi\|_{L_{p'(\cdot)}(\Omega)} \le 1$, therefore $\|c\varphi\|_{(p')} \le 1$ and we get $\left(\rho_{p'(\cdot)}(c\varphi)\right)^{1/(p')_+} \le \|c\varphi\|_{(p')}^{(p')_-^1/(p')_+}$ by (7.15). This entails that

$$\rho_{p'(\cdot)}(c\varphi) \le \|c\varphi\|_{(p')}^{(p')_-^1/(p')_+} + \|c\varphi\|_{L_\infty(\Omega_\infty(p'))}.$$

Since $A^\lambda + B \le 2^{1-\lambda}(A+B)^\lambda$, $0 \le \lambda \le 1$, $A \ge 0$, $0 \le B \le 1$, we get that $\delta_{p'(\cdot)}(c\varphi) \le 1$ and (7.48) is proved as the left-hand side inequality of (7.47).

To finish, if $m(\Omega_1(p)) = 0$ or $p(x) = const$ for $x \in \Omega\backslash(\Omega_\infty \cup \Omega_1)$, then we have $\|\varphi\|_{L_\infty(\Omega_\infty(p'))} = 0$ or $(p')_-^1/(p')_+ = 1$, respectively, and we obtain (7.48) with $c = 1$, which implies the coincidence of norms. □

The Luxemburg-Nakano norm is equivalent to the norm given in (7.43) in the following way.

Theorem 7.33. *Let $p_-^1 > 1$. The spaces $L_{p(\cdot)}(\Omega)$ and $\mathfrak{L}_{p(\cdot)}(\Omega)$ coincide modulo norm convergence:*

$$\frac{1}{3}\,\|f\|_{L_{p(\cdot)}(\Omega)} \leq \|f\|_p^* \leq \left(\frac{1}{p_-}+\frac{1}{(p')_-}\right)\|f\|_{L_{p(\cdot)}(\Omega)} \tag{7.49}$$

where $1/3$ can be replaced by 1 if $m(\Omega_1)=m(\Omega_\infty)=0$.

Proof. From the inclusion in (7.42) it suffices to show

$$\mathfrak{L}_{p(\cdot)}(\Omega) \subseteq L_{p(\cdot)}(\Omega). \tag{7.50}$$

Let $f\in\mathfrak{L}_{p(\cdot)}(\Omega)$ and let us take first the case $\|f\|_p^*\leq 1$. Take $\varphi_0(x)=|f(x)|^{p(x)-1}$ if $x\in\Omega\backslash(\Omega_1\cup\Omega_\infty)$ and $\varphi_0(x)=0$ otherwise. We now show that

$$\varphi_0\in L_{p'(\cdot)}(\Omega) \quad \text{and} \quad \rho_{p'(\cdot)}(\varphi_0)\leq 1. \tag{7.51}$$

Suppose that $\rho_{p'(\cdot)}(\varphi_0)>1$. Then

$$\rho_{p(\cdot)}(f) \geq \int\limits_{\Omega\backslash\Omega_\infty(p')} |\varphi_0(x)|^{p'(x)}\,\mathrm{d}x > 1. \tag{7.52}$$

Let

$$f_{N,k}(x)=\begin{cases} f(x), & \text{when } |f(x)|\leq N \text{ and } |x|\leq k;\\ 0, & \text{otherwise.}\end{cases}$$

Then $\varphi_{N,k}(x)=|f_{N,k}|^{p(x)-1}\in L_{p'(\cdot)}(\Omega)$. From (7.52) we derive the existence of an $N_0\to\infty$ and $k_0\to\infty$ such that

$$\int\limits_{\Omega\backslash\Omega_\infty(p)} |f_{N_0,k_0}|^{p(x)}\,\mathrm{d}x > 1. \tag{7.53}$$

In consequence, from (7.45) we obtain

$$1<\rho_{p(\cdot)}(f_{N_0,k_0})\leq \|f_{N_0,k_0}\|_p^*\,\|f_{N_0,k_0}^{p(\cdot)-1}\|_{L_{p'(\cdot)}(\Omega)}.$$

Henceforth, in virtue of (7.15)–(7.16)

$$1<\|f_{N_0,k_0}\|_p^*\max\left\{\left[\rho_{p(\cdot)}(f_{N_0,k_0})\right]^{\frac{1}{(p')_+}},\left[\rho_{p(\cdot)}(f_{N_0,k_0})\right]^{\frac{1}{(p')_-}}\right\}. \tag{7.54}$$

Then,

$$\min\left\{\left[\rho_{p(\cdot)}(f_{N_0,k_0})\right]^{1-\frac{1}{(p')_+}},\left[\rho_{p(\cdot)}(f_{N_0,k_0})\right]^{1-\frac{1}{(p')_-}}\right\}\leq \|f_{N_0,k_0}\|_p^*$$

which, from inequality (7.53) we conclude that $1<\|f_{N_0,k_0}\|_p^*$. This means that

$$\sup_{\rho_{p'(\cdot)}(\varphi)\leq 1}\left|\int_{\Omega} f(x)\varphi^{N,k}(x)\,\mathrm{d}x\right| > 1$$

where

$$\varphi^{N,k}(x) = \begin{cases} \varphi(x), & \text{when } |f(x)| \leq N \text{ and } |x| \leq k; \\ 0, & \text{otherwise.} \end{cases}$$

Nevertheless, since $\rho_{p'(\cdot)}(\varphi^{N,K}) \leq \rho_{p'(\cdot)}(\varphi)$, this contradicts the supposition that $\|f\|_p^* \leq 1$, from which we get (7.50).

As a result

$$\int_{\Omega\setminus(\Omega_1(p)\cup\Omega_\infty(p))} |f(x)|^{p(x)}\,\mathrm{d}x \leq 1$$

and to get the embedding (7.50) it is only necessary to show that $\int_{\Omega_1(p)} |f(x)|\,\mathrm{d}x < \infty$ and moreover that $\sup_{x\in\Omega_\infty(p)} |f(x)| < \infty$, which follows from the inequality

$$\int_{\Omega_i} |f(x)\varphi(x)|\,\mathrm{d}x \leq c\|\varphi\|_{L_{p'(\cdot)}(\Omega_i)}\ , \ i = 1,2,$$

(see (7.45)), where $\Omega_1 = \Omega_1(p), \Omega_2 = \Omega_\infty(p)$ and $f \in L_1,\ \varphi \in L_\infty\ (q = 1)$ in the first case and $f \in L_\infty, \varphi \in L_1\ (q = \infty)$ in the second one.

We now take $\|f\|_p^* > 1$. Then $f(x)/\|f\|_p^* \in L_{p(\cdot)}(\Omega)$ as was previously proved. Therefore, $f \in L_{p(\cdot)}(\Omega)$ by the linearity of the space $L_{p(\cdot)}(\Omega)$ under the condition $p_+ < \infty$. The embedding (7.50) is then proved.

It is only necessary to show the inequality(7.49) for the norms. The right-hand side inequality is a consequence of the Hölder inequality (7.29) and from the definition of the norm (7.44). To show the left-hand side of the inequality we write $f(x) = f_1(x) + f_2(x) + f_3(x)$ with $f_2(x) = f(x), x \in \Omega_1$ and $f_2(x) = 0, x \in \Omega\setminus\Omega_1$ and $f_3(x) = f(x), x \in \Omega_\infty$, and $f_3(x) = 0, x \in \Omega\setminus\Omega_\infty$. Let us show that

$$\|f_1\|_{L_{p(\cdot)}(\Omega)} \ \leq\ \|f\|^*_{p(\Omega\setminus(\Omega_1\cup\Omega_\infty))}\ . \tag{7.55}$$

We have that

$$\rho_{p(\cdot)}\left(\frac{f_1}{\lambda}\right) = \frac{1}{\lambda}\int_{\Omega\setminus\Omega_\infty} |f_1(x)|\varphi_\lambda(x)\,\mathrm{d}x\ ,\ \lambda > 0, \tag{7.56}$$

with $\varphi_\lambda(x) = \left|\frac{f_1(x)}{\lambda}\right|^{p(x)-1}$. Choosing $\lambda = \|f_1\|_{(p)}$, due to (7.45) and (7.55) we obtain

$$1 = \frac{1}{\|f_1\|_{(p)}}\int_{\Omega\setminus\Omega_\infty} |f_1(x)|\,\varphi_\lambda(x)\,\mathrm{d}x \leq \frac{\|f_1\|_p^*}{\|f_1\|_{(p)}}\|\varphi_\lambda\|_{L_{p'(\cdot)}(\Omega)}.$$

Since $\rho_{p'(\cdot)}(\varphi_\lambda) \leq \rho_{p(\cdot)}\left(\frac{f_1}{\lambda}\right) = 1$, we also conclude that $\|\varphi_\lambda\|_{L_{p'(\cdot)}(\Omega)} \leq 1$ due to (7.15)–(7.16) and we obtain the coincidence $\|\varphi_\lambda\|_{L_{p'(\cdot)}(\Omega)} = \|\varphi_\lambda\|_{(p')}$. Therefore (7.56) implies (7.54). Since $\|f_2\|_{L_{p(\cdot)}(\Omega)} = \|f\|^*_{L_1(\Omega_1)}$ and $\|f_3\|_{L_{p(\cdot)}(\Omega)} = \|f\|^*_{L_\infty(\Omega_\infty)}$, we obtain the left-hand side inequality. □

Corollary 7.34. *Let $f \in L_{p(\cdot)}(\Omega)$, $\varphi \in L_{p'(\cdot)}(\Omega)$, $1 \leq p(x) \leq \infty$. Regarding the norms (7.43)–(7.44) the Hölder inequality is valid with constant 1:*

$$\int_\Omega |f(x)\varphi(x)|\,\mathrm{d}x \leq \|f\|^*_p \|\varphi\|_{L_{p'(\cdot)}(\Omega)}\,,\ p^1_- > 1, \tag{7.57}$$

and

$$\int_\Omega |f(x)\varphi(x)|\,\mathrm{d}x \leq \|f\|^{**}_p \|\varphi\|_{L_{p'(\cdot)}(\Omega)}. \tag{7.58}$$

The inequality

$$\int_\Omega |f(x)\varphi(x)|\,\mathrm{d}x \leq \|f\|^*_p \|\varphi\|^*_{p'} \tag{7.59}$$

is valid in the case

$$p^1_- > 1,\ p_+ < \infty,\ m(\Omega_\infty(p)) = m(\Omega_1(p)) = 0. \tag{7.60}$$

In reality, the inequality (7.57) was already given in (7.45); meanwhile the inequality (7.58) follows directly from the definition (7.44). The inequality (7.59) is a consequence of (7.57) since $\|\varphi\|_{L_{p'(\cdot)}(\Omega)} \leq \|\varphi\|^*_{p'}$ under the condition (7.60) by Theorem 7.33.

7.1.8 More on the Space $L_{p(\cdot)}(\Omega)$ in the Case $p_+ = \infty$

The definition given in (7.41) is one of the possible ways to define the space $L_{p(\cdot)}(\Omega)$ in order to be linear in the case $p_+ = \infty$. It is also possible to define the spaces from the beginning as the convex hull of the space $L_{p(\cdot)}(\Omega)$ or as

$$\mathbf{L}_{p(\cdot)}(\Omega) := \left\{ f \in \mathfrak{F}(\Omega, \mathscr{L}) : \exists \lambda > 0 \text{ such that} \right.$$
$$\left. \int_{\Omega\setminus\Omega_\infty} \left|\frac{f(x)}{\lambda}\right|^{p(x)} \mathrm{d}x + \|f\|_{L_\infty(\Omega_\infty)} < \infty \right\}. \tag{7.61}$$

This space is always linear for $0 \leq p(x) \leq \infty$. The homogeneity is obvious, meanwhile the additivity is evident in the set $\{x \in \Omega : p(x) \leq 1\}$ due to the inequality

$(a+b)^p \le a^p + b^p, p \le 1$, meanwhile in the set $\{x \in \Omega : p(x) > 1\}$ it is verified by the convexity (7.10) of the function $t \mapsto t^p$, $p > 1$.

Therefore, in the case $p_+ = \infty$ we can use the three different versions of the definition, i.e., $\mathrm{span}(L_{p(\cdot)}), \mathbf{L}_{p(\cdot)}$, or $\mathfrak{L}_{p(\cdot)}$. We can see that

$$\mathrm{span}\left(L_{p(\cdot)}\right) = \mathbf{L}_{p(\cdot)} \subseteq \mathfrak{L}_{p(\cdot)}. \tag{7.62}$$

The norm in the space $\mathfrak{L}_{p(\cdot)}$ is given by (7.44) whereas the norm is given by (7.6) in the spaces $\mathrm{span}(L_{p(\cdot)}) = \mathbf{L}_{p(\cdot)}$.

7.1.9 Minkowski Integral Inequality

We now extend the Minkowski integral inequality, given in Theorem 3.25 for the classical Lebesgue spaces, into the variable framework.

Theorem 7.35. *Let $1 \le p(x) \le p_+ < \infty$ and $p^1_- > 1$. Then we have the Minkowski integral inequality in the variable exponent Lebesgue space*

$$\left\| \int_\Omega f(\cdot,y)\,\mathrm{d}y \right\|_p^{**} \le \int_\Omega \|f(\cdot,y)\|_p^{**}\,\mathrm{d}y. \tag{7.63}$$

Proof. Let J be the expression in the left-hand side. We get

$$J \le \sup_{\|\varphi\|_{L_{p'(\cdot)}(\Omega)} \le 1} \int_\Omega \left(\int_\Omega |\varphi(x) f(x,y)|\,\mathrm{d}x \right) \mathrm{d}y.$$

Using the definition of norm given in (7.44), we obtain the desired inequality. □

Corollary 7.36. *Let $1 \le p(x) \le p_+ < \infty$ and $p^1_- > 1$. Then*

$$\left\| \int_\Omega f(\cdot,y)\,\mathrm{d}y \right\|_p^{*} \le c_1 \int_\Omega \|f(\cdot,y)\|_p^{*}\,\mathrm{d}y, \tag{7.64}$$

and

$$\left\| \int_\Omega f(\cdot,y)\,\mathrm{d}y \right\|_{L_{p(\cdot)}(\Omega)} \le c_2 \int_\Omega \|f(\cdot,y)\|_{L_{p(\cdot)}(\Omega)}\,\mathrm{d}y \tag{7.65}$$

where $c_1 = 1$ if $m(\Omega_1) = 0$ and $c_1 = 2^{-1+(p')_+/(p')_-^1}$ in the other case. The constant $c_2 = kc_1$ if $m(\Omega_\infty) = m(\Omega_1) = 0$ and $c_2 = 3kc_1$ in the other case, where $k = \frac{1}{p_-} + \frac{1}{(p')_-}$.

Proof. The inequality (7.64) with the constant $c_1 = 2^{-1+(p')_+/(p')_-^1}$ is a consequence of (7.63) due to (7.47). In the same way (7.65) follows from (7.63) by virtue of (7.49) and (7.47). To prove that $c_1 = 0$ in (7.64) in the case $m(\Omega_1) = 0$, note that

$$\left\| \int_\Omega f(\cdot,y)\,\mathrm{d}y \right\|_p^* \leq \sup_{\delta_{p'(\cdot)}(\varphi)\leq 1} \int_\Omega \|\varphi\|_{L_{p'(\cdot)}(\Omega)} \|f(\cdot,y)\|_p^*\,\mathrm{d}y.$$

To finish the proof it is only necessary to see that the conditions $\delta_{p'(\cdot)}(\varphi) \leq 1$ and $\|\varphi\|_{L_{p'(\cdot)}(\Omega)} \leq 1$ are equivalent in the case $m(\Omega_1) = 0$, as a result of (7.15)–(7.16). □

7.1.10 *Some Differences Between Spaces with Variable Exponent and Constant Exponent*

Let us start with a property of the variable exponent Lebesgue spaces which is contrary to our intuition from the classical Lebesgue spaces. In this case, let us take the space $\mathbf{L}_{p(\cdot)}(\Omega)$ given in (7.61). Let $\Omega = [1,\infty)$, $p(x) = x$ and $f(x) \equiv a$ where $a > 0$. We have that $f \in \mathbf{L}_x(\Omega)$ since taking some $\lambda > a$ the integral $\int_1^\infty |f(x)/\lambda|^x\,\mathrm{d}x$ is finite but $f \notin L_p(\Omega)$ for any constant p.

We now show two more differences between the constant and the variable framework, namely in regards to the invariance under translation and the Young convolution.

7.1.10.1 Invariance Under Translations

An important result in the classical theory of Lebesgue spaces has to do with the boundedness of the translation operator , i.e., if $f \in L_p(\mathbb{R}^n)$ then we have that $\tau_h f \in L_p(\mathbb{R}^n)$, where $\tau_h f(x) := f(x-h)$. This result stems from the fact that the classical Lebesgue space is isotropic with respect to the exponent, since the power p is the same in any direction. On the other hand, the variable exponent Lebesgue space is, in general, anisotropic regarding the exponent. This anisotropy of the space generates problems for the translation operator. Let us give a simple example, taking $f(x) = |x|^{-\frac{1}{3}}$. This function $f \in L_{p(\cdot)}((-1,1))$ taking the following exponent

$$p(x) = \begin{cases} 2, \ x \in |x| < \varepsilon \\ 5, \ x \in |x| \geq \varepsilon \end{cases} \tag{7.66}$$

but $\tau_\delta f \notin L_{p(\cdot)}((-1,1))$, when $\delta > \varepsilon$, since we translated the singularity from 0 to δ but the exponent was not shifted. ($\tau_\delta f$ is understood as the zero extension whenever necessary). One could argue that the problem in this example is the non-smoothness nature of the exponent. From (7.66) we can construct a smooth function (for example, via Urysohn construction) and we will end up with the same problem. Our example is not an isolated incident, since Diening proved that this phenomenon is persistent, i.e., if $p_+ > p_-$, then there exists a $h \in \mathbb{R}\backslash\{0\}$ such that the translation operator τ_h is not continuous, cf. Diening, Harjulehto, Hästö, and Růžička [17].

7.1.10.2 Young Convolution Inequality in Variable Exponent Lebesgue Spaces

Let

$$Kf(x) = (k * f)(x) = \int_{\mathbb{R}^n} k(x-y)f(y)\,\mathrm{d}y = \int_{\mathbb{R}^n} k(y)f(x-y)\,\mathrm{d}y \tag{7.67}$$

where $*$ is called *convolution*, cf. § 11.1 for more details. The *Young's inequality for convolutions* states that

$$\|k * f\|_{L_r(\mathbb{R}^n)} \leq \|k\|_{L_q(\mathbb{R}^n)} \|f\|_{L_p(\mathbb{R}^n)}, \quad \frac{1}{p} + \frac{1}{q} - 1 = \frac{1}{r},$$

which can be proved, among other means, using the following decomposition $|f(x-y)k(y)| = |f(x-y)|^{1-s}|k(y)||f(x-y)|^s$, for $s = 1 - p/r$, the Hölder inequality and the integral Minkowski inequality. Since the convolution depends on the translation operator, which is not continuous, the natural question is: does the Young inequality for convolutions holds in general in the case of variable Lebesgue spaces? The answer is no, in general, although there are some particular cases where it is possible to have some version of the inequality. Let us start with a counter-example.

Example 7.37. The Young inequality in the form

$$\|k * f\|_{L_p(\mathbb{R}^n)} \leq \|k\|_{L_1(\mathbb{R}^n)} \|f\|_{L_p(\mathbb{R}^n)},$$

is not valid for an arbitrary kernel $k \in L_1(\mathbb{R}^n)$ and an arbitrary variable exponent p. For simplicity let us consider $n = 1$. Let

$$p(x) = \begin{cases} p_1, & \text{if } x < 0 \\ p_2, & \text{if } x > 0 \end{cases},$$

where $1 \leq p_1 < p_2 < \infty$. Let us define the kernel k in the following way

$$k(x) = \begin{cases} |x-2|^{\alpha-1}, & \text{if } |x| \leq 3 \\ 0, & \text{if } |x| > 3 \end{cases},$$

where $0 < \alpha < 1/p_1 - 1/p_2$, therefore $k \in L_1(\mathbb{R})$. Let us take f as

$$f(x) = \begin{cases} |x+1|^{-\nu}, & \text{if } x \in (-2,0) \\ 0, & \text{if } x \notin (-2,0) \end{cases},$$

which implies that $f \in L_{p(\cdot)}(\mathbb{R})$ if $0 < \nu < 1/p_1$. But the function $k * f \notin L_{p(\cdot)}(\mathbb{R})$ if $\nu > \alpha + 1/p_2$. This is a consequence of

$$(k * f)(x) = \int_{\max\{x-3,-2\}}^{\min\{x+3,0\}} |x-y-2|^{\alpha-1} |y+1|^{-\nu} \, dy,$$

and taking $1 < x < 3/2$ we have

$$\begin{aligned} (k * f)(x) &\leq \int_{x-3}^{-1} |x-y-2|^{\alpha-1} |y+1|^{-\nu} \, dy \\ &= \int_0^{2-x} s^{-\nu}(x-1+s)^{\alpha-1} \, ds \\ &= (x-1)^{\alpha-\nu} \int_0^{\frac{2-x}{x-1}} \xi^{-\nu} (1+\xi)^{\alpha-1} \, d\xi \\ &\geq \frac{c}{(x-1)^{\nu-\alpha}} \end{aligned}$$

where $c = \int_0^1 \xi^{-\nu}(1+\xi)^{\alpha-1} \, d\xi$. Therefore $k * f$ cannot be p_2-integrable in $[1, 3/2]$, since $(\nu - \alpha)p_2 > 1$.

⊘

Let us now show a very particular version of Young's inequality for convolutions.

Theorem 7.38. *Let p and q be variable exponents such that $\frac{1}{p(x)} + \frac{1}{q(x)} \equiv 1 + \frac{1}{r}$ where $r = \text{const} \geq 1$. If $k \in L_{q_-}(\mathbb{R}^n) \cap L_{(p')_+}(\mathbb{R}^n)$ then the convolution operator* (7.67)

$$k * \cdot : L_{p(\cdot)}(\mathbb{R}^n) \longrightarrow L_r(\mathbb{R}^n)$$

is bounded.

Proof. Let us take f such that $\|f\|_{L_{p(\cdot)}(\Omega)} \leq 1$. Then

$$\left|(k*f)(x)\right| \leq \int_{\mathbb{R}^n} A^{1-\mu(y)}|f(y)|^{\frac{p(y)}{r}} \left|k(x-y)\right|^{\mu(y)} |f(y)|^{1-\frac{p(y)}{r}} \left|\frac{k(x-y)}{A}\right|^{1-\mu(y)} \mathrm{d}y$$

where the constant $A > 0$ and the function $\mu(y)$, $0 < \mu(y) < 1$, will be chosen later. Using the generalized Hölder inequality (7.80) with the exponents

$$p_1(y) = r, \;\; p_2(y) = \frac{rp(y)}{r - p(y)}, \;\; p_3(y) = p'(y) = \frac{p(y)}{p(y)-1}$$

we obtain

$$\left|(k*f)(x)\right| \leq c \left\{ \int_{\mathbb{R}^n} A^{r-r\mu(y)}|f(y)|^{p(y)} \left|k(x-y)\right|^{r\mu(y)} \mathrm{d}y \right\}^{\frac{1}{r}}$$
$$\times \left\| |f(y)|^{1-\frac{p(y)}{r}} \right\|_{p_2(y)} \left\| \left|\frac{k(x-y)}{A}\right|^{1-\mu(y)} \right\|_{p'(y)}. \tag{7.68}$$

By the estimate (7.20) we get

$$\left\| |f(y)|^{1-\frac{p(y)}{r}} \right\|_{p_2(y)} \leq \|f\|_{p(y)}^{\operatorname{ess\,inf}\left[1-\frac{p(y)}{r}\right]} \leq 1 \tag{7.69}$$

since $\|f\|_{L_{p(\cdot)}(\Omega)} \leq 1$ and the fact that $p < r$.

To estimate the third factor in (7.68) it is natural to choose $\mu(y)$ in such a way that $\left[1-\mu(y)\right] p'(y) = q(y)$, i.e.

$$\mu(y) = \frac{q(y)}{r}.$$

We now want to use the inequality (7.20) in the third factor. We are now interested in

$$\left\| \frac{k(x-y)}{A} \right\|_{q(y)} = \frac{1}{A} \left\|k(x-y)\right\|_{q(y)} \leq 1. \tag{7.70}$$

To get (7.70) we choose

$$A = \|k\|_{q_-} + \|k\|_{(p')_+}.$$

In this way (7.70) is valid by Lemma 7.15. We can now apply (7.20) and obtain

$$\left\| \left|\frac{k(x-y)}{A}\right|^{1-\mu(y)} \right\|_{p'(y)} \leq 1. \tag{7.71}$$

From the inequalities (7.69) and (7.71) we get, via (7.68), the estimate

$$\|k*f\|_r \le cA^\nu \left(\int_{\mathbb{R}^n} dx \int_{\mathbb{R}^n} |f(y)|^{p(y)} |k(x-y)|^{q(y)} dy \right)^{\frac{1}{r}}$$

$$= cA^\nu \left(\int_{\mathbb{R}^n} |f(y)|^{p(y)} dy \int_{\mathbb{R}^n} |k(x)|^{q(x+y)} dx \right)^{\frac{1}{r}}$$

where $\nu = 1 - q_+/r$ if $A \le 1$ and $\nu = 1 - q_-/r$ if $A \ge 1$. Therefore

$$\|k*f\|_r \le cA^\nu \left(\|k\|_{q_-}^{\frac{q_-}{r}} + \|k\|_{(p')_+}^{\frac{(p')_+}{r}} \right) \int_{\mathbb{R}^n} |f(y)|^{p(y)} dy.$$

To finish the proof, we only need to take into account that the integral is bounded by 1 due to (7.15). □

7.2 Grand Lebesgue Spaces

In this section we will introduce the so-called grand Lebesgue spaces, a function space that was introduced in the 1990s to deal with the problem of the integrability of the Jacobian under minimal hypothesis. The best way to study this space is in the framework of *Banach Function Spaces*, since this gives clearer proofs and follows the historical development of the theory. As an additional benefit, it will be clear that many function spaces fall under the umbrella of Banach function spaces.

7.2.1 Banach Function Spaces

In the following, we give the definitions and list some results regarding Banach Function Spaces, see Bennett and Sharpley [1] and Pick, Kufner, John, and Fučík [56] for the proofs.

In the sequel, Ω denotes an open subset Ω in $\mathbb{R}^n$. Let M_0 be the set of all measurable functions whose values lie in $[-\infty, \infty]$ and are finite a.e. in Ω. Also, let M_0^+ be the class of functions in M_0 whose values lie in $(0, \infty)$.

Definition 7.39. A mapping $\rho : M_0^+ \longrightarrow [0, \infty]$ is called a *Banach function norm* if for all f, g, f_n in M_0^+, $n \in \mathbb{N}$, for all constants $a \ge 0$ and all measurable subsets $E \subset \Omega$, the following properties hold:

(P7) $\rho(f) = 0$ if and only if $f = 0$ a.e. in Ω;
(P7) $\rho(af) = a\rho(f)$;
(P7) $\rho(f+g) \le \rho(f) + \rho(g)$;

(P7) $0 \le g \le f$ a.e. in Ω implies that $\rho(g) \le \rho(f)$ (lattice property);
(P7) $0 \le f_n \uparrow f$ a.e. in Ω implies that $\rho(f_n) \uparrow \rho(f)$ (Fatou's property);
(P7) $m(E) < +\infty$ implies that $\rho(\chi_E) < +\infty$;
(P7) $m(E) < +\infty$ implies that $\int_E f\,\mathrm{d}x \le C_E \rho(f)$ (for some constant C_E, $0 < C_E < \infty$, depending on E and ρ but independent of f). ⊘

It is noteworthy to mention that the lattice property is a consequence of the Fatou property, see Problem 7.73.

Based upon the notion of Banach function norm, we introduce the Banach function space X_ρ.

Definition 7.40. If ρ is a Banach function norm, the Banach space

$$X(\rho) = X_\rho = X = \{f \in M_0 : \rho(|f|) < +\infty\} \tag{7.72}$$

is called a *Banach Function Space*. For each $f \in X$ define

$$\|f\|_X = \rho(|f|). \tag{7.73}$$

⊘

There is also a notion of rearrangement invariant Banach function space, namely:

Definition 7.41. Let ρ be a Banach function norm. We say that the norm is *rearrangement invariant* if

$$\rho(f) = \rho(g)$$

for all equimeasurable functions f and g. In this case the Banach function space $X(\rho)$ is said to be a *rearrangement invariant Banach function space*. ⊘

A very important property of the Lebesgue space is its dual characterization, for example, in $L_p[(0,1)]$ we have

$$\|f\|_{L_p([0,1])} = \sup_{\|g\|_{L_{p'}([0,1])}} \int_0^1 f(x)g(x)\,\mathrm{d}x$$

where p and p' are conjugate exponents. This characterization gives us immediately one of the key inequalities in the theory of Lebesgue spaces, namely the Hölder inequality which gives an upper bound for the integral of the product of two functions based upon their norms. The following notion is introduced to capture this "duality" in the framework of Banach function spaces.

Definition 7.42. If ρ is a Banach function norm, its *associative function norm* ρ' defined on M_0^+ is given by

$$\rho'(g) = \sup\left\{ \int_\Omega fg\,\mathrm{d}x : f \in M_0^+,\ \ \rho(f) \le 1 \right\}. \tag{7.74}$$

$\oslash$

As in the case of Banach function space, we can introduce the associate Banach function space based upon the concept of Banach associative function norm.

Definition 7.43. Let ρ be a function norm and let $X = X(\rho)$ be the Banach function space determined by ρ. Let ρ' be the Banach associate function norm of ρ. The Banach function space $X' = X'(\rho')$ determined by ρ' is called the *associate space of* X.

$\oslash$

In particular from the definition of $\|f\|_X$ it follows that the norm of a function g in the associate space X' is given by

$$\|g\|_{X'} = \sup\left\{ \int_{\Omega} fg\,\mathrm{d}x : f \in M^+,\ \|f\|_X \leq 1 \right\}.$$

We now give some results without proof, see Bennett and Sharpley [1] and Pick, Kufner, John, and Fučík [56] for the proofs.

Theorem 7.44. *Every Banach function space X coincides with its second associate space X''.*

This proposition tells us, in particular, that the notion of associate space is different from the notion of dual space, but under certain conditions both notions coincide, cf. Proposition 7.51.

Theorem 7.45. *If X and Y are Banach function spaces and $X \hookrightarrow Y$, then $Y' \hookrightarrow X'$.*

Definition 7.46. A function f in a Banach function space X is said to have *absolutely continuous norm* on X if

$$\lim_{n\to\infty} \|f\chi_{E_n}\|_X = 0$$

for every sequence $\{E_n\}_{n=1}^{\infty}$ satisfying $E_n \downarrow \emptyset$.

$\oslash$

Definition 7.47. The subspace of functions in X with absolutely continuous norm is denoted by X_a. If $X = X_a$, then the space X itself is said to have absolutely continuous norm.

$\oslash$

Definition 7.48. Let X be a Banach function space. The closure in X of the set of bounded functions is denoted by X_b.

$\oslash$

Theorem 7.49. *Let X be a Banach function space. Then $X_a \subseteq X_b \subseteq X$.*

Corollary 7.50. *If $X_a = X$, then $X_b = X$.*

Theorem 7.51. *The dual space X^* of a Banach function space X is canonically isometric to the associate space X' if and only if X has absolutely continuous norm.*

Theorem 7.52. *A Banach function space is reflexive if and only if both X and its associate space X′ have absolutely continuous norm.*

These last two theorems are very important, since they give necessary and sufficient condition to check when the associate space and the dual space are equal. Characterizing the dual space can be quite difficult, whereas the notion of associate space is more manageable in some sense.

7.2.2 Grand Lebesgue Spaces

In this section we give a brisk introduction to the so-called grand Lebesgue spaces, also known as Iwaniec-Sbordonne spaces. For simplicity we work only in the Euclidean space.

Let Ω be a bounded set on $\mathbb{R}^n$. Let M_0 be the set of all measurable functions whose value lies in $[-\infty,\infty]$ and are finite a.e. in Ω.

Definition 7.53. The *grand Lebesgue space* $L_{p)}(\Omega)$ is defined as the set of measurable functions on Ω for which

$$\|f\|_{p)} = \sup_{0<\varepsilon<p-1} \left(\frac{\varepsilon}{m(\Omega)} \int_{\Omega} |f|^{p-\varepsilon}\, \mathrm{d}x \right)^{\frac{1}{p-\varepsilon}},$$

is finite, i.e.

$$L_{p)}(\Omega) = \left\{ f \in \mathfrak{F}(\Omega,\mathscr{L}) : \|f\|_{p)} < \infty \right\},$$

where $1 < p < +\infty$. We stress that $m(\Omega) < \infty$. ⊘

The following theorem justifies the nomenclature of *grand* Lebesgue space.

Theorem 7.54. *For $p > 1$, we have*

$$L_p(\Omega) \subsetneq L_{p)}(\Omega).$$

Proof. Let us take $t = \frac{p}{p-\varepsilon}$ and $s = \frac{p}{\varepsilon}$ for $0 < \varepsilon < p-1$. Direct calculations show that

$$\frac{1}{t} + \frac{1}{s} = \left(1 - \frac{\varepsilon}{p}\right) + \frac{\varepsilon}{p} = 1,$$

or, in other words, that t and s are conjugate exponents. Taking $f \in L_p(\Omega)$ and by Hölder's inequality, we have

$$\left(\frac{\varepsilon}{m(\Omega)} \int_{\Omega} |f|^{p-\varepsilon}\, \mathrm{d}x \right)^{\frac{1}{p-\varepsilon}} \leq \left(\frac{\varepsilon}{m(\Omega)} \right)^{\frac{1}{p-\varepsilon}} \left(\int_{\Omega} |f|^{p}\, \mathrm{d}x \right)^{\frac{p-\varepsilon}{p}\cdot\frac{1}{p-\varepsilon}} \cdot \left(m(\Omega)^{\varepsilon/p} \right)^{\frac{1}{p-\varepsilon}}$$

$$= \left(\frac{\varepsilon}{m(\Omega)}\right)^{\frac{1}{p-\varepsilon}} \left(\int_{\Omega} |f|^p \, dx\right)^{\frac{1}{p}} \cdot m(\Omega)^{\frac{\varepsilon}{p(p-\varepsilon)}}$$

$$= (\varepsilon)^{\frac{1}{p-\varepsilon}} (m(\Omega))^{-\frac{1}{p-\varepsilon}} (m(\Omega))^{\frac{\varepsilon}{p(p-\varepsilon)}} \left(\int_{\Omega} |f|^p \, dx\right)^{\frac{1}{p}}$$

$$= (\varepsilon)^{\frac{1}{p-\varepsilon}} \left(\frac{1}{m(\Omega)} \int_{\Omega} |f|^p \, dx\right)^{\frac{1}{p}}$$

$$\leq p^{\frac{1}{p-\varepsilon}} \left(\frac{1}{m(\Omega)} \int_{\Omega} |f|^p \, dx\right)^{\frac{1}{p}},$$

which is finite, hence $f \in L_{p)}(\Omega)$, since

$$\sup_{0<\varepsilon<p-1} \left(\frac{\varepsilon}{m(\Omega)} \int_{\Omega} |f|^{p-\varepsilon} \, dx\right)^{\frac{1}{p-\varepsilon}} < \infty.$$

We now want to show that the inclusion is strict. We give an example with a particular Ω which can be used to construct further examples. Let $\Omega = (0,1)$ and $f(x) = x^{-1/p}$ for $p > 1$. Now, let us show that $f \in L_{p)}(0,1) \setminus L_p(0,1)$. Indeed,

$$\int_0^1 |f(x)|^p \, dx = \lim_{\varepsilon \to 0} \int_{\varepsilon}^1 \frac{dx}{x} = \lim_{\varepsilon \to 0} \log x \,\Big|_{\varepsilon}^1,$$

consequently $f \notin L_p(0,1)$. On the other hand

$$\varepsilon^{\frac{1}{p-\varepsilon}} \left(\int_0^1 x^{-\frac{1}{p}(p-\varepsilon)} \, dx\right)^{\frac{1}{p-\varepsilon}} = \varepsilon^{\frac{1}{p-\varepsilon}} \left(\int_0^1 x^{-1+\frac{\varepsilon}{p}} \, dx\right)^{\frac{1}{p-\varepsilon}}$$

$$= \varepsilon^{\frac{1}{p-\varepsilon}} \left(\frac{p}{\varepsilon} x^{\frac{\varepsilon}{p}} \Big|_0^1\right)^{\frac{1}{p-\varepsilon}}$$

$$= \varepsilon^{\frac{1}{p-\varepsilon}} \left(\frac{p}{\varepsilon}\right)^{\frac{1}{p-\varepsilon}}$$

$$= p^{\frac{1}{p-\varepsilon}}$$

$$< p.$$

From this we have

$$\sup_{0<\varepsilon<p-1} \left(\varepsilon \int_0^1 x^{-\frac{1}{p}(p-\varepsilon)}\, dx \right)^{\frac{1}{p-\varepsilon}} \leq p < \infty.$$

Hence $f \in L_{p)}(0,1) \setminus L_p(0,1)$. □

One of the most important property of the grand Lebesgue spaces is the so-called *nesting property*, namely for $p > 1$ and $0 < \varepsilon < p-1$ we have (see Problem 7.72) that

$$L_p(\Omega) \subsetneq L_{p)}(\Omega) \subsetneq L_{p-\varepsilon}(\Omega). \tag{7.75}$$

The nesting property (7.75) is one of the reasons for the usefulness of grand Lebesgue spaces, since it permits to enlarge the L_p scale of function with the property $L_\infty(\Omega) \subsetneq L_p(\Omega) \subsetneq L_q(\Omega) \subsetneq L_1(\Omega)$ for $1 < q < p < \infty$ whenever Ω has finite measure.

We know that weak Lebesgue spaces contain Lebesgue spaces, therefore from the nesting property it is natural to ask what is the relation between weak Lebesgue spaces and grand Lebesgue spaces.

Theorem 7.55. *Let $1 < p < \infty$. We have the inclusion*

$$L_{(p,\infty)}(\Omega) \subset L_{p)}(\Omega).$$

Proof. Let $f \in L_{(p,\infty)}$, then

$$\begin{aligned} \frac{\varepsilon}{m(\Omega)} \int_\Omega |f|^{p-\varepsilon}\, dx &= \frac{\varepsilon(p-\varepsilon)}{m(\Omega)} \int_0^\infty \lambda^{p-\varepsilon-1} D_f(\lambda)\, d\lambda \qquad (7.76) \\ &= \frac{\varepsilon(p-\varepsilon)}{m(\Omega)} \left[\int_0^a \lambda^{p-\varepsilon-1} D_f(\lambda)\, d\lambda + \int_a^\infty \lambda^{p-\varepsilon-1} D_f(\lambda)\, d\lambda \right]. \end{aligned}$$

We have that $\lambda^p D_f(\lambda) \leq \|f\|^p_{L_{(p,\infty)}}$, then $D_f(\lambda) \leq \lambda^{-p} \|f\|^p_{L_{(p,\infty)}}$, therefore from (7.76) we get

$$\begin{aligned} \frac{\varepsilon}{m(\Omega)} \int_\Omega |f|^{p-\varepsilon}\, dx &\leq \frac{\varepsilon(p-\varepsilon)}{m(\Omega)} \left[\frac{m(\Omega) a^{p-\varepsilon}}{p-\varepsilon} + \frac{a^{-\varepsilon}}{\varepsilon} \|f\|^p_{L_{(p,\infty)}} \right] \qquad (7.77) \\ &= \varepsilon a^{p-\varepsilon} + \frac{a^{-\varepsilon}}{\varepsilon} \frac{\varepsilon(p-\varepsilon)}{m(\Omega)} \|f\|^p_{L_{(p,\infty)}}. \end{aligned}$$

Let $a = \|f\|_{L_{(p,\infty)}}$, replacing a in (7.77) we have

$$\frac{\varepsilon}{m(\Omega)} \int_{\Omega} |f|^{p-\varepsilon} \, \mathrm{d}x \leq \varepsilon \|f\|^{p-\varepsilon}_{L_{(p,\infty)}} + \frac{p-\varepsilon}{m(\Omega)} \|f\|^{p-\varepsilon}_{L_{(p,\infty)}}$$

$$= \left(\varepsilon + \frac{p-\varepsilon}{m(\Omega)} \right) \|f\|^{p-\varepsilon}_{L_{(p,\infty)}}$$

and thus

$$\sup_{0<\varepsilon<p-1} \left(\frac{\varepsilon}{m(\Omega)} \int_{\Omega} |f|^{p-\varepsilon} \, \mathrm{d}x \right)^{\frac{1}{p-\varepsilon}} \leq C \|f\|_{L_{(p,\infty)}}$$

where $C = \sup_{0<\varepsilon<p-1} \left(\varepsilon + \frac{p-\varepsilon}{m(\Omega)} \right)^{\frac{1}{p-\varepsilon}}$, hence $L_{(p,\infty)} \subset L_{p)}$. □

We now show that the grand Lebesgue space is a Banach space under natural restrictions.

Theorem 7.56. *Let $1 < p < \infty$. The grand Lebesgue space $L_{p)}(\Omega)$ is a Banach space.*

Proof. Let $\{f_n\}_{n\in\mathbb{N}}$ be a Cauchy sequence in $L_{p)}(\Omega)$, i.e.

$$\lim_{\substack{m\to\infty \\ n\to\infty}} \sup_{0<\varepsilon<p-1} \left(\frac{\varepsilon}{m(\Omega)} \int_{\Omega} |f_m - f_n|^{p-\varepsilon} \, \mathrm{d}x \right)^{\frac{1}{p-\varepsilon}} = 0.$$

Hence for an arbitrary $\eta > 0$ there exists $n_0 \in \mathbb{N}$ such that

$$\left(\frac{\varepsilon}{m(\Omega)} \int_{\Omega} |f_m - f_n|^{p-\varepsilon} \, \mathrm{d}x \right)^{\frac{1}{p-\varepsilon}} < \frac{\eta}{3}$$

for an arbitrary ε, $0 < \varepsilon < p-1$, when $m > n_0$, $n > n_0$. Consequently $\{f_n\}_{n\in\mathbb{N}}$ is a Cauchy sequence in $L_{p-\varepsilon}(\Omega)$ for an arbitrary ε, $0 < \varepsilon < p-1$, and let f be its limit in $L_{p-\varepsilon}(\Omega)$.

Let $n > n_0$. According to the definition of the supremum there exists an ε_0 (depending generally speaking on n), $0 < \varepsilon_0(n) < p-1$, such that

$$\|f - f_n\|_{p)} = \sup_{0<\varepsilon<p-1} \left(\frac{\varepsilon}{|\Omega|} \int_{\Omega} |f - f_n|^{p-\varepsilon} \, \mathrm{d}x \right)^{\frac{1}{p-\varepsilon}}$$

$$\leq \left(\frac{\varepsilon_0(n)}{m(\Omega)} \int_{\Omega} |f - f_n|^{p-\varepsilon_0(n)} \, \mathrm{d}x \right)^{\frac{1}{p-\varepsilon_0(n)}} + \frac{\eta}{3}$$

Furthermore, there exists $n_1 \in \mathbb{N}$ such that $m > n_1$

$$\left(\frac{\varepsilon_0(n)}{m(\Omega)}\int_\Omega |f_m - f_n|^{p-\varepsilon_0(n)}\,\mathrm{d}x\right)^{\frac{1}{p-\varepsilon_0(n)}} < \frac{\eta}{3},$$

therefore

$$\begin{aligned}\|f - f_n\|_{p)} &\leq \left(\frac{\varepsilon_0(n)}{m(\Omega)}\int_\Omega |f_m - f_n|^{p-\varepsilon_0(n)}\,\mathrm{d}x\right)^{\frac{1}{p-\varepsilon_0(n)}} \\ &+ \left(\frac{\varepsilon_0(n)}{m(\Omega)}\int_\Omega |f_m - f|^{p-\varepsilon_0(n)}\,\mathrm{d}x\right)^{\frac{1}{p-\varepsilon_0(n)}} + \frac{\eta}{3} \\ &< \frac{\eta}{3} + \frac{\eta}{3} + \frac{\eta}{3} = \eta\end{aligned}$$

whenever $n > n_1$ and $m > n_1$. □

One of the drawbacks of grand Lebesgue spaces is the fact that the set of C_0^∞ functions is not a dense set. Fortunately we have a characterization of the closure of C_0^∞ functions in the grand Lebesgue norm given in a somewhat manageable way.

Theorem 7.57 *The set $C_0^\infty(\Omega)$ is not dense in $L_{p)}(\Omega)$. Its closure $\overline{C_0^\infty}\big|_{L_{p)}(\Omega)}$ consists of functions $f \in L_{p)}(\Omega)$ such that*

$$\lim_{\varepsilon\to 0} \varepsilon \int_\Omega |f|^{p-\varepsilon}\,\mathrm{d}x = 0. \tag{7.78}$$

Proof. Let $f \in \overline{C_0^\infty}\big|_{L_{p)}(\Omega)}$, then there is a sequence of functions $f_n \in C_0^\infty$ such that

$$\|f - f_n\|_{p)} \to 0$$

as $n \to \infty$.

Let us take $\delta > 0$. Choose n_0 such that

$$\left\|f - f_{n_0}\right\|_{p)} < \frac{\delta}{2} \quad \text{and } f_{n_0} \in C_0^\infty.$$

Now observe that for f_{n_0}, by Hölder's inequality, we have

$$\left(\frac{\varepsilon}{m(\Omega)}\int_\Omega |f_{n_0}|^{p-\varepsilon}\,\mathrm{d}x\right)^{\frac{1}{p-\varepsilon}} \leq \varepsilon^{\frac{1}{p-\varepsilon}}\left(\frac{1}{m(\Omega)}\int_\Omega |f_{n_0}|^{p}\,\mathrm{d}x\right)^{\frac{1}{p}} \to 0$$

as $\varepsilon \to 0$.

Hence there is an $\varepsilon_0 > 0$ such that when $\varepsilon < \varepsilon_0$, we have the bound

$$\left(\frac{\varepsilon}{m(\Omega)}\int_\Omega |f_{n_0}|^{p-\varepsilon}\,\mathrm{d}x\right)^{\frac{1}{p-\varepsilon}} < \frac{\delta}{2}.$$

Finally

$$\begin{aligned}\left(\frac{\varepsilon}{m(\Omega)}\int_\Omega |f|^{p-\varepsilon}\,\mathrm{d}x\right)^{\frac{1}{p-\varepsilon}} &\leq \left(\frac{\varepsilon}{m(\Omega)}\int_\Omega |f-f_{n_0}|^{p-\varepsilon}\,\mathrm{d}x\right)^{\frac{1}{p-\varepsilon}}\\ &\quad+\left(\frac{\varepsilon}{m(\Omega)}\int_\Omega |f_{n_0}|^{p-\varepsilon}\,\mathrm{d}x\right)^{\frac{1}{p-\varepsilon}}\\ &\leq \|f-f_{n_0}\|_{p)}+\frac{\delta}{2}\\ &\leq \frac{\delta}{2}+\frac{\delta}{2}\end{aligned}$$

when $\varepsilon < \varepsilon_0$. This ends the proof. □

We now use the Proposition 7.52 which gives information regarding reflexivity of the space based upon the absolute continuity of the norm.

Theorem 7.58. *The spaces $L_{p)}(\Omega)$ is not reflexive.*

Proof. The non-reflexivity follows from the fact that there exists a function Φ for which the norm $\|\Phi\|_{p)}$ is not absolute continuous. Indeed taking the function Φ as

$$\Phi(x) = x^{-\frac{1}{p}}, \quad x \in (0,1),$$

we obtain

$$\limsup_{a\to 0}\sup_{\varepsilon>0}\left(\varepsilon\int_0^a x^{-\frac{p-\varepsilon}{p}}\,\mathrm{d}x\right)^{\frac{1}{p-\varepsilon}} \neq 0,$$

and this ends the proof. □

From Fiorenza and Karadzhov [18], we give the following characterization of the grand Lebesgue spaces (in the case $\mu(\Omega) = 1$, for simplicity):

$$\|f\|_{L_{p)}(\Omega)} \asymp \sup_{0<t<1}(1-\log t)^{-\frac{1}{p}}\left(\int_t^1 |f^*(s)|^p\,\mathrm{d}s\right)^{\frac{1}{p}},$$

where f^* is a decreasing rearrangement of f defined as

$$f^*(t) = \sup_{m(E)=t} \inf_E f$$

with $t \in (0,1)$.

We can introduce a generalization of the grand Lebesgue spaces, namely the spaces $L_{p),\theta}(\Omega)$, $\theta > 0$, defined by

$$\|f\|_{p),\theta} = \sup_{0<\varepsilon<p-1} \left(\frac{\varepsilon^\theta}{m(\Omega)} \int_\Omega |f|^{p-\varepsilon}\, dx \right)^{\frac{1}{p-\varepsilon}}. \tag{7.79}$$

For $\theta = 0$ we have $\|f\|_{p),0} = \|f\|_p$ and for $\theta = 1$ such spaces reduce obviously to the spaces $L_{p)}(\Omega)$.

Many results of grand Lebesgue spaces are also valid for generalized grand Lebesgue spaces, we will just mention the following:

Theorem 7.59. *The subspace $C_0^\infty(\Omega)$ is not dense in $f \in L_{p),\theta}(\Omega)$. Its closure consists of functions $f \in L_{p),\theta}(\Omega)$ such that*

$$\lim_{\varepsilon\to 0} \varepsilon^{\frac{\theta}{p}} \|f\|_{p-\varepsilon} = 0.$$

7.2.3 Hardy's Inequality

We recall the classical Hardy inequality for Lebesgue spaces (see (3.40) for the weighted version in the Lebesgue spaces)

$$\left(\int_0^1 \left(\frac{1}{x} \int_0^x f(y)\, dy \right)^p dx \right)^{\frac{1}{p}} \le \frac{p}{p-1} \left(\int_0^1 f^p(x)\, dx \right)^{\frac{1}{p}}.$$

Here we discuss the Hardy inequality in grand Lebesgue spaces to show some common techniques used in the aforementioned spaces.

Theorem 7.60. *Let $1 < p < \infty$. There exists a constant $C(p) > 1$ such that*

$$\left\| \frac{1}{x} \int_0^x f(y)\, dy \right\|_{L_{p)}([0,1])} \le C(p) \|f\|_{L_{p)}([0,1])}$$

for nonnegative measurable functions f on $[0,1]$.

Proof. Let $0 < \sigma < p - 1$, then we have

$$\left\| \frac{1}{x} \int_0^x f(y)\,dy \right\|_{p)}$$

$$= \max \left\{ \sup_{0<\varepsilon<\sigma} \left(\frac{\varepsilon}{m(\Omega)} \int_\Omega \left(\frac{1}{x} \int_0^x f(y)\,dy \right)^{p-\varepsilon} dx \right)^{\frac{1}{p-\varepsilon}}, \right.$$

$$\left. \sup_{0<\varepsilon<p-1} \left(\frac{\varepsilon}{m(\Omega)} \int_\Omega \left(\frac{1}{x} \int_0^x f(y)\,dy \right)^{p-\varepsilon} dx \right)^{\frac{1}{p-\varepsilon}} \right\}$$

$$\leq \max \left\{ \sup_{0<\varepsilon<\sigma} \left(\frac{\varepsilon}{m(\Omega)} \int_\Omega \left(\frac{1}{x} \int_0^x f(y)\,dy \right)^{p-\varepsilon} dx \right)^{\frac{1}{p-\varepsilon}}, \right.$$

$$\left. \left(\sup_{\sigma\leq\varepsilon<p-1} \varepsilon^{\frac{1}{p-\varepsilon}} \right) \sigma^{-\frac{1}{p-\sigma}} \sigma^{\frac{1}{p-\varepsilon}} \left(\frac{1}{m(\Omega)} \int_\Omega \left(\frac{1}{x} \int_0^x f(y)\,dy \right)^{p-\sigma} dx \right)^{\frac{1}{p-\sigma}} \right\}$$

$$\leq (p-1)\sigma^{-\frac{1}{p-\sigma}} \sup_{0<\varepsilon\leq\sigma} \left(\frac{\varepsilon}{m(\Omega)} \int_\Omega \left(\frac{1}{x} \int_0^x f(y)\,dy \right)^{p-\varepsilon} dx \right)^{\frac{1}{p-\varepsilon}}.$$

Now take $0 < \varepsilon \leq \sigma$, so that $p - \varepsilon > 1$. Applying the Hardy inequality with the exponent p replaced by $p - \varepsilon$ and multiplying both sides by $\varepsilon^{\frac{1}{p-\varepsilon}}$, we get

$$\left(\frac{\varepsilon}{m(\Omega)} \int_\Omega \left(\frac{1}{x} \int_0^x f(y)\,dy \right)^{p-\varepsilon} dx \right)^{\frac{1}{p-\varepsilon}}$$

$$\leq \frac{p-\varepsilon}{p-\varepsilon-1} \left(\frac{\varepsilon}{m(\Omega)} \int_\Omega f^{p-\varepsilon}\,dx \right)^{\frac{1}{p-\varepsilon}}.$$

If we take the sup over $0 < \varepsilon \le \sigma$ on both sides, the previous inequality becomes

$$\sup_{0<\varepsilon\le\sigma}\left(\frac{\varepsilon}{m(\Omega)}\int_{\Omega}\left(\frac{1}{x}\int_{0}^{x}f(y)\,\mathrm{d}y\right)^{p-\varepsilon}\mathrm{d}x\right)^{\frac{1}{p-\varepsilon}}$$

$$\le \frac{p-\sigma}{p-\sigma-1}\sup_{0<\varepsilon\le\sigma}\left(\frac{\varepsilon}{m(\Omega)}\int_{\Omega}f^{p-\varepsilon}\,\mathrm{d}x\right)^{\frac{1}{p-\varepsilon}}.$$

And therefore

$$\left\|\frac{1}{x}\int_{0}^{x}f(y)\,\mathrm{d}y\right\|_{p)} \le (p-1)\sigma^{-\frac{1}{p-\sigma}}\frac{p-\sigma}{p-\sigma-1}\sup_{0<\varepsilon<p-1}\left(\frac{\varepsilon}{m(\Omega)}\int_{\Omega}f^{p-\varepsilon}\,\mathrm{d}x\right)^{\frac{1}{p-\varepsilon}}.$$

Letting

$$C(p) = \inf_{0<\sigma<p-1}(p-1)\sigma^{-\frac{1}{p-\sigma}}\frac{p-\sigma}{p-\sigma-1} > 1,$$

we get the desired inequality. □

7.3 Problems

7.61. Show that, under the conditions of Definition 7.1, the function $g(x) := |f(x)|^{p(x)}$ is indeed a measurable function.

7.62. Show directly from (7.6) that $\|\alpha f\|_{(p)} = |\alpha|\|f\|_{(p)}$.

7.63. Prove Lemma 7.16.

7.64. Show the validity of the generalized Hölder inequality

$$\int_{\Omega}|f_1(x)\cdots f_m(x)|\,\mathrm{d}x \le c\|f_1\|_{p^1(\cdot)}\cdots\|f_m\|_{p^m(\cdot)} \tag{7.80}$$

where $p^1(x) \ge 1, \dots, p^m(x) \ge 1$ and $\Sigma_{k=1}^m 1/p^k(x) \equiv 1$, for $x \in \Omega$, where $c = \Sigma_{k=1}^m 1/p_-^k$, $p_-^k = \min_{x\in\Omega} p^k(x)$.

Hint: Use pointwise the generalized Young inequality from Problem 1.47.

7.65. Demonstrate that

$$\|f\|^r_{L_{rp(\cdot)}(\Omega)} = \left\||f|^r\right\|_{L_{p(\cdot)}(\Omega)} \tag{7.81}$$

where r is constant and $r \ge 1/p_-$.

7.66. Show that

$$\int_{\Omega} |f(x)g(x)|\,\mathrm{d}x \leq \frac{\delta^{p_-}}{p_-}\rho_{p(\cdot)}(f) + \max\left\{\frac{1}{\delta^{(p_-)'}(p_-)'}, \frac{1}{\delta^{(p_+)'}(p_+)'}\right\}\rho_{p'(\cdot)}(g)$$

where $f \in L_{p(\cdot)}(\Omega)$, $g \in L_{p'(\cdot)}(\Omega)$, $0 < \delta < 1$, $p_- > 1$ and $p_+ < \infty$.

Hint: Use the *Peter-Paul inequality* (1.25) and the monotonicity of the function $p \mapsto \delta^p/p$ when $p \geq 1$.

7.67. If, instead of (7.6), we introduce the semi-norm $|f|_{(p)}$ as

$$|f|_{(p)} = \inf\left\{\lambda > 0 : \int_{\Omega\setminus\Omega_\infty} \frac{2}{p(x)}\left|\frac{f(x)}{\lambda}\right|^{p(x)} \mathrm{d}x \leq 1\right\}. \tag{7.82}$$

Show that the Hölder inequality (7.29) is valid but now with constant $k = 1$.

7.68. Show the completeness of the space $L_{p(\cdot)}(\Omega)$ based in the Riesz-Fischer theorem, namely:

Riesz-Fischer Theorem: *Let V be a vector space and q a semi-norm in V. The following are equivalent:*

1. *(V,q) is complete.*
2. *for all sequence $\{v_j\}_{j\in\mathbb{N}}$ such that $\sum_{j\in\mathbb{N}} q(v_j) < \infty$ the series $\sum_{j\in\mathbb{N}} v_j$ converges in V.*

7.69. Show that the relations in (7.62) are indeed true.

7.70. Prove that:

1. If $f \in L_p(\mathbb{R}^n)$ then we have that $\tau_h f \in L_p(\mathbb{R}^n)$, where $\tau_h f(x) := f(x-h)$, for all $h > 0$;
2. If $f \in L_p(\mathbb{R}^n)$ then $\lim_{h\to 0}\|f - \tau_h f\|_{L_p(\mathbb{R}^n)} = 0$.

7.71. Show that $L_{p)}(\Omega)$ is a rearrangement invariant Banach function space.

7.72. Prove that $L_{p)} \subset L_{p-\varepsilon}$ for $0 < \varepsilon < p-1$. This result, together with proposition 7.54, tells us that

$$L_p \subsetneq L_{p)} \subsetneq L_{p-\varepsilon}.$$

7.73. Show that property (P4) is a consequence of (P5) in Definition 7.39.

7.74. Show that the norms given in the Definition (7.14) are in fact norms.

7.75. Prove that $\|f\|_{p),0} = \|f\|_p$, where $\|\cdot\|_{p),\theta}$ is defined in (7.79)

7.76. Show that, for all measurable functions $p : \Omega \to \mathbb{R}$ with $m-1 \leq p(x) < m$, there exists a $c_m \in \mathbb{R}$ such that

$$\int_{\Omega} c_m^{p(x)}\,\mathrm{d}x = m^{-2}.$$

7.4 Notes and Bibliographic References

The content of the section related to variable exponent Lebesgue spaces is largely based on Samko [66, 65], Sharapudinov [67] and Rafeiro and Rojas [64].

The first reference to variable exponent Lebesgue spaces already appears in the 1930s in Orlicz [55].

The influential paper on variable exponent Lebesgue spaces is Kováčik and Rákosník [41] where many properties of the variable exponent Lebesgue spaces are studied. We mention some papers preceding Kováčik and Rákosník [41] in which variable exponent Lebesgue spaces were studied Sharapudinov [67, 68, 69]. For a more detailed historic exposition on variable exponent Lebesgue spaces see the book Diening, Harjulehto, Hästö, and Růžička [17] and Cruz-Uribe and Fiorenza [9].

Grand Lebesgue spaces were introduced by Iwaniec and Sbordone [35] when dealing with the problem of the integrability properties of the Jacobian. The generalization of grand Lebesgue spaces mentioned in the chapter appeared in Greco, Iwaniec, and Sbordone [23].

Part II
A Concise Excursion into Harmonic Analysis

Chapter 8
Interpolation of Operators

It is extremely difficult to imagine the program of singular integrals without the Marcinkiewicz Interpolation Theorem.
ROBERT FEFFERMAN

Abstract In this chapter we overview the technique of interpolation of operators, which is widely used in harmonic analysis in connection with Lebesgue spaces. The underlying idea is to obtain boundedness of an operator based on the available information in the endpoints. In the first section we will deal with the Riesz-Thorin interpolation theorem, also known as the complex method, and give some applications, viz. Hausdorff-Young inequality and Young's inequality for convolution. In the second section we prove the Marcinkiewicz interpolation theorem in Lebesgue spaces and also mention the theorem in its natural environment, namely in the framework of Lorentz spaces. We end the chapter with the Young's convolution inequality in Lorentz spaces.

We start with some concepts which will play a significant role in the subsequent sections.

Definition 8.1. We say that $T : X \longrightarrow Y$ is a linear operator if $T(\alpha f + \beta g) = \alpha T(f) + \beta T(g)$. Moreover, a linear operator is bounded if

$$\sup_{f \neq 0} \frac{\|Tf\|_Y}{\|f\|_X}$$

is finite and denote that value by $\|T\|_{X \longrightarrow Y} := \|T\|$, which is called the norm of the linear operator. ⊘

It will also be necessary to deal with operators T defined on several spaces simultaneously. We will define it for L_p spaces, although it can be defined, mutatis mutandis, for abstract function spaces.

Definition 8.2. We define $L_{p_1} + L_{p_2}$ to be the space of all functions f, such that $f = f_1 + f_2$, with $f_1 \in L_{p_1}$ and $f_2 \in L_{p_2}$. ⊘

R.E. Castillo, H. Rafeiro, *An Introductory Course in Lebesgue Spaces*, CMS Books in Mathematics, DOI 10.1007/978-3-319-30034-4_8

Suppose that $p_1 < p_2$. Then we observe that

$$L_p \subset L_{p_1} + L_{p_2},$$

for all $p \in [p_1, p_2]$.

In fact, let $f \in L_p$ and let γ be a fixed positive constant. Define

$$f_1(x) = \begin{cases} f(x), & |f(x)| > \gamma, \\ 0, & |f(x)| \leq \gamma, \end{cases}$$

and $f_2(x) = f(x) - f_1(x)$. Then

$$\int |f_1(x)|^{p_1} \mathrm{d}x = \int |f_1(x)|^p |f_1(x)|^{p_1 - p} \mathrm{d}x \leq \gamma^{p_1 - p} \int |f(x)|^p \mathrm{d}x,$$

since $p_1 - p \leq 0$. Similarly,

$$\int |f_2(x)|^{p_2} \mathrm{d}x = \int |f_2(x)|^p |f_2(x)|^{p_2 - p} \mathrm{d}x \leq \gamma^{p_2 - p} \int |f(x)|^p \mathrm{d}x,$$

so $f_1 \in L_{p_1}$ and $f_2 \in L_{p_2}$, with $f = f_1 + f_2$.

We will also rely on the following theorem from complex analysis in the proof of the Riesz-Thorin interpolation theorem.

Theorem 8.3 (Hadamard Three Lines Theorem). *Assume that f is an analytic function on $\mathrm{int}(\mathbb{S})$ and bounded and continuous on $\mathbb{S}$, where $\mathbb{S}$ stands for the strip $\mathbb{S} = \{z \in \mathbb{C} : 0 \leq \mathrm{Re}(z) \leq 1\}$. Then*

$$\sup_{t \in \mathbb{R}} |f(\theta + it)| \leq \left(\sup_{t \in \mathbb{R}} |f(it)| \right)^{1-\theta} \left(\sup_{t \in \mathbb{R}} |f(1 + it)| \right)^{\theta},$$

for every $\theta \in [0, 1]$.

Before showing Hadamard's three lines theorem we will prove a *weaker* version of the Phragmén-Lindelöf theorem from which we will derive the aforementioned Hadamard's theorem.

Theorem 8.4 (Phragmén-Lindelöf theorem). *Let f be an analytic function in the strip $\mathbb{S} = \{z \in \mathbb{C} : \alpha < \Re z < \beta\}$, continuous and bounded in the closed strip $\overline{\mathbb{S}}$. Suppose moreover that $|f(\alpha + \mathrm{i}y)| \leq M$ and $|f(\beta + \mathrm{i}y)| \leq M$. Then for all $x \in (\alpha, \beta)$ we have $|f(x + \mathrm{i}y)| \leq M$.*

Proof. We will first show a particular case, from which we will bootstrap the general result.

PARTICULAR CASE: Let us suppose moreover that $f(x + \mathrm{i}y) \to 0$ uniformly in $0 \leq x \leq 1$ when $|y| \to \infty$. By the assumption, it is possible to find a rectangle R_M such that $|f(x)| \leq M$ whenever $x \in \partial R_M$. By the maximum principle we get that $|f(x)| \leq M$ in the interior or R_M.

GENERAL CASE: Let us define the following auxiliary analytic function

$$f_m(z) := f(z)\exp\left(\frac{z^2}{m}\right),$$

from which it follows that

$$|f_m(z)| = |f(z)|\exp\left(\frac{x^2-y^2}{m}\right) \tag{8.1}$$

if $z = x + iy$. The equality (8.1) shows that f_m satisfies the conditions in the particular case, since f is bounded in $\overline{\mathbb{S}}$. Taking $M_1 = M\exp(\max\{\alpha^2, \beta^2\}/m)$ we have

$$|f_m(\alpha + iy)| \leq M_1, \quad |f_m(\beta + iy)| \leq M_1$$

from which it follows that $|f_m(x+iy)| \leq M_1$ for all $x \in (\alpha, \beta)$. This last inequality can be written as

$$\left|f(z)\exp\left(\frac{z^2}{m}\right)\right| \leq M \cdot \exp\left(\frac{\gamma}{m}\right) \tag{8.2}$$

where $\gamma = \max\{\alpha^2, \beta^2\}$. Passing to the limit in (8.2) we obtain the result. □

We now prove Hadamard's three lines theorem using the Pragmén-Lindelöf theorem via an appropriate auxiliary function Φ.

Proof (Hadamard's Three Lines Theorem). Let us define the analytic function

$$\Phi(z) = \frac{f(z)}{M_0^{1-z}M_1^z}$$

where $|f(iy)| \leq M_0$ and $|f(1+iy)| \leq M_1$. Direct calculations show that

$$|\Phi(iy)| = \frac{|f(iy)|}{M_0} \leq 1 \quad \text{and} \quad |\Phi(1+iy)| = \frac{|f(1+iy)|}{M_1} \leq 1,$$

from which, using Phragmén-Lindelöf theorem, we obtain that $\Phi(z) \leq 1$ whenever $z \in \mathbb{S}$. This means that

$$|\Phi(\theta + iy)| = \frac{|f(\theta + iy)|}{\left|M_0^{1-\theta-iy}M_0^{\theta+iy}\right|} \leq 1$$

which implies that $|f(\theta + iy)| \leq M_0^{1-\theta}M_1^{\theta}$. □

8.1 Complex Method

Let us start with the following result already stated in Problem 3.129.

Theorem 8.5. *Let $f \in L_p(X) \cap L_q(X)$. Then $f \in L_r(X)$ for all $p \le r \le q$ and we have the bound*

$$\|f\|_r \le \|f\|_p^{\theta} \|f\|_q^{1-\theta} \tag{8.3}$$

when $\frac{1}{r} = \frac{\theta}{p} + \frac{1-\theta}{q}$.

This theorem can be seen as a proto interpolation theorem, since we infer that f belongs to the L_r space due to the fact that it belongs to the *endpoint* spaces.

The next theorem generalizes Theorem 8.5, loosely stating that if an operator is, at the same time, $(L_{p_1} \to L_{q_1})$ and $(L_{p_2} \to L_{q_2})$ bounded, then it is $(L_p \to L_q)$ bounded, where p and q are *intermediate* points. In Riesz-Thorin Interpolation Theorem the L_s functions are complex-valued functions. The correct formulation is:

Theorem 8.6 (Riesz-Thorin Interpolation Theorem). *Let (X,μ) and (Y,ν) be two σ-finite measure spaces. Let T be a linear operator defined on all simple functions in X with range in the set of measurable functions in Y. Let $p_0, p_1, q_0, q_1 \in [1,\infty]$. Assume that for all simple functions f*

$$\|Tf\|_{L_{q_0}(Y,\mathrm{d}\nu)} \le A_0 \|f\|_{L_{p_0}(X,\mathrm{d}\mu)}, \qquad \|Tf\|_{L_{q_1}(Y,\mathrm{d}\nu)} \le A_1 \|f\|_{L_{p_1}(X,\mathrm{d}\mu)},$$

for some $p_0 \neq p_1$ and $q_0 \neq q_1$. For a fixed θ $(0 < \theta < 1)$ we define

$$\frac{1}{p} = \frac{1-\theta}{p_0} + \frac{\theta}{p_1}, \quad \frac{1}{q} = \frac{1-\theta}{q_0} + \frac{\theta}{q_1}. \tag{8.4}$$

Then if f is a simple function in X we have

$$\|Tf\|_{L_q(Y,\mathrm{d}\nu)} \le A_\theta \|f\|_{L_p(X,\mathrm{d}\mu)}, \tag{8.5}$$

with

$$A_\theta \le A_0^{1-\theta} A_1^{\theta}. \tag{8.6}$$

By density, the estimate (8.5) *can be extended to all function in $L^p(X,\mathrm{d}\mu)$.*

Remark 8.7. The geometric interpretation of (8.4) is that the points $(1/p, 1/q)$ belong to the line segment connecting $(1/p_0, 1/q_0)$ and $(1/p_1, 1/q_1)$. The schematic of this fact is called the *Riesz square*, as in Figure 8.1.

The inequality (8.6) means that A_θ is logarithmically convex, i.e., $\log A_\theta$ is convex.

In the case of real-valued functions, the constant appearing in (8.6) should be changed, namely $A_\theta \le 2A_0^{1-\theta} A_1^{\theta}$, see, e.g., Bennett and Sharpley [1]. ⊘

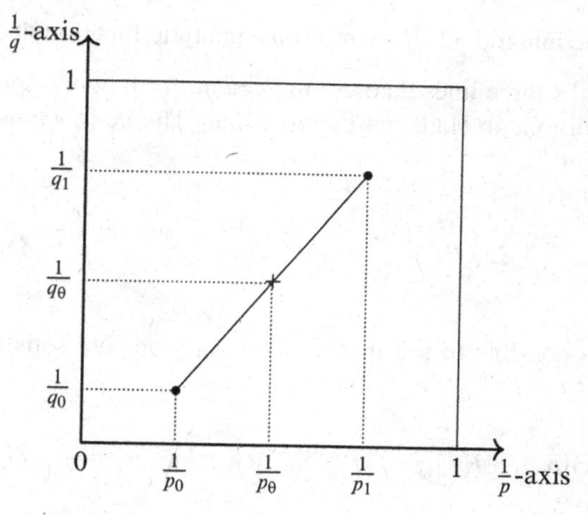

Fig. 8.1 Riesz square.

Proof (Riesz-Thorin Theorem 8.6). Let us take f has

$$f(x) = \sum_{k=1}^{m} a_k e^{i\alpha_k} \chi_{A_k}(x) \tag{8.7}$$

where A_k are finite measure pairwise disjoint sets, α_k real numbers and a_k strictly positive real numbers. By Riesz representation theorem, to estimate $\|Tf\|_q$ it suffices to estimate

$$\sup\left\{\int_Y (Tf) g \, d\nu : g \text{ simple function such that } \|g\|_{q'} \leq 1\right\}.$$

For definiteness we take a specific g in the form

$$g(x) = \sum_{s=1}^{n} b_s e^{i\beta_s} \chi_{B_s}(x), \tag{8.8}$$

where B_k are finite measure pairwise disjoint sets, β_k real numbers, and b_k strictly positive real numbers. We now introduce two auxiliary functions, namely

$$F_z = \sum_{k=1}^{m} a_k^{(1-z)\frac{p}{p_0} + z\frac{p}{p_1}} e^{i\alpha_k} \chi_{A_k}, \quad G_z = \sum_{s=1}^{n} b_k^{(1-z)\frac{q'}{q'_0} + z\frac{q'}{q'_1}} e^{i\beta_k} \chi_{B_s},$$

and finally we introduce

$$\Phi(z) = \int_Y T(F_z) G_z \, d\nu.$$

We embedded the integral $\int_Y (Tf)g$ into some analytic function Φ and now we want to use Hadarmard's three lines theorem to get a uniform bound for Φ. To see that Φ satisfies the conditions in Hadamard's three lines Theorem, we use linearity of the operator T and get

$$\Phi(z) = \sum_{\substack{k=1,\ldots,m\\ s=1,\ldots,n}} a_k^{(1-z)\frac{p}{p_0}+z\frac{p}{p_1}} b_s^{(1-z)\frac{q'}{q_0'}+z\frac{q'}{q_1'}} \mathrm{e}^{\mathrm{i}\alpha_k}\mathrm{e}^{\mathrm{i}\beta_s} \int_{B_s} T(\chi_{A_k})\,\mathrm{d}\nu,$$

from which it is possible to see that all the conditions are satisfied in the strip $\mathbb{S}$. Since

$$|F_{\mathrm{i}y}| = |f(x)|^{\frac{p}{p_0}}, \quad \|F_{\mathrm{i}}y\|_{p_0} = \|f\|_p^{\frac{p}{p_0}}, \quad |G_{\mathrm{i}y}| = |g(x)|^{\frac{q'}{q_0'}}, \quad \left\|G_{\mathrm{i}y}\right\|_{q_0'} = \|g\|_{q'}^{\frac{q'}{q_0'}}$$

we obtain

$$|\Phi(\mathrm{i}y)| \leq M_0\|f\|_p^{\frac{p}{p_0}}, \quad |\Phi(1+\mathrm{i}y)| \leq M_1\|f\|_p^{\frac{p}{p_1}}.$$

Since Φ satisfies all the conditions of the three lines theorem, we get that $\Phi(\theta) \leq M_0^{1-\theta}M_1^{\theta}$, which shows the result for simple functions. By a density argument we obtain the result. □

Now we give two applications of the Riesz-Thorin interpolation theorem.

Theorem 8.8 (Hausdorff-Young inequality). *Let* $1 \leq p \leq 2$ *and* $1/p + 1/p' = 1$. *Then the Fourier transform satisfies*

$$\|\widehat{f}\|_{p'} \leq \|f\|_p.$$

Proof. The proof can be given by a Riesz square,

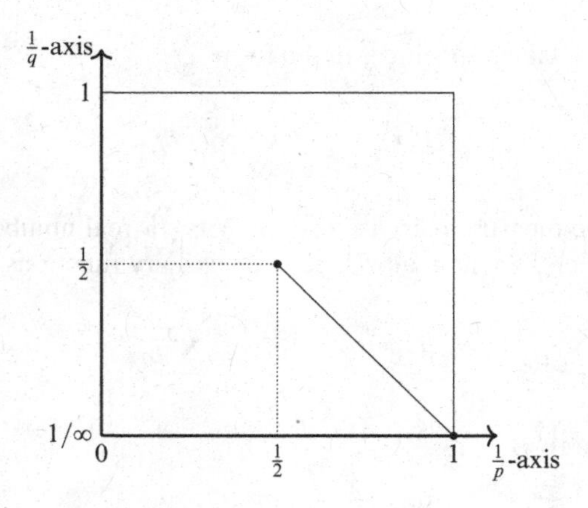

since $\|\widehat{f}\|_\infty \leq \|f\|_1$ and $\|\widehat{f}\|_2 = \|f\|_2$ (Parseval identity). □

Another application of the complex interpolation theorem is the so-called Young's inequality for convolution.

Theorem 8.9 (Young's inequality for convolution). *If $f \in L_p(\mathbb{R}^n)$ and $g \in L_q(\mathbb{R}^n)$, $1 \le p, q, r \le \infty$ and $\frac{1}{r} = \frac{1}{p} + \frac{1}{q} - 1$, then*

$$\|f * g\|_r \le \|f\|_p \|g\|_q .$$

Proof. Since a direct application of the interpolation theorem is not possible, we need to adapt our argument. Let us fix $f \in L_p(\mathbb{R}^n)$, $p \in [1, \infty]$ and then will apply the Riesz-Thorin interpolation theorem to the mapping $g \mapsto f * g$. From Hölder's inequality we have

$$|f * g(x)| \le \|f\|_p \|g\|_{p'}$$

and thus $g \mapsto f * g$ maps $L_{p'}(\mathbb{R}^n)$ to $L_\infty(\mathbb{R}^n)$. From the Young inequality (see Theorem 3.26) we have that if $g \in L_1$, then $\|f * g\|_p \le \|f\|_p \|g\|_1$, thus $g \mapsto f * g$ also maps L_1 to L_p. From the interpolation theorem we have that $g \mapsto f * g$ is $(L_q \to L_r)$ bounded where

$$\frac{1}{q} = \frac{1-\theta}{1} + \frac{\theta}{p'} \quad \text{and} \quad \frac{1}{r} = \frac{1-\theta}{p} + \frac{\theta}{\infty}.$$

Removing the parameter θ we arrive at $\frac{1}{r} = \frac{1}{p} + \frac{1}{q} - 1$. □

8.2 Real Method

We introduce some concepts that will play a fundamental role in the main theorem in this section.

Definition 8.10. An operator T is called *sublinear* if

(a) $|T(f+g)(x)| \le |Tf(x)| + |Tg(x)|$,
(b) $|T(\lambda f)(x)| \le |\lambda| |Tf(x)|$,

for all measurable functions f and g and all scalars λ. ⊘

We now give the notion of weak and strong type inequalities for sublinear operators.

Definition 8.11. A sublinear operator T defined in $L_1(\mathbb{R}^n)$ is said to be of *weak type* (p, q) with $1 \le p \le \infty$ and $1 \le q < \infty$, if exists a constant C, such that for each $f \in L_1(\mathbb{R}^n)$ and each $\lambda > 0$ we have

$$m\Big(\big\{x : |Tf(x)| > \lambda\big\}\Big) \le \left(\frac{C}{\lambda} \|f\|_{L_p}\right)^q$$

and it is said of *strong type* (p, q) if

$$\|Tf\|_{L_q} \le C \|f\|_{L_p}.$$

⊘

We now justify the above nomenclature.

Theorem 8.12 *Let T be an operator of strong type (p,q). Then T is of weak type (p,q).*

Proof. By the Markov inequality, with $g(\lambda) = \lambda^q$ we get

$$m\left(\left\{x : |Tf(x)| > \lambda\right\}\right) = m\left(\left\{x : |Tf(x)|^q > \lambda^q\right\}\right) \leq \frac{C}{\lambda^q}\int_{\mathbb{R}^n} |Tf(x)|^q \,\mathrm{d}m,$$

therefore $m\left(\left\{x : |Tf(x)| > \lambda\right\}\right) \leq \frac{C}{\lambda^q}\|Tf\|_{L_q}^q \leq \left(\frac{C}{\lambda}\|f\|_{L_p}\right)^q$. □

The following inequality, due to Kolmogorov, generalizes the simple fact

$$\left\|\mathrm{id}(f)\right\|_{L_r(E)}^r \leq m(E)^{1-\frac{r}{q}}\|f\|_{L_q(E)}^r, \tag{8.9}$$

where id is the identity operator and E is some finite measure set.

Theorem 8.13 (Kolmogorov's inequality). *Let E be a subset of $\mathbb{R}^n$ with finite measure. If T is of weak type (p,q) and $0 < r < q$, then*

$$\int_E |Tf(x)|^r \,\mathrm{d}m \leq C_{q,r}\left[m(E)\right]^{1-\frac{r}{q}}\|f\|_{L_p}^r. \tag{8.10}$$

Proof. For each $\lambda \in \mathbb{R}^+$ and Cavalieri's principle (3.34) we can write

$$\begin{aligned}
\int_E |Tf(x)|^r \,\mathrm{d}m &= r\int_0^\infty \lambda^{r-1} m\left(\left\{x\in E : |Tf(x)| > \lambda\right\}\right)\mathrm{d}\lambda \\
&= r\int_0^{\left(m(E)\right)^{-1/q}\|f\|_{L_p}} \lambda^{r-1} m\left(\left\{x\in E : |Tf(x)| > \lambda\right\}\right)\mathrm{d}\lambda \\
&\quad + r\int_{\left(m(E)\right)^{-1/q}\|f\|_{L_p}}^\infty \lambda^{r-1} m\left(\left\{x\in E : |Tf(x)| > \lambda\right\}\right)\mathrm{d}\lambda \\
&\leq r\int_0^{\left(m(E)\right)^{-1/q}\|f\|_{L_p}} \lambda^{r-1} m(E)\,\mathrm{d}\lambda
\end{aligned}$$

$$+r \int_{\left(m(E)\right)^{-1/q}\|f\|_{L_p}}^{\infty} \lambda^{r-1}\left(\frac{C^q}{\lambda^q}\|f\|_{L_p}^q\right) \mathrm{d}\lambda$$

$$= m(E)\left[m(E)\right]^{-r/q}\|f\|_{L_p}^r + rC^q\|f\|_{L_p}^r \frac{\left(m(E)\right)^{1-\frac{r}{q}}}{q-r}$$

$$= \left(1+\frac{rC^q}{q-r}\right)\left(m(E)\right)^{1-\frac{r}{q}}\|f\|_{L_p}^r.$$

from which we get (8.10). □

The reciprocal of Theorem 8.13 is also true, namely:

Theorem 8.14. *Suppose that for any measurable subset E of $\mathbb{R}^n$ with finite measure we have the inequality*

$$\int_E |Tf(x)|^r \,\mathrm{d}m \leq C_{r,q}\left(m(E)\right)^{1-\frac{r}{q}}\|f\|_{L_p}^r,$$

with $0 < r < q$. Then there exists $C > 0$ such that

$$m\left(\left\{x : |Tf(x)| > \lambda\right\}\right) \leq \left(\frac{C}{\lambda}\|f\|_{L_p}\right)^q.$$

Proof. Let $E = \left\{x : |Tf(x)| > \lambda\right\}$ for $\lambda \in \mathbb{R}^+$, then

$$\int_{\{x:|Tf(x)|>\lambda\}} |Tf(x)|^r \,\mathrm{d}m \leq C_{r,q}\left[m\left(\{x : |Tf(x)| > \lambda\}\right)\right]^{1-\frac{r}{q}}\|f\|_{L_p}^r.$$

On the other hand, it is clear that

$$\lambda^r m\left(\left\{x : |Tf(x)| > \lambda\right\}\right) = \int_{\{x:|Tf(x)|>\lambda\}} \lambda^r \,\mathrm{d}m$$

$$\leq \int_{\{x:|Tf(x)|>\lambda\}} |Tf(x)|^r \,\mathrm{d}m,$$

then

$$\lambda^r m\left(\left\{x : |Tf(x)| > \lambda\right\}\right) \leq C_{r,q}\left[m\left(\{x : |Tf(x)| > \lambda\}\right)\right]^{1-\frac{r}{q}}\|f\|_{L_p}^r$$

from which it follows the desired result. □

In the following result we get a kind of *improved* weak type (1,1) inequality in the form (8.11).

Theorem 8.15. *Let T be a sublinear operator defined in $L_1(\mathbb{R}^n)+L_\infty(\mathbb{R}^n)$, such that T is of weak type $(1,1)$ and moreover it satisfies $\|Tf\|_{L_\infty} \leq A\|f\|_{L_\infty}$, where A is a positive constant. Then for $f \in L_1(\mathbb{R}^n)+L_\infty(\mathbb{R}^n)$ and $\lambda > 0$, we have that*

$$m\Big(\Big\{x : |Tf(x)| > \lambda\Big\}\Big) \leq \frac{C}{\lambda} \int\limits_{\left[\frac{\lambda}{2A},\infty\right)} m\Big(\Big\{x : |f(x)| > t\Big\}\Big)\,\mathrm{d}t. \tag{8.11}$$

Proof. Let $f \in L_1(\mathbb{R}^n)$ and given $\lambda > 0$ we can write $f(x) = f_\lambda(x) + f^\lambda(x)$, where

$$f_\lambda(x) = f(x)\chi_{\{s:|f(s)|\leq\lambda/A\}}(x) \quad \text{and} \quad f^\lambda(x) = f(x)\chi_{\{s:|f(s)|>\lambda/A\}}(x).$$

From the sublinearity of T, we see that

$$\Big\{x : |Tf(x)| > \lambda\Big\} \subset \Big\{x : |Tf_{\lambda/2}(x)| > \lambda/2\Big\} \cup \Big\{x : |Tf^{\lambda/2}(x)| > \lambda/2\Big\}$$

then

$$\begin{aligned} m\Big(\{x : |Tf(x)| > \lambda\}\Big) \leq\; & m\Big(\{x : |Tf_{\lambda/2}(x)| > \lambda/2\}\Big) \\ & + m\Big(\{x : |Tf^{\lambda/2}(x)| > \lambda/2\}\Big). \end{aligned}$$

On the other hand, note that $\|f_{\lambda/2}\|_{L_\infty} \leq \dfrac{\lambda}{2A}$, then by hypothesis we have

$$\|Tf_{\lambda/2}\|_{L_\infty} \leq \lambda/2,$$

therefore

$$m\Big(\{x : |Tf_{\lambda/2}(x)| > \lambda/2\}\Big) = 0.$$

Gathering all estimates, we have

$$\begin{aligned} m\Big(\{x : |Tf(x)| > \lambda\}\Big) &\leq m\Big(\{x : |Tf^{\lambda/2}(x)| > \lambda/2\}\Big) \\ &\leq \frac{C}{\lambda} \int\limits_{\mathbb{R}^n} |f^{\lambda/2}(x)|\,\mathrm{d}m \\ &= \frac{C}{\lambda} \int\limits_{\{x:|f(x)|>\frac{\lambda}{2A}\}} |f(x)|\,\mathrm{d}m \end{aligned}$$

$$= \frac{C}{\lambda} \int_{[\lambda/(2A),\infty)} m\Big(\Big\{x : |f(x)| > t\Big\}\Big)\, \mathrm{d}t,$$

which ends the proof. □

We recall that a measurable function f is in the *Zygmund class* $L\log L(X)$, also denoted as Zygmund space, if

$$\int_X |f(x)| \log^+ |f(x)|\, \mathrm{d}x < \infty,$$

where $\log^+ t = \log t$ for $t > 1$ and 0 otherwise.

Theorem 8.16. *Let T be a sublinear operator defined in $L_1(\mathbb{R}^n) + L_\infty(\mathbb{R}^n)$, such that T is of weak type $(1,1)$ and moreover it satisfies $\|Tf\|_{L_\infty} \leq A\|f\|_{L_\infty}$, where A is a positive constant. For $B \subset \mathbb{R}^n$ such that $m(B) < \infty$, we have*

$$\int_B |Tf(x)|\, \mathrm{d}m \leq C \left(m(B) + \int_{\mathbb{R}^n} |f(x)| \log^+ |f(x)|\, \mathrm{d}m \right)$$

where $\log^+ t = \max(\log t, 0)$ and C is a constant independent of f.

Proof. By Theorem 3.54 with $p = 1$ we get

$$\int_B |Tf(x)|\, \mathrm{d}m = 2A \int_0^\infty m\Big(\Big\{x \in B : |Tf(x)| > 2A\lambda\Big\}\Big)\, \mathrm{d}\lambda$$

$$= C \left(\int_0^1 m\Big(\Big\{x \in B : |Tf(x)| > 2A\lambda\Big\}\Big)\, \mathrm{d}\lambda \right.$$

$$\left. + \int_1^\infty m\Big(\Big\{x \in B : |Tf(x)| > 2A\lambda\Big\}\Big)\, \mathrm{d}\lambda \right)$$

$$\leq C \left(m(B) + \int_1^\infty \left(\frac{C}{\lambda} \int_{\{x \in B : |f(x)| > \lambda\}} |f(x)|\, \mathrm{d}m \right) \mathrm{d}\lambda \right)$$

$$\leq C \left(m(B) + \int_{\mathbb{R}^n} \left(|f(x)| \int_1^{|f(x)|} \frac{\mathrm{d}\lambda}{\lambda} \right) \mathrm{d}m \right)$$

$$= C\left(m(B) + \int_{\mathbb{R}^n} |f(x)| \log^+ |f(x)| \,\mathrm{d}m \right),$$

where the first inequality is due to Theorem 8.15. □

We now study the so-called real method of interpolation, which is also known as Marcinkiewicz interpolation theorem. This theorem has the same working principle as the Riesz-Thorin interpolation theorem (showing boundedness using only information in the end points), but it requires less from the end points, it asks only that the operator is of weak type instead of strong type as in the case of Riesz-Thorin Theorem.

Theorem 8.17 (Marcinkiewicz's interpolation theorem). *Let $1 \leq p_0 \leq p_1 < \infty$. Suppose that T is a sublinear operator defined in $L_{p_0}(\mathbb{R}^n) + L_{p_1}(\mathbb{R}^n)$, which is, simultaneously, of weak type (p_0, p_0) and of weak type (p_1, p_1). Then T is of strong type (p, p) with $p_0 < p < p_1$.*

Proof. Let $f \in L_p(\mathbb{R}^n)$. For each $\lambda > 0$ we can write

$$f(x) = f_\lambda(x) + f^\lambda(x),$$

where

$$f_\lambda(x) = f(x)\chi_{\{x:|f(x)|\leq\lambda\}}(x)$$

and

$$f^\lambda(x) = f(x)\chi_{\{x:|f(x)|>\lambda\}}(x).$$

Let us consider only the case $p_1 < \infty$. Now, we want to show that $f^\lambda \in L_{p_0}(\mathbb{R}^n)$ and $f_\lambda \in L_{p_1}(\mathbb{R}^n)$, for that let us consider

$$\begin{aligned}
\int_{\mathbb{R}^n} |f^\lambda(x)|^{p_0} \,\mathrm{d}m &= \lambda^{p_0} \int_{\{x:|f(x)|>\lambda\}} \left|\frac{f(x)}{\lambda}\right|^{p_0} \mathrm{d}m \\
&< \lambda^{p_0} \int_{\{x:|f(x)|>\lambda\}} \left|\frac{f(x)}{\lambda}\right|^{p} \mathrm{d}m \\
&< \lambda^{p_0 - p} \int_{\mathbb{R}^n} |f(x)|^p \,\mathrm{d}m \\
&< \lambda^{p_0 - p} \|f\|_{L_p}^p,
\end{aligned}$$

then $f^\lambda \in L_{p_0}(\mathbb{R}^n)$. With a similar argument we can prove that $f_\lambda \in L_{p_1}(\mathbb{R}^n)$. Using the sublinearity of T it follows that

$$\left\{x : |Tf(x)| > \lambda\right\} \subset \left\{x : |Tf_{\lambda/2}(x)| > \lambda/2\right\} \cup \left\{x : |Tf^{\lambda/2}(x)| > \lambda/2\right\}.$$

On the other hand, due to Corollary 3.55 and the weak type estimate for the operator T, we obtain

$$\begin{aligned}
\|Tf\|_{L_p}^p &= \int_{\mathbb{R}^n} |Tf(x)|^p \,\mathrm{d}m \\
&= p\int_0^\infty \lambda^{p-1} m\Big(\big\{x : |Tf(x)| > \lambda\big\}\Big)\,\mathrm{d}\lambda \\
&\leq p\int_0^\infty \lambda^{p-1} m\Big(\big\{x : |Tf_{\lambda/2}(x)| > \lambda/2\big\}\Big)\,\mathrm{d}\lambda \\
&\quad + p\int_0^\infty \lambda^{p-1} m\Big(\big\{x : |Tf^{\lambda/2}(x)| > \lambda/2\big\}\Big)\,\mathrm{d}\lambda \\
&\leq p\int_0^\infty \lambda^{p-1}\frac{C^{p_0}}{\lambda^{p_0}}\left(\int_{\mathbb{R}^n} |f_{\lambda/2}(x)|^{p_0}\,\mathrm{d}m\right)\mathrm{d}\lambda \\
&\quad + p\int_0^\infty \lambda^{p-1}\frac{C^{p_1}}{\lambda^{p_1}}\left(\int_{\mathbb{R}^n} |f^{\lambda/2}(x)|^{p_1}\,\mathrm{d}m\right)\mathrm{d}\lambda \\
&\leq p\int_0^\infty \lambda^{p-p_0-1} C^{p_0}\left(\int_{\{x:|f(x)|\leq\lambda/2\}} |f(x)|^{p_0}\,\mathrm{d}m\right)\mathrm{d}\lambda \\
&\quad + p\int_0^\infty \lambda^{p-1}\frac{C^{p_1}}{\lambda^{p_1}}\left(\int_{\{x:|f(x)|>\lambda/2\}} |f(x)|^{p_1}\,\mathrm{d}m\right)\mathrm{d}\lambda.
\end{aligned}$$

By the Fubini theorem we get

$$\begin{aligned}
\|Tf\|_{L_p}^p &\leq pC^{p_0}\int_{\mathbb{R}^n}\left(|f(x)|^{p_0}\int_0^{2|f(x)|}\lambda^{p-p_0-1}\,\mathrm{d}\lambda\right)\mathrm{d}m \\
&\quad + pC^{p_1}\int_{\mathbb{R}^n}\left(|f(x)|^{p_1}\int_{2|f(x)|}^\infty \lambda^{p-p_1-1}\,\mathrm{d}\lambda\right)\mathrm{d}m \\
&= \frac{2^{p-p_0}pC^{p_0}}{p-p_0}\int_{\mathbb{R}^n} |f(x)|^{p_0}|f(x)|^{p-p_0}\,\mathrm{d}m
\end{aligned}$$

$$+\frac{2^{p-p_1}pC^{p_1}}{p_1-p}\int_{\mathbb{R}^n}|f(x)|^{p_1}|f(x)|^{p-p_1}\,\mathrm{d}m,$$

i.e.

$$\|Tf\|_{L_p}^p \le p\left(\frac{2^{p-p_0}C^{p_0}}{p-p_0}+\frac{2^{p-p_1}C^{p_1}}{p_1-p}\right)\int_{\mathbb{R}^n}|f(x)|^p\,\mathrm{d}m.$$

Then

$$\|Tf\|_{L_p} \le C(p_0,p,p_1)\|f\|_{L_p},$$

where

$$C(p_0,p,p_1)=\left(p\frac{2^{p-p_0}C^{p_0}}{p-p_0}+\frac{2^{p-p_1}C^{p_1}}{p_1-p}\right)^{1/p},$$

which ends the proof. □

The version of Marcinkiewicz theorem given in Theorem 8.17 is given only in the main diagonal of the Riesz square. The proof for the most general version is more involved and with a little extra effort it is possible to obtain the same theorem in the framework of Lorentz spaces, which is the natural environment for this theorem, since we ask that the operator is of weak type (p,q). Since the proof of this result is long and technical, we state it without proof, see Stein and Weiss [74] for a proof.

Definition 8.18. We will say T is of *restricted weak type* (r,p) if

$$\|Tf\|_{L_{(p,\infty)}} \le C\|f\|_{L_{(r,1)}}. \tag{8.12}$$

holds for every $f\in L_{(p,1)}\cap D$, where $1\le p<\infty$ and its domain D contains all the truncations of its members as well as all finite linear combinations of characteristic functions of sets of finite measure. In the case $r=\infty$, we define restricted weak type (∞,p) by the weak type estimate (∞,p) given by

$$\|Tf\|_{L_{(p,\infty)}} \le C\|f\|_{L_{(\infty,\infty)}}. \tag{8.13}$$

⊘

We are now in a position to state the Marcinkiewicz interpolation theorem in Lorentz spaces.

Theorem 8.19 (Marcinkiewicz Interpolation in Lorentz spaces). *Suppose that T is a subadditive operator of restricted weak types (r_j,p_j), $j=0,1$, with $r_0<r_1$ and $p_0\neq p_1$. Then there exists a constant $C=C_\theta$ such that*

$$\|Tf\|_{L_{(p,q)}} \le C\|f\|_{L_{(r,q)}},$$

for all f belonging to the domain of T and to $L_{r,q}$ where $1\le q\le\infty$.

$$\frac{1}{p}=\frac{1-\theta}{p_0}+\frac{\theta}{p_1},\quad \frac{1}{r}=\frac{1-\theta}{r_0}+\frac{\theta}{r_1},\quad 0<\theta<1.$$

We now obtain the Young inequality in weak Lebesgue spaces via Marcinkiewicz interpolation theorem.

Theorem 8.20 (Young Inequality in Weak Lebesgue Spaces). *Let $1 \leq r < \infty$ and $1 < p,q < \infty$ be such that*

$$1+\frac{1}{q}=\frac{1}{r}+\frac{1}{p}.$$

Then there is a constant $C = C$ such that, for every $f \in L_r(\mathbb{R}^n)$ and $g \in L_{(p,\infty)}(\mathbb{R}^n)$, we have

$$\|f*g\|_{L_{(q,\infty)}} \leq C\|g\|_{L_{(p,\infty)}}\|f\|_{L_r}.$$

Proof. Let λ be a real positive number to be chosen later and consider the decomposition $g = g_\lambda + g^\lambda$, where $g_\lambda = g\chi_{\{|g|\leq\lambda\}}$ and $g^\lambda = g\chi_{\{|g|>\lambda\}}$. We remember (see Theorem 5.15) the following relations between its distribution functions:

$$D_{g_\lambda}(\alpha)=\begin{cases}0, & \text{if } \alpha \geq \lambda;\\ D_g(\alpha)-D_g(\lambda), & \text{if } \alpha < \lambda;\end{cases}$$

and

$$D_{g^\lambda}(\alpha)=\begin{cases}D_g(\alpha), & \text{if } \alpha \geq \lambda;\\ D_g(\lambda), & \text{if } \alpha < \lambda.\end{cases}$$

We also know that

$$D_{f*g}(\beta) \leq D_{f*g_\lambda}\left(\frac{\beta}{2}\right)+D_{f*g^\lambda}\left(\frac{\beta}{2}\right),$$

hence, it is enough to evaluate the distributions of $f*g_\lambda$ and $f*g^\lambda$ separately. Since g_λ is the small part of g, we have $g_\lambda \in L_s$, if $s > p$. In fact,

$$\begin{aligned}\int_{\mathbb{R}^n}|g_\lambda(x)|^s\,\mathrm{d}x &= s\int_0^\infty \alpha^{s-1}D_{g_\lambda}(\alpha)\,\mathrm{d}\alpha\\ &= s\int_0^\lambda \alpha^{s-1}\{D_g(\alpha)-D_g(\lambda)\}\,\mathrm{d}\alpha\\ &\leq s\int_0^\lambda \alpha^{s-1-p}\|g\|_{p,\infty}^p\,\mathrm{d}\alpha - s\int_0^\lambda \alpha^{s-1}D_g(\lambda)\,\mathrm{d}\alpha\\ &= \frac{s}{s-p}\lambda^{s-p}\|g\|_{p,\infty}^p-\lambda^s D_g(\lambda),\end{aligned} \tag{8.14}$$

if $s < \infty$. In a similar fashion, $g^\lambda \in L_v$ with $v < p$. In fact,

$$
\begin{aligned}
\int_{\mathbb{R}^n} \left| g^{\lambda}(x) \right|^{\nu} \mathrm{d}x &= \nu \int_0^{\infty} \alpha^{\nu-1} D_{g^{\lambda}}(\alpha)\, \mathrm{d}\alpha \\
&= \nu \int_0^{\lambda} \alpha^{\nu-1} D_g(\lambda)\, \mathrm{d}\alpha + \nu \int_{\lambda}^{\infty} \alpha^{\nu-1} D_g(\alpha)\, \mathrm{d}\alpha \\
&\leq \lambda^{\nu} D_g(\lambda) + \nu \int_{\lambda}^{\infty} \alpha^{\nu-1-p} \|g\|_{p,\infty}^p \, \mathrm{d}\alpha \\
&\leq \lambda^{\nu-p} \|g\|_{p,\infty}^p + \frac{\nu}{p-\nu} \lambda^{t-p} \|g\|_{p,\infty}^p \\
&= \frac{p}{p-\nu} \lambda^{\nu-p} \|g\|_{p,\infty}^p .
\end{aligned}
\tag{8.15}
$$

Since $\frac{1}{p} = \frac{1}{r'} + \frac{1}{q}$, we have $1 < p < r'$. Using Hölder inequality and (8.14) we have

$$
\left| f * g_{\lambda}(x) \right| \leq \|f\|_r \|g_{\lambda}\|_{r'} \leq \|f\|_r \left(\frac{r'}{r'-p} \lambda^{r'-p} \|g\|_{p,\infty}^p \right)^{1/r'}, \tag{8.16}
$$

if $r' < \infty$. For the case $r' = \infty$ we have

$$
\left| f * g_{\lambda}(x) \right| \leq \lambda \|f\|_r . \tag{8.17}
$$

Take now λ in such a way that the right-hand side of (8.16) (or of (8.17) if $r' = \infty$) equals $C/2$. This is equivalent to

$$
\lambda = \left(C^{r'} 2^{-r'} \|f\|_r^{-r'} \frac{r'-p}{r'} \|g\|_{p,\infty}^{-p} \right)^{1/(r'-p)},
$$

if $r' < \infty$, or

$$
\lambda = \frac{C}{2\|f\|_1},
$$

if $r' = \infty$. In any case we have

$$
D_{f*g_{\lambda}}\left(\frac{C}{2} \right) = 0.
$$

By the Young inequality and (8.15) we have

$$
\left\| f * g^{\lambda} \right\|_r \leq \|f\|_r \left\| g^{\lambda} \right\|_1 \leq \|f\|_r \frac{p}{p-1} \lambda^{1-p} \|g\|_{p,\infty}^p . \tag{8.18}
$$

Using Chebyshev inequality and (8.18), for the chosen value of λ, we have

$$D_{f*g}(C) \leq D_{f*g^{\lambda}}\left(\frac{C}{2}\right) \leq \left(\frac{2\left\|f*g^{\lambda}\right\|_{r}}{C}\right)^{r}$$
$$\leq 2^{r}C^{-r}\left(\frac{p}{p-1}\right)^{r}\|f\|_{r}^{r}\lambda^{(1-p)r}\|g\|_{p,\infty}^{rp}$$
$$= 2^{q}\left(\frac{p}{p-1}\right)^{r}\left(\frac{r'-p}{r'}\right)^{(1-p)r/(r'-p)} C^{-q}\|f\|_{r}^{q}\|g\|_{p,\infty}^{q},$$

which is the desired result. □

We have a variant of the previous theorem, namely:

Theorem 8.21 (Young inequality in weak Lebesgue spaces). *Let $1 < p, q, r < \infty$ be such that*

$$1 + \frac{1}{q} = \frac{1}{r} + \frac{1}{p}.$$

Then there is a constant C such that, for every $f \in L_r(\mathbb{R}^n)$ and $g \in L_{(p,\infty)}(\mathbb{R}^n)$, we have

$$\|f*g\|_{L_q} \leq C\|g\|_{L_{(p,\infty)}}\|f\|_{L_r}.$$

8.3 Problems

8.22. Demonstrate the inequality given in (8.9).

8.23. Prove the Theorem 8.21.

8.24. Prove the following Theorem:

Theorem 8.25 (Hardy-Littlewood-Sobolev Inequality). *Let α be a real number with $0 < \alpha < n$ and let $1 < p, q < \infty$ be such that*

$$1 + \frac{1}{q} = \frac{1}{p} + \frac{\alpha}{n}.$$

Then there is a constant C such that for every $f \in L_p(\mathbb{R}^n)$ we have

$$\left\|f * |x|^{-\alpha}\right\|_{q} \leq C\|f\|_{p}.$$

Hint: Use Marcinkiewicz Interpolation Theorem in Lorentz spaces.

8.4 Notes and Bibliographic References

A version of the Riesz-Thorin theorem first appeared in Riesz [59]. The idea of the standard proof, as given in this chapter, appeared in Thorin [77] and was subsequently simplified in Tamarkin and Zygmund [76].

The Marcinkiewicz interpolation theorem was stated in Marcinkiewicz [50], but the complete proof appeared only in Zygmund [86]. The theorem was extended to the framework of Lorentz spaces by Hunt [33].

Chapter 9
Maximal Operator

Although Hardy and Littlewood invented the idea, it is only fair to give Zygmund and his students such as Calderón and Stein much credit for realizing its pervasive role in analysis.
ROBERT FEFFERMAN *(referring to the maximal function)*

Abstract In this chapter we study one of the most important operators in harmonic analysis, the maximal operator. In order to study this operator we need to have covering lemmas of Vitali type. After the covering lemmas we will study in some detail the maximal operator in Lebesgue spaces and show the Lebesgue differentiation theorem as well as a Theorem of Cotlar. We introduce and study the class of locally log-Hölder continuous functions in order to show the boundedness of the maximal operator in the space of variable exponent Lebesgue spaces whenever the exponent is in the aforementioned class. We end with a very short study of Muckenhoupt weights.

9.1 Locally Integrable Functions

Definition 9.1. A function $f : \mathbb{R}^n \longrightarrow \mathbb{C}$ is said to be *locally integrable* if

$$\int_K |f| \,\mathrm{d}\mu < \infty$$

for all compact sets $K \subset \mathbb{R}^n$. The space of all locally integrable function is denoted by $L_{1,\mathrm{loc}}(\mathbb{R}^n)$.

$\oslash$

Note that $L_1(\mathbb{R}^n) \subsetneq L_{1,\mathrm{loc}}(\mathbb{R}^n)$. In fact, if $f \in L_1(\mathbb{R}^n)$ and $K \subset \mathbb{R}^n$ is a compact set, then $\chi_K|f| \leq |f|$, from which we obtain

$$\int_{\mathbb{R}^n} \chi_K|f| \,\mathrm{d}\mu \leq \int_{\mathbb{R}} |f| \,\mathrm{d}\mu < \infty,$$

i.e., $f \in L_{1,\mathrm{loc}}(\mathbb{R}^n)$. On the other hand, any constant function $f(x) = c \in \mathbb{R}$ is locally integrable, but $f \notin L_1(\mathbb{R}^n)$.

R.E. Castillo, H. Rafeiro, *An Introductory Course in Lebesgue Spaces*, CMS Books in Mathematics, DOI 10.1007/978-3-319-30034-4_9

9.2 Vitali Covering Lemmas

In this section we study Vitali Covering Lemma (which is in the same spirit as the Borel-Lebesgue Covering Lemma or Lindelöf Covering Theorem), where from a collection of balls or cubes we can take a sub-collection of disjoint sets having some relation with the original collection. We start with a finite version, namely:

Lemma 9.2 (Finite Version of Vitali Covering Lemma). *Suppose that $\mathscr{B}$ is a finite collection of open balls in $\mathbb{R}^n$, i.e., $\mathscr{B} = \{B_1, B_2, \ldots, B_N\}$. Then, there exists a disjoint sub-collection $B_{j_1}, B_{j_2}, \ldots, B_{j_k}$ of $\mathscr{B}$ such that*

$$m\left(\bigcup_{\ell=1}^{N} B_\ell\right) \leq 3^n \sum_{i=1}^{k} m\left(B_{j_i}\right).$$

Proof. The argument given is constructive (based upon a greedy algorithm) and relies on the following simple observation: Suppose B and B' are a pair of balls that intersect, with the radius of B' being not greater than the radius of B. Then B' is engulfed by the ball $3B$ that is concentric with B but with 3 times its radius, as depicted in Fig. 9.1.

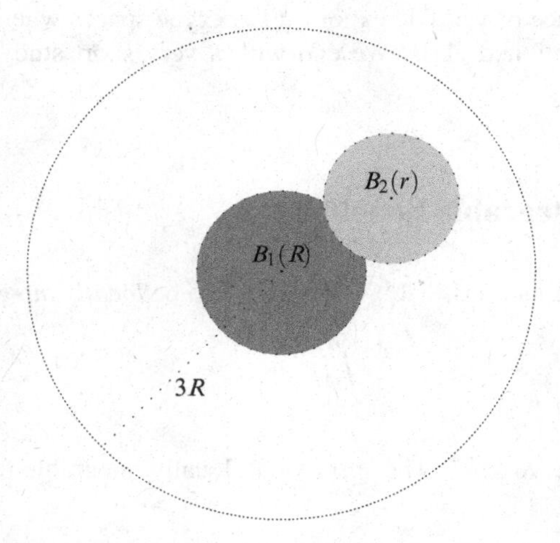

Fig. 9.1 Engulfing ball

As a first step, we pick a ball B_{j_1} in $\mathscr{B}$ with maximal (i.e., largest) radius, and then delete from $\mathscr{B}$ the ball B_{j_1} as well as any balls that intersect B_{j_1}. Thus all the balls that are deleted are contained in the ball $3B_{j_1}$ concentric with B_{j_1}, but with 3 times its radius.

The remaining balls form a new collection $\mathscr{B}'$, for which we repeat the procedure. We pick B_{j_2} and any ball that intersects B_{j_2}. Continuing this way, we find, after at most N steps, a collection of disjoint balls $B_{j_1}, B_{j_2}, \ldots, B_{j_k}$.

Finally, to prove that this disjoint collection of balls satisfies the inequality in the Lemma, we use the observation made at the beginning of the proof. Let $3B_{j_i}$ denote the ball concentric with B_{j_i}, but with 3 times its radius. Since any ball B in $\mathscr{B}$ must intersect a ball B_{j_i} and have equal or smaller radius than B_{j_i}, we must have $\cup_{B\cap B_{j_i}\neq\emptyset} B \subset 3B_{j_i}$, thus

$$m\left(\bigcup_{\ell=1}^{N} B_\ell\right) \le m\left(\bigcup_{i=1}^{k} \tilde{B}_{j_i}\right) \le \sum_{i=1}^{k} m\left(\tilde{B}_{j_i}\right) = 3^n \sum_{i=1}^{k} m\left(B_{j_i}\right).$$

□

The previous lemma can be generalized in several ways, for example we can take an arbitrary collection of balls, cubes, or even some arbitrary sets having some type of eccentricity. We use m_e for the Lebesgue exterior measure.

Lemma 9.3 (Vitali Covering Lemma). *Let $E \subset \mathbb{R}^n$, whose Lebesgue exterior measure satisfies $0 < m_e(E) < \infty$. Suppose that E is covered by a collection of cubes $\{Q\}$. Then there is a finite number of disjoint cubes $Q_1, \dots, Q_N$ in $\{Q\}$ and a constant $\gamma = \gamma(n) > 0$ such that $\sum_{i=1}^{n} m(Q_i) > \gamma m_e(E)$.*

Proof. We are going to index the cubes of the collection writing $Q = Q(t)$, where t is the length of the side of Q. Let $K_1 = \{Q\}$ and define

$$t_1^\star = \sup\{t; Q = Q(t) \in K_1\}.$$

If $t_1^\star = +\infty$, then K_1 contains a sequence of cubes Q with $m(Q) \to +\infty$. In this case, given $\beta > 0$, we simply choose a cube $Q \in K_1$ with $m(Q) \ge \beta m_e(E)$. If $t_1^\star < \infty$, the idea is to choose a "relatively" big cube: choose $Q_1 = Q_{t_1} \in K_1$ such that $t_1 > \frac{1}{2}t_1^\star$. Then, divide $K_1 = K_2 \cup K_2'$, where K_2 consists of the cubes in K_1 which are disjoint from Q_1, and K_2' of those which intercept Q_1. Denote by $Q_1^\star$ the concentric cube with Q_1 whose length of its side is $5t_1$. In this case, $m(Q_1^\star) = 5^n m(Q_1)$ and, since $2t_1 > t_1^\star$, every cube in K_2' is contained in $Q_1^\star$.

Starting with $j = 2$, continue the selection procedure for $j = 2, 3, \dots$ defining

$$t_j^\star = \sup\{t; Q(t) \in K_j\},$$

and choosing the cube $Q_j = Q_j(t_j) \in K_j$ with $t_j > \frac{1}{2}t_j^\star$. Now, write $K_j = K_{j+1} \cup K_{j+1}'$, where K_{j+1} consists of the cubes of K_j which are disjoint from Q_j. If K_{j+1} is empty, the process stops. We have that $t_j^\star \ge t_{j+1}^\star$ and furthermore, for every j, the cubes $Q_1, \dots, Q_j$ are mutually disjoint and disjoint from every cube in K_j. Moreover, every cube in K_{j+1}' is contained in the cube $Q_j^\star$ concentric with Q_j and whose side is $5t_j$. Notice that $m(Q_j^\star) = 5^n m(Q_j)$.

Consider the sequence $t_1^\star \ge t_2^\star \ge \dots$. If some K_{N+1} is empty (this is, $t_j^\star = 0$ for $j \ge N+1$) then, since

$$K_1 = K_2 \cup K_2' = \cdots = K_{N+1} \cup K_{N+1}' \cup \cdots \cup K_2'$$

and E is covered by cubes in K_1, it follows that E is covered by cubes in $K'_{N+1} \cup \cdots \cup K'_2$. Then, $E \subset \bigcup_{j=1}^{N} Q_j^\star$, so that

$$m_e(E) \leq \sum_{j=1}^{N} m(Q_j^\star) = 5^n \sum_{j=1}^{N} m(Q_j).$$

In this case, the Lemma is proven with $\beta = 5^{-n}$.

On the other hand, if no $t_j^\star$ is zero, then there is $\delta > 0$ such that $t_j^\star \geq \delta$ for every j, or $t_j^\star \to 0$. In the first case, $t_j \geq \frac{1}{2}\delta$ for every j and, hence, $\sum_{j=1}^{N} m(Q_j) \to +\infty$ when $N \to \infty$. Given any $\beta > 0$, the lemma follows, in this case, choosing N big enough.

Finally, if $t_j^\star \to 0$, it is easy to see that $K_1 \subset \cup_j Q_j^\star$, for otherwise, there would exist a cube $Q = Q(t)$ which would not intercept any Q_j. Since that cube would belong to K_j, t would satisfy $t \leq t_j^\star$ for every j and, hence, $t = 0$, which is a contradiction. Since E is covered by cubes in K_1, it follows that

$$m_e(E) \leq \sum_j m(Q_j^\star) = 5^n \sum_j m(Q_j).$$

Then, given β with $0 < \beta < 5^{-n}$, there is some N such that $\sum_{j=1}^{N} m(Q_j) \geq \beta m_e(E)$, which finishes the proof. □

Notice that the previous lemma does not assume that E is a measurable set. In the case where E is a measurable set, the proof of the previous lemma may be simplified. In fact, if E is measurable, we can suppose it is closed and bounded. We can also suppose that the cubes from the collection are open and, hence, it follows from the Heine-Borel Theorem that E can be covered by a finite number of cubes. Now we can follow the same argument used for balls.

We finally give the following version, the proof of which can be found in Jones [39].

Theorem 9.4. *Let $E \in \mathbb{R}^n$ be a bounded set. Let $\mathscr{B}$ be the collection of open balls centered in the points of E such that each point of E is the center of some ball in $\mathscr{B}$. Then there exists a sequence $B_1, B_2, \ldots$ of balls in $\mathscr{B}$ such that*

(a) The balls $B_1, B_2, \ldots$ are disjoint;
(b) $E \subset \bigcup_{\alpha \geq 1} 3B_\alpha$.

The set E is not covered by disjoint balls, nonetheless it is covered by concentric balls with radius 3 times bigger than the original balls.

9.3 Hardy-Littlewood Maximal Operator

Before introducing the maximal operator, we give the integral average operator which is related to the maximal operator.

Definition 9.5 (Integral Average Operator). For $f \in L^1(\mathbb{R}^n)$ and $r > 0$ we define de *integral average operator* $\mathrm{A}_r(f) : \mathbb{R}^n \longrightarrow \mathbb{R}_+$ as

$$\mathrm{A}_r(f)(x) := \frac{1}{m(B(x,r))} \int_{B(x,r)} |f(y)|\,\mathrm{d}y. \tag{9.1}$$

⊘

Using the integral average operator we can introduce the maximal operator in the following way.

Definition 9.6. Given a function $f \in L_{1,\mathrm{loc}}(\mathbb{R}^n)$, we define the *Hardy-Littlewood maximal function* for $x \in \mathbb{R}^n$ as

$$Mf(x) = M_B f(x) := \sup_{r>0} \mathrm{A}_r f(x) = \sup_{r>0} \frac{1}{m(B(x,r))} \int_{B(x,r)} |f(y)|\,\mathrm{d}y. \tag{9.2}$$

⊘

The adopted definition of maximal function is based in centered balls in x, but it is possible to define other Hardy-Littlewood maximal functions, for example $M_\square$

$$M_\square f(x) := \sup_{r>0} \frac{1}{(2r)^n} \int_{Q_r} |f(x-y)|\,\mathrm{d}y,$$

where $Q_r := [-r,r]^n$. The other possible definition is

$$\widetilde{M}_B f(x) := \sup_{B \ni x} \frac{1}{m(B)} \int_B |f(y)|\,\mathrm{d}y.$$

where the supremum is given over all balls $B \subset \mathbb{R}^n$ containing the point x. It is important to notice that all the above definitions are equivalent in the following sense:

$$Mf(x) \asymp M_\square f(x) \asymp \widetilde{M}_B f(x),$$

see Problem 9.29.

For $f \in L_{q,\mathrm{loc}}(\mathbb{R}^n)$, the q-th maximal operator is defined as

$$M_q f(x) := \sup_{Q \ni x} \left(\frac{1}{m(Q)} \int_Q |f(y)|^q\,\mathrm{d}y \right)^{1/q}. \tag{9.3}$$

From the Definition 9.6 the following properties of the operator M are almost immediate

(a) $0 \le Mf(x) \le \infty$;
(b) $M(f+g)(x) \le Mf(x) + Mg(x)$;
(c) $M(\lambda f)(x) = |\lambda| Mf(x)$.

We now show that the maximal function is measurable, namely:

Theorem 9.7 *The function Mf is lower semi-continuous and therefore measurable.*

Proof. To show that Mf is lower semi-continuous, we should verify that for each $\lambda > 0$, the set $\left\{x \in \mathbb{R}^n : Mf(x) > \lambda\right\}$ is open, and for that we show that the set $\left\{x \in \mathbb{R}^n : Mf(x) \le \lambda\right\}$ is closed. Let us fix $\lambda > 0$ and suppose that $x \in \overline{\left\{x \in \mathbb{R}^n : Mf(x) \le \lambda\right\}}$, then there exists a sequence $\{x_k\}_{k\in\mathbb{N}}$ in the set $\left\{x \in \mathbb{R}^n : Mf(x) \le \lambda\right\}$ such that $x_k \to x$ in $\mathbb{R}^n$ when $k \to \infty$. We first observe that, since $x_k \to x$, we will get $\lim_{k\to\infty} B(x_k, r)\Delta B(x, r) = \emptyset$ for all $r > 0$, where $A\Delta B$ is the symmetric difference, i.e., $A\Delta B := A\backslash B \cup B\backslash A$. Let $A_k = B(x_k, r)\Delta B(x, r)$ and $f_k = f\chi_{A_k}$, then we have that $|f_k| \le |f|$ and $\lim_{k\to\infty} f_k = 0$ almost everywhere. By the dominated convergence theorem, we have

$$\lim_{k\to\infty} \int_{\mathbb{R}^n} |f_k|\, dy = 0. \tag{9.4}$$

But

$$B(x, r) \subseteq B(x_k, r)\Delta B(x, r) \cup B(x_k, r)$$

and

$$m\Big(B(x_k, r)\Big) = m\Big(B(x, r)\Big),$$

therefore

$$\frac{1}{m\Big(B(x,r)\Big)} \int_{B(x,r)} |f(y)|\, dy \le$$

$$\frac{1}{m\Big(B(x,r)\Big)} \int_{B(x_k,r)\Delta B(x,r)} |f(y)|\, dy + \frac{1}{m\Big(B(x_k,r)\Big)} \int_{B(x_k,r)} |f(y)|\, dy.$$

Then, by (9.4) we have that

$$Mf(x) \le \lambda,$$

therefore $x \in \left\{x \in \mathbb{R}^n : Mf(x) \le \lambda\right\}$ and this finishes the proof. □

It is easy to show that taking the function $f(t) = |t|^\alpha$ with $\alpha > 0$ we get $Mf(x) = \infty$ for all $x \in \mathbb{R}^n$. Our next objective is to calculate Mf when $f \in L_p(\mathbb{R}^n)$. For $f \in L_\infty(\mathbb{R}^n)$ we see that

$$Mf(x) \le \|f\|_\infty$$

for each $x \in \mathbb{R}^n$, i.e., $Mf \in L_\infty(\mathbb{R}^n)$.

If M was $(L_1 \to L_1)$-bounded, we could apply the Riesz-Thorin Theorem 8.6 to get the $(L_p \to L_p)$-boundedness of M, unfortunately this is not true, since for $f \in L_1(\mathbb{R}^n)$ we have

$$Mf(x) \geq \frac{1}{m\Big(B(x,2|x|)\Big)} \int_{B(x,2|x|)} |f(y)|\,dy$$
$$\geq \frac{const}{|x|^n} \int_{B(0,a)} |f(y)|\,dy,$$

therefore for very large $|x|$, $Mf(x) \geq C|x|^{-n}\|f\|_1$, which implies that Mf cannot belong to $L_1(\mathbb{R}^n)$.

Let us now see that if $Mf \in L_1(\mathbb{R}^n)$ then $f = 0$. Let $a > 0$ be arbitrary and $|x| > a$, it results that

$$Mf(x) \geq \frac{1}{m\Big(B(x,2|x|)\Big)} \int_{B(x,2|x|)} |f(y)|\,dy$$
$$\geq \frac{1}{m\Big(B(0,2|x|)\Big)} \int_{B(0,a)} |f(y)|\,dy$$
$$= \frac{const}{|x|^n} \int_{B(0,a)} |f(y)|\,dy,$$

since $|x|^{-n}$ is not integrable for $|x| > a$, which yields that $\int_{B(0,a)} |f(y)|\,dy = 0$. From the arbitrariness of a, we conclude that $f = 0$. Let us give a particular example for the one-dimensional case.

Example 9.8. Take

$$f(x) = \frac{1}{x\log^2 x}\chi_{\left(0,1/2\right)}(x),$$

and let us use $r = x$. Note that $f \in L_1(\mathbb{R})$ and

$$Mf(x) \geq \frac{1}{2x}\int_0^{2x} |f(y)|\,dy$$
$$\geq \frac{1}{2x}\int_0^{x} \frac{dy}{y\log^2 y}$$
$$= \frac{-1}{2x\log x},$$

and since $\frac{-1}{2x\log x}$ is not integrable in the neighborhood of $x = 0$, we obtain that $Mf \notin L_1(\mathbb{R})$.

⊘

We now show that the range of the maximal function of L_1 functions is in weak Lebesgue space.

Theorem 9.9. *If $f \in L_1(\mathbb{R}^n)$, then*

$$m\Big(\{x \in \mathbb{R}^n : Mf(x) > \lambda\}\Big) \le \frac{3^n}{\lambda} \int_{\mathbb{R}^n} |f(y)|\,\mathrm{d}m,$$

i.e., $Mf \in L_{(1,\infty)}(\mathbb{R}^n)$.

Proof. For each $\lambda \in \mathbb{R}^+$, we define $A_\lambda = \Big\{x : Mf(x) > \lambda\Big\}$. We obtain that for each $x \in A_\lambda$, there exists $0 < r < \infty$ (which depends on x) such that

$$\frac{1}{m\Big(B(x,r)\Big)} \int_{B(x,r)} |f(y)|\,\mathrm{d}m > \lambda.$$

Note that we can write this last expression as

$$m\Big(B(x,r)\Big) < \frac{1}{\lambda} \int_{B(x,r)} |f(y)|\,\mathrm{d}m. \tag{9.5}$$

Suppose that $A_\lambda \neq \emptyset$, on the contrary the result is trivial. Note that to use Theorem 9.4 we should have that A_λ must be bounded, but a priori this is not clear. Nonetheless, we can consider the set $A_\lambda \cap B(0,k)$ (fixed k) instead of the set A_λ. Now, let $\mathscr{B}$ be a collection of open balls B with center in $A_\lambda \cap B(0,k)$ such that they satisfy (9.5). Observe that under this situation the hypothesis of Theorem 9.4 are satisfied, therefore, if $A_\lambda \cap B(0,k) \neq \emptyset$, there exists a sequence of balls $B_1, B_2, \ldots$ of $\mathscr{B}$ such that

(1) The balls $B_1, B_2, \ldots$ are disjoint;
(2) $A_\lambda \cap B(0,k) \subset \bigcup_{\alpha \ge 1} 3B_\alpha$.

Since $m\Big(3B_\alpha\Big) = 3^n m\Big(B_\alpha\Big)$, therefore by the inequality (9.5) we have

$$\begin{aligned} m(A_\kappa \cap B(0,k)) &\le \sum_{\alpha \ge 1} m(3B_\alpha) \\ &= \sum_{\alpha > 1} 3^n m(B_\alpha) \\ &< \sum_{\alpha \ge 1} 3^n \lambda^{-1} \int_{B_\alpha} |f(y)|\,\mathrm{d}m \end{aligned}$$

$$= 3^n \lambda^{-1} \int\limits_{\cup_{\alpha \geq 1}} |f(y)| \, dm$$

$$\leq 3^n \lambda^{-1} \int\limits_{\mathbb{R}^n} |f(y)| \, dm$$

and now letting $k \to \infty$, we get

$$m(A_\lambda) \leq \frac{3^n}{\lambda} \int\limits_{\mathbb{R}^n} |f(y)| \, dm,$$

i.e.

$$m\Big(\{x \in \mathbb{R}^n : Mf(x) > \lambda\}\Big) \leq \frac{3^n}{\lambda} \int\limits_{\mathbb{R}^n} |f(y)| \, dm.$$

and from this we get $Mf \in L_{(1,\infty)}(\mathbb{R}^n)$. □

Using the previous theorem, we show the so-called Lebesgue differentiation theorem, which is a generalization of the fundamental theorem of calculus.

Theorem 9.10 (Lebesgue Differentiation Theorem). *Let $f \in L_1(\mathbb{R}^n)$, then*

$$\lim_{r \to 0} \frac{1}{m\Big(B(x,r)\Big)} \int\limits_{B(x,r)} |f(y)| \, dm = f(x), \tag{9.6}$$

m-almost everywhere.

Proof. Let $f \in L_1(\mathbb{R}^n)$, then by Luzin's theorem, there exists a continuous function g such that $g \in L_1(\mathbb{R}^n)$ and for an arbitrary fixed $\varepsilon > 0$ we have

$$\int\limits_{\mathbb{R}^n} |f(x) - g(x)| \, dm < \varepsilon.$$

We have, for appropriate small $r > 0$, that

$$\left| \frac{1}{m\Big(B(x,r)\Big)} \int\limits_{B(x,r)} |g(y)| \, dm - g(x) \right| \leq \frac{1}{m\Big(B(x,r)\Big)} \int\limits_{B(x,r)} |g(y) - g(x)| \, dm < \varepsilon,$$

i.e.

$$\lim_{r \to 0} \frac{1}{m\Big(B(x,r)\Big)} \int\limits_{B(x,r)} |g(y)| \, dm = g(x).$$

On the other hand, observe that

$$\limsup_{r\to 0}\left|\frac{1}{m\Big(B(x,r)\Big)}\int_{B(x,r)} |f(y)|\,\mathrm{d}m - f(x)\right| =$$

$$\limsup_{r\to 0}\left|\frac{1}{m\Big(B(x,r)\Big)}\int_{B(x,r)} f(y)-g(y)\,\mathrm{d}m + \right.$$

$$\left.\left(\frac{1}{m\Big(B(x,r)\Big)}\int_{B(x,r)} g(y)\,\mathrm{d}m - g(x)\right) + (g(x)-f(x))\right|$$

$$\leq M(f-g)(x) + 0 + |g(x)-f(x)|.$$

Now, let us consider the following sets

$$E_\lambda = \left\{ x : \limsup_{r\to 0}\left|\frac{1}{m\Big(B(x,r)\Big)}\int_{B(x,r)} |f(y)|\,\mathrm{d}m - f(x)\right| > \lambda \right\}$$

$$F_\lambda = \{x : M(f-g)(x) > \lambda\}$$
$$H_\lambda = \{x : |f(x)-g(x)| > \lambda\}.$$

Note that

$$E_\lambda \subset F_{\lambda/2} \cup H_{\lambda/2}$$

and

$$\frac{\lambda}{2} m\Big(H_{\lambda/2}\Big) \leq \int_{\mathbb{R}^n} |g(x)-f(x)|\,\mathrm{d}m < \varepsilon,$$

therefore

$$m\Big(H_{\lambda/2}\Big) < \frac{2\varepsilon}{\lambda}.$$

Moreover, by Theorem 9.9 we get

$$m\Big(F_{\lambda/2}\Big) \leq \frac{2c}{\lambda}\int_{\mathbb{R}^n} |f(x)-g(x)|\,\mathrm{d}m < \frac{2c\varepsilon}{\lambda}$$

giving

$$m\Big(E_\lambda\Big) \leq m\Big(F_{\lambda/2}\Big) + m\Big(H_{\lambda/2}\Big) \leq 2(1+c)\frac{\varepsilon}{\lambda}$$

which tends to 0 whenever $\varepsilon \to 0$. Therefore

$$m\Big(E_\lambda\Big) = 0,$$

and we proved that

$$\limsup_{r\to 0}\left|\frac{1}{m\Big(B(x,r)\Big)}\int_{B(x,r)}|f(y)|\,dm - f(x)\right| = 0$$

m-almost everywhere. Since

$$0 \le \liminf_{r\to 0}\left|\frac{1}{m\Big(B(x,r)\Big)}\int_{B(x,r)}|f(y)|\,dm - f(x)\right|$$
$$\le \limsup_{r\to 0}\left|\frac{1}{m\Big(B(x,r)\Big)}\int_{B(x,r)}|f(y)|\,dm - f(x)\right| = 0$$

m-almost everywhere, therefore

$$\lim_{r\to 0}\left|\frac{1}{m\Big(B(x,r)\Big)}\int_{B(x,r)}|f(y)|\,dm - f(x)\right| = 0$$

m-almost everywhere, proving (9.6). □

It is noteworthy to mention that we can use other types of sets instead of balls in Theorem 9.10, the requirement is that those sets should have some type of eccentricity to guarantee a relation of type (9.6), see Stein [71, p. 10].

Theorem 9.11. *Let $f \in L_p(\mathbb{R}^n)$, $1 < p \le \infty$, then $Mf \in L_p(\mathbb{R}^n)$. Moreover, there exists a constant $C = C(p,n)$ such that*

$$\|Mf\|_{L_p} \le C(p,n)\|f\|_p.$$

Proof. For $p = \infty$, we observe that $Mf \in L_\infty(\mathbb{R}^n)$ and

$$\|Mf\|_{L_\infty} \le \|f\|_\infty.$$

Suppose that $1 < p < \infty$, then for each $\lambda \in \mathbb{R}^n$ let us define

$$f_\lambda(x) = \begin{cases} f(x) & \text{if } |f(x)| \ge \lambda/2 \\ 0 & \text{if } |f(x)| < \lambda/2 \end{cases}$$

therefore for all $x \in \mathbb{R}^n$

$$|f(x)| \leq |f_\lambda(x)| + \lambda/2,$$

from which we get

$$Mf(x) \leq Mf_\lambda(x) + \lambda/2,$$

hence

$$\Big\{x : Mf(x) > \lambda\Big\} \subset \Big\{x : Mf_\lambda(x) > \lambda/2\Big\}.$$

From this last relation we obtain

$$m\Big(\Big\{x : Mf(x) > \lambda\Big\}\Big) \leq m\Big(\Big\{x : Mf_\lambda(x) > \lambda/2\Big\}\Big),$$

By Theorem 9.9 we get

$$\begin{aligned} m\Big(\Big\{x : Mf(x) > \lambda\Big\}\Big) &\leq \frac{2.3^n}{\lambda} \int_{\mathbb{R}^n} |f_\lambda(x)| \, \mathrm{d}m \\ &= \frac{2.3^n}{\lambda} \int_{\{x:|f(x)|\geq\lambda/2\}} |f(x)| \, \mathrm{d}m \end{aligned} \tag{9.7}$$

Now using Corollary 3.54 and (9.7), we obtain

$$\begin{aligned} \int_{\mathbb{R}^n} |Mf(x)|^p \, \mathrm{d}m &= p \int_0^\infty \lambda^{p-1} m\Big(\Big\{x : Mf(x) > \lambda\Big\}\Big) \, \mathrm{d}\lambda \\ &\leq 2.3^n p \int_0^\infty \lambda^{p-2} \left(\int_{\left\{x: f(x)\geq\lambda/2\right\}} |f(x)| \, \mathrm{d}m \right) \mathrm{d}\lambda \\ &= 2.3^n p \int_{\mathbb{R}^n} |f(x)| \left(\int_0^{2|f(x)|} \lambda^{p-2} d\lambda \right) \mathrm{d}m \\ &= \frac{2^{p-1} 2.3^n p}{p-1} \int_{\mathbb{R}^n} |f(x)| |f(x)|^{p-1} \, \mathrm{d}m \\ &= \frac{2^{p-1} 2.3^n p}{p-1} \int_{\mathbb{R}^n} |f(x)|^p \, \mathrm{d}m, \end{aligned}$$

which ends the proof. □

For a more *elegant* proof of Theorem 9.11 see Problem 9.42.

Theorem 9.12 (Cotlar Theorem). *Suppose that S and T are sublinear operators and that the operator T is majorized by S in the following sense: if $A(x,r)$ is the annulus $A(x,r) = \left\{y \in \mathbb{R}^n : r \leq |x-y| \leq 2r\right\}$, for each $x \in \mathbb{R}^n$ and $f \in L_1(\mathbb{R}^n)$ there exists an $\tilde{r} = r(x)$ with $0 < \tilde{r} < \infty$, such that $Tf(x) \leq \inf_{y \in A(x,\tilde{r})} |Sf(y)|$. Then if S is of weak type (p,p) for some $p > 0$, T is also of weak type (p,p).*

Proof. Let $0 < q < p$, then

$$|Tf(x)|^q \leq \inf_{y \in A(x,\tilde{r})} |Sf(y)|^q,$$

therefore

$$\begin{aligned} |Tf(x)|^q &\leq \frac{1}{m\left(A(x,\tilde{r})\right)} \int_{A(x,\tilde{r})} |Sf(y)|^q \, dm \\ &\leq \frac{m\left(B(x,4\tilde{r})\right)}{m\left(A(x,\tilde{r})\right) m\left(B(x,4\tilde{r})\right)} \int_{B(x,4\tilde{r})} |Sf(y)|^q \, dm. \end{aligned} \tag{9.8}$$

Then by (9.8) we obtain that $|Tf(x)|^q \leq CM\left(|Sf|^q(x)\right)$, with C independent of x and f.

Note that in virtue of Theorem 9.9, the operator M is of weak type $(1,1)$ and by Theorem 9.11 we get that $M\left(|Sf|^q\right) \in L_\infty(\mathbb{R}^n)$ and moreover

$$\|M\left(|Sf|^q\right)\|_{L_\infty} \leq C\||Sf|^q\|_{L_\infty}.$$

To finish, by Theorem 8.15 we obtain

$$\begin{aligned} m\left(\left\{x : M(|Sf|^q(x)) > \lambda^q\right\}\right) &\leq \frac{C}{\lambda^q} \int_{[C\lambda^q,\infty)} m\left(\left\{x : |Sf(x)| > t^{1/q}\right\}\right) dt \\ &\leq \frac{C}{\lambda^q} \int_{[C\lambda^q,\infty)} \|f\|_{L_p}^p t^{-p/q} \, dt \\ &= C\lambda^{-p} \|f\|_{L_p}^p \end{aligned}$$

therefore

$$m\left(\left\{x : |Tf(x)| > \lambda\right\}\right) \leq C\left(\frac{\|f\|_{L_p}}{\lambda}\right)^p.$$

□

9.4 Maximal Operator in Nonstandard Lebesgue Spaces

In order to get the boundedness of the maximal operator in variable exponent Lebesgue spaces it is necessary to impose some condition on the exponent. For simplicity, we will only deal with the case of bounded subsets in $\mathbb{R}^n$.

Lemma 9.13. *Let $\Omega \subset \mathbb{R}^n$ be an open set and let $p : \Omega \longrightarrow [1,\infty)$ be an uniformly continuous function. Then the following conditions are equivalent:*

(a) There exists a constant C_0 such that for all $x,y \in \Omega$, $|x-y| < 1/2$, we have

$$|p(x)-p(y)| \leq \frac{C_0}{-\log|x-y|}. \tag{9.9}$$

(b) There exists a constant C_1 such that for all open ball $B \subset \mathbb{R}^n$ with $m(\Omega \cap B) > 0$, we get

$$m(B)^{p_-(B)-p_+(B)} \leq C_1.$$

Proof. Let us suppose that *(b)* is valid. Let $x,y \in \Omega$ such that $|x-y| < 1/2$ and let $B \subset \mathbb{R}^n$ be an open ball such that $x,y \in B$ and $\operatorname{diam} B \leq 2|x-y| < 1$. Since Ω is open, then we have that $m(\Omega \cap B) > 0$, therefore

$$m(B)^{p_-(B)-p_+(B)} \leq C_1.$$

Since $m(B) \leq \operatorname{diam}(B)^n \leq (2|x-y|)^n$, we have

$$\left((2|x-y|)^n\right)^{-|p(x)-p(y)|} \leq m(B)^{p_-(B)-p_+(B)} \leq C_1,$$

and

$$|x-y|^{-|p(x)-p(y)|} \leq C_1^{1/n} 2^{|p(x)-p(y)|} \leq C_1^{1/n} 2^{p_+-p_-},$$

and taking the logarithm, we get

$$|p(x)-p(y)| \leq \frac{\log\left(C_1^{1/n} 2^{p_+-p_-}\right)}{-\log|x-y|}.$$

which shows (a).

Let us assume *(a)* valid. Let $B \subset \mathbb{R}^n$ be an open ball with $m(\Omega \cap B) > 0$, which implies $1 \leq p_-(\Omega) \leq p_-(B) \leq p_+(B) \leq p_+(\Omega) < \infty$. If $\operatorname{diam}(B) \geq 1/2$ we get

$$\begin{aligned} m(B)^{p_-(B)-p_+(B)} &= \left(m(B(0,1))\left(\operatorname{diam}(B)/2\right)^n\right)^{p_-(B)-p_+(B)} \\ &\leq \left(m(B(0,1))4^{-n}\right)^{p_-(\Omega)-p_+(\Omega)}, \end{aligned}$$

which permits us to restrict to the case of $\operatorname{diam}(B) < 1/2$. Let us choose $x_0, x_\infty \in B \cap \Omega$ such that $0 \leq 1/2\left(p_+(B)-p_-(B)\right) \leq p(x_0)-p(x_\infty)$. Since we have that $\operatorname{diam}(B) < 1/2$ implies that $|x_0-x_\infty| < 1/2$, and by hypothesis in p

$$|p(x_0) - p(x_\infty)| \leq \frac{C_0}{-\log|x_0 - x_\infty|},$$

which implies that

$$\exp(C_0) \geq |x_0 - x_\infty|^{-|p(x_0)-p(x_\infty)|} \geq |x_0 - x_\infty|^{\frac{1}{2}(p_-(B)-p_+(B))}.$$

Since $2^n m(B) \geq |x_0 - x_\infty|^n m(B(0,1))$, we obtain

$$\exp(2C_0) \geq |x_0 - x_\infty|^{p_-(B)-p_+(B)} \geq \left(2\left(\frac{m(B)}{m(B(0,1))}\right)^{\frac{1}{n}}\right)^{p_-(B)-p_+(B)}$$

which entails

$$\begin{aligned}(2m(B))^{p_-(B)-p_+(B)} &\leq \exp(2nC_0)m(B(0,1))^{p_-(B)-p_+(B)} \\ &\leq \exp(2nC_0)\max\left\{1, m(B(0,1))^{p_-(\Omega)-p_+(\Omega)}\right\},\end{aligned}$$

which shows that (a) implies (b). □

Definition 9.14. When an exponent $p : \Omega \longrightarrow [1,\infty)$ satisfies the condition (9.9), we say that the exponent p is log-*Hölder continuous* and we write that $p \in LH(\Omega)$. ⊘

Corollary 9.15. *Let $\Omega \subset \mathbb{R}^n$ be an open and bounded set and let $p : \Omega \longrightarrow [1,\infty)$ be an uniformly Hölder continuous function with power $\alpha > 0$, i.e.*

$$|p(x) - p(y)| \leq H|x - y|^\alpha$$

for all $|x - y| < 1/2$. Then $p \in LH(\Omega)$.

We now obtain a pointwise inequality which is very useful in the framework of variable exponent spaces.

Theorem 9.16. *Let $\Omega \subset \mathbb{R}^n$ be an open set and $p \in LH(\Omega)$. Then there exists a constant $C = C(p)$ such that for all functions $f \in L^{p(\cdot)}(\Omega)$ with $\|f\|_{L^{p(\cdot)}(\Omega)} \leq 1$ we have*

$$|\mathrm{A}_r(f)(x)|^{p(x)} \leq C(p)\left(\mathrm{A}_r\left(|f(\cdot)|^{p(\cdot)}\right)(x) + 1\right)$$

for all $r > 0$.

Proof. We will prove the theorem by cases: when $r \geq 1/2$ and when $0 < r < 1/2$.
Let $r \geq 1/2$. Then

$$|\mathrm{A}_r(f)(x)|^{p(x)} = \left(\frac{1}{m(B_r(x))}\int_{B_r(x)\cap\Omega} |f(y)|\,\mathrm{d}y\right)^{p(x)}$$

$$\leq \left(\frac{1}{m(B_r(x))} \int\limits_{B_r(x)\cap\Omega} |f(y)|^{p(y)} \, \mathrm{d}y + 1 \right)^{p(x)}$$
$$\leq \left(\frac{r^{-n}}{m(B(0,1))} \rho_{p(\cdot)}(f) + 1 \right)^{p(x)} \leq \left(\frac{2^n}{m(B(0,1))} + 1 \right)^{p_+}$$

where the first inequality is a consequence of

$$f = f\chi_{\{|f|\leqslant 1\}} + f\chi_{\{|f|>1\}}. \tag{9.10}$$

Let now $0 < r < 1/2$. We have

$$|\mathrm{A}_r(f)(x)|^{p(x)} = \left(\frac{1}{m(B_r(x))} \int\limits_{B_r(x)\cap\Omega} |f(y)| \, \mathrm{d}y \right)^{p(x)}$$
$$\overset{(d_1)}{\leq} \left(\frac{1}{m(B_r(x))} \int\limits_{B_r(x)\cap\Omega} |f(y)|^{p_-(B_r(x))} \, \mathrm{d}y \right)^{\frac{p(x)}{p_-(B_r(x))}}$$
$$\overset{(d_2)}{\leq} \left(\frac{1}{m(B_r(x))} \int\limits_{B_r(x)\cap\Omega} |f(y)|^{p(y)} \, \mathrm{d}y + 1 \right)^{\frac{p(x)}{p_-(B_r(x))}}$$
$$\leq m(B_r(x))^{-\frac{p(x)}{p_-(B_r(x))}} 2^{\frac{p_+(B_r(x))}{p_-(B_r(x))}} \left(\frac{1}{2}\rho_{p(\cdot)}(f) + \frac{1}{2} m(B_r(x)) \right)^{\frac{p(x)}{p_-(B_r(x))}}$$

where the inequality (d_1) is a consequence of the Jensen integral inequality and for (d_2) we used (9.10). Since $\rho_{p(\cdot)}(f) \leq 1$ and $0 < r < \frac{1}{2}$, we have the inequality

$$\frac{1}{2} \int\limits_{B_r(x)\cap\Omega} |f(y)|^{p(y)} \, \mathrm{d}y + \frac{1}{2} m(B_r(x)) \leq \frac{1}{2}\rho_{p(\cdot)}(f) + \frac{1}{2}(2r)^n < 1,$$

which implies

$$|\mathrm{A}_r(f)(x)|^{p(x)} \leq m(B_r(x))^{-\frac{p(x)}{p_-(B_r(x))}} 2^{\frac{p_+(B_r(x))}{p_-(B_r(x))}} \left(\frac{1}{2}\rho_{p(\cdot)}(f) + \frac{1}{2} m(B_r(x)) \right)$$
$$\leq m(B_r(x))^{\frac{p_-(B_r(x)) - p_+(B_r(x))}{p_+(B_r(x))}} 2^{\frac{p_-(B_r(x))}{p_+(B_r(x))} - 1} \left(\mathrm{A}_r(|f(\cdot)|^{p(\cdot)})(x) + 1 \right)$$
$$\leq C(p) \left(\mathrm{A}_r(|f(\cdot)|^{p(\cdot)})(x) + 1 \right)$$

where the last inequality is a consequence of Lemma 9.13. □

A consequence of Theorem 9.16 is the next Corollary.

Corollary 9.17. *Let $\Omega \subset \mathbb{R}^n$ be an open set and $p \in LH(\Omega)$. Then there exists a constant $C = C(p)$ such that for all functions $f \in L^{p(\cdot)}(\Omega)$ with $\|f\|_{L^{p(\cdot)}(\Omega)} \leq 1$, we get*

$$|Mf(x)|^{p(x)} \leq C(p)\left(M\left(|f(\cdot)|^{p(\cdot)}\right)(x)+1\right).$$

The Corollary 9.17 is the key to get boundedness for the maximal operator in the variable exponent Lebesgue space when Ω is a bounded set.

Theorem 9.18. *Let Ω be an open and bounded set and $p : \Omega \longrightarrow [1,\infty)$ measurable.*

(a) If $f \in L^{p(\cdot)}(\Omega)$ with $1 \leq p(x) \leq p_+ < \infty$ in Ω, then Mf is finite almost everywhere in $\mathbb{R}^n$;
(b) Let $p \in LH(\Omega)$ and $1 < p_- \leq p(x) \leq p_+ < \infty$ in Ω. Then there exists a constant $C(\Omega, p) > 0$ such that for all $f \in L^{p(\cdot)}(\Omega)$ we get

$$\|Mf\|_{L^{p(\cdot)}(\Omega)} \leq C(\Omega,p)\|f\|_{L^{p(\cdot)}(\Omega)}.$$

Proof. Let us prove by cases:

(a) Since $L^{p(\cdot)}(\Omega) \hookrightarrow L^1(\Omega)$ due to the fact that Ω is a bound set, the result follows from the Proposition 9.11 with $p = 1$.
(b) Taking $q(x) := p(x)/p_-$, we have that $1 \leq q(x) \leq p(x) \leq p_+ < \infty$. Since $L^{p(\cdot)}(\Omega) \hookrightarrow L^{q(\cdot)}(\Omega)$ because Ω is bounded, we have that there exists a constant $C_e > 0$ such that $\|f\|_{L^{q(\cdot)}}(\Omega) \leq C_e\|f\|_{L^{p(\cdot)}}(\Omega)$ for all $f \in L^{p(\cdot)}(\Omega)$. Now, let $f \in L^{p(\cdot)}(\Omega)$ such that $\|f\|_{L^{p(\cdot)}(\Omega)} \leq 1/C_e$, which implies that $\|f\|_{L^{q(\cdot)}(\Omega)} \leq 1$. Let us show that $\rho_{q(\cdot)}(Mf)$ is bounded independently of f. Since the exponent $q \in LH(\Omega)$ we can use Corollary 9.17 and we obtain

$$\begin{aligned}\rho_{q(\cdot)}(Mf) &= \left\|M(f)^{q(\cdot)}\right\|_{L^{p_-}(\Omega)}^{p_-}\\ &\leq \left\|C(p)\left(M(|f(\cdot)^{q(\cdot)}|)+1\right)\right\|_{L^{p_-}(\Omega)}^{p_-}\\ &\leq (C(p))^{p_-}\left(\left\|M(|f(\cdot)^{q(\cdot)}|)\right\|_{L^{p_-}(\Omega)} + \|1\|_{L^{p_-}(\Omega)}\right)^{p_-}.\end{aligned}$$

Using the Proposition 9.11 with constant exponent $p_- > 1$ we obtain

$$\begin{aligned}\rho_{q(\cdot)}(Mf) &\leq C(p)^{p_-}\left(C(p_-)\left\||f(\cdot)^{q(\cdot)}|\right\|_{L^{p_-}(\Omega)} + \|1\|_{L^{p_-}(\Omega)}\right)^{p_-}\\ &= C(p)^{p_-}\left(C(p_-)[\rho_{p(\cdot)}(f)]^{\frac{1}{p_-}} + \|1\|_{L^{p_-}(\Omega)}\right)^{p_-}\\ &\leq C(\Omega,p).\end{aligned}$$

The previous estimate shows that $\rho_{q(\cdot)}(Mf)$ is bounded for all functions f with $\|f\|_{L^{p(\cdot)}(\Omega)} \leq 1/C_e$, therefore the norm $\|Mf\|_{L^{p(\cdot)}(\Omega)}$ is also bounded for these functions. Since $M(\lambda f) = |\lambda|M(f)$ and $\|\lambda f\| = |\lambda|\|f\|$ we obtain

$$\|M(f)\|_{L^{p(\cdot)}(\Omega)} = C_e \|f\|_{L^{p(\cdot)}(\Omega)} \left\| M\left(\frac{f}{C_e\|f\|_{L^{p(\cdot)}(\Omega)}}\right)\right\|_{L^{p(\cdot)}(\Omega)}$$
$$\lesssim C_e \|f\|_{L^{p(\cdot)}(\Omega)}.$$

□

We now introduce the notion of Hardy-Littlewood fractional maximal operator.

Definition 9.19. Given a function $f \in L^1_{\text{loc}}(\mathbb{R}^n)$, we define the *Hardy-Littlewood fractional maximal operator* for any $x \in \mathbb{R}^n$ as

$$\mathscr{M}_\alpha f(x) := \sup_{x\in B} \frac{1}{m(B)^{1-\frac{\alpha}{n}}} \int_B |f(y)|\, dy \tag{9.11}$$

where the supremum is taken over all balls that contain the point x. ⊘

The Hardy-Littlewood fractional maximal operator is a generalization of the maximal operators, since when $\alpha = 0$, we have that $\mathscr{M}_0 = M$, where M is the Hardy-Littlewood maximal operator (9.2). The classical result regarding the fractional maximal operator says that $\mathscr{M}_\alpha$ is $(L^p \to L^q)$-bounded, when $1 < p < n/\alpha$ and $1/q = 1/p - \frac{\alpha}{n}$. We now show that the same boundedness type result holds for the variable exponent Lebesgue space without resorting to the classical case, and in that way we also prove the classical result. We need the following pointwise estimate.

Lemma 9.20. *Let $0 < \alpha < n$ and p be an exponent function such that $1 < p_- \leq p(x) \leq p_+ < n/\alpha$ and the function q is defined pointwise by $1/q(x) = 1/p(x) - \frac{\alpha}{n}$. Then for all functions f we have the following pointwise inequality*

$$\mathscr{M}_\alpha(f)(x) \leq \left[M\left(|f(\cdot)|^{\frac{p(\cdot)}{q(\cdot)}\frac{n}{n-\alpha}} \right)(x) \right]^{1-\frac{\alpha}{n}} \left(\int_\Omega |f(y)|^{p(y)}\, dy \right)^{\frac{\alpha}{n}}. \tag{9.12}$$

Proof. Let B be any ball containing the point x. Since

$$\frac{p(x)}{q(x)} + \frac{\alpha p(x)}{n} = 1$$

we have

$$\frac{1}{m(B)^{1-\frac{\alpha}{n}}} \int_B |f(y)|\, dy = \frac{1}{m(B)^{1-\frac{\alpha}{n}}} \int_B |f(y)|^{\frac{p(x)}{q(x)}} |f(y)|^{\frac{\alpha p(x)}{n}}\, dy$$
$$\leq \left(\frac{1}{m(B)} \int_B |f(y)|^{\frac{p(x)}{q(x)}\frac{n}{n-\alpha}}\, dy \right)^{1-\frac{\alpha}{n}} \left(\int_\Omega |f(y)|^{p(y)}\, dy \right)^{\frac{\alpha}{n}}$$

$$\leq \left[M \left(|f(\cdot)|^{\frac{p(\cdot)}{q(\cdot)} \frac{n}{n-\alpha}} \right) (x) \right]^{1-\frac{\alpha}{n}} \left(\int_{\Omega} |f(y)|^{p(y)} \, dy \right)^{\frac{\alpha}{n}},$$

which shows the Lemma. □

Using the pointwise inequality (9.12) and the boundedness of the maximal operator we obtain the boundedness of the Hardy-Littlewood fractional maximal function.

Theorem 9.21. *Let $1 < p(x) < n/\alpha$ and define $q(x)$ by the pointwise equality $1/q(x) = 1/p(x) - \frac{\alpha}{n}$. Then $\mathscr{M}_\alpha : L^{p(\cdot)}(\Omega) \longrightarrow L^{q(\cdot)}(\Omega)$ is bounded.*

Proof. Let $f \in L^{p(\cdot)}(\Omega)$ be such that $\rho_{p(\cdot)}(f) = 1$. Then by the inequality (9.12) we have

$$\begin{aligned}
\|\mathscr{M}_\alpha(f)\|_{L^{q(\cdot)}(\Omega)} &\leq \left\| \left[M \left(|f(\cdot)|^{\frac{p(\cdot)}{q(\cdot)} \frac{n}{n-\alpha}} \right) \right]^{1-\frac{\alpha}{n}} \right\|_{L^{q(\cdot)}(\Omega)} \\
&= \left\| M \left(|f(\cdot)|^{\frac{p(\cdot)}{q(\cdot)} \frac{n}{n-\alpha}} \right) \right\|_{L^{(1-\frac{\alpha}{n})q(\cdot)}(\Omega)}^{1-\frac{\alpha}{n}} \\
&\lesssim \left\| |f|^{\frac{p(\cdot)}{q(\cdot)} \frac{n}{n-\alpha}} \right\|_{L^{(1-\frac{\alpha}{n})q(\cdot)}(\Omega)}^{1-\frac{\alpha}{n}} = \left\| |f|^{\frac{p(\cdot)}{q(\cdot)}} \right\|_{L^{q(\cdot)}(\Omega)} \\
&\leq 1.
\end{aligned}$$

The general result now follows from the homogeneity of the fractional maximal operator. □

We now want to investigate the proof of boundedness of Hardy-Littlewood maximal function in grand Lebesgue spaces. For simplicity we will deal only with the interval $(0,1)$. We will use an important relation between rearrangements and maximal operator

$$Mf(x) = \sup_{(0,1) \supset I \ni x} \frac{1}{|I|} \int_I f(y) \, dy$$

given by a well-known theorem through the notion of decreasing rearrangement f^* of f.

Let

$$f^{**}(t) = \frac{1}{t} \int_0^t f^*(s) \, ds, \quad t \in [0,1].$$

The following theorem is given in Bennett and Sharpley [1, Theorem 3.8].

Theorem 9.22. *There are absolute constants c and c' such that for all $f \in L^1(0,1)$,*

$$c(Mf)^*(t) \leq f^{**}(t) \leq c'(Mf)^*(t), \tag{9.13}$$

$t \in (0,1)$.

Theorem 9.23. *Let $1 < p < \infty$. There exists a constant $C(p) > 1$ such that*

$$\|Mf\|_{p)} \le C(p)\|f\|_{p)}$$

for all $f \in L^1(0,1)$.

Proof. Since

$$\|f\|_p = \|f^*\|_p,$$

from (9.13) and Theorem 9.23 applied to f^* we get

$$\|Mf\|_{p)} = \|(Mf)^*\|_{p)} \le C\|f^{**}\|_{p)} = C\|f\|_{p)}$$

from which the assertion follows. □

9.5 Muckenhoupt Weights

We now try to characterize all weights $w(x)$ such that the strong type (p,p) inequality

$$\int_{\mathbb{R}^n} [Mf(x)]^p w(x)\,dx \le C_p \int_{\mathbb{R}^n} |f(x)|^p w(x)\,dx \tag{9.14}$$

is valid for all $f \in L^p(w)$.

Suppose that (9.14) is valid for some weight w and all $f \in L^p(w)$ for some $1 < p < \infty$. Applying (9.14) to the function $f\chi_B$ supported in a ball B and we use that

$$M(f\chi_B)(x) \ge \frac{1}{m(B)} \int_B |f(y)|\,dy$$

for all $x \in B$, to obtain

$$\begin{aligned} w(B)\left[⨍_B |f(y)|\,dy \right]^p &\le \int_B [M(f\chi_B)(x)]^p w(x)\,dx \\ &\le C_p \int_B |f(x)|^p w(x)\,dx. \end{aligned}$$

where $w(B)$ is given by $\int_B w(x)\,d\mu(x)$ and $⨍_B f(x)dx = \frac{1}{m(B)} \int_B f(x)\,dx$. It follows then that

$$\left(⨍_B |f(y)|\,dy \right)^p \le \frac{C_p}{w(B)} \int_B |f(x)|^p w(x)\,dx \tag{9.15}$$

for all balls B and all functions f. At this point it is tempting to choose a function such that the two integrands are equal. We do so by setting $f = w^{-q/p}$ where $1/p + 1/q = 1$, which gives $f^p w = w^{1-q} = w^{-q/p}$. Under the assumption that $\inf_B w(B) > 0$ for all balls B, it would follow from (9.15) that

$$\left(\fint_B [w(x)]^{-q/p} \, dx \right)^p \leq \frac{C_p}{w(B)} \int_B [w(x)]^{-q/p} \, dx$$

and thus

$$w(B) \left(\fint_B [w(x)]^{-q/p} \right)^{p-1} \leq C_p^p.$$

Therefore

$$\sup_B \left(\fint_B w(x) \, dx \right) \left(\fint_B [w(x)]^{-\frac{1}{p-1}} \right)^{p-1} \leq C_p^p, \tag{9.16}$$

where the supremum is taken over all balls B.

If $\inf_B w(B) = 0$ for some ball B, we take $f = (w + \varepsilon)^{-q/p}$ to obtain

$$\sup_B \left(\fint_B w(x) \, dx \right) \left(\fint_B [w(x) + \varepsilon]^{-\frac{1}{p-1}} \right)^{p-1} \leq C_p^p, \tag{9.17}$$

from which we deduce (9.16) via the Lebesgue monotone convergence theorem by letting $\varepsilon \to 0$.

We have now obtained that every weight w that satisfies (9.14) must also satisfy the rather strange-looking condition (9.16) which we refer to in the sequel as the A_p condition, sometimes also called *Muckenhoupt condition.*

Definition 9.24. Let $1 < p < \infty$. A weight w is said to be of class A_p if

$$\sup_{B \text{ balls in } \mathbb{R}^n} \left(\fint_B w(x) \, dx \right) \left(\fint_B [w(x)]^{-\frac{1}{p-1}} \, dx \right)^{p-1} < \infty. \tag{9.18}$$

The expression in (9.18) is called the A_p Muckenhoupt characteristics constant of w and will be denoted by $[w]_{A_p}$. ⊘

For the case of $p = 1$ we define the notion of A_1 weight.

Definition 9.25. A weight belongs to the A_1 class if $Mw(x) \leq Cw(x)$ a.e. for some constant C. ⊘

We now state, for a proof cf. García-Cuerva and Rubio de Francia [20], one pivotal results in the theory of weighted Lebesgue spaces, namely if the weight belongs to the Muckenhoupt class, then the maximal operator is bounded in the weighted Lebesgue space.

Theorem 9.26 *Let $f \in L^p(w)$, $1 < p < \infty$ and $w \in A_p$ then $Mf \in L^p(w)$. Moreover*

$$\|Mf\|_{L^p(w)} \leq [w]_{A_p}^{1/p} \|f\|_{L^p(w)}. \tag{9.19}$$

Now, let us recall the definition of weighted Hardy-Littlewood maximal function on $\mathbb{R}^n$ over balls

$$M_w(f)(x) = \sup_{x \in B} \frac{1}{w(B)} \int_B |f(y)| w(y) \, dy$$

where w is any weight and $w(B) = \int_B w(y) \, dy$.

In the following theorem our proof avoids the Calderón-Zygmund decomposition. In place of it we use the Vitali covering Theorem 9.4 and the fact that w, as a measure, satisfies the doubling condition, i.e.,

$$w(\lambda B) \leq \lambda^{np} [w]_{A_p} w(B)$$

see Problem 9.41.

Theorem 9.27. *For $1 \leq p < \infty$, then the weak (p,p) inequality*

$$w\left(\left\{x \in \mathbb{R}^n : Mf(x) > \lambda\right\}\right) \leq \frac{C}{\lambda^p} \int_{\mathbb{R}^n} |f(x)|^p w(x) \, dx$$

holds if $w \in A_p$.

Proof. We have

$$\left(\fint_B |f(x)| \, dx \right)^p \leq \frac{1}{m(B)^p} \int_B |f(x)|^p w(x) \, dx \left(\int_B [w(x)]^{-q/p} \, dx \right)^{p/q}.$$

Next, the right-hand side term in the above inequality is

$$\left(\frac{1}{w(B)}\int_B |f(x)|^p w(x)\,\mathrm{d}x\right)\frac{w(B)}{m(B)}\left(⨍_B (w(x))^{-\frac{1}{p-1}}\,\mathrm{d}x\right)^{p-1}$$

$$\leq \left(\frac{1}{w(B)}\int_B |f(x)|^p w(x)\,\mathrm{d}x\right)[w]_{A_p}.$$

And thus

$$\left(⨍_B |f(x)|\,\mathrm{d}x\right)^p \leq [w]_{A_p}\left(\frac{1}{w(B)}\int_B |f(x)|^p w(x)\,\mathrm{d}x\right). \tag{9.20}$$

Fix $\lambda > 0$, from (9.20) and the definition of M_w we get

$$\{x \in \mathbb{R}^n : M(f)(x) > \lambda\} \subset \left\{x \in \mathbb{R}^n : M_w(f^p)(x) > \frac{\lambda^p}{[w]_{A_p}}\right\}.$$

Thus

$$w\left(\{x \in \mathbb{R}^n : M(f)(x) > \lambda\}\right) \leq w\left(\left\{x \in \mathbb{R}^n : M_w(f^p)(x) > \frac{\lambda^p}{[w]_{A_p}}\right\}\right).$$

Let $A_\lambda \cap B(0,k)$, k fix where

$$A_\lambda = \left\{x \in \mathbb{R}^n : M_w(f^p)(x) > \frac{\lambda^p}{[w]_{A_p}}\right\}.$$

We assume $A_\lambda \neq \emptyset$, of course, since the result is trivial otherwise. For each $x \in A_\lambda$ there exists an $r > 0$, depending on x

$$\frac{\lambda^p}{[w]_{A_p}} < \frac{1}{w(B_r)}\int_{B_r} |f(x)|^p w(x)\,\mathrm{d}x. \tag{9.21}$$

After we obtain an estimate for the measure of $A_\lambda \cap B(0,k)$, we can let $k \to \infty$. Now, let $\mathscr{B}$ be the collection of open balls B with center in $A_\lambda \cap B(0,k)$ and satisfying (9.21). Then the hypothesis of the Vitali covering Theorem 9.4 is satisfied. Now, if $A_\lambda \cap B(0,k) \neq \emptyset$, then there exist balls $B_1, B_2, \ldots$ in $\mathscr{B}$ such that

(1) $B_1, B_2, \ldots$ are disjoint,
(2) $A_\lambda \cap B(0,k) \subset \cup_{r\geq 1} 3B_r$.

All we have to do is to assemble this information. Here is the method: first we use the fact that $w(3Br) \leq 3^{np}[w]_{A_p} w(B_r)$. From the inequalities (9.21) and (9.14), the disjointness of the selected balls, we obtain

$$\begin{aligned} w(A_\lambda \cap B(0,k)) &\leq \sum_{r\geq 1} w(3B_r) \\ &\leq 3^{np}[w]_{A_p} \sum_{r\geq 1} w(B_r) \\ &\leq \frac{3^{np}[w]_{A_p}^2}{\lambda^p} \sum_{r\geq 1} \int_{B_r} |f(x)|^p w(x)\,\mathrm{d}x \\ &\leq \frac{3^{np}[w]_{A_p}^2}{\lambda^p} \int_{\cup_{r\geq 1} B_r} |f(x)|^p w(x)\,\mathrm{d}x \\ &\leq \frac{3^{np}[w]_{A_p}^2}{\lambda^p} \int_{\mathbb{R}^n} |f(x)|^p w(x)\,\mathrm{d}x \end{aligned}$$

Finally, let $k \to \infty$, to obtain

$$w(A_\lambda) \leq \frac{3^{np}[w]_{A_p}^2}{\lambda^p} \int_{\mathbb{R}^n} |f(x)|^p w(x)\,\mathrm{d}x,$$

that is, we obtain the weak (p,p) inequality. □

We now collect some almost immediate properties of the A_p weights following directly from the definition. For more properties see Problems 9.39–9.41.

Theorem 9.28. *Let $1 < p < \infty$. We have*

(a) $w \in A_p$ if and only if $w^{1-p'} \in A_{p'}$ and $\left[w^{1-p'}\right]_{p'} = [w]_p^{p'-1}$.
(b) If $q < p$, then $A_q \subset A_p$ and $[w]_p \leq [w]_q$.
(c) If $w \in A_p$ and $0 \leq \alpha \leq 1$, then $w^\alpha \in A_p$ and $[w^\alpha]_p \leq [w]_p^\alpha$.
(d) If $w_1, w_2 \in A_p$ and $0 \leq \alpha \leq 1$, then $w_1^\alpha w_2^{1-\alpha} \in A_p$ and

$$[w_1^\alpha w_2^{1-\alpha}]_p \leq [w_1]_p^\alpha [w_2]_p^{1-\alpha}.$$

(e) If $w_1, w_2 \in A_1$ then $w_1 w_2^{1-p} \in A_p$ and

$$[w_1 w_2^{1-p}]_p \leq [w_1]_1 [w_2]_1^{p-1}.$$

9.6 Problems

9.29. Prove that the various definitions of maximal function are equivalent in the sense:

$$Mf(x) \asymp M_{\square}f(x) \asymp \widetilde{M}_B f(x).$$

where $\asymp A \asymp B$ means that there exists $C > 0$ such that $C^{-1}B \leq A \leq CB$.

9.30. Prove that

$$f(x) = \frac{1}{|x|^{n-1}} \notin L_1(\mathbb{R}^n),$$

but f restricted to any closed ball with center in 0 and radius $r > 0$ is locally integral, i.e.

$$f(x) = \chi_{\overline{B}(0,1)}(x)\frac{1}{|x|^{n-1}} \in L_{1,\mathrm{loc}}(\mathbb{R}^n).$$

9.31. Given $f(x) = \frac{|x|^{-1/2}}{1+|x|^{-1/2}}$ for $x \in \mathbb{R}$, prove that $f \notin L_\infty(m)$ but $f \in \mathrm{Weak}L_2$.

9.32. Let $(X, \mathscr{A}, \mu)$ be a finite measure space. Prove that the dual space of $L_1(\mu)$ is $L_\infty(\mu)$.

9.33. If $f \geq 0$ is a nondecreasing function in $(0, \infty)$ and $0 < p \leq q \leq \infty$, $\alpha \in \mathbb{R}$. Show that

$$\left(\int_0^\infty (t^\alpha f(t))^q \frac{dt}{t}\right)^{1/q} \leq C\left(\int_0^\infty (t^\alpha f(t))^p \frac{dt}{t}\right)^{1/p}$$

where $C = C(p, q, \alpha)$.

9.34. Let f be a decreasing function in $[a, b]$ $(a \neq 0)$ such that $0 < \int_0^b f(x)\,\mathrm{d}x < \infty$ and $0 < \int_0^a f(x)\,dx < \infty$. Prove that

$$\log\left(\frac{\int_0^b f(x)\,\mathrm{d}x}{\int_0^a f(x)\,dx}\right) \leq \log\left(\frac{b}{a}\right)$$

which is equivalent to

$$a\int_0^b f(x)\,\mathrm{d}x \leq b\int_0^a f(x)\,dx.$$

9.35. Let $B(x,r) \subset \mathbb{R}^n$ be an open ball with center in x and radius r. Let us define the *Lebesgue set of f* as

$$L_f = \left\{ x : \lim_{r\to 0} \frac{1}{m\left(B(x,r)\right)} \int_{B(x,r)} |f(y) - f(x)|\,\mathrm{d}y = 0 \right\}.$$

Demonstrate that if $f \in L_1(\mathbb{R}^n, \mathscr{L}, m)$ then

$$|f(x)| \leq Mf(x) \qquad \text{for all } x \in L_f$$

where $Mf(x)$ is the Hardy-Littlewood maximal function.

9.36. Let $f \in L_1(\mathbb{R}^n, \mathscr{L}, m)$ and

$$\left| \int_E f\,\mathrm{d}m \right| \leq m(E)$$

where E is a Lebesgue measurable set. Prove that $|f| \leq 1$ almost everywhere.

9.37. Let $f : \mathbb{R} \to [0, \infty)$ be defined as

$$f(x) = \begin{cases} \frac{1}{x\log^2 x} & \text{if } x \in (0, 1/e) \\ 0 & \text{if } x \notin (0, 1/e). \end{cases}$$

Prove that

a) $\int_{(0,x)} f(t)\,\mathrm{d}t = -1/\log x$ para $x \in (0, 1/e)$.

b) $\int_0^r Mf(x)\,\mathrm{d}x = \infty$.

9.38. Show that the following conditions are equivalent:

(a) $p \in LH(\Omega)$;
(b) $1/p \in LH(\Omega)$;
(c) $p' \in LH(\Omega)$.

9.39. If $w \in A_p$ prove that $w(B_{2r}) \leq cw(B_r)$ (doubling condition).

9.40. Let $w \in A_p$ for some $1 \leq p < \infty$. Prove that:

1. $[\delta^\lambda(w)]_p = [w]_p$ where $\delta^\lambda(w)(x) = w(\lambda x_1, \cdots, \lambda x_n)$.
2. $[\tau^z(w)]_p = [w]_p$ where $\tau^z(w)(z) = w(x - z)$, $z \in \mathbb{R}^n$.
3. $[\lambda w]_p = [w]_p$ for all $\lambda > 0$.

4. $[w]_p \geq 1$ for all $w \in A_p$. Equality holds if and only if w is a constant.
5. For $a \leq p < q < \infty$, then $[w]_p \leq [w]_q$.
6. $\lim_{p \to 1+} [w]_p = [w]_1$ if $w \in A_1$.
7.

$$[w]_p = \sup_Q \sup_{f \in L_p(w); m\left(\{Q \cap \{|f|=0\}\}\right)=0} \left\{ \frac{\left(\frac{1}{m(Q)} \int_Q |f(x)|\, dx \right)^p}{\frac{1}{w(Q)} \int_Q |f(x)|^p w(x)\, dx} \right\}$$

9.41. The measure $w(x)\,dx$ is doubling, precisely, for all $\lambda > 1$ and all cubes Q show that

$$w(\lambda Q) \leq \lambda^{np} [w]_p w(Q).$$

9.42. Show, using Marcinkiewicz theorem, that the maximal operator M is a bounded operator in L_p spaces, for $1 < p < \infty$ (see Theorem 9.11 for other proof).

9.43. Show that, for radial weights $w(x) = w(|x|)$, the A_p condition is reduced to the following inequality

$$\left(\int_0^r \rho^{n-1} w(\rho)\, d\rho \right) \left(\int_0^r \rho^{n-1} w(\rho)^{-\frac{1}{p-1}}\, d\rho \right)^{p-1} \leq C r^{np}.$$

9.44. Let $w(x) = |x|^\alpha$ be a radial weight. Show that

(a) $w \in A_p$, $1 < p < \infty$ if $-n < \alpha < n(p-1)$.
(b) $w \in A_1$ if $-n < \alpha \leq 0$.

9.45. Show that the *geometric maximal operator*

$$M_0 f(x) := \sup_{r>0} \exp \left(\frac{1}{m(B(x,r))} \int_{B(x,r)} \log|f(y)|\, dy \right)$$

is obtained by

$$\lim_{q \to 0+} M_q f(x) = \lim_{q \to 0+} \left(\sup_{r>0} \frac{1}{m(B(x,r))} \int_{B(x,r)} |f(y)|^q\, dy \right)^{\frac{1}{q}}$$

Hint: Recall that $(x^\alpha - 1)/\alpha \xrightarrow[\alpha \to 0]{} \log(x)$.

9.7 Notes and Bibliographic References

The one-dimensional maximal operator was introduced in Hardy and Littlewood [28] and the multi-dimensional version was given in Wiener [84].

The first occurrence of Vitali covering type theorems was in Vitali [80]. The result of Theorem 9.11 can be stated without the dependence on the dimension, see Stein [72]. The boundedness of the maximal operator in variable exponent Lebesgue spaces was solved for bounded sets by Diening [16], see also Diening, Harjulehto, Hästö, and Růžička [17] and Cruz-Uribe and Fiorenza [9]. Muckenhoupt weights were studied in Muckenhoupt [53], for more on the topic of A_p weights see García-Cuerva and Rubio de Francia [20].

Chapter 10
Integral Operators

> *A large part of mathematics which becomes useful developed with absolutely no desire to be useful, and in a situation where nobody could possibly know in what area it would become useful; and there were no general indications that it ever would be so.*
> JOHN VON NEUMANN

Abstract Integral operator theory is a vast field on itself. In this chapter we briefly touch some questions that are related to Lebesgue spaces. We prove the Hilbert inequality, we show the Minkowski integral inequality, and with that tool we show a boundedness result of an integral operators having a homogeneous kernel of degree -1. We introduce the Hardy operator and study its adjoint operator. One of the sections of the chapter is dedicated to the L_2 space now focusing on the fact that this is the only Hilbert space in the L_p scale. We present a proof of the Radon-Nikodym theorem, due to J. von Neumann, which does not use the Hahn decomposition theorem.

10.1 Some Inequalities

In the following we obtain the so-called *Minkowski integral inequality* using a different approach from the one used in Theorem 3.25.

Theorem 10.1 (Minkowski integral inequality). *Let $(X, \mathscr{A}_1, \mu)$ and $(Y, \mathscr{A}_2, \nu)$ be σ-finite measure spaces. Suppose that f is a measurable $\mathscr{A}_1 \times \mathscr{A}_2$ function and $f(\cdot, y) \in L_p(\mu)$ for all $y \in Y$. Then for $1 \leq p < \infty$ we have that*

$$\left(\int_X \left| \int_Y f(x,y)\, \mathrm{d}\nu \right|^p \mathrm{d}\mu \right)^{1/p} \leq \int_Y \left(\int_X |f(x,y)|^p \, \mathrm{d}\mu \right)^{1/p} \mathrm{d}\nu.$$

R.E. Castillo, H. Rafeiro, *An Introductory Course in Lebesgue Spaces*, CMS Books in Mathematics, DOI 10.1007/978-3-319-30034-4_10

Proof. For $p = 1$, notice that since

$$\left| \int_Y f(x,y)\,d\nu \right| \leq \int_Y |f(x,y)|\,d\nu$$

we get

$$\int_X \left| \int_Y f(x,y)\,d\nu \right| d\mu \leq \int_X \int_Y |f(x,y)|\,d\nu\,d\mu,$$

and now using Fubini's theorem we obtain

$$\int_X \left| \int_Y f(x,y)\,d\nu \right| d\mu \leq \int_Y \int_X |f(x,y)|\,d\mu\,d\nu.$$

Now, for $p = \infty$, we get

$$\begin{aligned} \left| \int_Y f(x,y)\,d\nu \right| &\leq \int_Y |f(x,y)|\,d\nu \\ &\leq \int_Y \|f(\cdot,y)\|_\infty\,d\nu, \end{aligned}$$

again by Fubini's theorem we arrive at

$$\begin{aligned} \int_X \left| \int_Y f(x,y)\,d\nu \right| d\mu &\leq \int_X \left(\int_Y \|f(\cdot,y)\|_\infty\,d\nu \right) d\mu \\ &\leq \int_Y \left(\int_X \|f(\cdot,y)\|_\infty\,d\mu \right) d\nu. \end{aligned}$$

We now take $1 < p < \infty$. By Fubini's theorem and Theorem 3.20 (Hölder's inequality) we obtain

$$\begin{aligned} &\int_X \left| \int_Y f(x,y)\,d\nu \right|^p d\mu \\ &= \int_X \left(\left| \int_Y f(x,y)\,d\nu \right| \left| \int_Y f(x,y)\,d\nu \right|^{p-1} \right) d\mu \end{aligned}$$

$$\leq \int_X \int_Y |f(x,y)|\,\mathrm{d}\nu \left| \int_Y f(x,y)\,\mathrm{d}\nu \right|^{p-1} \mathrm{d}\mu$$

$$= \int_Y \left(\int_X |f(x,y)| \left| \int_Y f(x,y)\,\mathrm{d}\nu \right|^{p-1} \mathrm{d}\mu \right) \mathrm{d}\nu$$

$$\leq \int_Y \left[\left(\int_X |f(x,y)|^p \,\mathrm{d}\mu \right)^{1/p} \left(\int_X \left| \int_Y f(x,y)\,\mathrm{d}\nu \right|^{q(p-1)} \mathrm{d}\mu \right)^{1/q} \right] \mathrm{d}\nu$$

$$= \int_Y \left[\left(\int_X |f(x,y)|^p \,\mathrm{d}\mu \right)^{1/p} \left(\int_X \left| \int_Y f(x,y)\,\mathrm{d}\nu \right|^{p} \mathrm{d}\mu \right)^{1/q} \right] \mathrm{d}\nu,$$

therefore

$$\left(\int_X \left| \int_Y f(x,y)\,\mathrm{d}\nu \right|^p \mathrm{d}\mu \right)^{1-1/q} = \left(\int_X \left| \int_Y f(x,y)\,\mathrm{d}\nu \right|^p \mathrm{d}\mu \right)^{1/p}$$

$$\leq \int_Y \left(\int_X |f(x,y)|^p \,\mathrm{d}\mu \right)^{1/p} \mathrm{d}\nu$$

which ends the proof. □

We now study the boundedness of integral operators in Lebesgue spaces with a homogeneous kernel, which permits to separate variables simplifying the calculations. In this regard we have the following result.

Theorem 10.2. *Let K be a measurable function in $(0,\infty) \times (0,\infty)$ such that the kernel K is homogeneous of degree -1, i.e., $K(\lambda x, \lambda y) = \lambda^{-1} K(x,y)$ for all $\lambda > 0$, and the kernel K satisfies the following integral bound*

$$\int_0^\infty |K(x,1)| x^{-1/p}\,\mathrm{d}x = C < \infty.$$

For $f \in L_p(\mu)$ we define

$$Tf(y) = \int_0^\infty K(x,y) f(x)\,\mathrm{d}x.$$

Then

$$\|Tf\|_p \le C\|f\|_p.$$

Proof. Let

$$\|Tf\|_p = \left(\int_0^\infty |Tf(y)|^p \,\mathrm{d}y\right)^{1/p} = \left(\int_0^\infty \left|\int_0^\infty K(x,y)f(x)\,\mathrm{d}x\right|^p \mathrm{d}y\right)^{1/p}.$$

Writing $x = zy$, then $\mathrm{d}x = y\,\mathrm{d}z$, now by Minkowski's integral inequality

$$\begin{aligned}\left(\int_0^\infty \left|\int_0^\infty K(zy,y)f(zy)y\,\mathrm{d}z\right|^p \mathrm{d}y\right)^{1/p} &= \left(\int_0^\infty \left|\int_0^\infty y^{-1}K(z,1)f(zy)y\,\mathrm{d}z\right|^p \mathrm{d}y\right)^{1/p}\\ &= \left(\int_0^\infty \left|\int_0^\infty K(z,1)f(zy)\,\mathrm{d}z\right|^p \mathrm{d}y\right)^{1/p}\\ &\le \int_0^\infty K(z,1)\left(\int_0^\infty |f(zy)|^p \,\mathrm{d}y\right)^{1/p} \mathrm{d}z.\end{aligned}$$

Now, if $x = zy$, then $z^{-1}dx = dy$, and we get

$$\begin{aligned}\int_0^\infty K(z,1)\left(\int_0^\infty |f(zy)|^p dy\right)^{1/p} dz &= \int_0^\infty K(z,1)\left(\int_0^\infty |f(x)|^p z^{-1} dx\right)^{1/p} dz\\ &= \left(\int_0^\infty K(z,1)z^{-1/p} dz\right)\|f\|_p.\end{aligned}$$

from which it follows that $\|Tf\|_p \le C\|f\|_p$. □

We now show the integral analogue of the Hilbert inequality given in Theorem 2.19.

Theorem 10.3 (Hilbert inequality). *Let $f \in L_p(m)$ and $g \in L_q(m)$ with $1/p + 1/q = 1$. Then*

$$\left|\int_0^\infty \int_0^\infty \frac{f(y)g(x)}{x+y}\,\mathrm{d}y\,\mathrm{d}x\right| \le \frac{\pi}{\sin\left(\frac{\pi}{p}\right)}\|f\|_p\|g\|_q. \tag{10.1}$$

Proof. Note that

$$\left|\int_0^\infty\int_0^\infty \frac{f(y)g(x)}{x+y}\,\mathrm{d}y\,\mathrm{d}x\right| \le \int_0^\infty\int_0^\infty \frac{|f(y)g(x)|}{x+y}\,\mathrm{d}x\,\mathrm{d}y.$$

Let $y = xz$, then $\mathrm{d}y = x\,\mathrm{d}z$, and now by Fubini's theorem we have

$$\begin{aligned}\int_0^\infty\int_0^\infty \frac{|f(y)g(x)|}{x+y}\,\mathrm{d}y\,\mathrm{d}x &= \int_0^\infty\int_0^\infty \frac{|f(xz)g(x)|}{x(1+z)}\,x\,\mathrm{d}z\,\mathrm{d}x\\ &= \int_0^\infty\int_0^\infty \frac{|f(xz)g(x)|}{1+z}\,\mathrm{d}z\,\mathrm{d}x\\ &= \int_0^\infty \frac{1}{1+z}\int_0^\infty |f(xz)g(x)|\,\mathrm{d}x\,\mathrm{d}z.\end{aligned}$$

If $u = xz$, then $\mathrm{d}u = z\,\mathrm{d}x$, moreover $x = \frac{u}{z}$, and by Hölder's inequality we get

$$\begin{aligned}&\int_0^\infty \frac{1}{1+z}\int_0^\infty |f(xz)g(x)|\,\mathrm{d}x\,\mathrm{d}z\\ &= \int_0^\infty \frac{1}{1+z}\left(\int_0^\infty \left|f(u)g\left(\frac{u}{z}\right)\right| z^{-1}\,\mathrm{d}u\right)\mathrm{d}z\\ &\le \int_0^\infty \frac{1}{1+z}\left(\int_0^\infty |f(u)|^p\,\mathrm{d}u\right)^{1/p}\left(\int_0^\infty \left|g\left(\frac{u}{z}\right)\right|^q z^{-q}\,\mathrm{d}u\right)^{1/q}\mathrm{d}z\\ &\le \int_0^\infty \frac{1}{1+z}\left(\int_0^\infty |f(u)|^p\,\mathrm{d}u\right)^{1/p}\left(\int_0^\infty |g(w)|^q z^{-q+1}\,\mathrm{d}w\right)^{1/q}\mathrm{d}z\\ &= \int_0^\infty \frac{z^{\frac{1}{q}-1}}{1+z}\,\mathrm{d}z\,\|f\|_p\|g\|_q\\ &= \mathrm{B}\left(\frac{1}{q}, 1-\frac{1}{q}\right)\|f\|_p\|g\|_q\end{aligned}$$

where $\mathrm{B}(1/q, 1-1/q)$ is the Beta function and the result now follows from (C.7). □

We now introduce an operator which is widely used, the so-called Hardy operator.

Definition 10.4 (Hardy operator). Let f be a positive and measurable function in $(0,\infty)$. The Hardy operator is defined as

$$Hf(x) = \frac{1}{x}\int_0^x f(y)\,\mathrm{d}y. \tag{10.2}$$

The Hardy operator is an average operator. ⊘

The principal result in Lebesgue spaces regarding this operator is that it is L_p bounded, namely:

Theorem 10.5 (Hardy's inequality). *Let $f \in L_p(0,\infty)$ be positive and $1 < p < \infty$. Then*

$$\|Hf\|_p \le \frac{p}{p-1}\|f\|_p. \tag{10.3}$$

Proof. Observe that if $y = zx$, then $\mathrm{d}y = x\,\mathrm{d}z$, therefore

$$Hf(x) = \frac{1}{x}\int_0^x f(y)\,\mathrm{d}y = \frac{1}{x}\int_0^1 f(xz)\,x\,\mathrm{d}z = \int_0^1 f(xz)\,\mathrm{d}z.$$

Now using the integral Minkowski inequality, we obtain

$$\begin{aligned}
\left(\int_0^\infty \left(\frac{1}{x}\int_0^x f(y)\,\mathrm{d}y\right)^p \mathrm{d}x\right)^{1/p} &= \left(\int_0^\infty \left(\int_0^1 f(zx)\,\mathrm{d}z\right)^p \mathrm{d}x\right)^{1/p} \\
&\le \int_0^1 \left(\int_0^\infty (f(zx))^p\,\mathrm{d}x\right)^{1/p} \mathrm{d}z \\
&= \int_0^1 z^{-1/p} \left(\int_0^\infty (f(u))^p\,\mathrm{d}u\right)^{1/p} \mathrm{d}z \\
&= \frac{p}{p-1}\|f\|_p,
\end{aligned}$$

therefore (10.3) is true. □

To see a somewhat surprising proof of the boundedness of the Hardy operator result using convolution and Theorem 11.9, see details in p. 390.

Remark 10.6. The assertion of the Hardy inequality does not hold for $p = 1$. This can be observed by taking $f = \chi_{(0,1)}$, since then

$$\int_0^\infty f(x)\,\mathrm{d}x = 1,$$

but

$$\int_0^\infty \frac{1}{t}\int_0^t f(s)\,\mathrm{d}s\,\mathrm{d}t \geq \int_1^\infty \frac{1}{t}\int_0^1 \mathrm{d}s\,\mathrm{d}t = \int_1^\infty \frac{\mathrm{d}t}{t} = \infty.$$

For $p = 1$ the Hardy inequality is not true even when $(0,\infty)$ is replaced with a finite interval. For instance, there is no positive constant C that would render the inequality

$$\int_0^1 \frac{1}{t}\int_0^t f(s)\,\mathrm{d}s\,\mathrm{d}t \leq C\int_0^1 f(x)\,\mathrm{d}x$$

true for all positive functions on $(0,1)$. To see this, take for example

$$f(t) = \frac{1}{t(\log\frac{2}{t})^2}, \quad t \in (0,1).$$

Then, again, $\int_0^1 f(x)\,\mathrm{d}x < \infty$ but, with appropriate C

$$\int_0^1 \frac{1}{t}\int_0^t f(s)\,\mathrm{d}s\,\mathrm{d}t = C\int_0^1 \frac{\mathrm{d}t}{t(\log\frac{2}{t})} = \infty.$$

⊘

We now remember the notion of adjoint operator.

Definition 10.7. Let $T : X \longrightarrow Y$ be a linear and bounded operator. We say that the operator $T^* : Y^* \longrightarrow X^*$ is the *adjoint operator of* T if it satisfies the duality identity, it means that for all $x \in X$, $y \in Y^*$ where X and Y are Banach spaces, we have

$$\langle Tx, y\rangle = \langle x, T^*y\rangle$$

where $\langle \xi, \Lambda\rangle := \Lambda(\xi)$ with $\Lambda \in \Xi^*$ and $\xi \in \Xi$. ⊘

We can now obtain the adjoint operator of the Hardy operator.

Theorem 10.8. *The adjoint operator of the Hardy operator* (10.2), $H : L^p \longrightarrow L^p$, *at least formally, is given by*

$$H^*f(y) = \int_y^\infty f(x)\frac{\mathrm{d}x}{x}$$

for $f \geq 0$.

Proof. Using the Definition 10.7, the Riesz representation theorem and the Fubini theorem we get

$$\begin{aligned}
\langle Hf, g\rangle &= \int_0^\infty Hf(x)g(x)\,\mathrm{d}x \\
&= \int_0^\infty \left(\frac{1}{x}\int_0^x f(y)\,\mathrm{d}y\right) g(x)\,\mathrm{d}x \\
&= \int_0^\infty \left(\frac{1}{x}\int_0^\infty \chi_{(0,x)}(y)f(y)\,\mathrm{d}y\right) g(x)\,\mathrm{d}x \\
&= \int_0^\infty \left(\int_0^\infty \chi_{(y,\infty)}(x)f(y)\,\mathrm{d}y\right) \frac{g(x)}{x}\,\mathrm{d}x \\
&= \int_0^\infty f(y)\int_y^\infty g(x)\frac{\mathrm{d}x}{x}\,\mathrm{d}y \\
&= \langle f, H^* g\rangle,
\end{aligned}$$

which ends the proof. □

We now mention the concept of compact operator and obtain a result in this regard.

Definition 10.9 (Compact operator). Suppose that X and Y are Banach spaces and B is the unit ball in X. A linear operator $T : X \to Y$ is *compact* if the closure of the set $T(B)$ is compact in Y. ⊘

This definition is equivalent to say that T is compact if and only if the bounded sequence $\{x_n\}$ in X contains a subsequence $\{x_{n_k}\}$ such that $\{T(x_{n_k})\}$ converges pointwise in Y.

We say that a set $S \subset C(X)$ (where $C(X)$ is the space of continuous functions in the topological space X) is *equicontinuous* at $x \in X$ if for each $\varepsilon > 0$ there is a neighborhood U of x such that $|f(y) - f(x)| < \varepsilon$ for all $f \in E$ and all $y \in U$. If this condition is satisfied for all elements of S we say that E is equicontinuous.

A metric space X is said to be a *compact metric space* if it has the *Borel-Lebesgue property*, i.e., if every open cover of X has a finite subcover.

Theorem 10.10 (Arzelà-Ascoli Theorem). *Suppose that (X, d) is a compact metric spaces. Then a subset $F \subset C(X)$ is compact if and only if F is closed, bounded, and equicontinuous.*

Now we show compactness of an integral operator.

Theorem 10.11. *Let (X,d) be a compact metric space and μ a Borel measure in X. Let $K : X \times X \to \mathbb{R}$ be a continuous function such that*

$$\int_X |K(x,y)|\,\mathrm{d}\mu(x) \leq C \text{ a.e. } y \in X$$

and

$$\int_X |K(x,y)|\,\mathrm{d}\mu(y) \leq C \text{ a.e. } x \in X.$$

Then, for $1 \leq p \leq \infty$, the integral operator $T : L_p(\mu) \to L_p(\mu)$ given by

$$Tf(x) = \int_X K(x,y)f(y)\,\mathrm{d}\mu(y)$$

is compact.

Proof. Suppose that $1 < p < \infty$. Taking q as the conjugate exponent of p, and using Hölder's inequality in the product

$$|K(x,y)f(y)| = |K(x,y)|^{1/q}\left(|K(x,y)|^{1/p}|f(y)|\right)$$

we obtain

$$\begin{aligned}\int_X |K(x,y)f(y)|\,\mathrm{d}\mu(y) &\leq \left[\int_X |K(x,y)|\,\mathrm{d}\mu(y)\right]^{1/q}\left[\int_X |K(x,y)||f(y)|^p\,\mathrm{d}\mu(y)\right]^{1/p}\\ &\leq C^{1/q}\left[\int_X |K(x,y)||f(y)|^p\,\mathrm{d}\mu(y)\right]^{1/p}\end{aligned}$$

for almost all $x \in X$. By Tonelli's theorem we get

$$\begin{aligned}\int_X\left[\int_X |K(x,y)||f(y)|\,\mathrm{d}\mu(y)\right]^p \mathrm{d}\mu(x) &\leq C^{p/q}\int_X\int_X |K(x,y)||f(y)|^p\,\mathrm{d}\mu(y)\,\mathrm{d}\mu(x)\\ &\leq C^{p/q+1}\int_X |f(y)|^p\,\mathrm{d}\mu(y),\end{aligned}$$

i.e.

$$\|Tf\|_p \leq C\|f\|_p,$$

therefore Tf is bounded.

Now, let us fix a $y_0 \in X$ and let $\varepsilon > 0$. By the uniform continuity of K in $X \times X$ there exists $\delta > 0$ such that if $d(y, y_0) < \delta$ then $|K(x,y) - K(x,y_0)| < \varepsilon$ for each $x \in X$. Then, if $d(y,y_0) < \delta$ and $f \in L_p(\mu)$ satisfies $\|f\|_p \leq 1$, then Hölder's inequality implies

$$\begin{aligned}
|Tf(x) - Tf(x_0)| &= \left| \int_X [K(x,y) - K(x,y_0)] f(y) \, \mathrm{d}\mu(y) \right| \\
&< \varepsilon \int_X |f(y)| \, \mathrm{d}\mu(y) \\
&\leq \varepsilon [\mu(X)]^{1/q} \|f\|_p \\
&\leq \varepsilon [\mu(X)]^{1/q}.
\end{aligned}$$

Therefore, we proved that $\{Tf : \|f\|_p \leq 1\}$ is a subspace of $C(X)$ which is $\|\cdot\|_p$-bounded and equicontinuous, and now by Arzelà-Ascoli theorem we conclude that T is a compact operator. □

10.2 The Space L_2

We postponed the introduction of the L_2 space which has special nature, e.g., the space coincides with its dual. Moreover, in the Lebesgue scale it is the only Hilbert space, and by this reason is widely used in application, for example, in quantum mechanics the state of a particle is given by a wave-function which belongs to the L_2-space.

Definition 10.12 (Inner product). Let $(X, +, \cdot)$ be a vector space. A functional $\langle \cdot, \cdot \rangle : X \times X \to \mathbb{F}$, where $\mathbb{F} = \mathbb{R}$ or $\mathbb{F} = \mathbb{C}$, such that satisfies:

(a) $\langle f + g, h \rangle = \langle f, h \rangle + \langle g, h \rangle$ for all $f, g, h \in X$,
(b) $\langle cf, g \rangle = c\langle f, g \rangle$ for all $f, g \in X$ and any scalar c,
(c) $\langle f, g \rangle = \overline{\langle g, f \rangle}$ for all $f, g \in X$,
(d) $0 \leq \langle f, f \rangle < +\infty$ for all $f \in X$,
(e) $\langle f, f \rangle = 0$ if and only if $f = 0$,

it is called *inner product* or scalar product. ⊘

All inner product generates a norm defined by

$$\|f\| = \sqrt{\langle f, f \rangle}.$$

Moreover, we have the Cauchy-Schwarz inequality

$$|\langle f, g \rangle| \leq \|f\| \|g\|.$$

Definition 10.13 (Hilbert space). A vector space with an inner product is called an *Hilbert space* if it is complete with respect to the norm generated by the inner product.

⊘

The following theorem characterizes the vector spaces induced by an inner product.

Theorem 10.14. *A norm $\|\cdot\|$ in a vector space is induced by an inner product if and only if it satisfies the following identity (the so-called* law of parallelogram*)*

$$\|f+g\|^2+\|f-g\|^2=2(\|f\|^2+\|g\|^2) \tag{10.4}$$

for any vectors f and g.

From Theorem 3.29 we get that $(L_2(\mu),\|\cdot\|_2)$ is a complete space. Let us now consider in $L_2(\mu)$ the inner product

$$\langle f,g\rangle=\int_X f\bar{g}\,\mathrm{d}\mu,$$

$f,g\in L_2(\mu)$. Observe that the inner product generates the norm $\|\cdot\|_2$ and moreover $\|\cdot\|_2$ satisfies the parallelogram law, therefore:

Theorem 10.15. *The space$(L_2(\mu),\|\cdot\|_2)$ is an Hilbert space.*

Let X be a space with an inner product. If A is a nonempty subset of X, then the orthogonal complement $A^\perp$ of A is the set of all vectors which are orthogonal to any vector of A, i.e.

$$A^\perp=\{x\in X : x\perp y \text{ for all } y\in A\},$$

where $x\perp y$ means that $\langle x,y\rangle=0$.

From the linearity and continuity of the inner product (see Problem 10.62) it is clear that $A^\perp$ is a closed subspace of X such that $A^\perp=(\overline{A})^\perp$ and $A\cap A^\perp=\{0\}$.

We recall that a vector space X is a *direct sum* of two subspaces X_1 and X_2, denoted by $X=X_1\oplus X_2$, if for all $x\in X$ there is a unique representation $x=x_1+x_2$, where $X_1\in X_1$ and $x_2\in X_2$.

Theorem 10.16. *If M is a closed subspace of an Hilbert space H, then $H=M\oplus M^\perp$.*

Remark 10.17. Since all inner product is continuous, it follows that any vector y in the space X with an inner product defines a linear functional $f_y : X\to\mathbb{C}$ via the formula

$$f_y(x)=\langle x,y\rangle, \tag{10.5}$$

as we will see in the next theorem. If X is an Hilbert space, then all continuous and linear functions will be of the form (10.5).

Theorem 10.18 (F. Riesz Theorem). *If H is an Hilbert space and $f : H \to \mathbb{C}$ is a continuous and linear function, then there exists a unique vector $y \in H$ such that*

$$f(x) = \langle x, y \rangle$$

for all $x \in H$. Moreover $\|f\| = \|y\|$.

Proof. Let $F : H \to \mathbb{C}$ be a continuous and linear functional in H different from the null functional $x \mapsto 0$, since otherwise we can take $y = 0$. Let M be its kernel, i.e.

$$M = \ker(f) = f^{-1}(0) = \{x \in H \mid f(x) = 0\},$$

since f is a continuous and linear functional, we have that M is a closed subspace of H. By Theorem 10.16 we have $H = M \oplus M^{\perp}$, from which there exists an element $\xi \in M^{\perp}$ with $\|\xi\| = 1$. Since $(f(x)\xi - f(\xi)x) \in M$ for all $x \in H$ we obtain that

$$\langle \xi, f(x)\xi - f(\xi)x \rangle = 0$$

for all $x \in H$, from which it follows that

$$f(x) = \langle \overline{f(\xi)}\xi, x \rangle.$$

The uniqueness follows from noticing that if $\langle x, y \rangle = \langle x, y_1 \rangle$ for each $x \in H$, then taking in particular $x = y - y_1$ entails that $\langle y - y_1, y - y_1 \rangle = 0$, from which we get that $y = y_1$. Finally, in virtue of the Cauchy-Schwarz inequality $|f(x)| = |\langle x, y \rangle| \leq \|x\| \|y\|$ we have that $\|f\| \leq \|y\|$. On the other hand, if $y \neq 0$, then $x = \frac{y}{\|y\|}$ satisfies $\|x\| = 1$ and $\|f\| \geq |f(x)| = \langle y/\|y\|, y \rangle| = \|y\|$. Therefore $\|f\| = \|y\|$. $\square$

Remark 10.19. Since H and H^* are the same in an isometric way, it can be difficult to distinguish if $\varphi \in H$ is to be taken as an element or the *generator* of the linear continuous functional. To get around this inconvenience, the physicists following Dirac use the so-called *bra-ket* notation. If $\Phi \in H^*$, then there exists a $\varphi \in H$ such that

$$\Phi(f) = \langle \varphi, f \rangle = \langle \varphi | f \rangle,$$

where the last notation is the *bracket Dirac notation*. In this case the functional Φ is written as $\langle \varphi |$, what is called the *bra* and the vector f will be called *ket* and denoted by $| f \rangle$. Therefore

$$\Phi(f) = \langle \varphi || f \rangle = \langle \varphi | f \rangle.$$

⊘

If H is a Hilbert spaces, then the Theorem of F. Riesz shows that a function $y \mapsto f_y$ where $f_y(x) = \langle x, y \rangle$ can be defined from H into H^*. By the properties

(a) $f_y + f_z = f_{y+z}$
(b) $\alpha f_y = f_{\overline{\alpha} y}$
(c) $\|f_y\| = \|y\|$

it is easy to see that f_y is a linear "conjugate" application which is an isometry from H into H^*. Due to this isometry, we can show that all Hilbert spaces are reflexive.

Corollary 10.20 *All Hilbert spaces are reflexive.*

Proof. Let H be an Hilbert space and $F : H^* \to \mathbb{C}$ be a linear functional. Let us define $\phi : H \to \mathbb{C}$ by the formula $\phi(y) = \overline{F(f_y)}$, now note that:

(a) $\phi(y+z) = \overline{F(f_{y+z})} = \overline{F(f_y + f_z)} = \overline{F(f_y)} + \overline{F(f_z)} = \phi(y) + \phi(z)$
(b) $\phi(\alpha y) = \overline{F(f_{\overline{\alpha} y})} = \overline{F(\overline{\alpha} f_y)} = \overline{\overline{\alpha} F(f_y)} = \alpha \overline{F(f_y)} = \alpha \phi(y)$
(c) $|\phi(y) = |\overline{F(f_y)}| = |F(f_y)| \leq \|F\| \|f_y\| = \|F\| \|y\|$

By (a), (b), and (c) we have that $\phi \in H^*$. Therefore by the Theorem of F. Riesz, there exists a unique $x \in X$ such that $\langle y, x \rangle = \phi(y) = \overline{F(f_y)}$ for all $y \in H$. This implies that, for $\hat{x} \in H^{**}$ we have

$$\hat{x}(f_y) = f_y(x) = \langle x, y \rangle = F(f_y)$$

for each $y \in H$, from this we have that $\hat{x} = F$, and this tell us that the natural immersion is onto in H^{***}, therefore H is a reflexive space. □

10.2.1 Radon-Nikodym Theorem

In this section we present an alternative proof of the classical Radon-Nikodym theorem which is independent from the Hahn decomposition. The presented proof, due to J. von Neumann, is based on another existence theorem, the Theorem of F. Riesz.

Lemma 10.21. *Let ν, λ be finite measures on $(X, \mathscr{A})$ such that $\nu A \leq \lambda A$ for all $A \in \mathscr{A}$. Then there exists $f \geq 0$, $f \in L_2(\lambda)$ such that*

$$\nu A = \int_A f \, d\lambda$$

for all $A \in \mathscr{A}$.

Proof. Let us take $F : L_2(\lambda) \longrightarrow \mathbb{R}$ such that $g \mapsto \int_X g \, d\nu$. By the Riesz representation theorem, there exists $f \in L_2(\lambda)$ such that $F(g) = \int_X f g \, d\lambda$ for all $g \in L_2(\lambda)$. Taking $g = \chi_A$ we get $\nu A = F(\chi_A) = \int_A f \, d\lambda$. □

We now prove a version of the Radon-Nikodym theorem for finite measures.

Theorem 10.22 (Radon-Nikodym theorem). *Let μ, ν be finite measures on $(X, \mathscr{A})$ such that $\nu \ll \mu$. Then there exists a μ-almost everywhere unique positive φ function, $\varphi \in L_1(\mu)$, such that*

$$\nu(E) = \int_E \varphi \, d\mu, \tag{10.6}$$

for all $E \in \mathscr{A}$.

Proof. Let us define $\lambda = \mu + \nu$. By Lemma 10.21 there exists $f \in L_2(\lambda)$ such that

$$\nu E = \int_E f \, d\lambda.$$

The function f satisfy the following inequality $0 \leq f \leq 1$ λ-almost everywhere, due to

$$0 \leq \nu E = \int_E f \, d\lambda \leq \lambda E = \int_E 1 \, d\lambda.$$

Let $A = \{x \in X : f(x) = 1\}$. We have

$$\nu A = \int_A f \, d\lambda = \lambda A,$$

from which we get that $\mu A = 0$ and thus obtaining that $\nu A = \lambda A = 0$ since $\nu \ll \mu$. Now, for all sets $E \in \mathscr{A}$ we have

$$\int_X \chi_A f \, d\mu = \int_X \chi_A (1 - f) \, d\nu,$$

from which we obtain, via a limiting argument, that

$$\int_X g f \, d\mu = \int_X g(1 - f) \, d\nu,$$

for all $g \geq 0$ measurable functions. Taking $g = \frac{\chi_A}{1-f}$ and remembering that $0 \leq f < 1$ λ-almost everywhere, we get

$$\int_X \frac{f}{1-f} \chi_A \, d\mu = \int_X \chi_A \, d\nu$$

from which we take the function $\varphi = \frac{f}{1-f}$ which satisfies the equality (10.6).

The uniqueness of φ follows from the fact that $\int_X f = 0$ implies that $f = 0$ almost everywhere. $\square$

We now give a version of the Radon-Nikodym theorem for σ-finite measure spaces.

Theorem 10.23 (Radon-Nikodym theorem). *Let $(X, \mathscr{A}, \mu)$ be a σ-finite measure space and let ν be a measure defined in $\mathscr{A}$ which is absolutely continuous with respect to μ, i.e., $\nu \ll \mu$. Then there exists a nonnegative measurable function φ such that for each $E \in \mathscr{A}$ we have*

$$\nu(E) = \int_E \varphi \, d\mu. \tag{10.7}$$

The following example shows us that the hypothesis that the measure μ must be σ-finite in the Radon-Nikodym theorem cannot be dropped.

Example 10.24. Let $X = [0,1]$ and $\mathscr{A}$ be the class of all measurable subset of $[0,1]$. Let ν be the Lebesgue measure and μ the counting measure in $\mathscr{A}$. Then ν is finite and absolutely continuous with respect to μ, but there is no function f such that

$$\nu(E) = \int_E f \, d\mu$$

for all $E \in \mathscr{A}$.

If $\mu(E) = 0$, then $E = \emptyset$ since μ is the counting measure, then $\nu(E) = \nu(\emptyset) = 0$, ν the Lebesgue measure, therefore $\nu \ll \mu$.

Since $\nu([0,1]) = 1$ then ν is finite. On the other hand $X = [0,1]$ is not numerable and $\mu(\{x\}) = 1 \; \forall x \in [0,1]$ which tells us that μ is not a σ-finite measure.

Supposing now that there exists $f : [0,1] \to [0,\infty)$ defined by $F = f(x)\chi_{\{x\}}$ such that $\nu(E) = \int_E f \, d\mu$ for all $E \in \mathscr{A}$. Let $x \in [0,1]$, then

$$0 = \nu(\{x\}) = \int_{\{x\}} f \, d\mu = \int_X f(x)\chi_{\{x\}} \, d\mu = f(x)\mu(\{x\}) = f(x)$$

for all $x \in [0,1]$. But $\nu([0,1]) = \int_{[0,1]} f \, d\mu = 0$ which is a contradiction. ⊘

10.3 Problems

10.25. Prove that $\langle f, g \rangle = \int_X f\overline{g} \, d\mu$ with $f, g \in L_2(\mu)$ is an inner product.

10.26. Let $f, g \in L_2(\mu)$, show that

$$|\langle f, g \rangle| \leq \|f\|_2 \|g\|_2.$$

This inequality is known as the *Cauchy-Schwarz inequality* or *Cauchy-Bunyakovsky inequality*.

10.27. Prove that the equality in the Cauchy-Schwarz inequality, i.e.

$$|\langle f,g\rangle| = \|f\|_2\|g\|_2.$$

if and only if f and g are linearly dependents.

10.28. Show that $\|\cdot\|_2 : L_2(\mu) \to \mathbb{R}$ or $\mathbb{C}$ defined by

$$\|f\|_2 = \left(\int_X |f|^2\,\mathrm{d}\mu\right)^{\frac{1}{2}}$$

is a norm over $L_2(\mu)$.

10.29. Demonstrate that the norm $\|\cdot\|_2$ satisfies the parallelogram law (10.4).

10.30. If $f \in L_2(\mu)$, prove that$\|f\|_2 = \sup_{\|g\|_2=1} |\langle f,g\rangle|$.

10.31. Demonstrate that the following norms are not induced by an inner product:

(a) $\|x\| = \max_{1\leq k\leq n}\{|x_k|\}$ in $\mathbb{R}^n$
(b) $\|f\| = \sup_{x\in[a,b]} |f(x)|$ in $C[a,b]$
(c) $\|f\|_p = \left(\int_X |f|^p\,\mathrm{d}\mu\right)^{\frac{1}{p}}$ in $L_p(\mu)$ where $p \neq 2$.

10.32. Let $f_n, g_n \in L_2(\mu)$ with $n \in \mathbb{N}$. If

$$\lim_{n\to\infty}\int_X (f_n - f)^2\,\mathrm{d}\mu = \lim_{n\to\infty}\int_X (g_n - g)^2\,\mathrm{d}\mu = 0.$$

Prove that

$$\lim_{n\to\infty}\int_X f_n g_n\,\mathrm{d}\mu = \int_X fg\,\mathrm{d}\mu.$$

10.33. Let $I = [0,\pi]$ and $f \in L_2\left([0,\pi],\mathscr{L},m\right)$. Is it possible to have simultaneously

$$\int_I \left(f(x) - \sin x\right)^2 \mathrm{d}x \leq 4$$

and

$$\int_I \left(f(x) - \cos x\right)^2 \mathrm{d}x \leq \frac{1}{9}?$$

10.34. Let $I = [0,1]$ and f be a Lebesgue measurable function. Show that $f \in L_2(I,\mathscr{L},m)$ if and only if $f \in L_1(I,\mathscr{L},m)$ such that exists an increasing function g such that for all closed interval $[a,b] \subset [0,1]$ we have

$$\left| \int_a^b f(x)\,\mathrm{d}x \right|^2 \leq (g(b) - g(a))\,|b-a|.$$

10.35. Let $f \in L_2([0,1], \mathscr{L}, m)$ be such that $\|f\|_2 = 1$ and $\int_0^1 f\,\mathrm{d}m \geq \alpha > 0$. Also, for $\beta \in \mathbb{R}$, let $E_\beta = \{x \in [0,1] : f(x) \geq \beta\}$. If $0 < \beta < \alpha$, prove that

$$m(E_\beta) \geq (\beta - \alpha)^2.$$

This inequality is known in the literature as the *Peley-Zygmund inequality*.

10.36. Let us consider the measure space $(X, \mathscr{A}, \mu)$ with $\mu(X) = 1$ and let $f, g \in L_2(\mu)$. If $\int_X f\,\mathrm{d}\mu = 0$, show that

$$\left(\int_X fg\,\mathrm{d}\mu \right)^2 \leq \left[\int_X g^2\,\mathrm{d}\mu - \left(\int_X g\,\mathrm{d}\mu \right)^2 \right] \int_X f^2\,\mathrm{d}\mu.$$

10.37. Let $f \in L_1(\mu) \cap L_2(\mu)$. Demonstrate that

(a) $f \in L_p(\mu)$ for each $1 \leq p \leq 2$.
(b) $\lim_{p \to 1^+} \|f\|_p = \|f\|_1$.

10.38. If $\int_{-\infty}^{\infty} x^2 |f(x)|^2\,\mathrm{d}x < \infty$ and $\int_{-\infty}^{\infty} |f'(x)|^2\,\mathrm{d}x < \infty$, prove that if $x \geq 0$, then

$$x|f(x)|^2 \leq 4 \left(\int_x^{\infty} x^2 |f(x)|^2\,\mathrm{d}x \right)^{1/2} \left(\int_x^{\infty} |f'(x)|^2\,\mathrm{d}x \right)^{1/2}.$$

10.39. Let f be a function defined in $\mathbb{R}$ such that $f(x)$ and $xf(x)$ belong to $L_2(\mathbb{R})$. Prove that

$$\left(\int_{-\infty}^{\infty} |f(x)|\,\mathrm{d}x \right)^2 \leq 8 \left(\int_{-\infty}^{\infty} |f(x)|^2\,\mathrm{d}x \right)^{1/2} \left(\int_{-\infty}^{\infty} |x|^2 |f(x)|^2\,\mathrm{d}x \right)^{1/2}.$$

10.40. We remember that the Gamma function is defined as

$$\Gamma(\alpha) = \int_0^{\infty} t^{\alpha-1} \mathrm{e}^{-t}\,\mathrm{d}t \qquad \alpha \in (0, \infty).$$

From Example 3.23 we already know that this function is log-convex.

(a) If $\alpha, \beta \in (0,\infty)$ show that

$$\frac{\Gamma(\alpha)\Gamma(\beta)}{\Gamma(\alpha+\beta)} = \int_0^1 t^{\alpha-1}(1-t)^{\alpha-1}\,\mathrm{d}t := B(\alpha,\beta),$$

where B is the so-called *Beta function* or *Euler integral of the first kind*.

(b) Let f be continuous in $[0,\infty)$ for $\alpha \in (0,\infty)$ and $x \geq 0$, let us define

$$I_\alpha f(x) = \frac{1}{\Gamma(\alpha)} \int_0^x (x-t)^{\alpha-1} f(t)\,\mathrm{d}t.$$

Prove that $I_\alpha(I_\beta f)(x) = I_{\alpha+\beta} f(x)$.

(c) Let us define $J_\alpha f(x) = x^{-\alpha} I_\alpha f(x)$. Prove that for $1 < p < \infty$ we have

$$\|J_\alpha f\|_p \leq \frac{\Gamma(1-1/p)}{\Gamma(\alpha+1-1/p)} \|f\|_p.$$

10.41. Let $1 \leq p < \infty$, $r > 0$, and h be a nonnegative measurable function in $(0,\infty)$. Demonstrate that

(a) $\int_0^\infty x^{-r-1} \left[\int_0^x h(y)\,\mathrm{d}y\right]^p \mathrm{d}x \leq \left(\frac{p}{r}\right)^p \int_0^\infty x^{p-r-1} \left[h(x)\right]^p \mathrm{d}x$

(b) $\int_0^\infty x^{r-1} \left[\int_x^\infty h(y)\,\mathrm{d}y\right]^p \mathrm{d}x \leq \left(\frac{p}{r}\right)^p \int_0^\infty x^{p+r-1} \left[h(x)\right]^p \mathrm{d}x$

10.42. Let k be a nonnegative measurable function in $(0,\infty)$ such that

$$\int_0^\infty k(x) x^{s-1}\,\mathrm{d}x = \varphi(s),$$

for $0 < s < 1$, if $1 < p < \infty$ and $\frac{1}{p} + \frac{1}{q} = 1$, moreover if f, g are nonnegative measurable functions in $(0,\infty)$. Prove that

$$\int_0^\infty \int_0^\infty k(xy) f(x) g(y)\,\mathrm{d}x\,\mathrm{d}y$$

$$\leq \varphi(p^{-1}) \left[\int_0^\infty x^{p-2} \left[f(x)\right]^p \mathrm{d}x\right]^{1/p} \left[\int_0^\infty \left[g(x)\right]^q \mathrm{d}x\right]^{1/q}$$

10.43. Let $F(x) = \int\limits_0^\infty \frac{f(y)}{x+y}\,\mathrm{d}y$; $0 < x < \infty$. If $1 < p < \infty$ show that

$$\|F\|_p \leq \frac{\pi}{\sin\left(\frac{\pi}{p}\right)} \|f\|_p.$$

10.44. Show that

$$\left| \int\limits_0^\infty \int\limits_0^\infty \frac{f(x)g(y)}{x+y}\,\mathrm{d}x\,\mathrm{d}y \right| \leq \pi \|f\|_2 \|g\|_2$$

for $f, g \in L_2\left((0,\infty), \mathscr{L}, m\right)$.

10.45. Let $K : [0,1] \times [0,1] \to \mathbb{R}$ be defined by

$$K(s,t) = \begin{cases} 0 & \text{if } 0 \leq t \leq s \leq 1 \\ 1 & \text{if } 0 \leq s \leq t \leq 1 \end{cases}$$

and $V : L_p[0,1] \to L_p[0,1]$ $(1 \leq p \leq \infty)$ be the operator defined by

$$Vx(t) = \int\limits_0^1 K(s,t)x(s)\,\mathrm{d}s = \int\limits_0^t x(s)\,\mathrm{d}s$$

for $x \in L_p[0,1]$. This operator is known as the *Volterra operator*. Show that the adjoint operator of the Volterra operator is given by

$$V^* y(s) = \int\limits_s^1 y(t)\,\mathrm{d}t.$$

10.46. Let k be a nonnegative measurable function in $(0,\infty)$ such that

$$\int\limits_0^\infty k(x)x^{s-1}\,\mathrm{d}x = \varphi(s)$$

for $0 < s < 1$. Let f be a nonnegative measurable function in $(0,\infty)$. Let us define

$$Tf(x) = \int\limits_0^\infty k(xy)f(y)\,\mathrm{d}y.$$

Prove that

$$\|Tf\|_2 \leq \varphi\left(\frac{1}{2}\right) \|f\|_2.$$

What can be said about Tf and $\varphi(s)$ if $k(x) = \mathrm{e}^{-x}$?

10.47. Let $(X, \mathscr{A}, \mu)$ be a measure space and $f \in L_p(X, \mathscr{A}, \mu)$. If

$$\mu\left(\{x \in X : |F(x)| > \lambda\}\right) \leq \frac{1}{\lambda} \int_{\{x \in X : |F(x)| > \lambda\}} |f| \, \mathrm{d}\mu.$$

Show that

$$\|F\|_p \leq \frac{p}{p-1} \|f\|_p.$$

10.48. Let $f \in L_p((0,\infty), \mathscr{L}, m)$. For each $t > 0$, let us define

$$Sf(t) = \int_0^\infty \min\left(1, \frac{s}{t}\right) f(s) \frac{ds}{s},$$

prove that

$$\|Sf\|_p \leq \frac{p^2}{p-1} \|f\|_p.$$

10.49. Let $T : L_p(\mu) \to L_p(\mu)$ be a continuous operator where $1 < p < \infty$ and $0 \leq r \leq p$. Demonstrate that

(a) If $f \in L_p(\mu)$, then $|f|^{p-r}|Tf|^r \in L_1(\mu)$ and

$$\int |f|^{p-r}|Tf|^r \, \mathrm{d}\mu \leq \|T\|^r \left(\|f\|_p\right)^r.$$

(b) If for some $f \in L_p(\mu)$ with $\|f\|_p \leq 1$ we have that

$$\int |f|^{p-r}|Tf|^r \, \mathrm{d}\mu = \|T\|^r,$$

then

$$|Tf| = \|T\| |f|.$$

10.50. If $f \in L_1\left((0,\infty), \mathscr{L}, m\right)$, show that

(a)

$$\int_0^\infty \mathrm{e}^{\frac{1}{x}\int_0^x \log f(t)\,\mathrm{d}t} \leq \mathrm{e} \int_0^\infty f(x)\,\mathrm{d}x$$

(b) For $0 < p < 1$

$$\int_0^\infty \left(\mathrm{e}^{\frac{1}{x}\int_0^x \log f(t)dt}\right) x^p \,\mathrm{d}x \leq \frac{e}{1-p} \int_0^\infty f(x) x^p \,\mathrm{d}x$$

(c) Let f be a nonnegative measurable function in $(0,b)$, $0 < b \leq \infty$ such that $0 < \int_0^b [f(x)]^p \, dx < \infty$. Show that

(c1) for $p \geq 1$,

$$\int_0^b \left(\frac{1}{x} \int_0^x f(t) dt \right)^p \frac{dx}{x} \leq \int_0^b \left(1 - \frac{t}{b}\right) [f(t)]^p \frac{dt}{t}.$$

(c2) for $p \geq 1$,

$$\int_0^b x^{-\frac{1}{p}+1} \left(\frac{1}{x} \int_0^b f(t) \, dt \right)^p dx \leq \frac{p}{p-1} b^{\frac{p-1}{p}} \int_0^b \left[1 - \left(\frac{x}{b} \right)^{\frac{p-1}{p}} \right] [f(x)]^p \, dx.$$

10.51. Let $1 < q < \infty$ and p such that $\frac{1}{p} + \frac{1}{q} = 1$. Define

$$T(f)(x) = \int_{\mathbb{R}} k(x,t) f(t) \, dt$$

Prove that for all $f \in L_p(\mathbb{R})$, the operator T is linear and bounded from $L_p(\mathbb{R})$ into $L_q(\mathbb{R})$ and moreover

$$\|T\| \leq \left(\int \int |k(x,t)|^q \, dt \, dx \right)^{1/q}.$$

10.52. Let $1 < p < \infty$ and $Tf(x) = x^{-1/p} \int_0^x f(t) \, dt$ with $\frac{1}{p} + \frac{1}{q} = 1$. Show that T is a linear and bounded operator from $L_q(0,\infty)$ into $C_0((0,\infty))$.

10.53. Let $s < r - 1$ and $r > 1$. Let f be defined in $(0,\infty)$ such that

$$\int_0^\infty |f(x)|^r x^s \, dx < \infty.$$

Let $F(x) = \int_0^x f(t) \, dt$. Prove that

$$\left(\int_0^\infty \left| \frac{F(x)}{x} \right|^r x^s \, dx \right)^{1/r} \leq \frac{r}{r-s-1} \left(\int_0^\infty |f(x)|^r x^s \, dx \right)^{1/r}.$$

10.54. Let $s < r-1$ and $r > 1$. Let f be a differentiable function a.e. in $(0,\infty)$ such that

$$\int_0^\infty |f'(x)|^r x^s \,\mathrm{d}x < \infty,$$

and moreover f satisfies the following properties:

(a) $f(0) = 0$.
(b) $f(\infty) = \lim_{t\to\infty} f(t) = 0$.

Demonstrate that

$$\left(\int_0^\infty |f(x)|^r x^{s-r} \,\mathrm{d}x\right)^{1/r} \leq \frac{r}{r-s-1}\left(\int_0^\infty |f'(x)|^r x^s \,\mathrm{d}x\right)^{1/r}.$$

10.55. Let $\lambda > 0$, if the differential equation

$$\lambda \frac{d}{\mathrm{d}x}\left(y'(x)\right)^{q/p'} + g(x)\left[y(x)\right]^{q/p'} = 0$$

has solution y such that

(a) $y(0) = y(\infty) = 0$.
(b) $y(x) > 0$.
(c) $y'(x) > 0$.

$0 < x < \infty$. Prove that

$$\left(\int_0^\infty |u(x)|^q g(x) \,\mathrm{d}x\right)^{1/q} \leq \lambda^{1/q}\left(\int_0^\infty |u'(x)|^p \,\mathrm{d}x\right)^{1/p}$$

for all function $u(x)$ such that

$$u(x) \in \mathrm{AC}[0,\infty)$$
$$u(0) = \lim_{t\to\infty} u(t) = 0.$$

10.56. Suppose that f and g are nonnegative measurable functions in $(0,\infty)$ and

(a) $\int_0^\infty f(t) t^{-1/2} \,\mathrm{d}t < \infty$.
(b) $\int_0^\infty \left[g(t)\right]^2 \mathrm{d}t < \infty$.

Prove that

$$\int_0^\infty \int_0^x \frac{g(x)}{x} f(t) \,\mathrm{d}t\,\mathrm{d}x < \infty.$$

10.57. Let $(X, \mathscr{A}, \mu)$ be a measure space and u, v are $\mathscr{A}$-measurable nonnegative functions such that

$$t\mu\left(\{x \in X : u(x) \geq t\}\right) \leq \int_{\{u(x)\geq t\}} v\,\mathrm{d}\mu.$$

If $u, v \in L_p(X, \mathscr{A}, \mu)$ show that

$$\|u\|_p \leq \frac{p}{p-1}\|v\|_p.$$

10.58. Let $f \geq 0$ and $f(x) = \int_0^x f(t)\,\mathrm{d}t$. Prove that

$$\left(\int_0^1 [F(x)]^p\,\mathrm{d}x\right)^{1/p} \leq \frac{p}{p-1}\left(\int_0^1 [f(x)]^p\,\mathrm{d}x\right)^{1/p}$$

for $1 < p < \infty$.

10.59. Let $T : C(X) \to \mathbb{R}$ with $\mu(X) < \infty$ ($C(X)$ denotes the space of all continuous functions in X) such that $T(f) = \int_X f\,\mathrm{d}\mu$. Show that T is a linear operator and find $\|T\|$.

10.60. Let φ be a Lebesgue measurable function defined in $(0,1)$ such that $t\varphi(t) \in L_p\left((0,1), \mathscr{L}, \frac{dt}{t}\right)$. Prove that

$$\left\|(1+|\log t|)^{-1}\int_t^1 \varphi(s)\,\mathrm{d}s\right\|_{L_p\left(\frac{dt}{t}\right)} \leq \frac{p}{p-1}\left\|t\varphi(t)\right\|_{L_p\left(\frac{dt}{t}\right)}.$$

10.61. Let g be a positive measurable function in $(0,\infty)$. Let φ be a convex function in $(0,\infty)$. Show that

$$\int_0^\infty \varphi\left(\frac{1}{x}\int_0^x g(t)\,\mathrm{d}t\right)\frac{\mathrm{d}x}{x} \leq \int_0^\infty \varphi(g(x))\frac{\mathrm{d}x}{x}.$$

10.62. Prove that: *Let X be an Hilbert space. If $x_n, y_n, x, y \in X$, $x_n \to x$ and $y_n \to y$, then $\langle x_n, y_n\rangle \to \langle x, y\rangle$.*

10.63. Prove Theorem 10.23.

10.4 Notes and Bibliographic References

The Minkowski integral inequality already appears in Hardy, Littlewood, and Pólya [30] and Theorem 10.2 is due to Hardy, Littlewood, and Pólya [29]. The Hardy operator and Hardy's inequality (10.11) are from Hardy [26]. The proof of Radon-Nikodym based upon the Riesz representation of functionals on Hilbert spaces is due to von Neumann [81].

Chapter 11
Convolution and Potentials

Abstract In this chapter we study the convolution which is a very powerful tool and some operators defined using the convolution. We first start with a detailed study about the translation operator and after that we introduce the convolution operator and give some immediate properties of the operator. As an immediate application we show that the convolution with the Gauss-Weierstrass kernel is an approximate identity operator. We also study the Young inequality for the convolution operator. The definition of a support of a convolution is given based upon the definition of the support of a (class of) function which differs from the classical definition of support of a function. Approximate identity operators are studied in a general framework via Dirac sequences and Friedrich mollifiers. We end the chapter with a succinct study of the Riesz potential.

11.1 Convolution

Definition 11.1. Let $E \subset \mathbb{R}$ and let us define $\sigma(E) \subset \mathbb{R}^2$ as

$$\sigma(E) = \left\{(x,y) \in \mathbb{R}^2 : x - y \in E\right\}.$$

$\oslash$

Theorem 11.2 *Let $T : \mathbb{R}^2 \to \mathbb{R}$ defined by $T(x,y) = x + y$. Then T is continuous at the origin.*

Proof. In the first place, we want to show that $T(V \times V) = V + V$ for $V \subset \mathbb{R}$, where $V + V = \left\{x + y : x \in V, y \in V\right\}$. In fact, let $z \in T(V \times V)$ which is equivalent to the existence of $(x,y) \in V \times V$ such that $z = T(x,y)$ and this is equivalent to $z = x + y$ with $x \in V$ and $y \in V$. Then $T(V \times V) = V + V$.

R.E. Castillo, H. Rafeiro, *An Introductory Course in Lebesgue Spaces*, CMS Books in Mathematics, DOI 10.1007/978-3-319-30034-4_11

In the second place, we want to show that T is continuous at 0. To do this, let U be a neighborhood of 0 in $\mathbb{R}$ and V a neighborhood of 0 in $\mathbb{R}^2$, but

$$T(V \times V) = V + V \subset U,$$

showing in this way the continuity of T at 0. □

Remark 11.3. An alternative proof of Proposition 11.2 is to observe that $\mathbb{R}^2$, as a normed space, can be equipped with the norm

$$\|(x,y)\|_{\mathbb{R}^2} = \|x\|_{\mathbb{R}} + \|y\|_{\mathbb{R}},$$

therefore, for $T(x,y) = x+y$, we have

$$\begin{aligned}\|T(x,y) - T(x_0,y_0)\|_{\mathbb{R}} &= \|x - x_0 + y - y_0\|_{\mathbb{R}} \\ &\le \|x - x_0\|_{\mathbb{R}} + \|y - y_0\|_{\mathbb{R}} \\ &= \|(x - x_0, y - y_0)\|_{\mathbb{R}^2} \\ &= \sqrt{(x-x_0)^2 + (y-y_0)^2} < \delta = \varepsilon.\end{aligned}$$

The last result tells us that the continuity of T in (x_0, y_0) is uniform.

Theorem 11.4 *Let $h : \mathbb{R}^2 \to \mathbb{R}$ be defined by $h(x,y) = x - y$. Then*

(a) h is continuous at 0.
(b) $h^{-1}(E) = \sigma(E)$ for all $E \in \mathbb{R}$.
(c) If E is open in $\mathbb{R}$, then $\sigma(E)$ is open in $\mathbb{R}^2$.
(d) If E is closed in $\mathbb{R}$, then $\sigma(E)$ is closed in $\mathbb{R}^2$.
(e) $\sigma\left(\bigcup_{n=1}^{\infty} E_n\right) = \bigcup_{n=1}^{\infty} \sigma(E_n)$
(f) $\sigma\left(\bigcap_{n=1}^{\infty} E_n\right) = \bigcap_{n=1}^{\infty} \sigma(E_n)$.

Proof. Item (a) is an immediate consequence of Proposition 11.1.

(b) Let

$$\begin{aligned}E \in \mathbb{R} \text{ and } (x,y) \in h^{-1}(E) &\Leftrightarrow h(x,y) \in E \\ &\Leftrightarrow x - y \in E \\ &\Leftrightarrow (x,y) \in \sigma(E).\end{aligned}$$

Therefore $h^{-1}(E) = \sigma(E)$.

(c) and (d) are obtained from (b).

(e) $(x,y) \in \sigma\left(\bigcup_{n=1}^{\infty} E_n\right) \Leftrightarrow x - y \in \bigcup_{n=1}^{\infty} E_n \Leftrightarrow x - y \in E_n$ for some $n \Leftrightarrow (x,y) \in \sigma(E_n) \Leftrightarrow (x,y) \in \bigcup_{n=1}^{\infty} \sigma(E_n)$. In this way we proved that

$$\sigma\left(\bigcup_{n=1}^{\infty} E_n\right) = \bigcup_{n=1}^{\infty} \sigma(E_n).$$

(f) Exercise. □

Lemma 11.5. *If $E \subset \mathbb{R}$ is a Lebesgue measurable set, then $\sigma(E)$ is a measurable set in the product space.*

Proof. Let us suppose that E is a bounded set, then $m(E) < \infty$, and the Lemma is true.

If E is a G_δ or F_σ set. In fact, by known results in measure theory, we can find a set K which is F_σ and a set H which is G_δ such that

$$K \subset E \subset H \quad \text{and} \quad m(K) = m(E) = m(H). \tag{11.1}$$

Then $m(H \backslash K) = 0$. It is clear that $\sigma(K) \subset \sigma(E) \subset (H)$. Note that for all $A \subset \mathbb{R}$ we have that

$$\chi_{\sigma(A)}(x,y) = \chi_A(x-y).$$

Therefore, for each $x \in \mathbb{R}$ we get

$$\begin{aligned} \int_{\mathbb{R}} \chi_{\sigma(K)}(x,y)\,\mathrm{d}y &= \int_{\mathbb{R}} \chi_K(x-y)\,\mathrm{d}y \\ &= \int_{\mathbb{R}} \chi_K(-y)\,\mathrm{d}y \\ &= \int_{\mathbb{R}} \chi_K(y)\,\mathrm{d}y \\ &= m(K) \end{aligned} \tag{11.2}$$

Let $C \in \mathbb{R}$ be an arbitrary bounded set in $\mathbb{R}$, using the Tonelli theorem we get

$$\int_{C\times\mathbb{R}} \chi_{\sigma(K)}\,\mathrm{d}m\otimes m = \int_C \int_{\mathbb{R}} \chi_{\sigma(K)}\,\mathrm{d}y\,\mathrm{d}x = \int_C m(K)\,\mathrm{d}x = m(K)m(C).$$

Similarly, we can show that

$$\int_{C\times\mathbb{R}} \chi_{\sigma(H)}\,\mathrm{d}m\otimes m = m(H)m(C). \tag{11.3}$$

Therefore

$$\int_{C\times\mathbb{R}} \chi_{\sigma(H)} - \chi_{\sigma(K)}\,\mathrm{d}m\otimes m = 0. \tag{11.4}$$

Since $\sigma(K) \subset \sigma(H)$, then

$$\chi_{\sigma(H)} - \chi_{\sigma(K)} = \chi_{\sigma(H)\Delta\sigma(K)} = \chi_{\sigma(H)-\sigma(K)},$$

then

$$\int_{C\times\mathbb{R}} \chi_{\sigma(H)-\sigma(K)} \, \mathrm{d}m \times m = \int_{C\times\mathbb{R}} \chi_{\sigma(H)-\sigma(K)} \, \mathrm{d}m \otimes m$$

i.e.

$$\int_{C\times\mathbb{R}} \chi_{\sigma(H)-\sigma(K)} \, \mathrm{d}m \otimes m = m \otimes m\Big([\sigma(H) - \sigma(K)] \cap C \times \mathbb{R} \Big).$$

By (11.4) we get that

$$m \otimes m\Big([\sigma(H) - \sigma(K)] \cap C \times \mathbb{R} \Big) = 0.$$

In particular

$$m \otimes m\Big([\sigma(H) - \sigma(K)] \cap [-n,n] \times \mathbb{R} \Big) = 0,$$

for all $n \in \mathbb{N}$, but

$$\bigcup_{n=1}^{\infty} \Big([\sigma(H) - \sigma(K)] \cap [-n,n] \times \mathbb{R} \Big) = \sigma(H) - \sigma(K),$$

from which we conclude that

$$m \otimes m\Big(\sigma(H) - \sigma(K) \Big) = 0.$$

On the other hand, we know that

$$\sigma(E) - \sigma(K) \subset \sigma(H) - \sigma(K)$$

since $m \otimes m$ is a complete measure we have that $\sigma(E) - \sigma(K)$ is a measurable set, moreover $\sigma(K)$ is a set F_σ, therefore

$$\sigma(E) = \sigma(K) \cup \Big(\sigma(E) - \sigma(K) \Big)$$

is a measurable set, and this finishes the proof of Lemma 11.5. □

Corollary 11.6 *Let f be a measurable function. Let us define $F : \mathbb{R}^2 \to \mathbb{R}$ by $F(x,y) = f(x-y)$. Then F is measurable in $\mathbb{R}^2$.*

Proof. Let $h : \mathbb{R}^2 \to \mathbb{R}$ be defined by $h(x,y) = x - y$. Note that

$$F(x,y) = f(h(x,y)) = f \circ h(x,y).$$

Let $\alpha \in \mathbb{R}$, then

$$\begin{aligned}\Big\{(x,y) \in \mathbb{R}^2 : F(x,y) < \alpha\Big\} &= F^{-1}(-\infty,\alpha) \\ &= (f \circ h)^{-1}(-\infty,\alpha) \\ &= h^{-1}\Big(f^{-1}(-\infty,\alpha)\Big).\end{aligned}$$

Since f is measurable, then $f^{-1}(-\infty,\alpha)$ is also measurable.

Let $F(x,y) = f(x-y)$ and $G(x,y) = g(y)$.

In virtue of Corollary 11.6, the function F is measurable in $\mathbb{R}^2$, Now, observe that

$$\Big\{(x,y) \in \mathbb{R}^2 : G(x,y) < \alpha\Big\} = \mathbb{R} \times g^{-1}(-\infty,\alpha),$$

then G is measurable in $\mathbb{R}^2$. Then ϕ is measurable in $\mathbb{R}^2$.

On the other hand, by Tonelli's Theorem, we get

$$\begin{aligned}\int_{\mathbb{R}\times\mathbb{R}} |\phi(x,y)|\,\mathrm{d}m \times m &= \int_{\mathbb{R}} \left[\int_{\mathbb{R}} |f(x-y)|\,\mathrm{d}x\right] |g(y)|\,\mathrm{d}y \\ &= \int_{\mathbb{R}} \left[\int_{\mathbb{R}} |f(x)|\,\mathrm{d}x\right] |g(y)|\,\mathrm{d}y \\ &= \|f\|_1 \int_{\mathbb{R}} |g(y)|\,\mathrm{d}y \\ &= \|f\|_1 \|g\|_1 < \infty,\end{aligned}$$

if $E = f^{-1}(-\infty,\alpha)$, then E is measurable, then

$$\Big\{(x,y) \in \mathbb{R}^2 : F(x,y) < \alpha\Big\} = h^{-1}(E),$$

in other words, $\Big\{(x,y) \in \mathbb{R}^2 : F(x,y) < \alpha\Big\} = \sigma(E)$. By Lemma 11.5 the set $\sigma(E)$ is measurable, therefore F is measurable. □

We now introduce the convolution of two functions in $\mathbb{R}$.

Theorem 11.7. *Let $f,g \in L_1\Big(\mathbb{R},\mathscr{L},m\Big)$. For each $x \in \mathbb{R}$ let us define*

$$C(x) = (f * g)(x) := \int_{\mathbb{R}} f(x-y)g(y)\,\mathrm{d}y$$

Then $C \in L_1(m)$ and moreover $\|C\|_1 \le \|f\|_1 \|g\|_1$.

Proof. First we should prove that if $\varphi(x,y) = f(x-y)g(y)$, then φ is measurable in $\mathbb{R}^2$. This follows from the fact that the product of measurable functions is again a measurable function and Corollary 11.6. Therefore $\varphi \in L_1(\mathbb{R}^2)$. By Tonelli's theorem, we get

$$\begin{aligned}\int_{\mathbb{R}} |C(x)|\,\mathrm{d}x &\leq \int_{\mathbb{R}} \left[\int_{\mathbb{R}} |f(x-y)||g(y)|\,\mathrm{d}y\right] \mathrm{d}x \\ &= \int_{\mathbb{R}} \left[\int_{\mathbb{R}} |f(x-y)|\,\mathrm{d}x\right] |g(y)|\,\mathrm{d}y \\ &= \int_{\mathbb{R}} \left[\int_{\mathbb{R}} |f(x)|\,\mathrm{d}x\right] |g(y)|\,\mathrm{d}y \\ &= \|f\|_1 \|g\|_1.\end{aligned}$$

Therefore, we get $\|C\|_1 \leq \|f\|_1 \|g\|_1$. □

The notion of convolution can be introduced in more abstract frameworks, we will give only in the framework of n-dimensional Euclidean spaces.

Definition 11.8. Let $f : \mathbb{R}^n \longrightarrow \mathbb{R}$ and $g : \mathbb{R}^n \longrightarrow \mathbb{R}$. The *convolution of f and g*, denoted by $f * g$, is given *formally* by

$$(f * g)(x) = \int_{\mathbb{R}^n} f(x-y)g(y)\,\mathrm{d}y.$$

⊘

We will show some results regarding the convolution in the one-dimensional case, but many of the proofs can be adapted to the multidimensional case.

We start with the following properties, which are almost immediate from the definition of convolution.

Theorem 11.9. *Let $f, g \in L_1(m)$ and $\alpha, \beta \in \mathbb{C}$, then we have*

(a) $f * g = g * f$ (Commutativity)
(b) $(f * g) * h = f * (g * h)$ (Associativity)
(c) $f * (\alpha g + \beta h) = \alpha(f * g) + \beta(f * h)$ (Distributivity)

We now guarantee that the convolution belongs to the Lebesgue space under some hypothesis on the belongness of some Lebesgue spaces of the functions f and g.

Theorem 11.10. *Let $g \in L_1(m)$ and $f \in L_p(m)$ with $1 \leq p \leq \infty$. Then $\|f * g\|_p \leq \|g\|_1 \|f\|_p$*

Proof. If $p = \infty$ and $f \in L_\infty$, then

$$
\begin{aligned}
|(f*g)(x)| &\le \int_{\mathbb{R}} |f(x-y)||g(y)|\,\mathrm{d}y \\
&\le \|f\|_\infty \int_{\mathbb{R}} |g(y)|\,\mathrm{d}y \\
&= \|g\|_1 \|f\|_\infty,
\end{aligned}
$$

from which

$$\|(f*g)(x)\|_\infty \le \|g\|_1 \|f\|_\infty.$$

Let $1 < p < \infty$ and $1 < q < \infty$ the conjugate exponent of p, i.e., $\frac{1}{p} + \frac{1}{q} = 1$. Now, note that by the Hölder inequality, Tonelli's Theorem, and the translation invariance of the Lebesgue measure, we have

$$
\begin{aligned}
\|f*g\|_p^p &= \int_{\mathbb{R}} \left[\int_{\mathbb{R}} |f(x-y)| \big(g(y)\big)^{1/p} \big(g(y)\big)^{1/q} \mathrm{d}y \right]^p \mathrm{d}x \\
&\le \int_{\mathbb{R}} \left[\left(\int_{\mathbb{R}} |f(x-y)|^p |g(y)|\,\mathrm{d}y \right)^{1/p} \right]^p \left[\int_{\mathbb{R}} |g(y)|\,\mathrm{d}y \right]^{p/q} \mathrm{d}x \\
&= \int_{\mathbb{R}} \left(\int_{\mathbb{R}} |f(x-y)|^p |g(y)|\,\mathrm{d}y \right) \|g\|_1^{p/q}\,\mathrm{d}x \\
&= \|g\|_1^{p/q} \int_{\mathbb{R}} \left(\int_{\mathbb{R}} |f(x-y)|^p |g(y)|\,\mathrm{d}y \right) \mathrm{d}x \\
&= \|g\|_1^{p/q} \left(\int_{\mathbb{R}} |f(x-y)|^p\,\mathrm{d}x \right) \left(\int_{\mathbb{R}} |g(y)|\,\mathrm{d}y \right) \\
&= \|g\|_1^{p/q} \|f\|_p^p \|g\|_1 \\
&= \|g\|_1^{\frac{p}{q}+1} \|f\|_p^p.
\end{aligned}
$$

From which it follows that $\|f*g\|_p \le \|g\|_1^{\frac{1}{q}+\frac{1}{p}} \|f\|_p \le \|g\|_1 \|f\|_p$. □

Theorem 11.10 plays an important role in the theory of semi-groups. For example, let us define in $L_p(m)$ the following operator

$$T_t(f)(x) = \frac{1}{\sqrt{4\pi t}} \int_{\mathbb{R}} f(y) \exp\left(-\frac{|x-y|^2}{4t}\right) dy.$$

With $f \in L_p(m)$, then we can write

$$T_t(f)(x) = (G_t * f)(x), \quad \text{where} \quad G_t(x) = \frac{1}{\sqrt{4\pi t}} \exp\left(-\frac{|x|^2}{4t}\right).$$

By the Theorem 11.10 we have

$$\|T_t(f)\|_p = \|G_t * f\|_p \le \|G_t\|_1 \|f\|_p,$$

but

$$\|G_t\|_1 = \frac{1}{\sqrt{4\pi t}} \int_{\mathbb{R}} \exp\left(-\frac{|x|^2}{4t}\right) dx = \frac{1}{\sqrt{\pi}} \int_{\mathbb{R}} e^{-y^2} dy = \frac{\sqrt{\pi}}{\sqrt{\pi}} = 1.$$

Finally $\|T_t(f)\| \le 1$, and taking $T(0) = I$ the identity operator, it can be shown that $T_t T_s(f) = T_{t+s}(f)$. This semigroup is called the *Gauss-Weierstrass semigroup.*

The following application of the Theorem 11.10 is remarkable, since transforms an operator that is not defined via convolution and in this way we can apply Theorem 11.10.

Let H be the Hardy operator given in the Definition 10.4, i.e.

$$Hf(x) = \frac{1}{x} \int_0^x f(y)\, dy \quad \text{for} \quad 0 < x < \infty.$$

Let us do the following change of variables $x = e^s$ and $y = e^t$. Observe that

$$Hf(e^s) = e^{-s} \int_{-\infty}^{s} f(e^t) e^t \, dt.$$

On the other hand, note that

$$\|f\|_p^p = \int_0^\infty |f(x)|^p \, dx = \int_{-\infty}^{\infty} |f(e^s)|^p e^s \, ds = \|e^{s/p} f(e^s)\|_{L_p}^p.$$

Now, the equation

$$Hf(e^s) = e^{-s} \int_{-\infty}^{s} f(e^t) e^t \, dt$$

give us

$$\begin{aligned} e^{s/p} Hf(e^s) &= e^{-s/q} \int_{-\infty}^{s} e^{t/p} f(e^t) e^{t/q}\, dt \\ &= \int_{-\infty}^{s} e^{t/p} f(e^t) e^{-\frac{s-t}{q}}\, dt \\ &= \int_{-\infty}^{s} e^{t/p} f(e^t) g(s-t)\, dt, \end{aligned}$$

where

$$g(y) = \begin{cases} e^{-y/q} & \text{if } 0 < y < \infty \\ 0 & \text{if } -\infty < y < 0. \end{cases}$$

As we can see, we transformed (via change of variables) the Hardy operator into an operator defined via convolution, and now we can apply the Theorem 11.10 and obtain

$$\|Hf\|_p = \|e^{s/p} Hf(e^s)\|_p \leq \|g\|_1 \|e^{t/p} f(e^t)\|_p = \|g\|_1 \|f\|_p.$$

Since

$$\|g\|_1 = \int_{-\infty}^{\infty} |g(y)|\, dy = \int_{0}^{\infty} |g(y)|\, dy = \int_{0}^{\infty} e^{-y/q}\, dy = q = \frac{p}{p-1},$$

we finally obtain

$$\|Hf\|_p \leq \frac{p}{p-1} \|f\|_p.$$

If we fix $g \in L_1(m)$ and define

$$T(f) = f * g.$$

then we can interpret the Theorem 11.10 in the following way. For $1 \leq p \leq \infty$ the operator $T : L_p(m) \to L_p(m)$ is a linear bounded operator.

We now obtain the so-called Young inequality for the convolution operator in the one-dimensional case for simplicity, but the result is valid also in $\mathbb{R}^n$ as already proved in Theorem 8.9 via interpolation theory.

Theorem 11.11 (Young's Inequality for Convolution). *Let p, q and r be real numbers such that $p > 1$, $q > 1$ and $\frac{1}{p} + \frac{1}{q} - 1 = \frac{1}{r} > 0$. Let $f \in L_p(m)$ and $g \in L_q(m)$. Then $f * g \in L_r(m)$ and*

$$\|f * g\|_r \leq \|f\|_p \|g\|_q.$$

Proof. Let a, b and c be real numbers such that $\frac{1}{p} = \frac{1}{a} + \frac{1}{b}$, $\frac{1}{q} = \frac{1}{a} + \frac{1}{c}$ and $a = r$. Note that

$$\frac{1}{a} + \frac{1}{b} + \frac{1}{c} = \left(\frac{1}{a} + \frac{1}{b}\right) + \left(\frac{1}{a} + \frac{1}{c}\right) - \frac{1}{a} = \frac{1}{p} + \frac{1}{q} - \frac{1}{r} = 1.$$

Now, we can write

$$\begin{aligned} |f(x-y)g(y)| &= |f(x-y)||g(y)| \\ &= \left(|f(x-y)|^{p/a}|g(y)|^{q/a}\right)\left(|f(x-y)|^{p(\frac{1}{p}-\frac{1}{a})}|g(y)|^{q(\frac{1}{q}-\frac{1}{a})}\right). \end{aligned}$$

By Corollary 3.22 (generalized Hölder's inequality), we obtain

$$\begin{aligned} \int_{\mathbb{R}} |f(x-y)g(y)|\,\mathrm{d}y \leq & \left(\int_{\mathbb{R}} \left[|f(x-y)|^{p/a}|g(y)|^{q/a}\right]^a \mathrm{d}y\right)^{1/a} \\ & \times \left(\int_{\mathbb{R}} |f(x-y)|^{pb(\frac{1}{p}-\frac{1}{a})}\,\mathrm{d}y\right)^{1/b} \left(\int_{\mathbb{R}} |g(y)|^{qc(\frac{1}{q}-\frac{1}{a})}\,\mathrm{d}y\right)^{1/c}, \end{aligned}$$

but $\dfrac{1}{p} - \dfrac{1}{a} = \dfrac{1}{b}$ and $\dfrac{1}{q} - \dfrac{1}{a} = \dfrac{1}{c}$, therefore

$$\begin{aligned} &\int_{\mathbb{R}} |f(x-y)g(y)|\,\mathrm{d}y \leq \\ &\left(\int_{\mathbb{R}} |f(x-y)|^p|g(y)|^q\,\mathrm{d}y\right)^{1/a} \left(\int_{\mathbb{R}} |f(x-y)|^p\,\mathrm{d}y\right)^{1/b} \left(\int_{\mathbb{R}} |g(y)|^q\,\mathrm{d}y\right)^{1/c} \end{aligned}$$

i.e.

$$\int_{\mathbb{R}} |f(x-y)g(y)|\,\mathrm{d}y \leq \left(\int_{\mathbb{R}} |f(x-y)|^p|g(y)|^q\,\mathrm{d}y\right)^{1/r} \|f\|_p^{p/b}\|g\|_q^{q/c}.$$

Let us define

$$h(x) = \int_{\mathbb{R}} |f(x-y)g(y)|\,\mathrm{d}y,$$

then

$$|h(x)|^r \leq \left(\int_{\mathbb{R}} |f(x-y)|^p|g(y)|^q\,\mathrm{d}y\right) \|f\|_p^{r\frac{p}{b}}\,\|g\|_q^{r\frac{q}{c}},$$

therefore

$$\left(\int_{\mathbb{R}} |h(x)|^r \, dx\right)^{1/r} \leq \|f\|_p^{p(\frac{1}{b}+\frac{1}{r})} \|g\|_q^{q(\frac{1}{c}+\frac{1}{r})}.$$

Note that

$$p(\frac{1}{b}+\frac{1}{r}) = p(\frac{1}{b}+\frac{1}{a}) = 1$$

and

$$q(\frac{1}{c}+\frac{1}{r}) = q(\frac{1}{c}+\frac{1}{a}) = 1.$$

Then

$$\left(\int_{\mathbb{R}} |h(x)|^r \, dx\right)^{1/r} \leq \|f\|_p \|g\|_q,$$

from this last inequality it is easy to see that $f * g \in L_r(m)$, therefore

$$\|f * g\|_r \leq \|f\|_p \|g\|_q,$$

which ends the proof. □

11.2 Support of a Convolution

The classical definition of support of a function is well known, for example for $f : \mathbb{R}^n \to \mathbb{R}$ the support is given by

$$\text{supp}(f) = \overline{\{x \in \mathbb{R}^n : f(x) \neq 0\}} \tag{11.5}$$

which is always a closed set. From the definition (11.5) we have that if $x \notin \text{supp}(f)$, then there exists an open neighborhood of the point x where the function f is zero. This notion is not robust enough when we deal with equivalent classes of functions, since taking different representations of the same class can give different results, e.g., let $f(x) = \chi_{\mathbb{Q}}(x)$ and $g(x) = 0$, which belong to the same equivalence class with the Lebesgue measure. Using the definition of support from (11.5) we get

$$\text{supp}(f) = \overline{\{x \in \mathbb{R} : f(x) \neq 0\}} = \mathbb{R}$$

but

$$\text{supp}(g) = \overline{\{x \in \mathbb{R} : g(x) \neq 0\}} = \emptyset$$

which is clearly different.

Since the notion of support is not robust enough for functions defined almost everywhere, we introduce the following notion of support in a negative way.

Definition 11.12. Let $x \in \mathbb{R}^n$. We say that $x \notin \operatorname{supp}(f)$ if and only if there exists an open V such that $x \in V$ and $f = 0$ in almost every point of V. $\oslash$

In this new notion of support, let us see what happens with the previous counterexample.

Example 11.13. Let $f(x) = \chi_{\mathbb{Q}}(x)$ and $g(x) = 0$. Using the classical definition of support (11.5) we already know that $\operatorname{supp}(f) \neq \operatorname{supp}(g)$ even so $f = g$ m-almost everywhere. Let us see what happens with the new definition of support given in Definition 11.12. Let $V = \big(|[x]| - 1, |[x]| + 1\big)$ be an open set. Since $f(x) = 0$ almost everywhere in $\mathbb{R}$ in particular in V, then

$$x \notin \operatorname{supp}(f) = \overline{\{x \in \mathbb{R} : f(x) \neq 0\}}$$

if $x \notin \mathbb{R}$, therefore $\operatorname{supp}(f) = \emptyset$, which now is equal to the support of g. $\oslash$

If f is a continuous function in $\mathbb{R}^n$ it is not difficult to show that this new definition coincides with the notion given in (11.5).

On the other hand, for $f = g$ a.e. in $\mathbb{R}^n$ with the new definition we have that $\operatorname{supp}(f) = \operatorname{supp}(g)$. In this sense, we can talk about the support of a measurable function.

Theorem 11.14 *If f and g have compact support, then $f * g$ has compact support. Moreover*

$$\operatorname{supp}(f * g) \subseteq \operatorname{supp}(f) + \operatorname{supp}(g).$$

Proof. Note that

$$(f * g)(x) = \int_{\mathbb{R}^n} f(t) g(x - t)\, dt = \int_{\operatorname{supp}(f)} f(t) g(x - t)\, dt$$

since if $t \notin \operatorname{supp}(f)$, then $f(t) = 0$.

Analogously if $x - t \in \operatorname{supp}(g)$, then $t \in x - \operatorname{supp}(g)$, which means that $g(x - t) = 0$ if $t \notin x - \operatorname{supp}(g)$. From this we get that

$$(f * g)(x) = \int_{\operatorname{supp}(f) \cap (x - \operatorname{supp}(g))} f(t) g(x - t)\, dt.$$

If $(f * g)(x) \neq 0$, then $\operatorname{supp}(f) \cap (x - \operatorname{supp}(g)) \neq \emptyset$ therefore there exists $y \in \operatorname{supp}(f) \cap (x - \operatorname{supp}(g))$. Since $y \in x - \operatorname{supp}(g)$, we have that $y = x - w$ with $w \in \operatorname{supp}(g)$. Since $x = y + w$ with $y \in \operatorname{supp}(f)$, $w \in \operatorname{supp}(g)$.

We have proved that

$$\{(f * g)(x) \neq 0\} \subseteq \operatorname{supp}(f) + \operatorname{supp}(g),$$

but the sum of two compact sets is compact, therefore $\operatorname{supp}(f * g) \subset \operatorname{supp}(f) + \operatorname{supp}(g)$. $\square$

The next result shows that, in some sense, the convolution function preserves the best properties from the convoluted functions.

Theorem 11.15. *If $f \in L_1(\mathbb{R}^n, \mathscr{L}, m)$ and k is uniformly continuous and bounded in $\mathbb{R}^n$, then $f * k$ is bounded and uniformly bounded.*

Proof. Let us see that $f * k$ is uniformly continuous. If $f = 0$, then $f * k = 0$. Suppose that $f \neq 0$. Given $\varepsilon > 0$, since k is uniformly continuous we can find a $\delta > 0$ such that if $|x - y| < \delta$, then $|k(x) - k(y)| < \frac{\varepsilon}{\|f\|_1}$. We now have

$$\begin{aligned}
|f * k(x) - f * k(y)| &= \left| \int_{\mathbb{R}^n} f(t)k(x-t)\,\mathrm{d}t - \int_{\mathbb{R}^n} f(t)k(y-t)\,\mathrm{d}t \right| \\
&\leq \int_{\mathbb{R}^n} |f(t)||k(x-t) - k(y-t)|\,\mathrm{d}t \\
&< \frac{\varepsilon}{\|f\|_1} \int_{\mathbb{R}^n} |f(t)|\,\mathrm{d}t \\
&= \varepsilon
\end{aligned}$$

since $|(x-t) - (y-t)| = |x - y| < \delta$, which shows that $f * k$ is the uniformly continuous. Now using Theorem 11.10 we have $\|f * g\|_\infty \leq \|g\|_1 \|k\|_\infty$. □

11.3 Convolution with Smooth Functions

We recall some standard notation regarding the space of differentiable functions.

For each $m \in \mathbb{N}$ we denote by C^m the class of functions having continuous partial derivatives up to order m. By C^∞ we mean the set of all infinite differentiable functions. By C_0^m we denote the subset of C^m where the functions have compact support and in a similar fashion we define C_0^∞.

If $\alpha = (\alpha_1, \alpha_2, \ldots, \alpha_n)$ is a multi-index, where $\alpha_i \in \mathbb{N} \cup \{0\}$, we denote the partial derivative as

$$(D^\alpha f)(x) = \frac{\partial^{|\alpha|} f}{\partial x_1^{\alpha_1} \partial x_2^{\alpha_2} \cdots \partial x_n^{\alpha_n}},$$

where $|\alpha| = \alpha_1 + \alpha_2 + \cdots + \alpha_n$.

We now study the behavior of the convolution when we convolve L_p functions with smooth functions. The convolution inherits the best properties of each *parent* function.

Theorem 11.16. *Let $1 \leq p \leq \infty$, $f \in L_p(\mathbb{R}^n, \mathscr{L}, m)$ and $k \in C_0^m$. Then $f * k \in C_0^m$ and moreover*

$$D^\alpha (f * k)(x) = (f * D^\alpha k)(x)$$

whenever $|\alpha| = \alpha_1 + \alpha_2 + \cdots + \alpha_n \leq m$.

Proof. Let us show first that if k is continuous with compact support then $f*k$ is continuous. We have

$$
\begin{aligned}
|(f*k)(x+h)-(f*k)(x)| \\
&= \left| \int_{\mathbb{R}^n} f(t)k(x+h-t)\,\mathrm{d}t - \int_{\mathbb{R}^n} f(t)k(x-t)\,\mathrm{d}t \right| \\
&= \left| \int_{\mathbb{R}^n} f(t)[k(x+h-t)-k(x-t)]\,\mathrm{d}t \right| \\
&= \left| \int_{\mathbb{R}^n} f(t)[k(u+h)-k(u)]\,\mathrm{d}u \right| \\
&\le \left(\int_{\mathbb{R}^n} |f(x-u)|^p\,\mathrm{d}u \right)^{\frac{1}{p}} \left(\int_{\mathbb{R}^n} |k(u+h)-k(u)|^q\,\mathrm{d}u \right)^{\frac{1}{q}}.
\end{aligned}
$$

We affirm that

$$
\lim_{h\to 0} \left(\int_{\mathbb{R}^n} |k(u+h)-k(u)|^q\,\mathrm{d}u \right)^{\frac{1}{q}} = 0.
$$

Since k is continuous and have compact support, then it is uniformly continuous in $\mathbb{R}^n$, hence, for given $\varepsilon > 0$ there exists a $\delta > 0$ such that for all $u \in \mathbb{R}^n$, if $|h| < \delta$, then

$$
|k(u+h)-k(u)| < \varepsilon.
$$

We can suppose moreover that $\delta < 1$. Therefore, if $|h| < \delta$, we have

$$
\begin{aligned}
\int_{\mathbb{R}^n} |k(u+h)-k(u)|^q\,\mathrm{d}u &= \int_I |k(u+h)-k(u)|^q\,\mathrm{d}u \\
&< \int_I \varepsilon^q\,\mathrm{d}u \\
&= \varepsilon^q m(I)
\end{aligned}
$$

where $I = \{x \in \mathbb{R}^n : d(x, \operatorname{supp}(k)) \le 1\}$ (which is compact and therefore have finite measure).

Let $k \in C_0^m$ $(m \ge 1)$, fix i with $1 \le i \le m$ and let e_i be the usual unit vector from the canonical base, then

$$\frac{(f*k)(x+h\cdot e_i)-(f*k)(x)}{h} = \int_{\mathbb{R}^n} f(t)\left[\frac{k(x-t+h\cdot e_i)-k(x-t)}{h}\right]\mathrm{d}t$$
$$= \int_{\mathbb{R}^n} f(t)\left[\frac{\partial k}{\partial x_i}(x-t+h^*)\right]\mathrm{d}t$$

by the mean value theorem, for some $h^* = \zeta \cdot e_i$ depending on x and t, where ζ is between 0 and h.

Therefore, when $|h| \to 0$, $\frac{\partial k}{\partial x_i}(x-t+h^*)$ converges to $\frac{\partial k}{\partial x_i}(x-t)$ uniformly in t.

Since $\frac{\partial k}{\partial x_i}$ has compact support, we deduce from the theorem on the uniform convergence that the last integral converges to

$$\int_{\mathbb{R}^n} f(t)\left[\frac{\partial k}{\partial x_i}(x-t)\right]\mathrm{d}t.$$

Consequently, $\frac{\partial}{\partial x_i}(f*k)(x)$ exists and is equal to $\left(f*\frac{\partial k}{\partial x_i}\right)(x)$ which is continuous by previous arguments. This shows the theorem for the case $m=1$, the proof for any m is obtained by induction. If follows that $f*k \in C^\infty$ if $f \in L_p$ $(1 \le p \le \infty)$ and $k \in C_0^\infty$. By the Theorem 11.15 $f*k$ has compact support. □

11.3.1 Approximate Identity Operators

One of the main applications of the convolution operator is the construction of the so-called *approximate identity operators*.

Definition 11.17. A sequence $\{\varphi_k\}$ of real-valued continuous functions in $\mathbb{R}^n$ is called a *Dirac sequence* if satisfies:

DIR1 $\varphi_k \ge 0$ for all k;

DIR2 For each k we have $\int_{\mathbb{R}^n} \varphi_k(x)\,\mathrm{d}x = 1$;

DIR3 Given $\varepsilon, \delta > 0$ there exists a k_0 such that

$$\int_{|x|\ge\delta} \varphi_k(x)\,\mathrm{d}x < \varepsilon$$

for all $k \ge k_0$.

⊘

In other words, all the functions φ_k have constant mass, the mass is concentrated around the origin and the functions are positive (Fig. 11.1).

The usefulness of the Dirac sequences steams from the following result.

Theorem 11.18. *Let f be a measurable and bounded function in $\mathbb{R}^n$, K be a compact set on which f is continuous and $\{\varphi_k\}$ be a Dirac sequence. Then $\varphi_k * f$ converges to f uniformly in the set K, i.e.*

$$\varphi_k * f \rightrightarrows f.$$

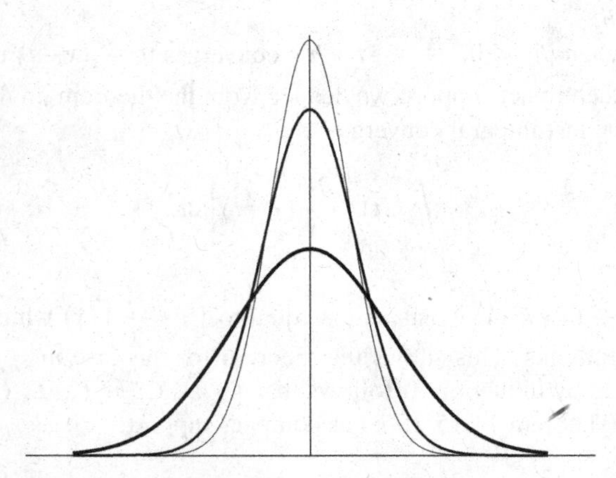

Fig. 11.1 Example of a Dirac sequence

Proof. Let $x \in K$. Then

$$\begin{aligned}|(\varphi_k * f)(x) - f(x)| &= \left| \int_{\mathbb{R}^n} \varphi_k(y) f(x-y)\,\mathrm{d}y - \int_{\mathbb{R}^n} f(x)\varphi_k(y)\,\mathrm{d}y \right| \\ &\leqslant \int_{\mathbb{R}^n} \varphi_k(y)\,|f(x-y) - f(x)|\,\mathrm{d}y \\ &\leqslant \left(\int_{|y|<\delta} + \int_{|y|\geq\delta} \right) \varphi_k(y)\,|f(x-y) - f(x)|\,\mathrm{d}y \\ &= I_{1,\delta} + I_{2,\delta},\end{aligned}$$

where the first equality follows from **DIR2**. Given $\varepsilon > 0$, let us choose $\delta > 0$ such that $|y| < \delta$, therefore for all $x \in K$ we have $|f(x-y) - f(x)| < \varepsilon$. Due to the choice of $\delta > 0$ we have that $I_{1,\delta} < \varepsilon$. To bound the integral $I_{2,\delta}$, we observe that **DIR3** guarantee that $I_{2,\delta} < 2\|f\|_\infty \varepsilon$ for k sufficiently large. □

We can construct Dirac sequences via *Friedrich mollifiers*.

Definition 11.19 (Friedrich Mollifier). Let $\varphi : \mathbb{R}^n \longrightarrow \mathbb{R}_+$ be a function with the following conditions:

(a) $\varphi \in C_0^\infty(\mathbb{R}^n)$;

(b) $\varphi(x) = 0$ when $|x| > 1$; and

(c) $\int\limits_{\mathbb{R}^n} \varphi(x)\,\mathrm{d}x = 1$.

We define the *Friedrich mollifier* φ_ε has

$$\varphi_\varepsilon(x) := \varepsilon^{-n}\varphi\left(\frac{x}{\varepsilon}\right)$$

for all $\varepsilon > 0$ and $x \in \mathbb{R}^n$. ⊘

Using Friedrichs mollifiers we can construct approximate identity operators with "smooth" properties, which can be obtained by convolving f with an appropriate Friedrichs mollifier. We have the following.

Theorem 11.20. *Let φ_ε be a Friedrichs mollifier, $1 \leqslant p < \infty$ and $f \in L^p(\mathbb{R}^n)$. Therefore:*

(a) $\varphi_\varepsilon * u \in C^\infty(\mathbb{R}^n)$;
(b) $\|\varphi_\varepsilon * u\|_{L^p} \leq \|u\|_{L^p}$;
(c) $\lim_{\varepsilon\to 0}\|\varphi_\varepsilon * u - u\|_{L^p} = 0$.

The proof of the theorem is not difficult. The idea of the proof of the item (c) is similar to the one given in Theorem 11.18 with the respective changes.

Remark 11.21. The function

$$(\varphi_\varepsilon * f)(x) = \frac{1}{\varepsilon^n}\int\limits_{\mathbb{R}^n} \varphi\left(\frac{x-u}{\varepsilon}\right) f(u)\,\mathrm{d}u = \frac{1}{\varepsilon}\int\limits_{\mathbb{R}^n} \varphi\left(\frac{u}{\varepsilon}\right) f(x-u)\,\mathrm{d}u$$

is sometimes denoted, especially in the Russian literature, the *Sobolev ε-average*.

The problem to extend the Theorem 11.20 for the variable exponent Lebesgue spaces is the fact that the proof relies on the continuity of the translation operator

$$\|\varphi_\varepsilon * u - u\|_p \leq \int\limits_\Omega \|\tau_{\varepsilon y}u - u\|_p\,|\varphi(y)|\,\mathrm{d}y,$$

which is not valid in general, cf. § 7.1.10.1. Fortunately it is possible to show a similar result for variable Lebesgue spaces using the boundedness of the maximal operator. We need some auxiliary lemmas, see Stein [71] for more details.

Lemma 11.22. *Let $\psi : \mathbb{R}^n \longrightarrow \mathbb{R}$ be a radial function (i.e., $\varphi(x) = \varphi(|x|)$), positive and decreasing such that $\psi \in L^1(\mathbb{R}^n)$. Then $\psi(r) = o(r^{-n})$ when $r \to 0$ and $r \to \infty$.*

Proof. The result follows from the following estimate

$$\|\psi\|_{L^1(\mathbb{R}^n)} \geq \int\limits_{r/2 \leq |x| \leq r} \psi(x)\,\mathrm{d}x \geq \psi(r) \int\limits_{r/2 \leq |x| \leq r} \mathrm{d}x = c\psi(r)r^n.$$

□

Lemma 11.23. *Let $f \in L^1_{\mathrm{loc}}(\mathbb{R}^n)$. Then we have the following estimate*

$$\int\limits_{B(0,r)} f(x)\,\mathrm{d}x \leqslant V_n r^n Mf(0)$$

where M is the maximal function (9.2) *and V_n is the volume of the unit ball in $\mathbb{R}^n$ given by $V_n = \frac{\pi^{n/2}}{\Gamma(n/2+1)}$.*

Proof. The proof is direct, since

$$\int\limits_{B(0,r)} f(x)\,\mathrm{d}x = m(B(0,r))|\frac{1}{m(B(0,r))|} \int\limits_{B(0,r)} f(x)\,\mathrm{d}x \leq m(B(0,r))Mf(0).$$

□

Lemma 11.24. *Let φ be a positive, decreasing and radial function in $\mathbb{R}_+$ and integrable. Then*

$$\sup_{\varepsilon>0} |(\varphi_\varepsilon * f)(x)| \leq \|\varphi\|_{L^1} Mf(x).$$

Proof. We will first prove that $(\varphi * f)(0) \leq \|\varphi\|_{L^1} Mf(0)$ and will show the general case based on this particular case. Let $Mf(0) < \infty$ and let us define the following functions

$$\lambda(r) = \int\limits_{\mathbb{S}^{n-1}} f(rx)\,\mathrm{d}\sigma(x)$$

and

$$\Lambda(r) = \int\limits_{\mathbb{B}(0,r)} f(x)\,\mathrm{d}x,$$

where $\mathbb{S}^{n-1}$ is the unit sphere in $\mathbb{R}^n$ and $\mathbb{B}(0,r)$ denotes the ball centered at the origin and radius r in $\mathbb{R}^n$. By a change of variables in spherical coordinates we have

$$\Lambda(r) = \int_0^r \lambda(t) t^{n-1}\,\mathrm{d}t.$$

Now, we have

$$
\begin{aligned}
(\varphi * f)(0) &= \int_{\mathbb{R}^n} f(x)\varphi(x)\,\mathrm{d}x \\
&= \int_0^\infty \lambda(r)\varphi(r)r^{n-1}\,\mathrm{d}r \\
&= \lim_{\substack{\varepsilon\to 0\\ N\to\infty}} \int_\varepsilon^N \varphi(r)\,\mathrm{d}(\Lambda(r)) \\
&= \lim_{\substack{\varepsilon\to 0\\ N\to\infty}} [\Lambda(N)\varphi(N) - \Lambda(\varepsilon)\varphi(\varepsilon)] - \lim_{\substack{\varepsilon\to 0\\ N\to\infty}} \int_\varepsilon^N \Lambda(r)\,\mathrm{d}(\varphi(r)).
\end{aligned}
$$

Using Lemmas 11.22 and 11.23, the previous estimate and (11.24) we obtain

$$
\begin{aligned}
(\varphi * f)(0) &= \int_0^\infty \Lambda(r)\,\mathrm{d}(-\varphi(r)) \\
&\leq V_n Mf(0) \int_0^\infty r^n\,\mathrm{d}(-\varphi(r)) \\
&\leq \|\varphi\|_{L^1} Mf(0).
\end{aligned}
$$

since $V_n = \omega_{n-1}/n$, which proves the result for $x = 0$ and $\varepsilon = 1$. Let $x \in \mathbb{R}^n$ be arbitrary, then taking τ_h has the translation operator: $\tau_h \circ f(x) = f(x-h)$ and $\check{f}(x) = f(-x)$, we get $(\varphi_\varepsilon * f)(x) = \left(\varphi_\varepsilon * \left(\tau_x \circ \check{f}\right)\right)(0) \leq \|\varphi_\varepsilon\|_{L^1} M(\tau_x \circ \check{f})(0) = \|\varphi\|_{L^1} Mf(x)$. □

After some preparation we arrive at the following important theorem.

Theorem 11.25. *Let $\Omega \subset \mathbb{R}^n$ be a bounded open set, and $p \in LH(\Omega)$. Let $\varphi : \mathbb{R}^n \longrightarrow \mathbb{R}$ be an integrable function and let φ_ε be a Friedrichs mollifier. Moreover, let us suppose that the least decreasing radial majoran of φ is integrable, i.e., $\psi(x) = \sup_{|y|\geq|x|} |\varphi(y)|$ then $\int_{\mathbb{R}^n} \psi(x)\,\mathrm{d}x = A < \infty$. We then have*

(a) $\sup_{\varepsilon>0} |(f * \varphi_\varepsilon)(x)| \leq AMf(x)$ for all function $f \in L^{p(\cdot)}(\Omega)$;
(b) $\lim_{\varepsilon\to 0+} (f * \varphi_\varepsilon)(x) = f(x)$ almost everywhere in Ω for all $f \in L^{p(\cdot)}(\Omega)$;
(c) For all $f \in L^{p(\cdot)}(\Omega)$ we have $f * \varphi_\varepsilon \to f$ in $L^{p(\cdot)}(\Omega)$ whenever $\varepsilon \to 0+$;
(d) For all $f \in L^{p(\cdot)}(\Omega)$ we have the following estimate (uniform with respect to $\varepsilon > 0$)

$$
\|f * \varphi_\varepsilon\|_{L^{p(\cdot)}(\Omega)} \lesssim \|Mf\|_{L^{p(\cdot)}(\Omega)} \lesssim \|f\|_{L^{p(\cdot)}(\Omega)}.
$$

Proof. Since Ω is a bounded set, we have that $L^{p(\cdot)}(\Omega) \hookrightarrow L^1(\Omega)$, which implies the pointwise estimates (a) and (b) via Theorem 11.20 and Lemma 11.24. To prove (c), for fixed $x \in \Omega$, we have

$$\begin{aligned}|(f*\varphi_\varepsilon)(x) - f(x)|^{p(x)} &\lesssim (|(f*\varphi_\varepsilon)(x)| + |f(x)|)^{p(x)}\\ &\lesssim (AMf(x) + |f(x)|)^{p(x)}\end{aligned}$$

which implies that $|(f*\varphi_\varepsilon)(x) - f(x)|^{p(x)} \in L^1(\Omega)$. Using (b) with the Dominated Convergence Theorem, we obtain

$$\begin{aligned}\lim_{\varepsilon\to 0+} \rho_{p(\cdot)}(f*\varphi_\varepsilon - f) &= \lim_{\varepsilon\to 0+} \int_\Omega |(f*\varphi_\varepsilon)(x) - f(x)|^{p(x)}\,\mathrm{d}x\\ &= \int_\Omega \lim_{\varepsilon\to 0+} |(f*\varphi_\varepsilon)(x) - f(x)|^{p(x)}\,\mathrm{d}x\\ &= 0,\end{aligned}$$

from which we get the convergence in norm due to the fact $\|f*\varphi_\varepsilon - f\|_{L^{p(\cdot)}(\Omega)} \to 0$ whenever $\varepsilon \to 0$. To show (d) we take into consideration the Theorem 11.20(a) and the fact that the Lebesgue space is ideal (cf. Remark 7.10). □

11.4 Riesz Potentials

The inequalities that involve the Riesz potential provide us with an important tool which permits to estimate functions in terms of the norm of its derivative. We will use the Fourier transform. Let $f \in L_1(\mathbb{R}^n)$, let us define $\widehat{f}$ by

$$\widehat{f}(\xi) = \int_{\mathbb{R}^n} \mathrm{e}^{-2i\pi x\cdot\xi} f(x)\,\mathrm{d}x, \quad \xi \in \mathbb{R}^n. \tag{11.6}$$

The function $\widehat{f}$ is called the *Fourier transform* of the function f, sometimes we will denote it by $\mathscr{F}(f)$ see Appendix D for general properties of the Fourier transform. Let us consider the Laplacian of f, i.e.

$$\Delta f = \sum_{k=1}^n \frac{\partial^2 f}{\partial x_k^2}.$$

Now, let us take the Fourier transform of the minus Laplacian

$$\widehat{-\Delta f}\,(x) = -\widehat{\sum_{k=1}^n \frac{\partial^2 f}{\partial x_k^2}}\,(x) = -\sum_{k=1}^n (\mathrm{i}2\pi x_k)^2 \widehat{f}(x) = 4\pi^2 \sum_{k=1}^n x_k^2 \widehat{f}(x) = 4\pi^2 |x|^2 \widehat{f}(x).$$

Now, we want to substitute the exponent 2 in $|x|^2$ by an arbitrary exponent β and in this way to define, at least formally, the fractional Laplacian by

$$\widehat{(-\Delta)^{\frac{\beta}{2}} f}\,(x) = (2\pi|x|)^{\beta}\widehat{f}(x). \tag{11.7}$$

Looking to equation (11.7) we see that, formally, it can be obtained as the Fourier transform of $k_\alpha * f$ where $k_\alpha(x) = \mathscr{F}^{-1}(|\xi|^{-\alpha}(2\pi)^{-\alpha})$ since $\widehat{k_\alpha * f}(\xi) = \widehat{k_\alpha}(\xi)\widehat{f}(\xi)$. We now try to compute the function k_α by some *operational* way, in the sense that we will not care about rigor following instead formal rules. In Lemma 11.27 we will return to orthodoxy and a rigorous prove will be given regarding the nature of k_α. We first notice that the Fourier transform of a radial function is again a radial function (see Appendix D). Using a scalar argument we get

$$\mathscr{F}(|\cdot|^{-\alpha})(t\xi) = |t|^{\alpha-n}\mathscr{F}(|\cdot|^{-\alpha})(\xi)$$

from which it follows

$$\mathscr{F}(|\cdot|^{-\alpha})(\xi) = |\xi|^{\alpha-n}\mathscr{F}(|\cdot|^{-\alpha})\left(\frac{\xi}{|\xi|}\right) =: |\xi|^{\alpha-n}C(n,\alpha).$$

We operated in a purely formal way without taking care of the fact that we were working with an improper integral. To try to compute the constant $C(n,\alpha)$ we will use the fact that the function $\exp(-\pi|x|^2)$ is invariant under the Fourier transform and we will also use the multiplication formula (D.8) disregarding the fact that the functions $|\cdot|^{-\alpha}$ and $|\cdot|^{n-\alpha}$ do not belong to $L^1(\mathbb{R}^n)$. From the multiplication formula we have

$$\int_{\mathbb{R}^n} \exp(-\pi|x|^2)|x|^{-\alpha}\,\mathrm{d}x = \int_{\mathbb{R}^n} \exp(-\pi|x|^2)C(n,\alpha)|x|^{\alpha-n}\,\mathrm{d}x$$

equality that can be transformed, via polar coordinates, to

$$\int_0^\infty \exp(-\pi r^2)r^{n-\alpha-1}\,\mathrm{d}r = C(n,\alpha)\int_0^\infty \exp(-\pi r^2)r^{\alpha-1}\,\mathrm{d}r, \tag{11.8}$$

from which we get, after some routine calculations,

$$C(n,\alpha) = \pi^{\alpha-\frac{n}{2}}\frac{\Gamma(\frac{n-\alpha}{2})}{\Gamma(\frac{\alpha}{2})}.$$

Gathering all the previous facts we get , at least formally, that

$$k_\alpha(x) = \frac{|x|^{\alpha-n}}{\gamma_n(\alpha)}, \quad \gamma_n(\alpha) = \frac{2^\alpha\pi^{\frac{n}{2}}\Gamma\left(\frac{\alpha}{2}\right)}{\Gamma\left(\frac{n-\alpha}{2}\right)}.$$

We will now define the so-called *Riesz potential operator* as

$$I_\alpha := k_\alpha * f(x) = \frac{1}{\gamma_n(\alpha)} \int_{\mathbb{R}^n} \frac{f(y)}{|x-y|^{n-\alpha}}\, \mathrm{d}y, \tag{11.9}$$

where $\gamma_n(\alpha) = \frac{2^\alpha \pi^{\frac{n}{2}} \Gamma(\frac{\alpha}{2})}{\Gamma(\frac{n-\alpha}{2})}$ is the normalizing factor given in such a way that $\widehat{I_\alpha f}(\xi) = |\xi|^{-\alpha}\widehat{f}(\xi)$.

In the next Lemma we will verify, in a rigorous way, that the Fourier transform of $k_\alpha(x)$ is indeed the function $|x|^{-\alpha}$, but first an auxiliary result.

Lemma 11.26. *Let $f(x) = \exp(-\pi\delta|x|^2)$ with $\delta > 0$. Then $\widehat{f}(\xi) = \delta^{-\frac{n}{2}} f(\xi/\delta)$.*

Proof. We have

$$\begin{aligned}
\widehat{f}(\xi) &= \int_{\mathbb{R}^n} \exp(-2\pi i \xi \cdot x) \exp(-\pi\delta|x|^2)\, \mathrm{d}x \\
&= \exp\left(\frac{-\pi|\xi|^2}{\delta}\right) \int_{\mathbb{R}^n} \exp\Big(-\pi\delta(y + i\xi/\delta)\cdot(x + i\xi/\delta)\Big)\, \mathrm{d}x \\
&= \exp\left(-\frac{\pi|\xi|^2}{\delta}\right) \int_{\mathbb{R}^n} \exp(-\pi\delta x \cdot x)\, \mathrm{d}x \\
&= \delta^{-\frac{n}{2}} \exp\left(-\frac{\pi|\xi|^2}{\delta}\right),
\end{aligned}$$

where the last equality is due to the Euler-Poisson formula. □

Lemma 11.27. *Let $0 < \alpha < n$ and $\mathscr{S}$ stands for the Schwartz class (see Definition D.6), then:*

(a) The Fourier transform of $|x|^{\alpha-n}$ is the function

$$(2\pi)^{-\alpha}\gamma_n(\alpha)|x|^{-\alpha},$$

in the following sense

$$\int_{\mathbb{R}^n} |x|^{\alpha-n}\widehat{\varphi}(x)\, \mathrm{d}x = \int_{\mathbb{R}^n} (2\pi)^{-\alpha}\gamma_n(\alpha)|x|^{-\alpha}\varphi(x)\, \mathrm{d}x,$$

for all $\varphi \in \mathscr{S}$.

(b) The identity

$$\widehat{I_\alpha f}(x) = |x|^{-\alpha}\widehat{f}(x),$$

is obtained in the following sense

$$\int_{\mathbb{R}^n} I_\alpha(f)(x)\widehat{g}(x)\,\mathrm{d}x = \int_{\mathbb{R}^n} \widehat{f}(x)|x|^{-\alpha} g(x)\,\mathrm{d}x,$$

for $f, g \in \mathscr{S}$.

Proof. (a) By Lemma 11.26 and the multiplication formula we have

$$\int_{\mathbb{R}^n} \exp(-\pi\delta|x|^2)\widehat{\varphi}(x)\,\mathrm{d}x = \delta^{-\frac{n}{2}} \int_{\mathbb{R}^n} \exp(-\pi|x|^2/\delta)\varphi(x)\,\mathrm{d}x$$

for all $\varphi \in \mathscr{S}$. We now multiply the previous equality by $\delta^{\frac{n-\alpha}{2}-1}$ and integrating with respect to δ we obtain

$$\int_{\mathbb{R}^n} \left(\int_{\mathbb{R}} \exp(-\pi\delta|x|^2)\delta^{\frac{n-\alpha}{2}-1}\,\mathrm{d}\delta \right) \widehat{\varphi}(x)\,\mathrm{d}x$$

$$= \int_{\mathbb{R}^n} \left(\int_{\mathbb{R}} \exp\left(\frac{-\pi\delta|x|^2}{\delta} \right) \delta^{-\frac{\alpha}{2}-1}\,\mathrm{d}\delta \right) \varphi(x)\,\mathrm{d}x. \quad (11.10)$$

Taking into account that

$$\int_{\mathbb{R}} \exp(-\pi\delta|x|^2)\delta^{\frac{n-\alpha}{2}-1}\,\mathrm{d}\delta = (\pi|x|^2)^{-\frac{n-\alpha}{2}}\Gamma\left(\frac{n-\alpha}{2}\right),$$

we obtain from (11.10) the equality

$$\pi^{-\frac{n-\alpha}{2}}\Gamma\left(\frac{n-\alpha}{2}\right)\int_{\mathbb{R}^n} |x|^{\alpha-n}\widehat{\varphi}(x)\,\mathrm{d}x = \pi^{-\frac{\alpha}{2}}\Gamma\left(\frac{\alpha}{2}\right)\int_{\mathbb{R}^n} |x|^{-\alpha}\varphi(x)\,\mathrm{d}x.$$

(b) Using (a) we obtain

$$\frac{1}{\gamma_n(\alpha)}\int_{\mathbb{R}^n} f(x-y)|y|^{\alpha-n}\,\mathrm{d}y = \frac{1}{\gamma_n(\alpha)}\int_{\mathbb{R}^n} \widehat{\Phi}(y)|y|^{\alpha-n}\,\mathrm{d}y = \int_{\mathbb{R}^n} |y|^{-\alpha}\Phi(y)\,\mathrm{d}y,$$

where $\widehat{\Phi}(y) = f(x-y)$, from which we get that $\Phi(y) = \mathrm{e}^{2\pi\mathrm{i}x\cdot y}\widehat{f}(y)$. The previous equality can be written as

$$\frac{1}{\gamma_n(\alpha)}\int_{\mathbb{R}^n} f(x-y)|y|^{\alpha-n}\,\mathrm{d}y = \int_{\mathbb{R}^n} \mathrm{e}^{2\pi\mathrm{i}x\cdot y}\widehat{f}(y)|y|^{-\alpha}\,\mathrm{d}y.$$

Multiplying both sides by $\widehat{g}(x)$ and integrating over $\mathbb{R}^n$ we obtain

$$\int_{\mathbb{R}^n} I_\alpha f(x)\widehat{g}(x)\,\mathrm{d}x = \int_{\mathbb{R}^n}\left(\int_{\mathbb{R}^n} \mathrm{e}^{2\pi i x\cdot y}\widehat{f}(y)|y|^{-\alpha}\,\mathrm{d}y\right)\widehat{g}(x)\,\mathrm{d}x,$$

which entails the result via Fubini's theorem. □

Lemma 11.28. If $f \in \mathscr{S}$, then

(a) $I_\alpha(I_\beta f) = I_{\alpha+\beta}(f)$ where $\alpha > 0, \beta > 0$ and $\alpha+\beta < n$.
(b) $\Delta(I_\alpha f) = I_\alpha(\Delta f) = -I_{\alpha-2}(f)$ with $n > 3, n \geq \alpha \geq 2$.

Proof. (a) Applying the Fourier transform we obtain

$$\begin{aligned}
\widehat{I_\alpha(I_\beta f)}(\xi) &= (2\pi|\xi|)^{-\alpha}\widehat{(I_\beta f)}(\xi) \\
&= (2\pi|\xi|)^{-\alpha}(2\pi|\xi|)^{-\beta}\widehat{f}(\xi) \\
&= (2\pi|\xi|)^{-(\alpha+\beta)}\widehat{f}(\xi) \\
&= \widehat{I_{\alpha+\beta} f}\,(\xi).
\end{aligned}$$

Therefore, taking the inverse Fourier transform, we get $I_\alpha(I_\beta f) = I_{\alpha+\beta}(f)$.

(b)

$$\begin{aligned}
\widehat{\Delta(I_\alpha f)}(\xi) &= \Sigma_{k=1}^n \widehat{\partial_k^2(I_\alpha f)}(\xi) \\
&= -\sum_{k=1}^n 4\pi^2\xi_k^2\widehat{I_\alpha f}\,(\xi) \\
&= -(2\pi|\xi|)^2(2\pi|\xi|)^{-\alpha}\widehat{f}(\xi) \\
&= -\widehat{(I_{2-\alpha} f)}(\xi).
\end{aligned} \tag{11.11}$$

On the other hand

$$\begin{aligned}
\widehat{I_\alpha(\Delta f)}(\xi) &= (2\pi|\xi|)^{-\alpha}\widehat{\Delta f}(\xi) \\
&= -(2\pi|\xi|)^{-\alpha}(2\pi|\xi|)^2\widehat{f}(\xi) \\
&= -(2\pi|\xi|)^{2-\alpha}\widehat{f}(\xi) \\
&= -\widehat{(I_{\alpha-2} f)}(\xi).
\end{aligned} \tag{11.12}$$

From (11.11) and (11.12), we obtain the result. □

11.5 Potentials in Lebesgue Spaces

In Section 11.4, we consider the Riesz potential from a formal point of view. In particular we operated with smooth functions which behave quite well in the infinity. Since the Riesz operator is an integrable operator, it is natural to study its action in the Lebesgue spaces. By this reason, we formulate the following problem:

Given a real α with $0 < \alpha < n$ for what pairs of (p,q) is the operator $I_\alpha : L_p(\mathbb{R}^n) \longrightarrow L_q(\mathbb{R}^n)$ bounded?

In other words, when the following inequality

$$\|I_\alpha(f)\|_q \leq C\|f\|_p \tag{11.13}$$

holds?

To answer the question regarding the boundedness of the operator I_α, we consider the dilation operator δ_λ defined by

$$\delta_\lambda(f)(x) = f(\lambda x), \qquad \lambda > 0.$$

On the one hand, we have

$$\begin{aligned} \|\delta_\lambda(f)\|_p &= \left(\int_{\mathbb{R}^n} |f(\lambda x)|^p \, dx\right)^{\frac{1}{p}} \\ &= \lambda^{-\frac{n}{p}}\|f\|_p. \end{aligned} \tag{11.14}$$

On the other hand,

$$\begin{aligned} \delta_{\lambda^{-1}}(I_\alpha(\delta_\lambda f))(x) &= I_\alpha(\delta_\lambda f)(\lambda^{-1}x) \\ &= \frac{1}{\gamma_n(\alpha)} \int_{\mathbb{R}^n} \frac{\delta_\lambda(f)(y)}{|\lambda^{-1}x - y|^{n-\alpha}} \, dy \\ &= \lambda^{-\alpha} I_\alpha f(x), \end{aligned}$$

from which it follows

$$\lambda^{-\alpha}\|I_\alpha(f)\|_q = \|\delta_{\lambda^{-1}}(I_\alpha(\delta_\lambda(f)))\|_q = \lambda^{\frac{n}{q}}\|I_\alpha(\delta_\lambda f)\|_q. \tag{11.15}$$

Gathering the above considerations and supposing (11.13) we obtain

$$\|I_\alpha f\|_q \leq C\lambda^{\alpha + \frac{n}{q} - \frac{n}{p}}\|f\|_p$$

which, being valid for all $\lambda > 0$, implies that

$$\frac{1}{q} = \frac{1}{p} - \frac{\alpha}{n}. \tag{11.16}$$

We will see that relation (11.16) fails for the limiting cases $p = 1$ and $q = \infty$.

$\boxed{p = 1}$ From (11.16) we get that $q = \frac{n}{n-\alpha}$. Taking a Dirac sequence φ_k, for $x \neq 0$, we have

$$I_\alpha \varphi_k(x) \sim \frac{1}{|x|^{n-\alpha}}$$

for k sufficiently large, from which we get

$$\left|I_\alpha \varphi_k(x)\right|^{\frac{n}{n-\alpha}} \sim \frac{1}{|x|^n}$$

and this entails a contradiction if we suppose (11.13) to be true for $p = 1$. See Problem 11.46 for more details.

$\boxed{p = \frac{n}{\alpha}}$ We will show directly with a counterexample that this case is also not possible. Let

$$f(x) = \begin{cases} |x|^{-\alpha}\left(\log \frac{1}{|x|}\right)^{-\frac{\alpha}{n}(1+\varepsilon)} & \text{if } |x| \leq \frac{1}{2} \\ 0 & \text{if } |x| > \frac{1}{2}. \end{cases}$$

where ε is a positive number sufficiently small. Now $f \in L_{\frac{n}{\alpha}}(\mathbb{R}^n)$ since

$$\begin{aligned} \int_{\mathbb{R}^n} |f(x)|^{\frac{n}{\alpha}}\, dx &= \int_{|x|\leq\frac{1}{2}} |x|^{-n}\left(\log\frac{1}{|x|}\right)^{-(1+\varepsilon)} dx \\ &= c\int_0^{\frac{1}{2}} \frac{dr}{r\left(\log\frac{1}{r}\right)^{(1+\varepsilon)}} \\ &< \infty. \end{aligned}$$

Nonetheless, $I_\alpha(f)$ is essentially not bounded near the origin. This is

$$\begin{aligned} I_\alpha(f)(0) &= \frac{1}{\gamma_n(\alpha)} \int_{|y|\leq\frac{1}{2}} |y|^{-n}\left(\log\frac{1}{|x|}\right)^{-\frac{\alpha}{n}(1+\varepsilon)} dy \\ &= C(\alpha)\int_0^{\frac{1}{2}} \frac{dr}{r\left(\log\frac{1}{r}\right)^{\frac{\alpha}{n}(1+\varepsilon)}} \\ &= \infty. \end{aligned}$$

if $\alpha(1+\varepsilon) \leq n$.

Now, if we take a subset $\Omega \subseteq \mathbb{R}^n$ ($\alpha < n$), such that $0 < m(\Omega) < \infty$, we obtain the following result.

Lemma 11.29. Let $\Omega \subseteq \mathbb{R}^n$ be a measurable set with $0 < m(\Omega) < \infty$ and $0 < \alpha < n$. Then there exists a constant $C > 0$, such that

$$\|I_\alpha(\chi_\Omega)\|_\infty \leq C\|\chi_\Omega\|_{\frac{n}{\alpha}}.$$

Proof. We can suppose that $x = 0$, and find an $R > 0$, for which $m(B(0,R)) = m(\Omega)$. Let us denote $B = B(0,R)$; therefore, if $y \in \Omega \setminus B$, then $|y| > R$ as $\alpha - n < 0$:

$$\int_{\Omega\setminus B} |y|^{\alpha-n}\,\mathrm{d}y \le R^{\alpha-n} m(\Omega \setminus B)$$

$$= R^{\alpha-n} m(B \setminus \Omega)$$

$$\int_{\Omega\setminus B} |y|^{\alpha-n}\,\mathrm{d}y \le \int_{B\setminus\Omega} |y|^{\alpha-n}\,\mathrm{d}y. \tag{11.17}$$

On the other hand

$$\int_{\Omega} |y|^{\alpha-n}\,\mathrm{d}y = \int_{\Omega\setminus B} |y|^{\alpha-n}\,\mathrm{d}y + \int_{\Omega\cap B} |y|^{\alpha-n}\,\mathrm{d}y,$$

and by (11.17), we have that

$$\int_{\Omega} |y|^{\alpha-n}\,\mathrm{d}y \le \int_{B\setminus\Omega} |y|^{\alpha-n}\,\mathrm{d}y + \int_{\Omega\cap B} |y|^{\alpha-n}\,\mathrm{d}y,$$

from which

$$\int_{\Omega} |y|^{\alpha-n}\,\mathrm{d}y \le \int_{B} |y|^{\alpha-n}\,\mathrm{d}y.$$

Since

$$\begin{aligned}
\int_{B} |y|^{\alpha-n}\,\mathrm{d}y &= \sigma_n m(B(0,1)) \int_0^R r^{n-1} r^{\alpha-n}\,dr \\
&= \frac{\sigma_n}{\alpha} m(B(0,1)) R^{\alpha} \\
&= R^{\alpha} m(B(0,1)) m(B(0,1))^{-\alpha/n} m(B(0,1))^{\alpha/n} \\
&= m(B(0,1))^{1-\alpha/n} (R^n m(B(0,1)))^{\alpha/n} \\
&= m(B(0,1))^{1-\alpha/n} m(B(0,R))^{\alpha/n}.
\end{aligned}$$

Finally we obtain

$$\int_{\Omega} |y|^{\alpha-n}\,\mathrm{d}y \le m(B(0,1))^{1-\alpha/n} m(B(0,R))^{\alpha/n};$$

i.e.

$$\|I_\alpha(\chi_\Omega)\|_\infty \leq C\|\chi_\Omega\|_{\frac{n}{\alpha}}.$$

□

We remember the concept of a *lower semi-continuous function*, namely f is said to be a lower semi-continuous function at x if

$$\liminf_{x_n \to x} f(x_n) \geq f(x).$$

Theorem 11.30 *The $I_\alpha f$ is semi-continuous when $f \geq 0$.*

Proof. If $x_n \to x$, then $|x_n - y|^{\alpha-n} f(y) \to |x-y|^{\alpha-n} f(y)$. By the Fatou Lemma we obtain

$$\int_{\mathbb{R}^n} \liminf |x_n - y|^{\alpha-n} f(y)\,\mathrm{d}y \leq \liminf \int_{\mathbb{R}^n} |x_n - y|^{\alpha-n} f(y)\,\mathrm{d}y$$

which implies that $I_\alpha f(x) \leq \liminf I_\alpha f(x_n)$. □

We now obtain a type of Cavalieri's principle for the Riesz potential $I_\alpha \mu$ of a measure μ which is sometimes called α-potentials.

Lemma 11.31. *Let μ be a Radon measure in $\mathbb{R}^n$ and $\alpha < n$. Then*

$$\int_{\mathbb{R}^n} \frac{\mathrm{d}\mu(y)}{|x-y|^{n-\alpha}} = (n-\alpha) \int_0^\infty r^{\alpha-n-1} \mu(B(x,r))\,\mathrm{d}r.$$

Proof. Using Corollary 3.55, we can write

$$\begin{aligned}
\int_{\mathbb{R}^n} \frac{\mathrm{d}\mu(y)}{|x-y|^{n-\alpha}} &= \int_0^\infty \mu\left(\{y : |x-y|^{\alpha-n} > \lambda\}\right) \mathrm{d}\lambda \\
&= \int_0^\infty \mu\left(\left\{y : |x-y| < \left(\frac{1}{\lambda}\right)^{\frac{1}{n-\alpha}}\right\}\right) \mathrm{d}\lambda \\
&= \int_0^\infty \mu\left(B\left(x, \left(\frac{1}{\lambda}\right)^{\frac{1}{n-\alpha}}\right)\right) \mathrm{d}\lambda,
\end{aligned}$$

making the following change of variable $\left(\frac{1}{\lambda}\right)^{\frac{1}{n-\alpha}} = r$, results that

$$\int_0^\infty \mu\left(B\left(x, \left(\frac{1}{\lambda}\right)^{\frac{1}{n-\alpha}}\right)\right) \mathrm{d}\lambda = (n-\alpha) \int_0^\infty r^{\alpha-n-1} \mu\Big(B(x,r)\Big)\,\mathrm{d}r.$$

Finally

$$\int_{\mathbb{R}^n} \frac{d\mu(y)}{|x-y|^{n-\alpha}} = (n-\alpha) \int_0^\infty r^{\alpha-n-1} \mu\Big(B(x,r)\Big)\, dr.$$

□

Theorem 11.32. *Let $1 \leq p < \infty$. Let Ω be a finite measure set and let $f \in L_p(\Omega)$, then*

$$\|I_\alpha f\|_{L_p(\Omega)} \leq C\|f\|_{L_p(\Omega)} \tag{11.18}$$

where $C = m(B(0,1))^{1-\alpha/n} m(\Omega)^{\frac{\alpha}{n}}$.

Proof. By Lemma 11.29 with $f = 1$, we get

$$\int_\Omega |x-y|^{\alpha-n}\, dy \leq Cm(\Omega)^{\alpha/n}.$$

Then, if $p \geq 1$, using Hölder's inequality we get

$$\begin{aligned}
\int_\Omega |f(y)||x-y|^{\alpha-n}\, dy &\leq \left(\int_\Omega |f(y)|^p|x-y|^{\alpha-n}\, dy\right)^{1/p} \left(\int_\Omega |x-y|^{\alpha-n}\, dy\right)^{1/q} \\
&\leq C^{1-1/p} \left(\int_\Omega |f(y)|^p|x-y|^{\alpha-n}\, dy,\right)^{1/p}
\end{aligned}$$

from which we get

$$\begin{aligned}
\int_\Omega |I_\alpha f(x)|^p dx &\leq C^{p-1} \int_\Omega \left(\int_\Omega |f(y)|^p|x-y|^{\alpha-n}\, dy\right) dx \\
&= C^{p-1} \int_\Omega |f(y)|^p\, dy \int_\Omega |x-y|^{\alpha-n}\, dx \\
&\leq C^p \int_\Omega |f(y)|^p\, dy.
\end{aligned}$$

Therefore

$$\|I_\alpha f\|_{L_{p(\Omega)}} \leq C\|f\|_{L_{p(\Omega)}}.$$

□

Now, we will estimate the norm of the Riesz potential in a more general way.

Theorem 11.33. *If $0 < \alpha < n$, $\beta > 0$ and $\delta > 0$, then for $x \in \mathbb{R}^n$*

$$\int\limits_{B(x,\delta)} \frac{|f(y)|}{|x-y|^{n-\alpha}}\,\mathrm{d}y \leq C_\alpha \delta^\alpha M f(x),$$

where $C_\alpha = \frac{n}{\alpha} m(B(0,1))$.

Proof. For $x \in \mathbb{R}^n$ and $\delta > 0$ we use the Lemma 11.31, then we obtain

$$\begin{aligned}
&\int\limits_{B(x,\delta)} \frac{|f(y)|}{|x-y|^{n-\alpha}}\,\mathrm{d}y \\
&= (n-\alpha)\int\limits_0^\infty \left(\int\limits_{B(x,r)\cap B(x,\delta)} |f(y)|\,\mathrm{d}y \right) \frac{\mathrm{d}r}{r^{n-\alpha+1}} \\
&\leq (n-\alpha)\left(m(B(0,1))\int\limits_0^\delta Mf(x) r^n \frac{\mathrm{d}r}{r^{n-\alpha+1}} + m(B(0,1))\int\limits_\delta^\infty Mf(x)\delta^n \frac{\mathrm{d}r}{r^{n-\alpha+1}} \right) \\
&= \frac{n}{\alpha} m(B(0,1)) Mf(x)\delta^\alpha,
\end{aligned}$$

which ends the proof. □

We now obtain a very important inequality, the so-called *Hedberg inequality*.

Theorem 11.34 (Hedberg inequality). *Let $0 < \alpha < n$ and $f \in L_p(\mathbb{R}^n)$. Then for $1 \leq p < \frac{n}{\alpha}$ we have the following pointwise inequality*

$$|I_\alpha f(x)| \leq C \|f\|_p^{\frac{p\alpha}{n}} (Mf(x))^{1-\frac{p\alpha}{n}}. \tag{11.19}$$

Proof. For $x \in \mathbb{R}^n$ and $\delta > 0$ we have

$$|I_\alpha f(x)| \leq \int\limits_{B(x,\delta)} \frac{|f(y)|}{|x-y|^{n-\alpha}}\,\mathrm{d}y + \int\limits_{\mathbb{R}^n\setminus B(x,\delta)} \frac{|f(y)|}{|x-y|^{n-\alpha}}\,\mathrm{d}y$$

by Theorem 11.33, we obtain

$$\begin{aligned}
\int\limits_{B(x,\delta)} \frac{|f(y)|}{|x-y|^{n-\alpha}}\,\mathrm{d}y &= (n-\alpha)\int\limits_0^\infty \left(\int\limits_{B(x,r)\cap B(x,\delta)} |f(y)| dy \right) \frac{\mathrm{d}r}{r^{n-\alpha+1}} \\
&\leq (n-\alpha)\left(m(B(0,1))\int\limits_0^\delta Mf(x) r^n \frac{\mathrm{d}r}{r^{n-\alpha+1}} \right.
\end{aligned}$$

$$+m(B(0,1))\int_{\delta}^{\infty} Mf(x)\delta^n \frac{\mathrm{d}r}{r^{n-\alpha+1}}\Bigg)$$

$$\leq \frac{n}{\alpha} m(B(0,1))Mf(x)\delta^{\alpha}. \tag{11.20}$$

Now, for $\delta > 0$ the Hölder inequality implies that

$$\int_{\mathbb{R}^n \setminus B(x,\delta)} \frac{|f(y)|}{|x-y|^{n-\alpha}}\,\mathrm{d}y \leq \|f\|_p \left(\int_{\mathbb{R}^n \setminus B(x,\delta)} |x-y|^{(\alpha-n)q}\,\mathrm{d}y \right)^{\frac{1}{q}}$$

$$= \|f\|_p \left(m(B(0,1)) \int_{\delta}^{\infty} r^{n-1-q(n-\alpha)}\,\mathrm{d}r \right)^{\frac{1}{q}}$$

$$\leq \frac{m(B(0,1))}{n-q(n-\alpha)} \|f\|_p \delta^{\alpha-\frac{n}{p}}. \tag{11.21}$$

Finally, from (11.20) and (11.21), we get

$$\left| \int_{\mathbb{R}^n} \frac{|f(y)|}{|x-y|^{n-\alpha}}\,\mathrm{d}y \right| \leq C\left(\delta^{\alpha} Mf(x) + \|f\|_p \delta^{\alpha-\frac{n}{p}}\right). \tag{11.22}$$

If we choose $\delta = \left(\frac{Mf(x)}{\|f\|_p}\right)^{-\frac{p}{n}}$, then (11.22) transforms into

$$|I_{\alpha} f(x)| \leq C(Mf(x))^{1-\frac{\alpha p}{n}} \|f\|_p^{\frac{\alpha p}{n}}.$$

□

We now obtain mapping properties of the Riesz operator.

Theorem 11.35. *Let* $0 < \alpha < n$.

(a) If $1 < p < \frac{n}{\alpha}$ *and* $q = \frac{np}{n-\alpha p}$, *then* $I_{\alpha} : L_p \longrightarrow L_q$ *is bounded, i.e.,* $\|I_{\alpha} f\|_q \leq C\|f\|_p$.
(b) If $q = \frac{n}{n-\alpha}$, *then* $I_{\alpha} : L_1 \longrightarrow L_{(1,\frac{n}{n-\alpha})}$ *is bounded, i.e.,* $m(\{x : I_{\alpha} f(x) > \lambda\}) \leq c\left(\frac{\|f\|_1}{\lambda}\right)^{\frac{n}{n-\alpha}}$.

Proof. We have:

(a) Observe that the Hedberg inequality together with Theorem 9.11 implies that

$$\left(\int_{\mathbb{R}^n} |I_\alpha f(x)|^q \,\mathrm{d}x\right)^{\frac{1}{q}} \le C\|f\|_p^{\frac{\alpha p}{n}} \left(\int_{\mathbb{R}^n} |Mf(x)|^{q(1-\frac{\alpha p}{n})} \,\mathrm{d}x\right)^{\frac{1}{q}}$$

$$= C\|f\|_p^{\frac{\alpha p}{n}} \left(\int_{\mathbb{R}^n} |Mf(x)|^p \,\mathrm{d}x\right)^{\frac{1}{p}}$$

$$\le C\|f\|_p.$$

(b) For the case $p = 1$, the Hedberg inequality transforms into

$$|I_\alpha f(x)| \le C\|f\|_1^{\frac{\alpha}{n}} (Mf(x))^{1-\frac{\alpha}{n}}. \tag{11.23}$$

Observe that by Theorem 9.9 and (11.23) we have that

$$m(\{x : I_\alpha f(x) > \lambda\}) \le m\left(\left\{x : Mf(x) > \left(\frac{\lambda}{c\|f\|_1^{\frac{\alpha}{n}}}\right)^{\frac{n}{n-\alpha}}\right\}\right)$$

$$\le c\left(\frac{\|f\|_1}{\lambda}\right)^{\frac{n}{n-\alpha}},$$

therefore

$$m(\{x : I_\alpha f(x) > \lambda\}) \le c\left(\frac{\|f\|_1}{\lambda}\right)^{\frac{n}{n-\alpha}}.$$

□

11.6 Problems

11.36. Calculate $(f * f)(x)$ in each case

a) $f : \mathbb{R} \to \mathbb{R}$ given by $f = \chi_{[-1,1]}$.
b) $f : \mathbb{R}^2 \to \mathbb{R}$ given by $f = \chi_{B(0,1)}$.

11.37. Suppose that $f \in L_p\left((\mathbb{R}^n, \mathscr{L}, m)\right)$ and $g \in L_q\left((\mathbb{R}^n, \mathscr{L}, m)\right)$ with $\frac{1}{p} + \frac{1}{q} = 1$. Prove that for each $\varepsilon > 0$ there exists $R > 0$ such that

$$(f * g)(x) < \varepsilon \qquad \forall\, |x| > R.$$

11.38. Let

$$\ell_1(L_\infty)(\mathbb{R}^n) = \left\{ f \in L_{\text{loc},\infty} : \|f\|_{\infty,1} = \sum_{k\in\mathbb{Z}^n} \|f\|_{L_\infty(Q_k)} < \infty \right\}$$

$$\ell_\infty(L_1)(\mathbb{R}^n) = \left\{ f \in L_{\text{loc},1} : \|f\|_{1,\infty} = \sup_{k\in\mathbb{Z}^n} \|f\|_{L_1(Q_k)} < \infty \right\}$$

where $Q_k = Q_0 + k$ and $Q_0 = [-\frac{1}{2}, \frac{1}{2}]^n$. If $f \in \ell_1(L_\infty)(\mathbb{R}^n)$ and $g \in L_\infty(L_1)(\mathbb{R}^n)$ show that $f * g \in L_\infty(\mathbb{R}^n, \mathcal{L}, m)$ and

$$\|f * g\|_\infty \leq 2^n \|f\|_{\infty,1} \|g\|_{1,\infty}.$$

11.39. Let $\varphi : \mathbb{R} \to \mathbb{R}$ be a function defined by

$$\varphi(x) = \begin{cases} \exp(\frac{1}{x^2-1}) & \text{if } |x| < 1, \\ 0 & \text{if } |x| \geq 1. \end{cases}$$

a) Prove that φ belongs to the class C^∞ with $\operatorname{supp}\varphi = [-1, 1]$,
b) For each $\varepsilon > 0$ and $a \in \mathbb{R}$, show that the function $f(x) = \varphi\left(\frac{x-a}{\varepsilon}\right)$ belongs to the class C^∞ with $\operatorname{supp} f = [a - \varepsilon, a + \varepsilon]$.

11.40. Let $[a, b] \subset \mathbb{R}$ and $\varepsilon > 0$ be such that $a + \varepsilon < b - \varepsilon$ where φ is defined as in the Problem 11.36. Let us define $h : \mathbb{R} \to \mathbb{R}$ by $h(x) = \int_a^b \varphi\left(\frac{t-x}{\varepsilon}\right) dt$ for all $x \in \mathbb{R}$. Prove that

a) $\operatorname{supp} h \subset [a - \varepsilon, a + \varepsilon]$,
b) $h(x) = c$ (constant function) for all $x \in [a + \varepsilon, b - \varepsilon]$,
c) h belongs to the class C^∞ and $h^{(n)}(x) = \int_a^b \frac{\partial^n}{\partial x^n} \varphi\left(\frac{t-x}{\varepsilon}\right) dt$ for all $x \in \mathbb{R}$, and
d) The function $f = h/c$ of C^∞ class satisfies $0 \leq f(x) \leq 1$ for all $x \in \mathbb{R}$, $f(x) = 1$ and for all $x \in [a + \varepsilon, b - \varepsilon]$ and $\int_{\mathbb{R}} |\chi_{(a,b)} - f|\, dm < 4\varepsilon$.

11.41. Let $f : \mathbb{R} \to \mathbb{R}$ be an integral function with respect to the Lebesgue measure. Given $\varepsilon > 0$, show that there exists a function g belonging to the class C^∞ such that $\int_{\mathbb{R}} |f - g| dm < \varepsilon$.

11.42. Let us consider the vector space of functions

$$E = \left\{ f : \mathbb{R} \to \mathbb{R} \mid f \in C^\infty \text{ and } \int_{\mathbb{R}} f dm = 0 \right\}$$

Prove that for each $1 < p < \infty$, the vector space E is dense in $L_p(\mathbb{R})$. Is E dense in $L_1(\mathbb{R})$?

11.43. The *Poisson integral* and the *Gauss-Weierstrass integral* are given, respectively, by:

(a) $P_t\varphi(x) = \int\limits_{\mathbb{R}^n} P(y,t)\varphi(x-y)\,\mathrm{d}y\; t > 0$,

(b) $W_t\varphi(x) = \int\limits_{\mathbb{R}^n} W(y,t)\varphi(x-y)\,\mathrm{d}y$,

where

$$P(y,t) = \frac{C_n t}{\left(|y|^2+t^2\right)^{(n+1)/2}}, \quad C_n = \pi^{-(n+1)/2}\Gamma\left(\frac{n+1}{2}\right)$$

and

$$W(x,t) = (4\pi t)^{-n/2}\exp\left(-\frac{|x|^2}{4t}\right).$$

Let $\varphi \in L_p(\mathbb{R}^n, \mathscr{L}, m)$ with $1 < p < n/\alpha$. Show that

$$\begin{aligned} I_\alpha\varphi(x) &= \frac{1}{2^{\alpha-1}\Gamma(\alpha)}\int\limits_0^\infty t^{\alpha-1}P_t\varphi(x)\,\mathrm{d}t \\ &= \frac{1}{\Gamma(\alpha/2)}\int\limits_0^\infty t^{\alpha/2-1}W_t\varphi(x)\,\mathrm{d}t. \end{aligned}$$

11.44. Prove that the following sequences are indeed Dirac sequences:

1. The *Landau function*

$$\varphi_k(x) = \begin{cases} \frac{1}{c_k}\left(1-x^2\right)^k, & \text{if } |x| \leqslant 1; \\ 0, & \text{if } |x| > 1; \end{cases}$$

   where $c_k = \int\limits_{-1}^{1}(1-x^2)^k\,\mathrm{d}x$.

2. The *Gauss kernel*

$$\varphi_k(x) := k^{-n}\left(4\pi\right)^{-n/2}\mathrm{e}^{-|x/k|^2/4}$$

   for $x \in \mathbb{R}^n$ and $k > 0$.

11.45. Show that, under the hypothesis of Lemma 11.24, we have

$$\int\limits_0^\infty r^n\,\mathrm{d}(-\varphi(r)) = \frac{n}{\omega_{n-1}}\int\limits_{\mathbb{R}^n}\varphi(x)\,\mathrm{d}x, \tag{11.24}$$

where $\omega_{n-1} = \frac{2\pi^{n/2}}{\Gamma(n/2)}$ is the surface area of the unit sphere $\mathbb{S}^{n-1}$.

11.46. Prove that the relation $\|I_\alpha f\|_{L_{n/(n-\alpha)}(\mathbb{R}^n)} \leq C\|f\|_{L_1(\mathbb{R}^n)}$ cannot be true for all $f \in L_1(\mathbb{R}^n)$.

Hint: Use a Dirac sequence φ_k and show that, passing to the limit, we get

$$\left\| \frac{1}{\gamma_n(\alpha)} \frac{1}{|x|^{n-\alpha}} \right\|_{L_{n/(n-\alpha)}(\mathbb{R}^n)} \leq C < \infty$$

which gives a contradiction.

11.7 Notes and Bibliographic References

The Riesz potential was studied for the first time by Frostman [19] but many properties were obtained by Riesz [60], see also the monograph of Landkof [42].

The Hedberg inequality (11.19) appeared in Hedberg [31].

Appendix A
Measure and Integration Theory Toolbox

What I don't like about measure theory is that you have to say "almost everywhere" almost everywhere.
KURT FRIEDRICHS

In this appendix we collect all the necessary information regarding measure and integration theory[1].

A.1 Measure Spaces

Definition A.1. A collection $\mathscr{A}$ of subsets of a given set X is called a *σ-algebra* if

(a) $X \in \mathscr{A}$;
(b) if $A \in \mathscr{A}$, then $X \setminus A \in \mathscr{A}$;
(c) if $A_n \in \mathscr{A}$, then $\cup_{n=1}^{\infty} A_n \in \mathscr{A}$.

The pair $(X, \mathscr{A})$ is called a *measurable space*. ⊘

Not every collection of sets is a σ-algebra. However, if $\mathfrak{T}$ is an arbitrary family of subsets of X, then there exists the *smallest* σ-algebra $\sigma(\mathfrak{T})$ which contains $\mathfrak{T}$. Such a σ-algebra is simply the intersection of all σ-algebras (in X) which contain $\mathfrak{T}$. It surely exists, since there is at least one such a σ-algebra (the σ-algebra $\mathfrak{P}(X)$ of *all* subsets of X) and the intersection of any collection of σ-algebras is again a σ-algebra. The collection $\sigma(\mathfrak{T})$ is called the *σ-algebra generated by* $\mathfrak{T}$.

Definition A.2. Let P be a topological space. The σ-algebra $\mathfrak{B}(P)$ generated by the family of all open subsets of P is called the *Borel σ-algebra of P*; its elements are called *Borel sets*. ⊘

Definition A.3. Let $\mathscr{A}$ be a collection of subsets of a set X. A nonnegative set function $\mu : \mathscr{A} \to [0, \infty]$ is called a *measure* if

(a) $\mathscr{A}$ is a σ-algebra;
(b) $\mu(\emptyset) = 0$;

[1] We follow very closely Lukeš and Malý [46].

R.E. Castillo, H. Rafeiro, *An Introductory Course in Lebesgue Spaces*, CMS Books in Mathematics, DOI 10.1007/978-3-319-30034-4

(c) for each sequence $\{A_n\}$ of pairwise disjoint sets from $\mathscr{A}$,

$$\mu(\bigcup_{n=1}^{\infty} A_n) = \sum_{n=1}^{\infty} \mu A_n.$$

The triplet $(X, \mathscr{A}, \mu)$ is termed a *measure space*. ⊘

We say that a measure is:

(a) *finite* if $\mu X < +\infty$,
(b) σ*-finite* if there exist sets $M_n \in \mathscr{A}$ such that $\mu M_n < +\infty$ and $X = \cup_{n=1}^{\infty} M_n$;
(c) a *probability measure* if $\mu X = 1$;
(d) *complete* if whenever $B \in \mathscr{A}$ is a null set and $A \subset B$, then also $A \in \mathscr{A}$.

By $(\mathbb{R}, \mathscr{L}, m)$ we denote the Lebesgue measure space, with the Lebesgue σ-algebra and the Lebesgue measure m. The Lebesgue measure is the natural extension of the notion of length of intervals since it is the only translation invariant measure on $\mathfrak{A}(\mathbb{R})$ such that $\mu\left(0,1]\right)) = 1$. For a more detailed construction of the Lebesgue measure, the reader should consult the references given at the end of Appendix A.

A.2 Measurable Functions

If $\Sigma \subset \mathscr{A}$, then we say that a function $f : \Sigma \longrightarrow \overline{\mathbb{R}}$ is $\mathscr{A}$*-measurable*, sometimes denoted only by measurable when the underlying σ-algebra is understood, if $\{x \in \Sigma : f(x) > \alpha\} \in \mathscr{A}$ for each $\alpha \in \mathbb{R}$. We denote by $\mathfrak{F}(X, \mathscr{A})$ the set of all measurable functions. The following theorem encapsulates the main properties of measurable functions.

Theorem A.4. *Let f, g, f_n be $\mathscr{A}$-measurable functions (with possibly different domains in $\mathscr{A}$), $\lambda \in \mathbb{R}$ and φ be a continuous function on an open set $G \subset \mathbb{R}$. Then the following functions are $\mathscr{A}$-measurable where defined (and their definition domain in $\mathscr{A}$):*

(a) $\lambda f, f+g, \max(f,g), \min(f,g), |f|, fg, f/g$;
(b) $\sup f_n, \inf f_n, \limsup f_n, \liminf f_n$ *and* $\lim f_n$;
(c) $\varphi \circ f$.

We will need one more result regarding measurable functions, namely:

Theorem A.5. *Let $f(x)$ be measurable in $\mathbb{R}^n$. Then $(x,y) \mapsto f(x-y)$ is measurable in $\mathbb{R}^n \times \mathbb{R}^n = \mathbb{R}^{2n}$.*

A.3 Integration and Convergence Theorems

If s is a nonnegative simple function expressed as $s = \sum_{j=1}^{n} \beta_j \chi_{B_j}$, where B_j are pairwise disjoint sets and β_j are nonnegative coefficients, define

$$\int_X s\,\mathrm{d}\mu = \int_X \left(\sum_{j=1}^{n} \beta_j \chi_{B_j}\right) \mathrm{d}\mu := \sum_{j=1}^{n} \beta_j \mu B_j.$$

Next, if $f \geq 0$ is a μ-measurable function, define

$$\int_X f\,\mathrm{d}\mu := \sup\left\{\int_X s\,\mathrm{d}\mu : 0 \leq s \leq f,\, s \text{ simple}\right\}.$$

For an arbitrary $\mathscr{A}$-measurable function f, we define its integral as a difference of two positive functions, namely $f^+ = \max\{f, 0\}$ and $f^- = \min\{-f, 0\}$. Namely,

$$\int_X f\,\mathrm{d}\mu := \int_X f^+\,\mathrm{d}\mu - \int_X f^-\,\mathrm{d}\mu$$

provided that at least one of the integrals is finite.

By $L^*(X,\mu) = L^*(\mu)$ we denote the family of all μ-measurable functions defined μ-almost everywhere on X for which the Lebesgue integral is defined.

Lemma A.6 (Fatou's Lemma). *Let $\{f_n\}$ be a sequence of μ-measurable functions and $g \in L^1$. If $f_n \leq g$ almost everywhere for all $n \in \mathbb{N}$, then*

$$\int_X \liminf_{n\to\infty} f_n\,\mathrm{d}\mu \leq \liminf_{n\to\infty} \int_X f_n\,\mathrm{d}\mu.$$

Theorem A.7 (Beppo-Levi's Theorem also known as Lebesgue Monotone Convergence Theorem). *Let $\{f_n\}$ be a sequence of μ-measurable functions, $f_n \uparrow f$ almost everywhere and let $\int_X f_1\,\mathrm{d}\mu > -\infty$. Then $\int_X f\,\mathrm{d}\mu = \lim_{n\to\infty} \int_X f_n\,\mathrm{d}\mu$.*

Theorem A.8 (Lebesgue Dominated Convergence Theorem). *Let $\{f_n\}$ be a sequence of μ-measurable functions, $f_n \to f$ almost everywhere. If there exists a function $h \in L^1$ such that $|f_n| \leq h$ almost everywhere for all n, then $f \in L^1$ and $\int_X f\,\mathrm{d}\mu = \lim \int_X f_n\,\mathrm{d}\mu$.*

A function f defined on a measurable set A has the *property* $\mathscr{C}$ on the set A if given $\varepsilon > 0$ there exists a closed set $F \subset A$ such that

(a) $\mu(A \backslash F) < \varepsilon$;
(b) f is continuous relative to F.

Theorem A.9 (Luzin's Theorem). *Let f be defined and finite on a measurable set A. Then f is measurable if and only if f has the property $\mathscr{C}$ on A.*

Loosely speaking, the Luzin theorem states that a measurable function is *almost* a continuous function.

A.4 Absolutely Continuous Norms

We say that a measure ν on $\mathscr{A}$ is *absolutely continuous* with respect to μ, and write $\nu \ll \mu$, if $\nu E = 0$ for every $E \in \mathscr{A}$ with $\mu E = 0$.

Theorem A.10 (Radon-Nikodým Theorem). *Let μ, ν be finite measures on $(X, \mathscr{A})$, $\nu \ll \mu$. Then there exists a nonnegative function $h \in L^1(\mu)$ such that*

$$\nu A = \int_A h \, \mathrm{d}\mu$$

for all $A \in \mathscr{A}$. This function h is unique up to μ-almost everywhere equality.

A proof of Radon-Nikodym Theorem is given in Theorem 10.23 using an approach based on the Riesz representation theorem, whereas the classical approach is using the Hahn decomposition theorem.

A.5 Product Spaces

If $\mathscr{A}$ and $\mathfrak{A}$ are σ-algebras, the *product σ-algebra* $\mathscr{A} \otimes \mathfrak{A}$ is defined as the smallest σ-algebra which contains all sets of the form $A \times B$ where $A \in \mathscr{A}$ and $B \in \mathfrak{A}$.

Let $(X, \mathscr{A}, \mu)$ and $(Y, \mathfrak{A}, \nu)$ be σ-finite measure spaces. A measure τ on $\mathscr{A} \otimes \mathscr{B}$ is called a *product measure of μ and ν* (denoted by $\mu \otimes \nu$) if

$$\tau(A \times B) = \mu(A)\nu(B)$$

whenever $A \in \mathscr{A}$ and $B \in \mathfrak{A}$.

With the previous definitions at hand, we state the important Fubini's theorem and some of its variants.

Theorem A.11 (Fubini's Theorem). *Let $(X, \mathscr{A}, \mu)$, $(Y, \mathfrak{A}, \nu)$ be σ-finite measure spaces, and $h \in L^*(\mu \otimes \nu)$. Then*

$$\int_{X \times Y} h \, \mathrm{d}\mu \otimes \nu = \int_X \left(\int_Y h(x,y) \, \mathrm{d}\nu \right) \mathrm{d}\mu = \int_Y \left(\int_X h(x,y) \, \mathrm{d}\mu \right) \mathrm{d}\nu.$$

Theorem A.12 (Tonelli's Theorem). *Let $(X, \mathscr{A}, \mu)$, $(Y, \mathfrak{A}, \nu)$ be σ-finite complete measure space. Let $f \geq 0$ be a $\mathscr{A} \otimes \mathfrak{A}$-measurable function. Then:*

$$\int_{X\times Y} f \, \mathrm{d}\mu \otimes \nu = \int_X \left(\int_Y f(x,y) \, \mathrm{d}\nu(y) \right) \mathrm{d}\mu(x) = \int_Y \left(\int_X f(x,y) \, \mathrm{d}\mu(x) \right) \mathrm{d}\nu(y).$$

A.6 Atoms

Definition A.13. Given a measurable space $(X, \mathscr{A})$ and a measure μ on that space, a set $A \in \mathscr{A}$ is called an atom if $\mu(A) > 0$ and for any measurable subset B of A with $\mu(A) > \mu(B)$, one has $\mu(B) = 0$. A measure which has no atoms is called nonatomic or atomless. ⊘

Example A.14. Let us consider two examples:

(a) Consider the set $X = \{1,2,3,4,5,6,7,8,9,10\}$ and let the σ-algebra be the power set of X. Define the measure μ of a set to be its cardinality, that is, the number of elements in the set. Then each of the singletons $\{k\}$ for $k = 1,2,\ldots,9,10$ is an atom.
(b) Consider the Lebesgue measure on the real line. This measure has no atoms.

Remark A.15. The following important properties of nonatomic measures are used:

(a) A nonatomic measure with at least one positive value has an infinite number of distinct values, as starting with a set A with $\mu(A) > 0$ one can construct a decreasing sequence of measurable sets

$$A = A_1 \supset A_2 \supset A_3 \supset \ldots$$

such that

$$\mu(A) = \mu(A_1) > \mu(A_2) > \mu(A_3) > \ldots > 0.$$

This may not be true for measure having atoms, see the first example above.
(b) It turns out that nonatomic measures actually have a continuum of values. It can be proved that if μ is a nonatomic measure and A is a measurable set with $\mu(A) > 0$, then for any real number b satisfying

$$\mu(A) \geq b \geq 0,$$

there exists a measurable subset B of A such that $\mu(B) = b$.

A.7 Convergence in Measure

The following notion is of importance in probability theory.

Definition A.16. Let f, f_n be measurable functions on the measurable space (X,μ). The sequence $\{f_n\}_{n\in\mathbb{N}}$ is said to converge in measure to f, denoted by $f_n \xrightarrow{\mu} f$, if for all $\varepsilon > 0$ there exists an $n_0 \in \mathbb{N}$ such that

$$\mu\left(\{x \in X : |f_n(x) - f(x)| > \varepsilon\}\right) < \varepsilon \quad \text{for all} \quad n \geq n_0. \tag{A.1}$$

⊘

Remark A.17. The preceding definition is equivalent to the following statement. For all $\varepsilon > 0$,

$$\lim_{n\to\infty} \mu\left(\{x \in X : |f_n(x) - f(x)| > \varepsilon\}\right) = 0. \tag{A.2}$$

Definition A.18. We say that a sequence of measurable functions $\{f_n\}_{n\in\mathbb{N}}$ on the measure space $(X,\mathscr{A},\mu)$ is Cauchy in measure if for every $\varepsilon > 0$ there exists an $n_0 \in \mathbb{N}$ such that for n, $m > n_0$ we have

$$\mu\left(\{x \in X : |f_n(x) - f_m(x)| > \varepsilon\}\right) < \varepsilon.$$

⊘

A.8 σ-Homomorphism

Let $(X,\mathscr{A})$ be a measurable space. If f is a real $\mathscr{A}$-measurable function in X. Therefore the set function φ defined by

$$\varphi(A) = f^{-1}(A) = \{x \in X : f(x) \in A\}$$

is a function defined in the σ-algebra $\mathscr{L}$ from $\mathbb{R}$ into the σ-algebra $\mathscr{A}$ such that

$$\varphi(A \cup B) = \varphi(A) \cup \varphi(B) \quad \text{if } A \cap B = \emptyset \tag{A.3}$$

and

$$\varphi(A \backslash B) = \varphi(A) \backslash \varphi(B) \quad A, B \in \mathscr{L}. \tag{A.4}$$

Definition A.19. A function φ which satisfies (A.3) and (A.4) is said to be a homomorphism from $\mathscr{L}$ into $\mathscr{A}$. The homomorphism is said to be a σ-homomorphism if $\varphi(\mathbb{R}) = X$ and for all sequence $\{A_j\}_{j\in\mathbb{N}}$ of disjoint sets in $\mathscr{L}$ we have that

$$\varphi\left(\cup_{j=1}^{\infty}\right) = \cup_{j=1}^{\infty} \varphi(A_j).$$

⊘

Theorem A.20 (Sikorki). *Let $(X,\mathscr{A})$ be a measurable space and φ a σ-homomorphism defined in the σ-algebra of Borel in $\mathscr{A}$. Therefore there exists an $\mathscr{A}$-measurable function f such that $\varphi(A) = f^{-1}(A)$ for any Borel sets in A.*

Proof. For each real number r, let $A_r = \varphi([-\infty, r])$, note that $A_\infty = X$ and $A_{r_1} \subset A_{r_2}$ if $r_1 \geq r_2$.

Now, for each $x \in X$ let us define

$$f(x) = \inf\{r \in \mathbb{R} \mid x \in A_r\}.$$

We get that $f : X \to \mathbb{R}$ and for each $t \in \mathbb{R}$

$$\{x \in X \mid f(x) \leq t\} = \cup_{r \leq t} A_r = \cup\{A_S \mid s \leq t, s \in \mathbb{Q}\}.$$

It is clear that f is an $\mathscr{A}$-measurable function, since all Borel set can be expressed as a countable union and intersection of closed and open subintervals of $\mathbb{R}$. Therefore

$$A_r = f^{-1}\left([-\infty, r]\right).$$

Finally, since φ is a σ-homomorphism it is easy to see that $\varphi(A) = f^{-1}(A)$. $\square$

A.9 References

Classical references regarding measure theory and integration are, among others, Rudin [62], Wheeden and Zygmund [83]. For a detailed introduction to measure theory in the framework of the Euclidean space, see Jones [39].

Appendix B
A Glimpse on Functional Analysis

Mathematics is as old as Man.
STEFAN BANACH

In this appendix we gather some definitions and results from functional analysis that are used throughout the book.

Definition B.1. A *normed space* X is a vector space over the field of real or complex numbers (denoted by $\mathbb{F}$) endowed with a function $\mathscr{N} : X \to [0, \infty)$, which satisfies the following properties:

(N1) $\mathscr{N}(x) = 0$ if and only if $x = 0$ (positive definite);
(N2) $\mathscr{N}(\alpha x) = |\alpha| \mathscr{N}(x)$ for all $x \in X$ and $\alpha \in \mathbb{F}$ (homogeneity);
(N3) $\mathscr{N}(x+y) \leq \mathscr{N}(x) + \mathscr{N}(y)$ for all $x, y \in X$ (triangle inequality).

The function $\mathscr{N}$ (also denoted as *functional*) is designated by *norm* when it satisfies the properties (N1)-(N3). In that case, we use the notations $\|\cdot\|_X$, $\|\cdot\|$, and $\|\cdot \mid X\|$ for the functional $\mathscr{N}$. ⊘

Sometimes the functional $\mathscr{N}$ does not satisfy all properties. For example, when we have (N1) and (N2) but (N3) is replaced by

$$\mathscr{N}(x+y) \leq C(\mathscr{N}(x) + \mathscr{N}(y))$$

for all $x, y \in X$, the functional $\mathscr{N}$ is said to be a *quasi-norm*.

Let $\|\cdot\|_1 : X \to [0, \infty)$ and $\|\cdot\|_2 : X \to [0, \infty)$ be two norms in X such that there exists $C > 0$ and

$$C^{-1}\|x\|_1 \leq \|x\|_2 \leq C\|x\|_1$$

for all $x \in X$, then we say that the *norms are equivalent*.

Using the concept of norm we can introduce the notion of *norm convergence* and *Cauchy sequences*. A sequence $\{x_n\}_{n\in\mathbb{N}}$ is said to *converge in norm* or simply *converge* to $x \in X$, and denoted by $\lim_{n\to\infty} x_n = x$ or $s\text{-}\lim_{n\to\infty} x_n = x$, if

$$\lim_{n\to\infty} \|x_n - x\|_X = 0.$$

R.E. Castillo, H. Rafeiro, *An Introductory Course in Lebesgue Spaces*, CMS Books in Mathematics, DOI 10.1007/978-3-319-30034-4

A sequence $\{x_n\}_{n\in\mathbb{N}}$ is a Cauchy sequence if for all $\varepsilon > 0$, there exists $N(\varepsilon)$ such that, for $n,m > N(\varepsilon)$ we have $\|x_n - x_m\|_X < \varepsilon$.

We now arrive at a fundamental concept in the theory of normed spaces.

Definition B.2. A normed space $(X, \|\cdot\|)$ is said to be a *complete space* or a *Banach space* if very Cauchy sequence in X converges in X. ⊘

Let X be a normed space. A function $F : X \to \mathbb{R}$ is called a *linear functional* if it satisfies

$$F(\alpha f + \beta g) = \alpha F(f) + \beta F(g)$$

and F is called *bounded* if there is a real number $M > 0$ such that

$$\|F(f)\| \leq M\|f\|$$

for all f in X. The smallest constant M for which the above inequality is true is called the *norm* of F. That is,

$$\|F\| = \sup_{f \neq 0} \frac{|F(f)|}{\|f\|}.$$

Let X and Y be vector spaces. A linear functional $T : X \to Y$ is an *isomorphism*, if T is 1-1 and onto. Moreover, if X and Y are normed spaces such that $\|T(x)\| = \|x\|$ for each $x \in X$, then we say that T is an *isometric isomorphism* and X and Y are *isometrically isomorphic*.

We now introduce the definition of a Schauder basis, namely:

Definition B.3. A sequence $\{x_n\}_{n\in\mathbb{N}}$ in a normed space X is said to be a *Schauder basis*, if for all $x \in X$ there exists a unique sequence of scalars $\{\alpha_n\}_{n\in\mathbb{N}}$ such that $x = \sum_{n=1}^{\infty} \alpha_n x_n$, where the convergence of the series is understood with respect to the norm. ⊘

In finite dimension vector spaces X it is a known fact that if $\{e_1, \ldots, e_n\}$ is a basis of X, then the dual space X', called the *algebraic dual of* X defined by

$$X' = \{F : X \to \mathbb{R} : F \text{ is linear}\}$$

has dimension n and the set $\{f_1, \ldots, f_n\}$, where $f_i(e_k) = \delta_{ik}$ is a basis of X'. A similar result can be demonstrated in infinite dimension using Schauder basis.

This fact is the starting point to define the concept of dual space in arbitrary normed spaces.

Definition B.4. Let $(X, +, \cdot, \|\cdot\|)$ be a normed space. We call the *dual space of* X to

$$X^* = \{F : X \to \mathbb{R} : F \text{ is linear and bounded}\}.$$

The dual space is sometimes denoted as *continuous dual space* to emphasize the fact that the linear functionals are also continuous. ⊘

An observation is that whenever $\dim(X) < \infty$, the continuous dual concept coincides with the algebraic dual. For our purposes, we will only work with the continuous dual space throughout the book the expression "dual space" means continuous dual space.

Definition B.5. A normed linear space X is said to be *reflexive* if X may be identified with its *second dual* or the *bidual* $X^{**} = (X^*)^*$ by the canonical isomorphism given by

$$\begin{aligned} \phi : & X \to X^{**} \\ & x \mapsto \phi(x) \end{aligned}$$

such that $\phi(x) : X^* \to \mathbb{R}$ is given by the equality $\phi(x)(x^*) = x^*(x)$. In other words, X is reflexive if $\phi(X) = X^{**}$. ⊘

Be warned that there are *non-reflexive* normed spaces X which are isometrically isomorphic to X^{**}, see James [36] for the classical example with *James spaces*.

Appendix C
Eulerian Integrals

The Gamma function ... is simple enough for juniors in college to meet, but deep enough to have called forth contributions from the finest mathematicians.
PHILIP J. DAVIS

In this appendix we give a terse introduction to the special functions known as Eulerian integrals in the case of *real variables*. It should be pointed out that the most profound applications of these special functions are in the framework of the complex plane, which is outside the scope of the applications given in this book. In our short exposition we avoid, as much as possible, direct calculations, giving instead the ideas of the procedures, which the reader is welcomed to fill out with more details.

C.1 Beta Function

The special function $\mathrm{B}(a,b)$

$$\mathrm{B}(a,b) = \int_0^1 x^{a-1}(1-x)^{b-1}\,\mathrm{d}x, \tag{C.1}$$

is known as *Eulerian integral of the first kind* and also as *Beta function* (it seems that the former designation was coined by Legendre and the latter by Binet). An examination shows that the Beta function is well defined for $a, b > 0$.

The Beta function can be given in several equivalent ways, for example

$$\mathrm{B}(a,b) = \int_0^{+\infty} \frac{x^{a-1}}{(1+x)^{a+b}}\,\mathrm{d}x \tag{C.2}$$

which follows by a simple change of variables $x = s/(1+s)$. Due to the fundamental trigonometric identity, we can further obtain

R.E. Castillo, H. Rafeiro, *An Introductory Course in Lebesgue Spaces*, CMS Books in Mathematics, DOI 10.1007/978-3-319-30034-4

$$B(a,b) = 2\int_0^{\pi/2} \cos^{2a-1}\vartheta \cos^{2b-1}\vartheta \mathrm{d}\vartheta. \tag{C.3}$$

The Beta function enjoys some interesting properties, for example:

Symmetry: $B(a,b) = B(b,a)$, which follows by a simple change of variables $x = 1 - s$;

Reduction formulas: Integration by parts allows us to obtain a reduction formula, namely

$$B(a,b) = \frac{b-1}{a+b-1}B(a,b-1) \tag{C.4}$$

and in the case $b = n \in \mathbb{N}$, we further obtain

$$B(a,n) = \frac{n-1}{a+n-1}\frac{n-2}{a+n-2}\cdots\frac{1}{a+1}B(a,1).$$

Together with the fact that

$$B(a,1) = \int_0^1 x^{a-1}\mathrm{d}x = \frac{1}{a}$$

we get the equality

$$B(a,n) = \frac{(n-1)!}{a(a+1)(a+2)\ldots(a+n-1)}. \tag{C.5}$$

In the case $a = m \in \mathbb{N}$ we further obtain

$$B(m,n) = \frac{(n-1)!(m-1)!}{(m+n-1)!}. \tag{C.6}$$

Special values of $B(a,b)$: In the particular case $b = 1 - a$ with $0 < a < 1$, we get

$$B(a,1-a) = \int_0^{+\infty} \frac{x^{a-1}}{1+x}\mathrm{d}x$$

and it is possible to show that

$$B(a,1-a) = \frac{\pi}{\sin(a\pi)} \tag{C.7}$$

when $0 < a < 1$. We will show (C.7) using complex integration theory. We start by recalling the well-known residue theorem.

Theorem C.1 (Residue theorem). *Let f be an analytic function inside and over a simple closed curve γ except in the points $z_1, \ldots, z_n$ which are in the interior. Therefore*

$$\oint_{\gamma} f(z)\mathrm{d}z = 2\pi\mathrm{i} \sum_{k=1}^{n} \mathrm{Res}(f, z = z_k).$$

Lemma C.2. *Let $p > 1$, then*

$$\int_0^{\infty} \frac{x^{-\frac{1}{p}}}{1+x}\mathrm{d}x = \frac{\pi}{\sin(\pi/p)}.$$

Proof. We consider the integral of complex variable given by (C.8) where C is the region given in the next figure.

$$\oint_C \frac{z^{-\frac{1}{p}}}{1+z}\mathrm{d}z \tag{C.8}$$

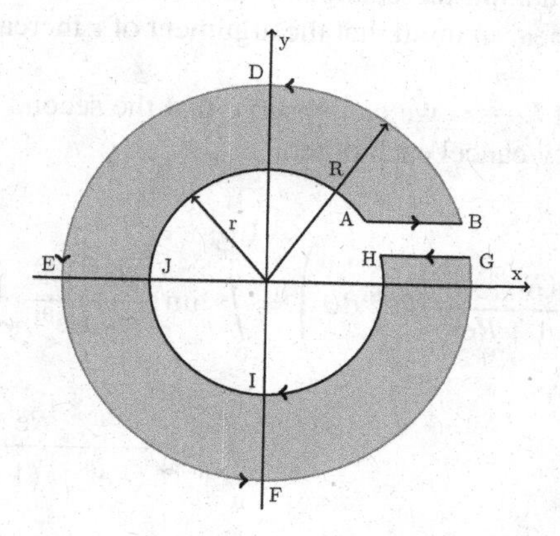

The segments AB and GH are parallel between each other and with the real axis. Let

$$f(z) = \frac{z^{-\frac{1}{p}}}{1+z}.$$

It is not difficult to deduce that f has a simple pole in $z = -1$ inside the region C. If $z = -1$, then

$$z = \cos(\pi) + \mathrm{i}\sin(\pi) = \mathrm{e}^{\pi\mathrm{i}}.$$

Therefore

$$\mathrm{Res}(f, z = -1) = \lim_{z \to -1}(z+1)f(z) = \lim_{z \to -1}(z+1)\frac{z^{-\frac{1}{p}}}{1+z} = \lim_{z \to -1} z^{-\frac{1}{p}} = \mathrm{e}^{-\frac{\pi\mathrm{i}}{p}}.$$

Now invoking the residue theorem we get

$$\oint_C \frac{z^{-\frac{1}{p}}}{1+z}\mathrm{d}z = 2\pi \mathrm{i} e^{-\frac{\pi \mathrm{i}}{p}}.$$

On the other hand, using the path integrals we get

$$\begin{aligned}
e^{-\frac{\pi \mathrm{i}}{p}} &= \int_{AB} f(z)\mathrm{d}z + \int_{BDEFG} f(z)\mathrm{d}z + \int_{GH} f(z)\mathrm{d}z + \int_{HJA} f(z)\mathrm{d}z \\
&= \int_r^R \frac{x^{-\frac{1}{p}}}{1+x}\mathrm{d}x + \int_0^{2\pi} \frac{(Re^{\theta \mathrm{i}})^{-\frac{1}{p}}}{1+Re^{\theta \mathrm{i}}} iRe^{\theta \mathrm{i}}\mathrm{d}\theta \\
&\quad + \int_R^r \frac{(xe^{2\pi \mathrm{i}})^{-\frac{1}{p}}}{1+xe^{2\pi \mathrm{i}}}\mathrm{d}x + \int_{2\pi}^0 \frac{(Re^{\theta \mathrm{i}})^{-\frac{1}{p}}}{1+Re^{\theta \mathrm{i}}} iRe^{\theta \mathrm{i}}\mathrm{d}\theta.
\end{aligned}$$

The last one we get making the change of variables $z = xe^{2\pi \mathrm{i}}$ in the third integral; moreover we should bear in mind that the argument of z increases 2π going through the $BDEFG$ circle.

Now, if $r \to 0$ and $R \to \infty$ we can observe that the second and third integrals in the previous inequality cancel each other.

Indeed

$$\begin{aligned}
\lim_{R\to\infty}\left(\int_0^{2\pi} \frac{(Re^{\theta \mathrm{i}})^{-\frac{1}{p}}}{1+Re^{\theta \mathrm{i}}} iRe^{\theta \mathrm{i}}\mathrm{d}\theta\right) &= \int_0^{2\pi} \lim_{R\to\infty} \frac{(Re^{\theta \mathrm{i}})^{-\frac{1}{p}}}{1+Re^{\theta \mathrm{i}}} iRe^{\theta \mathrm{i}}\mathrm{d}\theta \\
&= \int_0^{2\pi} \lim_{R\to\infty} \frac{iRe^{\theta \mathrm{i}}}{(Re^{\theta \mathrm{i}})^{\frac{1}{p}}(1+Re^{\theta \mathrm{i}})}\mathrm{d}\theta \\
&= \int_0^{2\pi} \lim_{R\to\infty} \frac{i}{(Re^{\theta \mathrm{i}})^{\frac{1}{p}}}\mathrm{d}\theta \\
&= \int_0^{2\pi} 0\mathrm{d}\theta \\
&= 0.
\end{aligned}$$

To calculate the other limit, we proceed in a similar way, first making the change of variables $u = r^{\frac{1}{p}}$ where $u \to 0$. Therefore

$$\lim_{r\to 0}\left(\int_{2\pi}^{0}\frac{(re^{\theta i})^{-\frac{1}{p}}}{1+re^{\theta i}}ire^{\theta i}d\theta\right)=\int_{2\pi}^{0}\frac{ie^{\theta i}}{e^{\frac{\theta i}{p}}}\left(\lim_{r\to 0}\frac{r}{r^{\frac{1}{p}}(1+re^{\theta i})}\right)d\theta$$
$$=\int_{2\pi}^{0}\frac{ie^{\theta i}}{e^{\frac{\theta i}{p}}}\left(\lim_{u\to 0}\frac{u^{p}}{u(1+ue^{\theta i})}\right)d\theta$$
$$=\int_{2\pi}^{0}\frac{ie^{\theta i}}{e^{\frac{\theta i}{p}}}(0)d\theta$$
$$=0.$$

Finally, we get

$$2\pi ie^{-\frac{\pi i}{p}}=\int_{0}^{\infty}\frac{x^{-\frac{1}{p}}}{1+x}dx+\int_{\infty}^{0}\frac{(xe^{2\pi i})^{-\frac{1}{p}}}{1+xe^{2\pi i}}dx$$
$$=\int_{0}^{\infty}\frac{x^{-\frac{1}{p}}}{1+x}dx-\int_{0}^{\infty}\frac{x^{-\frac{1}{p}}e^{-\frac{2\pi i}{p}}}{1+xe^{2\pi i}}dx$$
$$=\int_{0}^{\infty}\frac{x^{-\frac{1}{p}}}{1+x}dx-e^{-\frac{2\pi i}{p}}\int_{0}^{\infty}\frac{x^{-\frac{1}{p}}}{1+x(\cos(2\pi)+i\sin(2\pi))}dx$$
$$=(1-e^{-\frac{2\pi i}{p}})\int_{0}^{\infty}\frac{x^{-\frac{1}{p}}}{1+x}dx.$$

Therefore

$$\int_{0}^{\infty}\frac{x^{-\frac{1}{p}}}{1+x}dx=\frac{2\pi ie^{-\frac{\pi i}{p}}}{1-e^{-\frac{2\pi i}{p}}}=\frac{2\pi i}{e^{\frac{\pi i}{p}}}\left(1-\frac{1}{e^{\frac{2\pi}{p}}}\right)=\frac{\pi}{\left(\frac{e^{\frac{\pi i}{p}}-e^{-\frac{\pi i}{p}}}{2i}\right)}=\frac{\pi}{\sin\left(\frac{\pi}{p}\right)},$$

which finishes the proof. □

Taking $a=1/2$ in (C.7) we obtain

$$B\left(\frac{1}{2},\frac{1}{2}\right)=\pi. \tag{C.9}$$

C.2 Gamma Function

The *Eulerian integral of the second kind*, also known as *Gamma function*, is given commonly has the improper integral

$$\Gamma(a) = \int_0^{+\infty} x^{a-1} \mathrm{e}^{-x} \mathrm{d}x, \tag{C.10}$$

which has sense only for $a > 0$. From (C.10) it is immediate that the Gamma function has no zeros. It is possible to prove that the Gamma function is continuous, and moreover, it is infinite differentiable.

Euler gave the following integral definition for the Gamma function

$$\Gamma(a) = \int_0^1 \left(\log \frac{1}{x} \right)^{a-1} \mathrm{d}x \tag{C.11}$$

which is obtained by a simple change of variable $x = \log(1/s)$ in (C.10). Formula (C.11) is useful to obtain another representation for the Gamma function, the so-called *Euler-Gauss formula*

$$\Gamma(a) = \lim_{n \to +\infty} \frac{1 \cdot 2 \cdot 3 \cdot \ldots \cdot (n-1)}{a(a+1)(a+2)\ldots(a+n-1)}. \tag{C.12}$$

To obtain (C.12) from (C.11) we note that

$$\log \frac{1}{x} = \lim_{n \to +\infty} n \left(1 - x^{1/n} \right) \tag{C.13}$$

and now replacing (C.13) into (C.11) and *formally* interchanging the limit with the integral we get

$$\Gamma(a) = \lim_{n \to +\infty} n^a \int_0^1 x^{n-1} (1-x)^{a-1} \mathrm{d}x = \lim_{n \to +\infty} n^a \mathrm{B}(a,n)$$

and now taking (C.5) we obtain the *Euler-Gauss formula* (C.12). The permissibility of interchanging the limit with the integral is given by the fact that the sequence of functions $n(1 - x^{1/n})$ is monotonically increasing.

We now list some of the most important properties of the Gamma function.

Reduction formula: Taking integration by parts in (C.10) we immediately obtain the following recursion formula

$$\Gamma(a+1) = a \cdot \Gamma(a)$$

which, when iterated, gives

$$\Gamma(a+n) = (a+n-1)(a+n-2)\dots(a+1)a\Gamma(a), \tag{C.14}$$

and since $\Gamma(1) = 1$ we get

$$\Gamma(n+1) = n!$$

which simply states the fact that the Gamma function is an extension of the factorial function.

Extension to negative values: From (C.10) we know that $\Gamma(a)$ is meaningful only for $a > 0$. To extend the function to the negative half-axis, we use the reduction formula (C.14). Taking the so-called *Pochhammer symbol*

$$(x)_n = x(x+1)\dots(x+n-1), \quad n \in \mathbb{N},$$

we can write the reduction formula (C.14) simply as

$$\Gamma(a) = \frac{\Gamma(a+n)}{(a)_n}. \tag{C.15}$$

Formula (C.15) is used to *define* the Gamma function for negative values, except for negative integers!

Link between the Gamma and Beta function: There is a relation between the Eulerian integrals, namely

$$\mathrm{B}(a,b) = \frac{\Gamma(a)\Gamma(b)}{\Gamma(a+b)} \tag{C.16}$$

whenever $a, b > 0$, which was already obtained for natural numbers a and b in (C.6). To obtain (C.16) we notice that, by a change of variables $x = (1+t)y$, $t > 0$, we obtain

$$\frac{\Gamma(s)}{(1+t)^s} = \int_0^{+\infty} y^{s-1}\mathrm{e}^{-(1+t)y}\mathrm{d}y$$

and taking $s = a+b$, $a, b > 0$, we have

$$\frac{\Gamma(a+b)}{(1+t)^{a+b}} = \int_0^{+\infty} y^{a+b-1}\mathrm{e}^{-(1+t)y}\mathrm{d}y. \tag{C.17}$$

Now, integrating with respect to t between 0 and $+\infty$ both sides of (C.17), due to (C.2), we obtain

$$\Gamma(a+b)\mathrm{B}(a,b) = \int_0^{+\infty} t^{q-1} \int_0^{+\infty} y^{a+b-1}\mathrm{e}^{-(1+t)y}\mathrm{d}y\mathrm{d}t$$

and now it is only necessary to use Fubini's theorem, obtaining (C.16).

Complement formula: Taking (C.7), (C.16), and the fact that $\Gamma(1) = 1$ we obtain the so-called *complement formula for the Gamma function*

$$\Gamma(a)\Gamma(1-a) = \frac{\pi}{\sin(\pi a)} \tag{C.18}$$

whenever $0 < a < 1$. It is also possible to use the Euler-Gauss (C.12) to obtain (C.18).

Graph of the Gamma function: From the fact that the Gamma function is continuous and the reduction formula (C.14) we get

$$\lim_{a\to 0+} \Gamma(a) = \lim_{a\to 0+} \frac{\Gamma(a+1)}{a} = +\infty.$$

On the other hand,

$$\lim_{a\to +\infty} \Gamma(a) = +\infty$$

since $\Gamma(a) > n!$ whenever $a > n+1$.

We now want to calculate some values of the Gamma function, e.g., $\Gamma(\frac{1}{2})$. In other words, we need to calculate

$$\Gamma\left(\frac{1}{2}\right) = 2\int_0^{+\infty} \mathrm{e}^{-x^2}\,\mathrm{d}x.$$

The integral $\int_0^{+\infty} \mathrm{e}^{-x^2}\,\mathrm{d}x$ is the so-called *Euler-Poisson integral*. We can calculate the Euler-Poisson integral using series expansion, but we rely on the following observation

$$\left(\int_{-\infty}^{+\infty} \mathrm{e}^{-x^2}\,\mathrm{d}x\right)^2 = \left(\int_{-\infty}^{+\infty} \mathrm{e}^{-x^2}\,\mathrm{d}x\right)\left(\int_{-\infty}^{+\infty} \mathrm{e}^{-y^2}\,\mathrm{d}y\right) = \int_{-\infty}^{+\infty}\int_{-\infty}^{+\infty} \mathrm{e}^{-(x^2+y^2)}\,\mathrm{d}x\mathrm{d}y. \tag{C.19}$$

The improper integral $\int_{-\infty}^{+\infty}\int_{-\infty}^{+\infty} \mathrm{e}^{-(x^2+y^2)}\,\mathrm{d}x\mathrm{d}y$ can be calculated as $\lim_{n\to+\infty} a_n$ where

$$a_n = \iint_{x^2+y^2<n^2} \mathrm{e}^{-x^2-y^2}\,\mathrm{d}x\mathrm{d}y = \int_0^{2\pi}\int_0^{n} \mathrm{e}^{-r^2}\, r\mathrm{d}r\mathrm{d}\vartheta = \pi(1-\mathrm{e}^{-n^2}) \tag{C.20}$$

where we passed to polar coordinates. Therefore $\lim_{n\to+\infty} a_n = \pi$ and the Euler-Poisson integral $\int_0^{+\infty} \mathrm{e}^{-x^2}\,\mathrm{d}x = \frac{\sqrt{\pi}}{2}$. We thus obtained that

$$\Gamma\left(\frac{1}{2}\right) = \sqrt{\pi}.$$

To obtain the value of $\Gamma(\frac{3}{2})$ we use the reduction formula (C.14) and get

$$\Gamma\left(\frac{3}{2}\right) = \Gamma\left(1+\frac{1}{2}\right) = \frac{1}{2}\Gamma\left(\frac{1}{2}\right) = \frac{\sqrt{\pi}}{2}.$$

Now, from the complement formula (C.18) we obtain the value of $\Gamma(-\frac{1}{2}) = -2\sqrt{\pi}$. In this manner we can obtain further half-integer values of the Gamma function (Fig. C.1). The graph of the Gamma function is

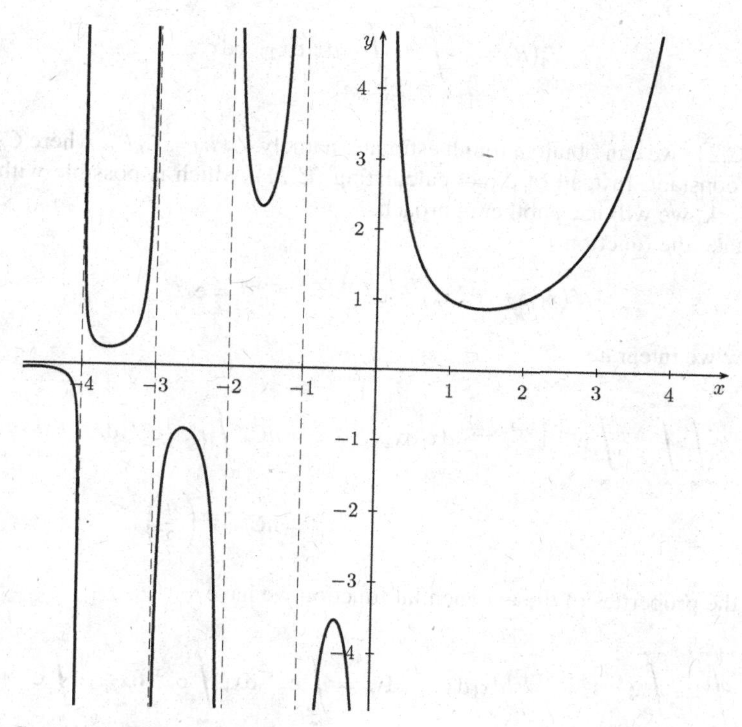

Fig. C.1 Graph of the Gamma function

C.3 Some Applications

In this section we want to obtain the volume of an n-dimensional ball and the surface area of an $(n-1)$-dimensional sphere. These problems are intimately related to the Beta and Gamma functions.

Volume of an n-Dimensional Ball

An n-dimensional ball centered at the origin with radius R is given by the condition

$$B^n(R) := \left\{(x_1, x_2, \ldots, x_n) \in \mathbb{R}^n : x_1^2 + x_2^2 + \ldots + x_n^2 \leq R^2\right\}.$$

By $V_n(R)$ we denote the volume of the $B^n(R)$. To calculate $V_n(R)$ we can use multi-dimensional integration, namely

$$V_n(R) = \int\limits_{x_1^2+x_2^2+\ldots+x_n^2 \leq R^2} \cdots \int \mathrm{d}x_1 \mathrm{d}x_2 \ldots \mathrm{d}x_n. \tag{C.21}$$

From (C.21) we can obtain a rough estimate, namely $V_n(R) = C_n R^n$, where C_n is the sought constant. Instead of direct calculating (C.21), which is possible with some extra work, we will use another approach.

We take the function

$$f(x_1, x_2, \ldots, x_n) = \mathrm{e}^{-(x_1^2+x_2^2+\ldots+x_2^2)} = \mathrm{e}^{-R^2}$$

and now we integrate

$$\begin{aligned} \int\limits_{-\infty}^{+\infty}\int\limits_{-\infty}^{+\infty} \ldots \int\limits_{-\infty}^{+\infty} \mathrm{e}^{-(x_1^2+x_2^2+\ldots x_n^2)} \mathrm{d}x_1 \mathrm{d}x_2 \ldots \mathrm{d}x_n &= nC_n \int\limits_0^{+\infty} r^{n-1} \mathrm{e}^{-r^2} \mathrm{d}r \\ &= nC_n \frac{1}{2} \Gamma\left(\frac{n}{2}\right). \end{aligned} \tag{C.22}$$

Using the properties of the exponential function we have

$$\begin{aligned} \int\limits_{-\infty}^{+\infty}\int\limits_{-\infty}^{+\infty} \ldots \int\limits_{-\infty}^{+\infty} \mathrm{e}^{-(x_1^2+x_2^2+\ldots x_n^2)} \mathrm{d}x_1 \mathrm{d}x_2 \ldots \mathrm{d}x_n &= \int\limits_{-\infty}^{+\infty} \mathrm{e}^{-x_1^2} \mathrm{d}x_1 \int\limits_{-\infty}^{+\infty} \mathrm{e}^{-x_2^2} \mathrm{d}x_2 \ldots \int\limits_{-\infty}^{+\infty} \mathrm{e}^{-x_n^2} \mathrm{d}x_n \\ &= \left(\int\limits_{-\infty}^{+\infty} \mathrm{e}^{-x^2} \mathrm{d}x\right)^n \end{aligned} \tag{C.23}$$

Taking (C.19) and (C.20), (C.22) and (C.23) we get that

$$C_n = \frac{\pi^{n/2}}{\Gamma\left(1 + \frac{n}{2}\right)},$$

therefore the formula for the volume of an n-dimensional ball of radius R is given by

$$V_n(R) = \frac{\pi^{n/2}}{\Gamma\left(1+\frac{n}{2}\right)} R^n. \tag{C.24}$$

Surface Area of An n-Dimensional Ball

We note that, by *Cavalieri's principle*, we can obtain $V_n(R)$ by

$$V_n(R) = \int_0^R S_{n-1}(r)\mathrm{d}r \tag{C.25}$$

where $S_{n-1}(R)$ denotes the *area* surface of the n-dimensional ball $B^n(R)$, i.e.

$$S_{n-1} = \left\{(x_1,x_2,\ldots,x_n) \in \mathbb{R}^n : x_1^2 + x_2^2 + \ldots x_n^2 = R^2\right\}.$$

From (C.25), Barrow's formula and (C.24) we obtain

$$S_{n-1}(R) = \frac{\mathrm{d}V_n(R)}{\mathrm{d}R} = \frac{2\pi^{n/2}}{\Gamma\left(\frac{n}{2}\right)} R^{n-1} \tag{C.26}$$

using the reduction formula (C.14).

Appendix D
Fourier Transform

The profound study of nature is the most fertile source of mathematical discoveries.
JOSEPH FOURIER

In this appendix we give the working bare minimum regarding the theory of Fourier transform. We will stress mainly operational rules and will avoid a rigorous study of the Fourier transform. The Fourier transform is a powerful tool which is widely exploited in probability theory, partial differential equations, theory of signal, just to name a few. For a more detailed account, the reader should consult Bracewell [2].

The Fourier transform is a tool to express a function as a "continuous" superposition of complex exponentials $\left\{e^{2\pi i x\cdot v}\right\}_{v\in\mathbb{R}}$ generalizing the Fourier series of a periodic function.

Definition D.1. For $f \in L_1(\mathbb{R}^n)$ we define the *Fourier transform*

$$\widehat{f}(\xi) := \mathscr{F}[f](\xi) := \int_{\mathbb{R}^n} f(x)e^{-2\pi i x\cdot\xi}\,dx \tag{D.1}$$

for all $\xi \in \mathbb{R}^n$, where $x\cdot\xi$ denotes the inner product in $\mathbb{R}^n$, viz. $x\cdot\xi = \sum_{k=1}^{n} x_k\xi_k$. ⊘

Warning: There is no universal agreement regarding the definition of the Fourier transform, therefore it is always necessary to check the definition being used, specially when using Fourier transform tables. Alternative definitions are

$$\widehat{f}(\xi) = \frac{1}{\sqrt[n]{2\pi}}\int_{\mathbb{R}^n} e^{-i\pi x\cdot\xi} f(x)dx, \quad \widehat{f}(\xi) = \int_{\mathbb{R}^n} e^{-i\pi x\cdot\xi} f(x)dx, \quad \widehat{f}(\xi) = \int_{\mathbb{R}^n} e^{-ix\cdot\xi} f(x)dx,$$

among others.

We now introduce the so-called multi-index notation, which is very useful to get very compact formulas resembling the one-dimensional versions. The use of this notation should be used with great care by the novice.

R.E. Castillo, H. Rafeiro, *An Introductory Course in Lebesgue Spaces*, CMS Books in Mathematics, DOI 10.1007/978-3-319-30034-4

Definition D.2. For $x = (x_1, \dots, x_n) \in \mathbb{R}^n$ and a multi-index $\alpha = (\alpha_1, \dots, \alpha_n) \in \mathbb{N}_0^n$, where $\mathbb{N}_0$ stands for $\mathbb{N} \cup \{0\}$, we define

$$x^\alpha := x_1^{\alpha_1} \cdots x_n^{\alpha_n} \quad \text{and} \quad |\alpha| := \alpha_1 + \cdots + \alpha_n. \tag{D.2}$$

Moreover, for a multi-index α, we define

$$D^\alpha := D_1^{\alpha_1} \cdots D_n^{\alpha_n}, \tag{D.3}$$

where D_j stands for the j-th partial derivative. ⊘

Caution: Some authors, mainly in Fourier theory, use the normalized notation $D_j f(x) = \frac{1}{\mathrm{i}} \partial_j f(x)$ to obtain cleaner formulas, but we will refrain from doing so.

The following properties of the Fourier transform follow almost immediately from the definition.

Bounded linear mapping: The operator $\mathscr{F} : L_1(\mathbb{R}^n) \longrightarrow L_\infty(\mathbb{R}^n)$ is a linear operator which satisfies $\|\widehat{f}\|_\infty \le \|f\|_1$.

Uniformly continuous: The function $\widehat{f}$ is uniformly continuous on $\mathbb{R}^n$.

Norm of the Fourier transform: If $f \ge 0$, then $\|\widehat{f}\|_\infty = \|f\|_1 = \widehat{f}(0)$.

Fourier transform of the derivative: If $f \in L_1(\mathbb{R}^n)$ and $D_k f \in L_1(\mathbb{R}^n)$ then

$$\widehat{(D_k f)}(\xi) = 2\pi \mathrm{i} \xi_k \widehat{f}(\xi). \tag{D.4}$$

Product of the Fourier transform by a monomial: If $f \in L_1(\mathbb{R}^n)$ and $x_k f(x) \in L_1(\mathbb{R}^n)$ then

$$D_k \widehat{f}(\xi) = (\widehat{-2\pi \mathrm{i} x_k f})(\xi). \tag{D.5}$$

Fourier transform of a translated function: If $f \in L_1(\mathbb{R}^n)$ and $\tau_y f(x) = f(x+y)$ denotes the translation of f, then

$$\widehat{(\tau_y f)}(\xi) = \mathrm{e}^{2\pi \mathrm{i} y \cdot \xi} \widehat{f}(\xi). \tag{D.6}$$

Fourier transform of a dilated function: If $f \in L_1(\mathbb{R}^n)$ and $\delta_\varepsilon f(x) = f(\varepsilon x)$ denotes the dilation of f, then

$$\widehat{(\delta_\varepsilon f)}(\xi) = \varepsilon^{-n} \widehat{f}\left(\frac{\xi}{\varepsilon}\right) = \varepsilon^{-n} \left(\delta_{\varepsilon^{-1}} \widehat{f}\right)(\xi). \tag{D.7}$$

The property (D.4) is widely used in differential equations, since it permits in some cases to convert some linear partial differential equations into algebraic equations, see Example D.8 at the end of section.

Riemann-Lebesgue lemma: If $f \in L_1(\mathbb{R}^n)$ then $\widehat{f}(x) \to 0$ when $|x| \to +\infty$.

Multiplication formula: If $f, g \in L_1(\mathbb{R}^n)$ then

$$\int_{\mathbb{R}^n} \widehat{f}(x)g(x)\mathrm{d}x = \int_{\mathbb{R}^n} f(x)\widehat{g}(x)\mathrm{d}x. \tag{D.8}$$

Definition D.3. We define the *convolution* of two function $f \in L_1(\mathbb{R}^n)$ and $g \in L^p(\mathbb{R}^n)$, with $1 \leq p \leq +\infty$, by

$$(f * g)(x) := \int_{\mathbb{R}^n} f(x-y)g(y)\mathrm{d}y,$$

for almost all $x \in \mathbb{R}^n$.
⊘

By Fubini's theorem, we know that $(f * g)(x)$ exists almost everywhere and it belongs to $L^p(\mathbb{R}^n)$.

The next property is widely used in harmonic analysis, see, e.g., § 11.

Fourier transform of convolution: If $f, g \in L_1(\mathbb{R}^n)$, then $f * g \in L_1(\mathbb{R}^n)$ and

$$\widehat{f * g}(\xi) = \widehat{f}(\xi) \cdot \widehat{g}(\xi). \tag{D.9}$$

A very useful property of the Fourier transform is the fact that it has an inverse.

Definition D.4. Let $f \in L_1(\mathbb{R}^n)$. We define the *inverse Fourier transform* by

$$\overline{f}(x) := \mathscr{F}^{-1}[f](x) := \int_{\mathbb{R}^n} \mathrm{e}^{2\pi i x \cdot \xi} f(\xi)\mathrm{d}\xi, \tag{D.10}$$

for all $x \in \mathbb{R}^n$.
⊘

Note that $\overline{f}$ is not standard notation, it is customary to use $\check{f}$. An immediate relation is $\overline{f}(x) = \widehat{f}(-x)$.

Fourier inversion formula: If $f \in L_1(\mathbb{R}^n)$ and $\widehat{f} \in L_1(\mathbb{R}^n)$, then

$$f(x) = \overline{(\widehat{f})}(x) = \widehat{(\overline{f})}(x) \tag{D.11}$$

for almost every x.

The proof of the Fourier inversion formula (D.11) is not straightforward and it should be noted that Fubini's theorem is not applicable.

If $f \in L_1(\mathbb{R}^n)$ it does not follow that $\widehat{f} \in L_1(\mathbb{R}^n)$, as the next example illustrates.

Example D.5. Taking $\chi_{[-1,1]} : \mathbb{R} \longrightarrow \mathbb{R}$, we have

$$\widehat{\chi_{[-1,1]}}(\xi) = \frac{\mathrm{e}^{2\pi\mathrm{i}\xi} - \mathrm{e}^{-2\pi\mathrm{i}\xi}}{2\pi\mathrm{i}\xi} = 2\,\mathrm{sinc}(2\pi\xi),$$

which is not integrable in $\mathbb{R}$. ⊘

A noticeable fact from the previous example is that even if a function has compact support, the support of its Fourier transform can be all $\mathbb{R}$. This fact is related to the so-called Heisenberg uncertainty principle, see Stein and Shakarchi [73] for more details.

It is then reasonable to ask for the natural set for which the Fourier inversion formula holds, or in other words, which is invariant under Fourier transforms. We are lead to the Schwartz class.

Definition D.6. The *Schwartz class* of functions, denoted by $\mathscr{S}(\mathbb{R}^n)$, is defined as the set of C^∞ functions such that the function and all its derivatives are rapidly decreasing, i.e., $f \in \mathscr{S}(\mathbb{R}^n)$ if

$$\varkappa(\ell, \alpha, f) := \sup_{x\in\mathbb{R}^n} \left(1 + |x|^\ell\right) |D^\alpha f(x)| < +\infty,$$

for all $\ell \in \mathbb{N}_0$ and all the multi-index $\alpha \in \mathbb{N}_0^n$. ⊘

An immediate observation is that $\mathscr{S}(\mathbb{R}^n) \supsetneq C_0^\infty(\mathbb{R}^n)$, since $\varphi(x) = \mathrm{e}^{-|x|^2} \in \mathscr{S}(\mathbb{R}^n)$ but does not belong to $C_0^\infty(\mathbb{R}^n)$.

With the notion of Schwartz class we now have the so-called

Fourier inversion theorem: The Fourier transform is a one-to-one mapping of $\mathscr{S}(\mathbb{R}^n)$ on $\mathscr{S}(\mathbb{R}^n)$. Moreover, if $f \in \mathscr{S}(\mathbb{R}^n)$, then $\widehat{f} \in \mathscr{S}(\mathbb{R}^n)$ and

$$\overline{(\widehat{f})}(x) = \widehat{(\overline{f})}(x) = f(x). \tag{D.12}$$

The previous theorem can be seen in the following diagram

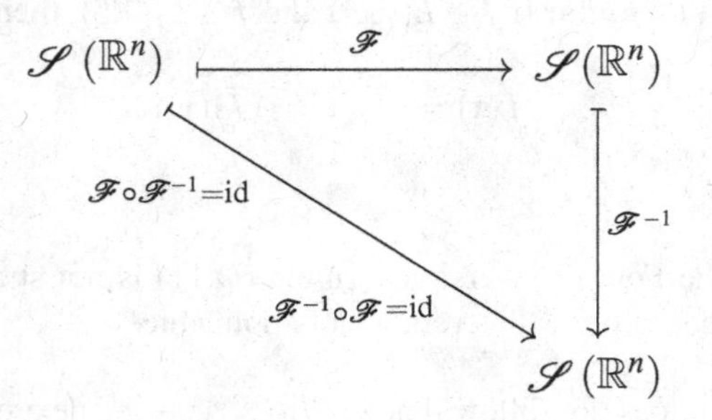

As a corollary of the inversion formula (D.12) we get

Plancherel identity: Let $f \in \mathscr{S}(\mathbb{R}^n)$, then $\|\varphi\|_2 = \|\widehat{\varphi}\|_2$.
We end with the most general case regarding the relation between the derivative and the Fourier transform.

Theorem D.7. *Let $f \in \mathscr{S}(\mathbb{R}^n)$. Then*

$$D^\alpha \widehat{f}(\xi) = (-2\pi \mathrm{i})^{|\alpha|} \widehat{(x^\alpha f)}(\xi)$$

and

$$\xi^{|\alpha|} \widehat{f}(\xi) = (-2\pi \mathrm{i})^{|\alpha|} \widehat{(D^\alpha f)}(\xi),$$

where $\alpha \in \mathbb{N}_0^n$ is a multi-index and $\xi \in \mathbb{R}^n$.

We now end with an example of how the Fourier transform can be used to solve partial differential equations with constant coefficients.

Example D.8. Let $p(x)$ be some polynomial. We want to solve the following partial differential equation

$$p\left(\frac{\partial}{\partial x}\right)\varphi = f \tag{D.13}$$

where f is the given function and φ is the sought one. Applying the Fourier transform in the expression (D.13), we obtain $\tilde{p}(-\mathrm{i}\xi)\widehat{\varphi}(\xi) = \widehat{f}(\xi)$ where $\tilde{p}$ is the polynomial p multiplied by the constant from the Fourier transform of the derivative. This entail

$$\widehat{\varphi}(\xi) = \frac{\widehat{f}(\xi)}{\tilde{p}(-\mathrm{i}\xi)}$$

and we get, at least formally,

$$\varphi(x) = \mathscr{F}^{-1}\left(\frac{\widehat{f}(\cdot)}{\tilde{p}(-\mathrm{i}\cdot)}\right)(x).$$

Since the product of Fourier transforms is simply the Fourier transform of a convolution, we finally get

$$\varphi(x) = \left(\mathscr{F}^{-1}\left(\frac{1}{\tilde{p}(-\mathrm{i}\cdot)}\right) * f\right)(x).$$

Therefore if we can calculate $\mathscr{F}^{-1}\left(1/\tilde{p}(-\mathrm{i}\cdot)\right)(x)$ we obtain the solution, at least in a formal way, since it is necessary to justify all the operations. ⊘

Appendix E
Greek Alphabet

α	alpha	A
β	beta	B
γ	gamma	Γ
δ	delta	Δ
ε	epsilon	E
ζ	zeta	Z
η	eta	H
θ, ϑ	theta	Θ
ι	iota	I
κ	kappa	K
λ	lambda	Λ
μ	mu	M
ν	nu	N
ξ	xi	Ξ
o	omicron	O
π	pi	Π
ρ	rho	P
σ, ς	sigma	Σ
τ	tau	T
υ	upsilon	Υ
ϕ, φ	phi	Φ
χ	xi	X
ψ	psi	Ψ
ω	omega	Ω

R.E. Castillo, H. Rafeiro, *An Introductory Course in Lebesgue Spaces*, CMS Books in Mathematics, DOI 10.1007/978-3-319-30034-4

References

[1] C. Bennett and R. Sharpley. *Interpolation of operators*, volume 129 of *Pure and Applied Mathematics*. Academic Press, Inc., Boston, MA, 1988.
[2] R. N. Bracewell. *The Fourier transform and its applications*. McGraw-Hill International Book Company, 2nd ed. edition, 1983.
[3] H. Brezis. *Functional Analysis, Sobolev Spaces and Partial Differential Equations*. Universitext. Springer, 2010.
[4] A.-P. Calderón. Intermediate spaces and interpolation, the complex method. *Studia Math.*, 24:113–190, 1964.
[5] R. E. Castillo, F. Vallejo Narvaez, and J.C. Ramos Fernández. Multiplication and composition operators on weak l_p spaces. *Bull. Malays. Math. Sci. Soc.*, 38(3):927–973, 2015.
[6] A. L. Cauchy. *Cours d'Analyse de l'Ecole Royale Polytechnique: Analyse Algébrique*. Debure, 1821.
[7] K. M. Chong and N. M. Rice. *Equimeasurable rearrangements of functions*. Queen's University, Kingston, Ont., 1971. Queen's Papers in Pure and Applied Mathematics, No. 28.
[8] J. A. Clarkson. Uniformly convex spaces. *Trans. Am. Math. Soc.*, 40:396–414, 1936.
[9] D. V. Cruz-Uribe and A. Fiorenza. *Variable Lebesgue spaces. Foundations and harmonic analysis*. New York, NY: Birkhäuser/Springer, 2013.
[10] M. Cwikel. The dual of weak L^p. *Ann. Inst. Fourier*, 25(2):81–126, 1975.
[11] M. Cwikel and C. Fefferman. Maximal seminorms on Weak L^1. *Stud. Math.*, 69:149–154, 1980.
[12] M. Cwikel and C. Fefferman. The canonical seminorm on weak L^1. *Stud. Math.*, 78:275–278, 1984.
[13] M. M. Day. The spaces L^p with $0 < p < 1$. *Bull. Am. Math. Soc.*, 46:816–823, 1940.
[14] M. M. Day. Reflexive Banach spaces not isomorphic to uniformly convex spaces. *Bull. Am. Math. Soc.*, 47:313–317, 1941.
[15] P. W. Day. *Rearrangements of measurable functions*. ProQuest LLC, Ann Arbor, MI, 1970. Thesis (Ph.D.)–California Institute of Technology.

R.E. Castillo, H. Rafeiro, *An Introductory Course in Lebesgue Spaces*, CMS Books in Mathematics, DOI 10.1007/978-3-319-30034-4

[16] L. Diening. Maximal function on generalized Lebesgue spaces $L^{p(\cdot)}$. *Math. Inequal. Appl.*, 7(2):245–253, 2004.

[17] L. Diening, P. Harjulehto, Hästö, and M. Růžička. *Lebesgue and Sobolev spaces with variable exponents*. Springer-Verlag, Lecture Notes in Mathematics, vol. 2017, Berlin, 2011.

[18] A. Fiorenza and G.E. Karadzhov. Grand and small Lebesgue spaces and their analogs. *Z. Anal. Anwend.*, 23(4):657–681, 2004.

[19] O. Frostman. Potentiel d'équilibre et capacité des ensembles avec quelques applications à la théorie des fonctions. Meddelanden Mat. Sem. Univ. Lund 3, 115 s (1935)., 1935.

[20] J. García-Cuerva and J. L. Rubio de Francia. *Weighted norm inequalities and related topics*. North-Holland Elsevier, 1985.

[21] G.G. Gould. On a class of integration spaces. *J. Lond. Math. Soc.*, 34:161–172, 1959.

[22] L. Grafakos. *Classical Fourier analysis*, volume 249 of *Graduate Texts in Mathematics*. Springer, New York, third edition, 2014.

[23] L. Greco, T. Iwaniec, and C. Sbordone. Inverting the p-harmonic operator. *Manuscr. Math.*, 92(2):249–258, 1997.

[24] A. Grothendieck. Réarrangements de fonctions et inégalités de convexité dans les algèbres de von Neumann munies d'une trace. In *Séminaire Bourbaki, Vol. 3*, pages Exp. No. 113, 127–139. Soc. Math. France, Paris, 1995.

[25] J. Hadamard. Étude sur les propriétés des fonctions entières et en particulier d'une fonction considérée par Riemann. *Journ. de Math. (4)*, 9:171–215, 1893.

[26] G. H. Hardy. Note on a theorem of Hilbert. *Math. Z.*, 6:314–317, 1920.

[27] G. H. Hardy. Note on a theorem of Hilbert concerning series of positive terms. *Proc. Lond. Math. Soc. (2)*, 23:xlv–xlvi, 1925.

[28] G. H. Hardy and J. E. Littlewood. A maximal theorem with function-theoretic applications. *Acta Math.*, 54(1):81–116, 1930.

[29] G. H. Hardy, J. E. Littlewood, and G. Pólya. The maximum of a certain bilinear form. *Proc. Lond. Math. Soc. (2)*, 25:265–282, 1926.

[30] G. H. Hardy, J. E. Littlewood, and G. Pólya. *Inequalities*. Cambridge, at the University Press, 1952. 2d ed.

[31] L. I. Hedberg. On certain convolution inequalities. *Proc. Am. Math. Soc.*, 36: 505–510, 1973.

[32] O. Hölder. Ueber einen Mittelwertsatz. *Gött. Nachr.*, 1889:38–47, 1889.

[33] R.A. Hunt. An extension of the Marcinkiewicz interpolation theorem to Lorentz spaces. *Bull. Am. Math. Soc.*, 70:803–807, 1964.

[34] R.A. Hunt. On L(p,q) spaces. *Enseign. Math. (2)*, 12:249–276, 1966.

[35] T. Iwaniec and C. Sbordone. On the integrability of the Jacobian under minimal hypotheses. *Arch. Ration. Mech. Anal.*, 119(2):129–143, 1992.

[36] R. C. James. A non-reflexive Banach space isometric with its second conjugate space. *Proc. Nat. Acad. Sci. U. S. A.*, 37:174–177, 1951.

[37] J. L. W. V. Jensen. Om konvexe Funktioner og Uligheder mellem Middelvaerdier. *Nyt Tidss. for Math.*, 16:49–68, 1905.

[38] J. L. W. V. Jensen. Sur les fonctions convexes et les inégalités entre les valeurs moyennes. *Acta Math.*, 30:175–193, 1906.

[39] F. Jones. *Lebesgue Integration on Euclidean Space*. Jones and Bartlett, 2001.

[40] G. Köthe. Topological vector spaces I. Berlin-Heidelberg-New York: Springer Verlag 1969. XV, 456 p. (1969)., 1969.

[41] O. Kováčik and J. Rákosník. On spaces $L^{p(x)}$ and $W^{k,p(x)}$. *Czech. Math. J.*, 41 (4):592–618, 1991.

[42] N.S. Landkof. *Foundations of modern potential theory.* Springer-Verlag, 1972.

[43] G. G. Lorentz. Some new functional spaces. *Ann. Math. (2)*, 51:37–55, 1950.

[44] G.G. Lorentz. Some new functional spaces. *Ann. Math. (2)*, 51:37–55, 1950.

[45] G.G. Lorentz. On the theory of spaces Λ. *Pac. J. Math.*, 1:411–429, 1951.

[46] J. Lukeš and J. Malý. *Measure and integral.* Prague: Matfyzpress, 2nd ed. edition, 2005.

[47] W.A.J. Luxemburg and A.C. Zaanen. Some examples of normed Köthe spaces. *Math. Ann.*, 162:337–350, 1966.

[48] L. Maligranda. Why Hölder's inequality should be called Rogers' inequality. *Math. Inequal. Appl.*, 1(1):69–83, 1998.

[49] L. Maligranda. Equivalence of the Hölder-Rogers and Minkowski inequalities. *Math. Inequal. Appl.*, 4(2):203–207, 2001.

[50] J. Marcinkiewicz. Sur l'interpolation d'opérations. *C. R. Acad. Sci., Paris*, 208: 1272–1273, 1939.

[51] H. Minkowski. Geometrie der Zahlen. I. Reprint. New York: Chelsea Co., 256 p. (1953)., 1953.

[52] D.S. Mitrinović, J.E. Pečarić, and A.M. Fink. *Classical and new inequalities in analysis.* Dordrecht: Kluwer Academic Publishers, 1993.

[53] B. Muckenhoupt. Weighted norm inequalities for the Hardy maximal function. *Trans. Am. Math. Soc.*, 165:207–226, 1972.

[54] C. Niculescu and L.-E. Persson. *Convex functions and their applications. A contemporary approach.* New York, NY: Springer, 2006.

[55] W. Orlicz. Über konjugierte Exponentenfolgen. *Stud. Math.*, 3:200–211, 1931.

[56] L. Pick, A. Kufner, O. John, and S. Fučík. *Function spaces. Volume 1. 2nd revised and extended ed.* Berlin: de Gruyter, 2013.

[57] F. Riesz. Untersuchungen über Systeme integrierbarer Funktionen. *Math. Ann.*, 69:449–497, 1910.

[58] F. Riesz. Les systèmes d'équations linéaires à une infinite d'inconnues. Paris: Gauthier-Villars, VI + 182 S. 8°. (Collection Borel.) (1913)., 1913.

[59] M. Riesz. Sur les maxima des formes bilinéaires et sur les fonctionelles linéaires. *Acta Math.*, 49:465–497, 1927.

[60] M. Riesz. Intégrales de Riemann-Liouville et potentiels. *Acta Litt. Sci. Szeged*, 9:1–42, 1938.

[61] L.J. Rogers. An extension of a certain theorem in inequalities. *Messenger of mathematics*, XVII(10):145–150, 1888.

[62] W. Rudin. *Real and complex analysis*. New York, NY: McGraw-Hill, 1987.

[63] J. V. Ryff. Measure preserving transformations and rearrangements. *J. Math. Anal. Appl.*, 31:449–458, 1970.

[64] H. Rafeiro and E. Rojas, Espacios de Lebesgue con exponente variable. *Un espacio de Banach de funciones medibles.* Caracas: Ediciones IVIC, xxii–134, 2014.

[65] S. G. Samko. Convolution and potential type operators in $L^{p(x)}(\mathbb{R}^n)$. *Integral Transforms Spec. Funct.*, 7(3-4):261–284, 1998.

[66] S. G. Samko. Convolution type operators in $L^{p(x)}$. *Integral Transforms Spec. Funct.*, 7(1-2):123–144, 1998.

[67] I.I. Sharapudinov. Topology of the space $\mathscr{L}^{p(t)}([0,t])$. *Math. Notes*, 26: 796–806, 1979.

[68] I.I. Sharapudinov. Approximation of functions in the metric of the space $L^{p(t)}$ ([a,b]) and quadrature formulae. Constructive function theory, Proc. int. Conf., Varna/Bulg. 1981, 189-193 (1983)., 1983.

[69] I.I. Sharapudinov. On the basis property of the Haar system in the space $\mathscr{L}^{p(t)}([0,1])$ and the principle of localization in the mean. *Math. USSR, Sb.*, 58:279–287, 1987.

[70] G. Sinnamon. The Fourier transform in weighted Lorentz spaces. *Publ. Mat.*, 47(1):3–29, 2003.

[71] E. M. Stein. *Singular integrals and differentiability properties of functions.* Princeton University Press, 1970.

[72] E. M. Stein. The development of square functions in the work of A. Zygmund. *Bull. Am. Math. Soc., New Ser.*, 7:359–376, 1982.

[73] E. M. Stein and R. Shakarchi. *Fourier analysis. An Introduction.* Princeton, NJ: Princeton University Press, 2003.

[74] E. M. Stein and G. Weiss. *Introduction to Fourier analysis on Euclidean spaces.* Princeton University Press, 1971.

[75] O. Stolz. Grundzüge der Differential- und Integralrechnung. Erster Teil: Reelle Veränderliche und Functionen. Leipzig. B. G. Teubner. X + 460 S. gr. 8° (1893)., 1893.

[76] J.D. Tamarkin and A. Zygmund. Proof of a theorem of Thorin. *Bull. Am. Math. Soc.*, 50:279–282, 1944.

[77] G.O. Thorin. An extension of a convexity theorem due to M. Riesz. Fysiogr. Sällsk. Lund Förh. 8, 166-170 (1939)., 1939.

[78] E. C. Titchmarsh. On conjugate functions. *Proc. Lond. Math. Soc. (2)*, 29:49–80, 1928.

[79] E. C. Titchmarsh. Additional note on conjugate functions. *J. Lond. Math. Soc.*, 4:204–206, 1929.

[80] G. Vitali. Sui gruppi di punti e sulle funzioni di variabili reali. *Torino Atti*, 43: 229–246, 1908.

[81] J. von Neumann. On rings of operators. III. *Ann. Math. (2)*, 41:94–161, 1940.

[82] H. Weyl. Singuläre Integralgleichungen mit besonderer Berücksichtigung des *Fourier*schen Integraltheorems. Göttingen, 86 S (1908)., 1908.

[83] R. L. Wheeden and A. Zygmund. *Measure and integral. An introduction to real analysis*. CRC Press, 2015.

[84] N. Wiener. The ergodic theorem. *Duke Math. J.*, 5:1–18, 1939.

[85] A. Zygmund. Sur les fonctions conjuguées. *C. R. Acad. Sci., Paris*, 187: 1025–1026, 1928.

[86] A. Zygmund. On a theorem of Marcinkiewicz concerning interpolation of operations. *J. Math. Pures Appl. (9)*, 35:223–248, 1956.

Symbol Index

Function Spaces

Operators

R.E. Castillo, H. Rafeiro, *An Introductory Course in Lebesgue Spaces*, CMS Books in Mathematics, DOI 10.1007/978-3-319-30034-4

Norms

Miscellaneous

Subject Index

R.E. Castillo, H. Rafeiro, *An Introductory Course in Lebesgue Spaces*, CMS Books in Mathematics, DOI 10.1007/978-3-319-30034-4

Printed in the United States
By Bookmasters